PYRETHRUM FLOWERS

PYRETHRUM FLOWERS

Production, Chemistry, Toxicology, and Uses

Edited by

JOHN E. CASIDA
GARY B. QUISTAD

New York Oxford
OXFORD UNIVERSITY PRESS
1995

Oxford University Press

Oxford New York Toronto
Delhi Bombay Calcutta Madras Karachi
Kuala Lumpur Singapore Hong Kong Tokyo
Nairobi Dar es Salaam Cape Town
Melbourne Auckland Madrid

and associated companies in
Berlin Ibadan

Published by Oxford University Press, Inc.,
200 Madison Avenue, New York, New York 10016

Library of Congress Cataloging-in-Publication Data

Pyrethrum flowers : production, chemistry, toxicology, and uses /
edited by John E. Casida and Gary B. Quistad.
p. cm.
Papers based on an International Symposium on "Pyrethrum Flowers: Production, Chemistry, Toxicology and Uses" held in Aug. 1992, in Honolulu, Hawaii.
Includes bibliographical references and index.
ISBN 0-19-508210-9
1. Pyrethrum (Insecticide) 2. Pyrethrum (Plant) 3. Pyrethrins.
4. Pyrethrins—Toxicology. I. Casida, John E., 1929– .
II. Quistad, Gary Bennet. III. International Symposium on "Pyrethrum Flowers: Production, Chemistry, Toxicology and Uses (1992 : Honolulu, Hawaii)
SB952.P9P93 1995
668'.651 dc20 93-48567

9 8 7 6 5 4 3 2
Printed in the United States of America
on acid-free paper

Dedicated to the producers, users, and researchers
who have been fascinated with pyrethrum
and particularly to
William D. Gullickson, Sr.
for helping us all to keep the faith

Preface

Pyrethrum is the most important botanical insecticide. The flowers of this daisy, at the base of the petals, contain a mixture of pleasant-smelling esters called the pyrethrins with highly unusual insecticidal properties. Many countries are involved in the production and use of pyrethrum. The major producers are Kenya and Tasmania, but there are also extensive fields of the flowers in Tanzania, Rwanda, and Papua New Guinea. Although pyrethrum culture is a farming practice, the natural product is used only to a limited extent to control crop pests but primarily to protect people, households, and domestic animals from pest infestation.

Pyrethrum has been used for more than 160 years establishing the longest safety record of any major pest control agent. However, the criteria of safety and effectiveness are constantly changing with social and political perceptions and this botanical insecticide is continually being reevaluated. Despite its chemical complexity, it must be thoroughly studied in the same manner as structurally-simpler pesticides. Knowledge from these new investigations further verifies the unique built-in safety features of the pyrethrins. They are knockdown and killing agents for pest insects yet are of very low toxicity when ingested by mammals. They are nonpolluting since they break down quickly on exposure to light and air and they are quickly metabolized. Pyrethrum is therefore a natural insecticide which does its job effectively and is then rapidly degraded.

Pyrethroids is a generic term for the pyrethrins and their synthetic analogs. Pesticide researchers have focused for many decades on improving the properties of this botanical material. They made analogs that are more potent, cheaper and with broader uses. However, they did not fully reproduce the properties of the marvelous botanical material and so there is a need and market for all the pyrethrum that can be produced currently. It is clear that pyrethrum has been the insecticide of choice for a variety of use conditions and will continue to be so for many years to come.

The continuing interest of the editors in pyrethrum research and their preparation of this volume were made possible by grant number PO1 ES00049 from the National Institute of Environmental Health Sciences, NIH. In addition, one of us (JEC) also acknowledges the Rockefeller Foundation for the opportunity to focus on pyrethrum while a scholar-in-residence at the Bellagio Study and Conference Center at Lake Como, Italy.

This book is based in part on an International Symposium on "Pyrethrum Flowers: Production, Chemistry, Toxicology and Uses" held in August 1992 in Hawaii as part of the American Institute of Biological Sciences annual meeting. Each contribution has been revised and expanded with information available through September, 1993. The compilation is an updating of "Pyrethrum, The Natural Insecticide" published in 1973. All chapters are by experts on this multi-faceted subject. Each author is thanked for a thorough and critical analysis of the relevant topic. Frederick Preiss and Mary Brennan (McLaughlin Gormley King Co. or MGK) and Sharon Wood, Betsy Kopmar, and Yolanda Leung (Environmental Chemistry and Toxicology Laboratory at Berkeley) skillfully coordinated the organizational, correspondence and word processing aspects of the conference and manuscripts. William D. Gullickson, Sr. and the Governing Board of MGK provided the impetus and resources to bring together all of the current information on pyrethrum both published and unpublished.

John E. Casida and Gary B. Quistad

Environmental Chemistry and Toxicology Laboratory
Department of Environmental Science, Policy, and Management
University of California
Berkeley, CA 94720

Contents

List of Authors

B. K. Bhat, Botanical Resources, Inc., Independence, OR

David J. Carlson, McLaughlin Gormley King Co., Minneapolis, MN

John E. Casida, Department of Environmental Science, Policy, and Management, University of California, Berkeley, CA

Donald G. Cochran, Department of Entomology, Virginia Polytechnic Institute and State University, Blacksburg, VA

Leslie Crombie, Department of Chemistry, University of Nottingham, Nottingham, England

Donald G. Crosby, Department of Environmental Toxicology, University of California, Davis, CA

Michael Elliott, Rothamsted Experimental Station, AFRC Institute of Arable Crops Research, Harpenden, Hertfordshire, England

Karl L. Gabriel, Biosearch, Incorporated, Philadelphia, PA

Eugene J. Gerberg, Gainesville, FL

William D. Gullickson, Sr., McLaughlin Gormley King Co., Minneapolis, MN

Robert L. Hamilton, Product Development Insect Control, S. C. Johnson Wax, Racine, WI

M. Keith Kennedy, Product Development Insect Control, S. C. Johnson Wax, Racine, WI

William L. MacDonald, CIG Pyrethrum, Hobart, Tasmania, Australia

Donald R. Maciver, Roussel-Uclaf Corporation, Montvale, NJ

Raymond Mark, Biosearch, Incorporated, Philadelphia, PA

Frederick J. Preiss, McLaughlin Gormley King Co., Minneapolis, MN

Gary B. Quistad, Department of Environmental Science, Policy, and Management, University of California, Berkeley, CA

Edwin S. Roth, Roussel-Uclaf Corporation, Montvale, NJ

Gerald P. Schoenig, Toxicology/Regulatory Services, Charlottesville, VA

Charles A. Silcox, Roussel-Uclaf Corporation, Montvale, NJ

David M. Soderlund, Department of Entomology, Cornell University, New York State Agricultural Experiment Station, Geneva, NY

Job M. G. Wainaina, Pyrethrum Board of Kenya, Nakuru, Kenya

Nomenclature of Pyrethrum Extract and the Six Natural Pyrethrins

Pyrethrum extract contains three naturally-occurring, closely-related insecticidal esters of chrysanthemic acid (pyrethrins I) and three corresponding esters of pyrethric acid (pyrethrins II). The alcohol moiety has three natural variations which are pyrethrolone in pyrethrin I and pyrethrin II, cinerolone in cinerin I and cinerin II, and jasmolone in jasmolin I and jasmolin II. The sum of pyrethrins I and pyrethrins II collectively designated as "the pyrethrins" constitutes 45 to 55% of pyrethrum extract.

"the pyrethrins"	R	R′
"pyrethrins I"—esters of chrysanthemic acid (chrysanthemates)		
pyrethrin I	CH_3	$CH{=}CH_2$
cinerin I	CH_3	CH_3
jasmolin I	CH_3	CH_2CH_3
"pyrethrins II"—esters of pyrethric acid (pyrethrates)		
pyrethrin II	$CH_3OC(O)$	$CH{=}CH_2$
cinerin II	$CH_3OC(O)$	CH_3
jasmolin II	$CH_3OC(O)$	CH_2CH_3

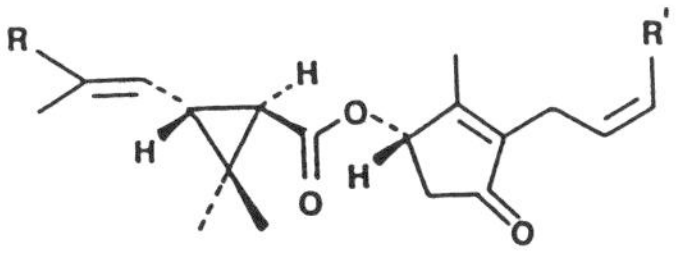

The Chemical Abstracts Service (CAS) numbers and chemical names for the six pyrethrins are as follows:

pyrethrin I: [121-21-1]; cyclopropanecarboxylic acid, 2,2-dimethyl-3-(2-methyl-1-propenyl)-2-methyl-4-oxo-3-(2,4-pentadienyl)-2-cyclopenten-1-yl ester [1R-[1α[S*(Z)],3β]].

cinerin I: [25402-06-6]; cyclopropanecarboxylic acid, 2,2-dimethyl-3-(2-methyl-1-propenyl)-3-(2-butenyl)-2-methyl-4-oxo-2-cyclopenten-1-yl ester [1R-[1α[S*(Z)],3β]].

jasmolin I: [4466-14-2]; cyclopropanecarboxylic acid, 2,2-dimethyl-3-(2-methyl-1-propenyl)-2-methyl-4-oxo-3-(2-pentenyl)-2-cyclopenten-1-yl ester [1R-[1α[S*(Z)],3β]].

pyrethrin II: [121-29-9]; cyclopropanecarboxylic acid, 3-(3-methoxy-2-methyl-3-oxo-1-propenyl)-2,2-dimethyl-2-methyl-4-oxo-3-(2,4-pentadienyl)-2-cyclopenten-1-yl ester [1R-[1α[S*(Z)],3β(E)]].

cinerin II: [121-20-0]; cyclopropanecarboxylic acid, 3-(3-methoxy-2-methyl-3-oxo-1-propenyl)-2,2-dimethyl-3-(2-butenyl)-2-methyl-4-oxo-2-cyclopenten-1-yl ester [1*R*-[1α[*S**(*Z*)],3β(*E*)]].

jasmolin II: [1172-63-0]; cyclopropanecarboxylic acid, 3-(3-methoxy-2-methyl-3-oxo-1-propenyl)-2,2-dimethyl-2-methyl-4-oxo-3-(2-pentenyl)-2-cyclopenten -1-yl ester [1*R*-[1α[*S**(*Z*)],3β(*E*)]].

Pyrethroids are insecticides based on the pyrethrins as prototypes.

I

Introduction

1

Chemicals in Insect Control

MICHAEL ELLIOTT

I. INTRODUCTION

About half (*ca.* 1 million) of the known species of living creatures are insects and 10,000 of these influence mankind adversely and are, therefore, considered to be pests (Metcalf and Metcalf, 1993). Insects consume about 30% of food grown and about $6,000 million is spent annually protecting crops from the havoc they cause. Diptera are vectors of malaria, filariasis, schistosomiasis, trypanosomiasis, and onchocerciasis, which debilitate and kill people and domestic animals on a vast scale; there are between 250 and 500 million cases of malaria annually with an estimated mortality of 1–2.5 million, mainly children. Insects such as the spruce budworm attack both seedling and mature forest trees and migrant pests (e.g., locusts and armyworms) are responsible for enormous crop losses. In the home, human and animal lice, fleas, bugs, and ticks cause much irritation and their bites can spread devastating bacterial diseases such as bubonic plague (Busvine, 1993). To maintain and extend the present standards of civilization insects must controlled, as benevolently as possible, and with minimum impact on the environment.

Throughout the world broad-ranging research in industrial and academic laboratories has aimed, for up to 10 decades, to discover the most effective compounds, formulations, and procedures for insect control. Considerable attention has recently been directed at alternatives to insecticides, e.g., semiochemicals (Pickett *et al.*, 1991) including pheromones and host recognition compounds (Pickett, 1988), antifeedants (Ley *et al.*, 1993), genetically engineered baculoviruses (Wood and Granados, 1991; Leisy and van Beek, 1992), *Bacillus thuringiensis* (Gill *et al.*, 1992), and neuropeptides (O'Shea 1985; O'Shea and Schaffer, 1985). However, the enormous scale of the problem determines that, for the foreseeable future, insecticides, on which the main emphasis is placed in this review, will remain of outstanding importance.

This survey covers practical achievements since an earlier conference (International Symposium on Recent Advances with Pyrethrum, the Natural Insecticide; Casida, 1973); in certain respects the present article updates an earlier one (Elliott, 1979).

Table 1-1. Classes of Insecticides

Class	Insecticides used or being developed		Change
	in 1972	in 1991	
Organochlorine	20	6	−14
Organophosphorus	87	82	− 5
Carbamate	22	25	+ 3
Pyrethroid	4	38	+34
Other	15	23	+ 8

II. MAIN CLASSES OF INSECTICIDES OTHER THAN THE SYNTHETIC PYRETHROIDS

To review the changes in use of insecticides in the period under consideration (i.e., between the two symposia in 1972 and 1992) an appropriate basis is a comparison of the compounds listed in the third (Martin, 1972) and ninth (Worthing and Hance, 1991) editions of the Pesticide Manual, respectively (Table 1-1). As these World Compendia cite pesticides "understood (by the authors) to be in current use or under active development" this will reflect realistically the trends being examined. A few other compounds that represent important stages in the evolution of a particular group, or that have special properties, are also included.

A. DDT

The most significant change in insecticide use in these 20 years has been the replacement, apart from minor uses, of the organochlorine insecticides, frequently by synthetic pyrethroids. In 1991, 221 million hectares worldwide were treated with organophosphorus insecticides, 100 million hectares with synthetic pyrethroids and 49 million hectares with carbamate insecticides, and except in underdeveloped countries very little DDT or other organochlorine insecticide was used on agricultural crops. Of the 20 organochlorine compounds listed in 1972 only 6 (Fig. 1-1) (DDT and methoxychlor; endosulfan, heptachlor and chlordane; gamma HCH) remained in the 1991 Pesticide Manual.

The remarkable properties of the first organochlorine insecticide DDT were discovered in 1939 and the compound was spectacularly successful in wartime, for example in a 1943 typhus epidemic in Naples and subsequently in many other applications (for a comprehensive and authoritative account of DDT and its application, see Mellanby, 1992). In the USA some 15,000 tons of DDT, probably about half the world total, were produced in 1945, and nearly 100,000 tons per annum in the late 1950s. Production had declined to 20,000 tons in 1971 at the start of the period under review (Green *et al.*, 1987). Mellanby (1992) estimates that a total of 2 million tons of DDT may have been produced worldwide and that much still exists. DDT was particularly effective (Mellanby, 1992) in the World Health Organization malaria eradication program in 124

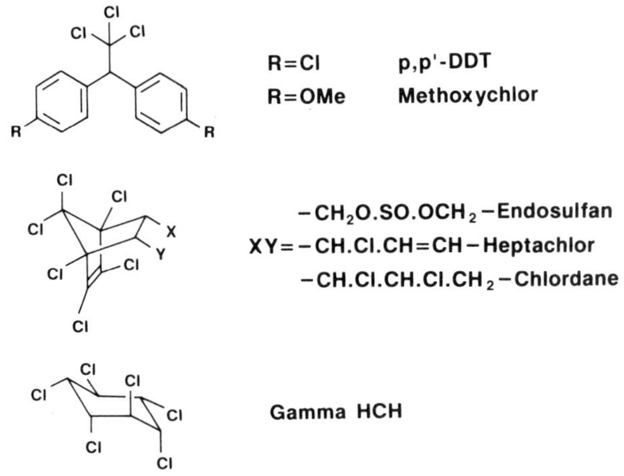

R=Cl p,p'-DDT
R=OMe Methoxychlor

$-CH_2O.SO.OCH_2-$Endosulfan
XY= $-CH.Cl.CH{=}CH-$Heptachlor
$-CH.Cl.CH.Cl.CH_2-$Chlordane

Gamma HCH

Figure 1-1. Chlorinated insecticides, 1991.

countries with a total population of 1,724 million. In the decade from 1935 before DDT was available there were, for example, over 105 million cases of malaria in 17 countries. In 1969 when DDT had been used fewer than 300,000 cases were recorded.

Despite the initial achievements with DDT, and when the special circumstances associated with the Second World War had passed, properties judged to be adverse began to be recognized. The stability of DDT was at first considered very valuable; moreover, DDT was easy to produce, had low acute toxicity to mammals and was active against many insect species, particularly as residual films. However, much disquiet developed over the scale of production and about the accumulation of residues in the environment. Although harm to man or other mammals from residues which could be detected in their bodies has never been demonstrated unequivocally, DDT accumulation in food chains, its purported oncogenicity, and definite impairment of the reproductive capacities of some species of birds led to very severe restrictions on manufacture and use, from 1972 in the USA and subsequently throughout the more developed nations. Production of DDT was transferred to countries where it was desperately needed because more advanced substitutes were too expensive. Green *et al.* (1987) observe that risk:benefit analyses give different results in poor, starving populations and in well fed affluent societies and the balance between hazards to wild and human life assumes a different perspective.

The discovery and application of DDT has probably brought greater benefit to more people than any other single compound, but with experience two specific disadvantages of DDT became apparent. First, because it is a broad spectrum insecticide, not only pests but also their predators and parasites are killed, and some natural control is lost. Second, and probably the overriding drawback, its ubiquitous presence promoted resistance both in direct target species and in others where no direct control was intended or necessary (Sawicki, 1985). Such limitations led the World Health Organization and many national and international bodies to discontinue outdoor applications of DDT. However, developing nations short of food and resources desperately need crop protection and pest control but often cannot afford the more recently developed alternatives to the overly persistent organochlorine compounds. In its favor DDT itself has been estimated to have saved 50 million human lives and averted 1 billion human diseases.

B. Other Organochlorine Insecticides

Other halogenated insecticides were discovered in the decade after 1942 (summary: Elliott, 1979), but after experience with DDT, their use (18 listed in Worthing and Hance, 1991) was severely restricted. Now only gamma HCH and the cyclodiene insecticides endosulfan, heptachlor and chlordane, which have special properties not shared by other members of the group, survive. Endosulfan, which is degraded in mammals and in the environment, is permitted in the USA for application to cotton, especially in association with organophosphorus insecticides. Treatments are permitted in orchards and on vegetables and in Africa on many crops. Heptachlor and chlordane are valuable for various domestic applications on animals and in the soil, whereas chlordane assists management of earthworms in gardens. Gamma HCH is considerably more volatile than DDT and significantly more soluble in water; it is thus a valuable seed dressing and has some special uses such as control of head lice. This restriction in the number of applications for DDT and other organochlorine compounds (only 8% of the worldwide market in 1992; Smith, 1992) reflects a very significant decrease in demand for them; their role in pest control has largely been assumed by synthetic pyrethroids (see below).

C. Organophosphorus Insecticides

Although as noted above, at first only DDT, and then the other organochlorine insecticides, appeared likely to be required for highly effective worldwide, safe, insect control, their limitations and disadvantages including resistance development (Mellanby, 1992) were soon recognized. As the need to protect crops developed, particularly those grown in monoculture, other classes of insecticides were examined. In Germany, the UK, and the USA chemists had investigated the toxicity of organophosphorus compounds and established that products such as dimefox, tabun and sarin (Fig. 1-2) were highly toxic not only to insects, but to mammals as well, a severe impediment to practical application. With extensive knowledge of structure–activity relationships in this class of compounds derived from work originated by Lange and Krueger (1932), Schrader (from 1937) and a succession of other investigators, the first commercial compound schradan (Fig. 1-2) (subsequently withdrawn) was marketed in Germany in 1943 (Schrader, 1963). The basic structural requirements for activity gave much scope for variation in physical properties and for manipulating activity to both insects and mammals. This was exploited

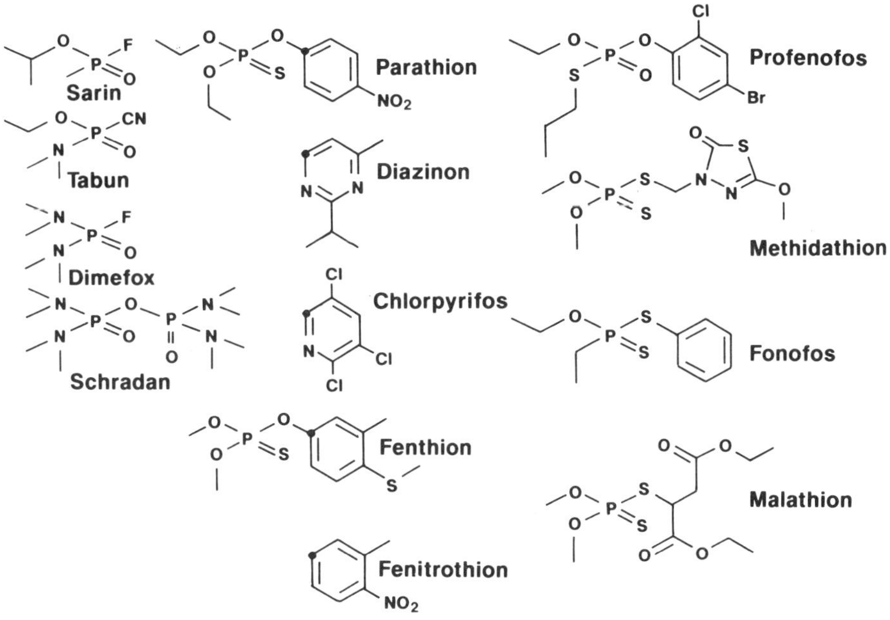

Figure 1-2. Organophosphorus insecticides I.

in research and development so effectively by up to 51 international companies that organophosphorus insecticides are now the most used group (45% of the world market; 222 million ha treated in 1991). Figs. 1-2 and 1-3 show some of the more important compounds now available; the range has not changed significantly since a previous survey (Elliott, 1979), the earlier investigations having disclosed so many valuable variations on the basic theme. Unlike the lipophilic organochlorine insecticides, there is scope in the organophosphorus

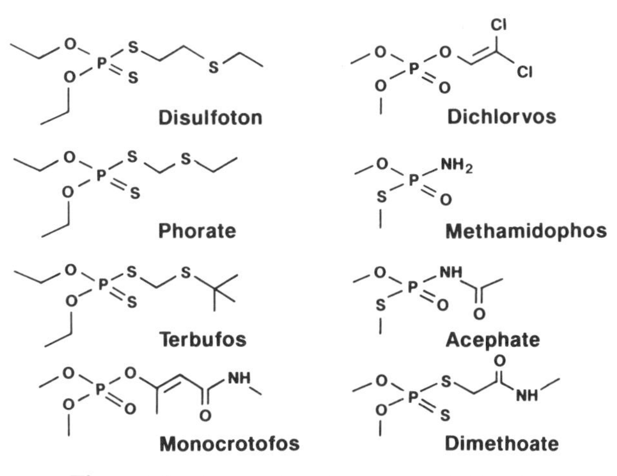

Figure 1-3. Organophosphorus insecticides II.

group to vary physical properties from polar, water-soluble, systemic compounds such as dimethoate to lipophilic, soil-active compounds (phorate and terbuphos) (reviewed by Naumann, 1989). However, because they were readily available and so effective, injudicious overapplication led to the emergence of high levels of resistance in some insects infesting crops of major economic importance such as cotton and to the rapid promotion of the photostable synthetic pyrethroids (see below) as substitutes. Despite some isolated problems with a few organophosphorus compounds such as delayed neurotoxicity (Johnson, 1988) and alleged mutagenicity, this class of compounds has proved of outstanding value in economic pest control.

D. Carbamate Insecticides

The need for further new types of insecticides led by 1947 to the exploitation of the long-established toxicity of carbamic acid esters which, like the organophosphorus compounds, were known anticholinesterases. The first compounds investigated, dimethylcarbamates, are now represented by pirimicarb (Baranyovits and Ghosh, 1969), a fast acting aphicide with fumigant and translaminar properties. Fig. 1-4 shows some of the most important carbamates in the Pesticide Manual (Worthing and Hance, 1991). The carbamate insecticides are generally more expensive than the organophosphorus

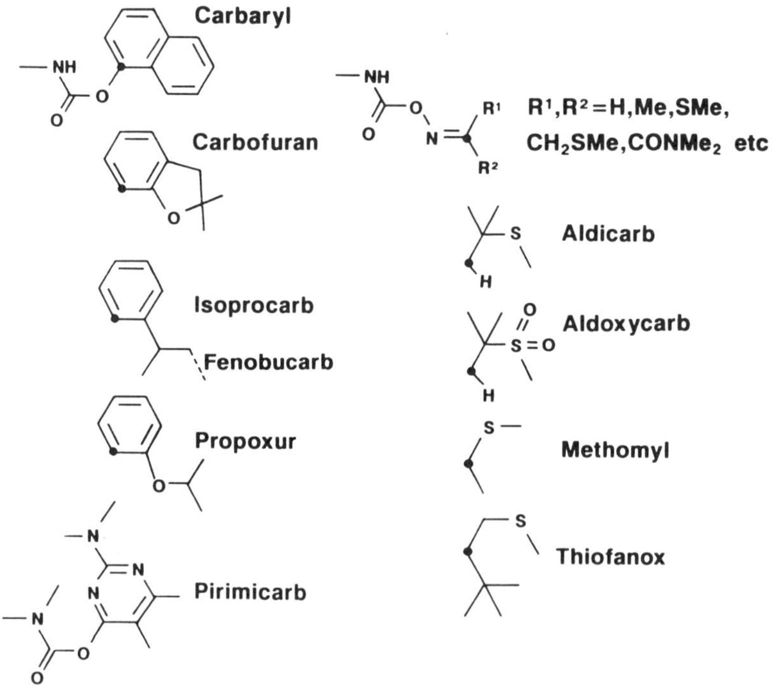

Figure 1-4. Carbamates and carbamoyloximes.

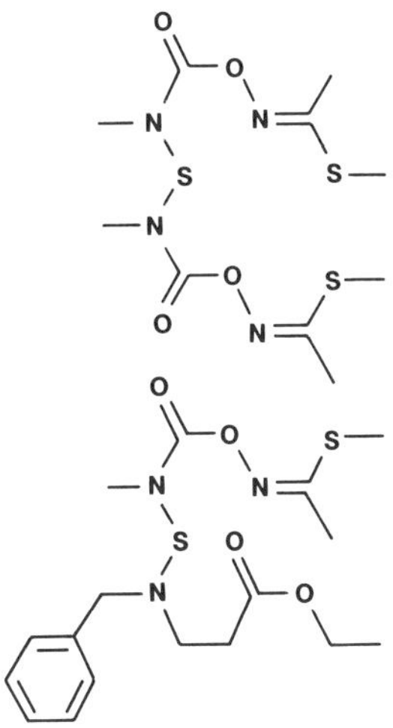

Thiodicarb

Alanycarb

Figure 1-5. More recent carbamoyloximes.

group and some, although excellent insecticides, have mammalian toxicities too high for general application. Carbaryl (Haynes *et al.*, 1957), the most widely used carbamate, is less toxic to mammals than other commercially important compounds such as: carbofuran, a systemic soil insecticide (Homeyer, 1975); propoxur, a fumigant for greenhouse and domestic applications (Unterstenhöfer, 1963); fenobucarb (Metcalf *et al.*, 1962); and isoprocarb (Anonymous, 1969), for control of rice, fruit and vegetable pests. Carbamoyloximes with various substituents (R^1 and R^2 in Fig. 1-4) have a range of physical properties combined with valuable potencies to various insect species. Aldicarb (Weiden *et al.*, 1965) in particular is translocated in plants after soil application, although high mammalian toxicity precludes use of all formulations except granules. Unlike the organophosphorus compounds, of which few examples have been introduced since 1972, new carbamates, e.g., thiodicarb (Sousa *et al.*, 1977) and alanycarb (Umetsu, 1984) (Fig. 1-5), continue to be developed. The scope for structural variation within this group of compounds (e.g., *N*-sulphenylated derivatives which are less toxic to mammals, but retain insect potency) (Black *et al.*, 1973) is probably not yet exhausted.

E. Botanical Insecticides and Their Analogs

Except pyrethrum, nicotine, and rotenone, no botanical insecticides are recognized in the current Pesticide Manual (Worthing and Hance, 1991) as important for practical insect control although ryanodine has garden uses and neem tree extract (azadirachtin) (Ley *et al.*, 1993) may have potential. Nicotine, an effective nonpersistent, nonsystemic contact insecticide has been used for more than 200 years. High mammalian toxicity limits use to enclosed spaces such as glasshouses; no related compounds with improved properties have been discovered (Elliott, 1979; Corbett *et al.*, 1984; Tomizawa and Yamamoto, 1992). Rotenone, isolated from the roots of *Derris* and *Lonchocarpus spp.*, has been

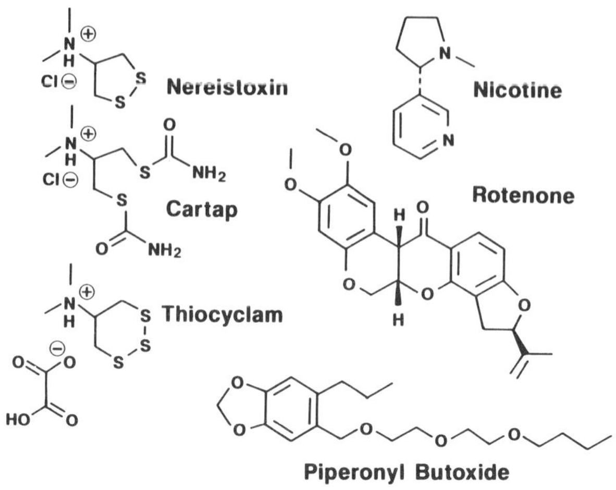

Figure 1-6. Natural insecticides, related compounds and a synergist.

recognized as a fish poison with low mammalian toxicity for many years; the Chinese used it as an insecticide. It is still used horticulturally, but larger scale agricultural applications are precluded by cost and by instability in air and light. The structural requirements for activity have been discussed recently (Crombie *et al.*, 1992), but no more active alternative to the natural product has been introduced.

Two synthetic compounds, cartap hydrochloride (Sakai *et al.*, 1967) and thiocyclam (Berg and Knutti, 1975), related to the natural compound nereistoxin (Fig. 1-6), are used, respectively, against rice and vegetable pests and for control of Lepidoptera and Coleoptera on cotton, rape, rice, sugar cane, and vegetables.

Unlike other botanical insecticides, the use of pyrethrum has greatly expanded during the period reviewed because it has an exceptionally favorable combination of properties, fully discussed in other chapters of this book. The pyrethrins are frequently formulated with the synergist piperonyl butoxide (Fig. 1-6) (Wachs, 1947; Casida, 1970) which enhances the activity of these relatively expensive insecticides and, therefore, minimizes the amount needed for efficient insect control. However, piperonyl butoxide and other synergists are not so effective with the more photostable synthetic pyrethroids (below) because, in general, they are insufficiently stable in air and light and formulation to ensure that toxicant and synergist are together where required is difficult. The natural pyrethrins are also important because their chemical constitution (Crombie, 1994) has provided the basis for developing more photostable synthetic pyrethroid insecticides, as discussed later.

F. Insecticides Affecting Insect Growth

Several classes of compounds influencing the growth of insects in various ways have been developed during the past two decades.

Disruption of the moulting process of insect larvae by the compound diflubenzuron (Fig. 1-7) was discovered unexpectedly during work on herbicides (Post and Mulder, 1974). Much research has since expanded the range of commercially available compounds with a mode of action similar to that of diflubenzuron. These are stomach and contact poisons which disrupt chitin deposition and thence cuticle formation. Diflubenzuron itself is a high-melting solid only slightly soluble in most solvents and therefore formulated as a wettable powder (Maas *et al.*, 1981). Later, more potent compounds (chlorfluazuron, teflubenzuron, hexaflumuron and decarafluron), still derived from 2,6-difluorobenzoic acid, became easier to formulate. They are retained in the insect as a result of more rapid transport from the gut into larval tissues where

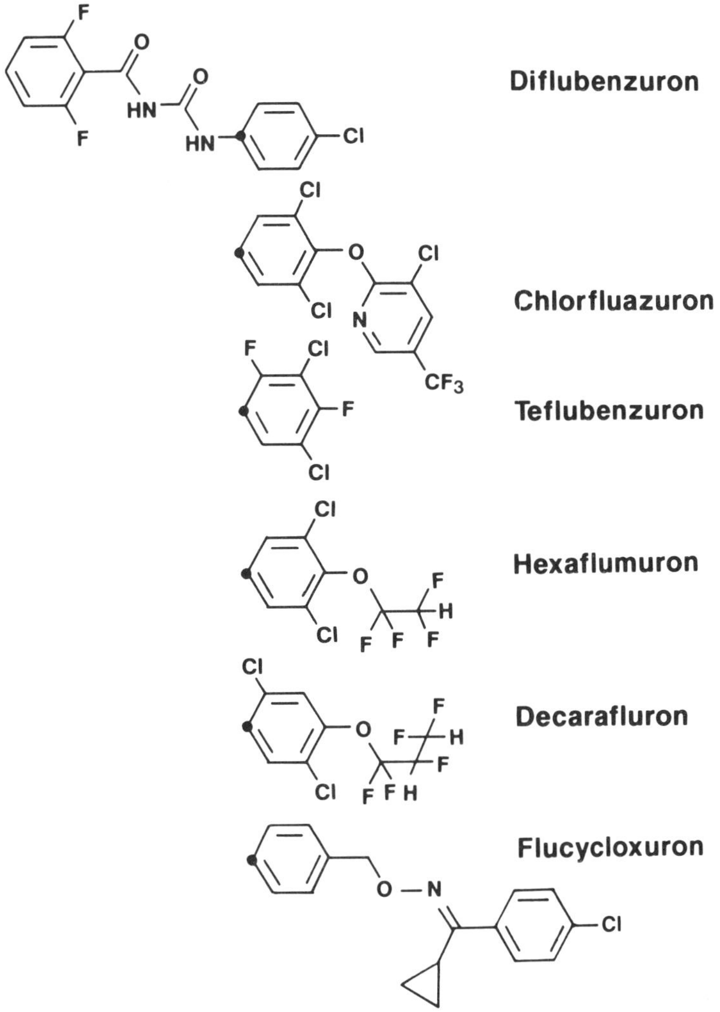

Diflubenzuron

Chlorfluazuron

Teflubenzuron

Hexaflumuron

Decarafluron

Flucycloxuron

Figure 1-7. Insecticides derived from urea: I.

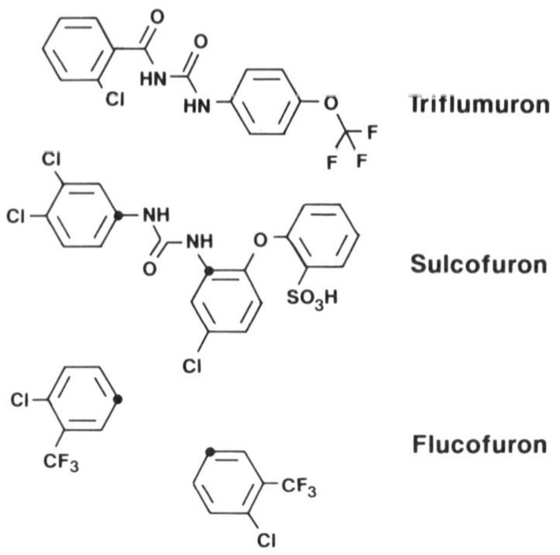

Triflumuron

Sulcofuron

Flucofuron

Figure 1-8. Insecticides derived from urea: II.

detoxification is slower (review with leading references: Ishaaya, 1990). Flucycloxuron (Grosscurt *et al.*, 1988) has both acaricidal and insecticidal properties. Another urea related compound, triflumuron (Fig. 1-8), is effective against pests of cotton, forests, fruit, and soybeans; sulcofuron and flucofuron (Ciba-Geigy AG), substituted diarylureas, were developed to control larvae which attack cotton fabrics.

In another approach to insect control by influencing growth, the synthetic compound methoprene (Fig. 1-9) (Henrick *et al.*, 1973) was derived from

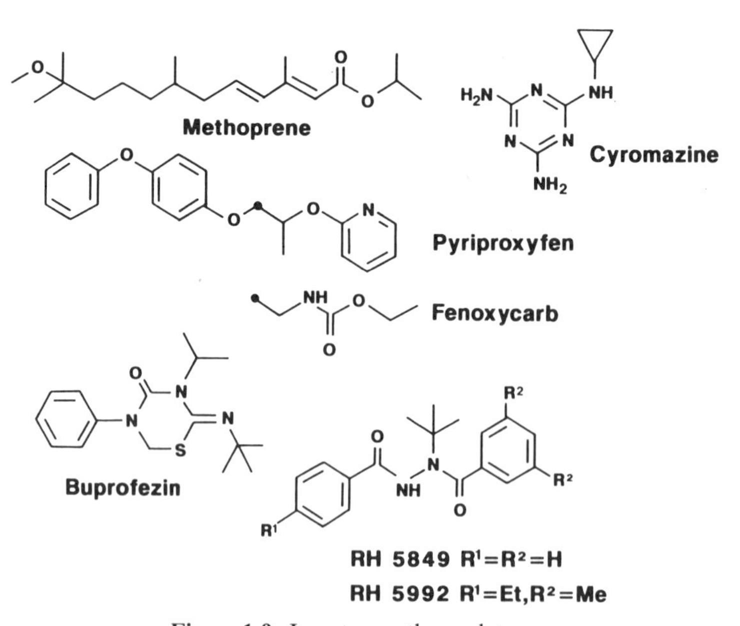

Figure 1-9. Insect growth regulators.

structures of the natural juvenile hormones. Methoprene is more stable than the natural hormones and has been used in nonagricultural control of mosquitoes and fleas. Other juvenile hormone mimics such as fenoxycarb (Dorn *et al.*, 1981) and pyriproxifen are intrinsically more active with unexpected acaricidal activity. Pyriproxyfen is 350 times more effective than methoprene against the last instar larvae of *Aedes aegypti* (Miyamoto, 1990).

Buprofezin (Kanno *et al.*, 1981), another agent disrupting chitin deposition, acts specifically against Homoptera that transmit viruses such as plant hoppers and whiteflies which are important pests of cotton and vegetables. Cyromazine (Hall and Fochse, 1980), a 1,3,5-triazine, represents another class of insect growth regulator, especially effective against Diptera by affecting ecdysis rather than chitin syntheses. RH 5849 (Fig. 1-9) is the first nonsteroidal ecdysone mimic (Wing and Ramsay, 1989); a later related compound, RH 5992, is highly specific for lepidopterous insects and has a moderately rapid action (Heller *et al.*, 1992).

G. Nitromethylene, Nitroguanidine, and Related Compounds

The insecticidal activity of nitromethylene derivatives (Fig. 1-10) was reported in 1978 (Soloway *et al.*, 1978), but the most active member of the group (nithiazine, R=H) was not adequately stable in sunlight for agricultural applications. A related compound (R=CHO) was more stable but not manufactured. However, another product (imidacloprid, Fig. 1-10) (Diehr *et al.*, 1991) in which the nitro group is on nitrogen rather than carbon is more photostable and has been developed for commercial application. It has excellent systemic activity against aphids, leafminers, plant hoppers, thrips and whiteflies, combined with low mammalian toxicity (Elbert *et al.*, 1990). Other compounds such as nitenpyram, TI 304 (Takeda), less persistent than imidacloprid, and the dichloro analogue

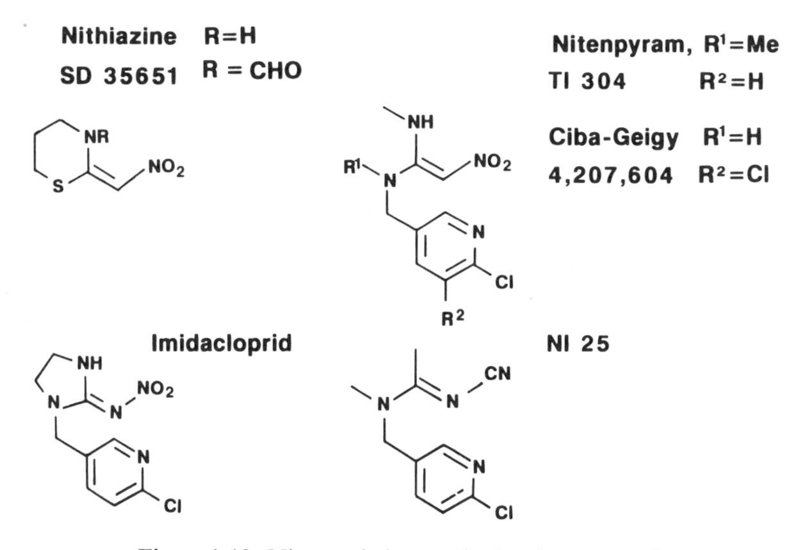

Figure 1-10. Nitromethylene and related compounds.

(Ciba-Geigy AG, European Patent Application 4,207,604) have structures with features in common with those of imidacloprid. NI 25 (Takehashi *et al.*, 1992) acts systemically by foliar application and in the soil. Imidacloprid and related compounds are considered to act on the nicotinic acetylcholine receptor and constitute a valuable new group of insecticides (Tomizawa and Yamamoto, 1992).

H. Other Active Compounds

For a significant period after the potential of synthetic pyrethroids for practical insect control had been demonstrated, much research was aimed to exploit the leads to new compounds in this group and, consequently, few other new types of insecticide were discovered. However in the past five years a number of original compound types, some indicated in Fig 1-11, have been synthesized.

Fipronil (Colliot *et al.*, 1992) controls soil and foliar insects, including those resistant to pyrethroids or carbamates, on a wide variety of crops. Only GABA-regulated chloride channels, and not acetylcholinesterases, are reported to be influenced.

Pymetrozine (CGA 215'944) (Flückiger *et al.*, 1992) is active against susceptible and resistant aphids and whiteflies (juvenile and adult stages) which are increasingly serious pests in many crops, including vegetables and cotton. At recommended rates of application only Homoptera, and not their natural enemies, are affected.

Fenazaquin (Dreikorn *et al.*, 1991) is claimed to have acaricidal activity comparable with that of the pyrethroids bifenthrin and fenproparthrin (compounds **6** and **27**, Fig. 1-16 referred to later) but not to be cross resistant to mites no longer susceptible to these pyrethroids. An additional advantage is that the activity is not affected by temperature, in contrast to some pyrethroids.

AC 303,630 (Lovell *et al.*, 1990) is a broad spectrum insecticide/acaricide,

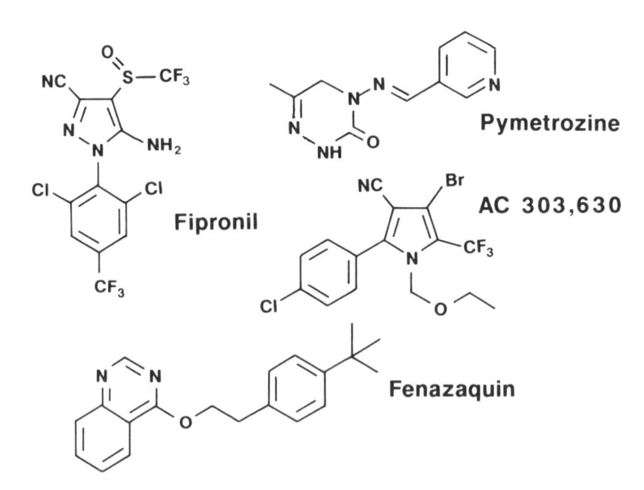

Figure 1-11. Other active compounds.

active by foliar application and by some systemic uptake via roots. It is a member of a new group of insecticidal pyrroles (Addor *et al.*, 1992)

Despite much research on insecticidal lipid amides (Blade, 1990) and on trioxabicyclooctanes and related compounds (Casida *et al.*, 1990) no commercial compounds of these types have yet been reported.

III. SYNTHETIC PYRETHROIDS

A. Introduction

Among the classes of insecticides, the six natural pyrethrins are unique for the intensity of their very rapid action against many species of insects combined with minimal hazard to mammals under normal conditions. However, all six esters are decomposed in air and light with loss of insecticidal activity and only limited stabilization is feasible (Allan and Miller, 1990). The greatly expanded application of pyrethroids referred to above has been associated with synthetic variants of the natural esters which are relatively more stable in air and light (for detailed accounts of natural and synthetic pyrethroids and their application: Casida, 1973; Leahey, 1985; Jackson, 1989; Naumann, 1990). Table 1-2 and Figs. 1-12 to 1-20 show the structures of pyrethroids available commercially at present and a few others. In 1991 pyrethroids were used to treat approximately

Table 1-2. Pyrethroids in Figs. 1-12 to 1-19[a]

Compound	Compound
1 Acrinathrin (16, 18)	**23** Empenthrin [(1*R*) isomer] (16, 18)
2 Allethrin (12, 16)	**24** Esfenvalerate (14, 17, 18)
3 Benfluthrin[b] (16, 18)	**25** Etofenprox (Nonester) (19)
4 Benzylnorthrin[b] (12, 16)	**26** Fenfluthrin[b] (16, 18)
5 Kadethrin (16, 18)	**27** Fenpropathrin (16, 18)
6 Bifenthrin (16, 18)	**28** Fenvalerate (17, 19)
7 Bioallethrin (12, 16)	**29** Flucythrinate (17, 18)
8 Bioallethrin (*S*-cyclopentenyl isomer) (12, 16)	**30** Flumethrin (*trans*)[b,d] (16, 18)
9 Bioethanomethrin (16, 18)	**31** Tau-fluvalinate (17, 18)
10 Bioresmethrin (13, 15, 16, 18)	**32** Jasmolin I[c] (12, 16)
11 Cinerin I[c] (12, 16)	**33** Jasmolin II[c] (12, 16)
12 Cinerin II[c] (12, 16)	**34** Permethrin (13, 16, 18)
13 Cycloprothrin (17, 18)	**35** Phenothrin [(1*R*) isomers] (16, 18)
14 Cyfluthrin (16, 18)	**36** Prallethrin (12, 16)
15 Beta-cyfluthrin (16, 18)	**37** Pyrethrin I (12, 13, 16)
16 Cyhalothrin (16, 18)	**38** Isopyrethrin I[b] (12, 16)
17 Lamda-cyhalothrin (16, 18)	**39** Pyrethrin II (12, 16)
18 Cypermethrin (16, 18)	**40** Resmethrin (16, 18)
19 Alpha-cypermethrin (16, 18)	**41** Tefluthrin (16, 18)
20 Beta-cypermethrin (16, 18)	**42** Tetramethrin (16, 18)
21 Cyphenothrin [(1*R*) isomers] (16, 18)	**43** Tetramethrin [(1*R*) isomers] (16, 18)
22 Deltamethrin (13, 16, 18)	**44** Tralomethrin (16, 18)

[a] The structures of each compound (or acid and alcohol components) are given in figures designated in parentheses.
[b] Not in the Pesticide Manual, 9th Ed.
[c] Not listed separately in the Pesticide Manual.
[d] Flumethrin is not a BSA-approved common name.

100 million ha, over 25% of agricultural areas worldwide. This greater use of synthetic pyrethroids is reflected by the increase of their numbers in Table 1-1 from 4 to 38 during the period reviewed.

B. Early Synthetic Pyrethroids

Following pioneering structure-activity studies of pyrethroids (Staudinger and Ruzicka, 1924; Staudinger *et al.*, 1924; Schechter *et al.*, 1949; Barthel, 1961), pyrethrin I (**37**) (Elliott, 1958, 1964b) was used as the standard to assess the influence on potency of modifying the side chain of cyclopentenolone chrysanthemates (Fig. 1-12). The benzyl side chain in compound **4** was thus found to be the most effective, and synthetically accessible, substitute for the *Z*-pentadienyl system in pyrethrins I and II (**37** and **39**); the ester (**38**) with an *E,Z*-pentadiene in conjugation with the cyclopentenolone ring was inactive (Elliott, 1964a). Examining chrysanthemates of aliphatic, alicyclic and aromatic alcohols with allyl and benzyl side chains identified the most active compound as 5-benzyl-3-furylmethyl chrysanthemate, bioresmethrin (**10**) (Fig. 1-13) (Elliott and Janes, 1978). Although more active against insects than previous pyrethroids and even less toxic to mammals than the natural esters and allethrin, bioresmethrin was not significantly more stable in air and light than earlier compounds. However, the furan alcohol was superior to cyclopentenolones such as allethrolone, which had been used previously (Barthel, 1961) for establishing the relative effectiveness of acid components of pyrethroid esters such as (**9**) (Fig. 1-16) (Velluz *et al.*, 1969). For example, the significance for increased activity and photostability of the dichloro isostere of chrysanthemic acid (Farkas *et al.*, 1959) was overlooked when it was first examined as an allethrolone ester (see Fig. 1-14). When the furan ring and the isobutenyl side

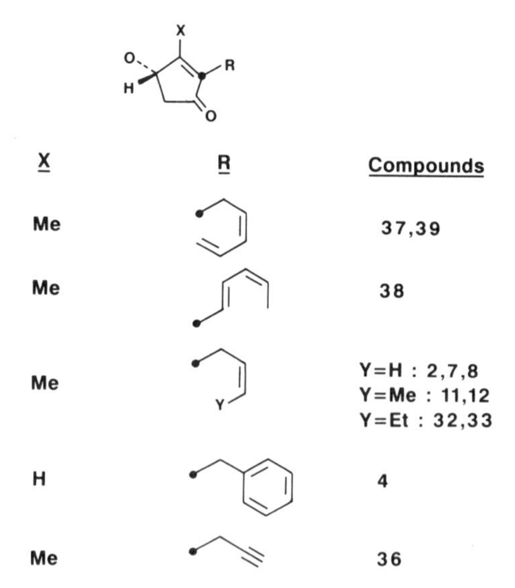

Figure 1-12. Alcohol components of pyrethroids: I.

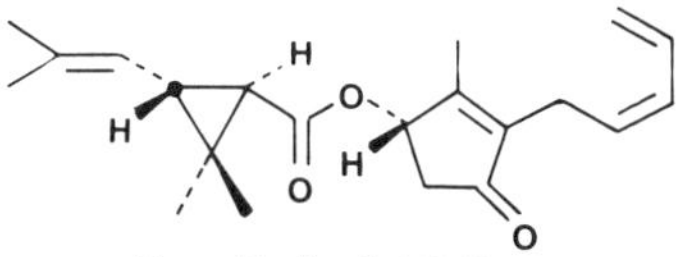

Pyrethrin I (37)

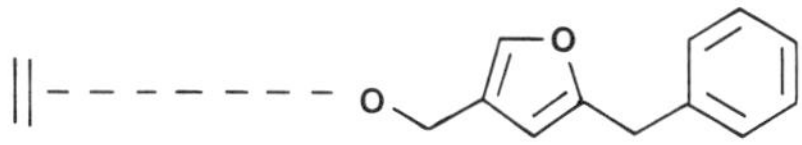

Bioresmethrin (10)

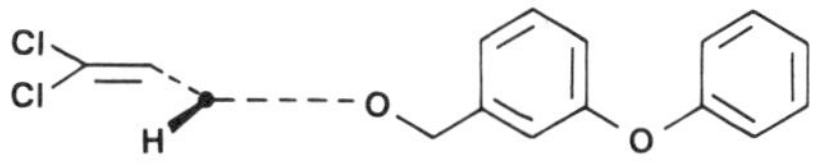

(1R)-trans-Permethrin (34)

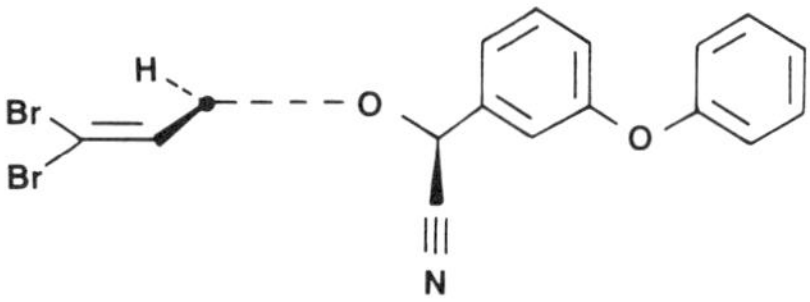

Deltamethrin (22)

Figure 1-13. Synthetic pyrethroids—evolution of the present range: I.

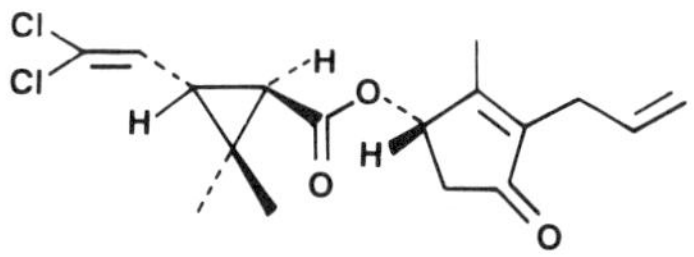

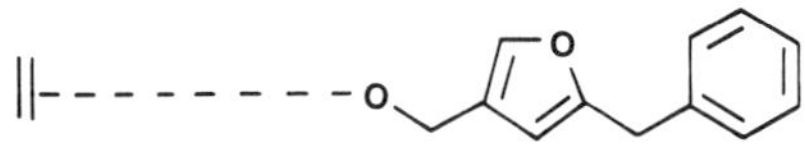

Furan ester over 20 times more effective than allethronyl

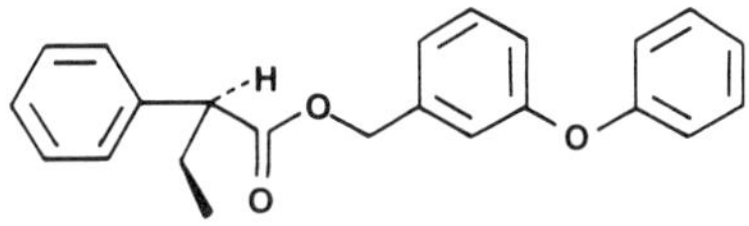

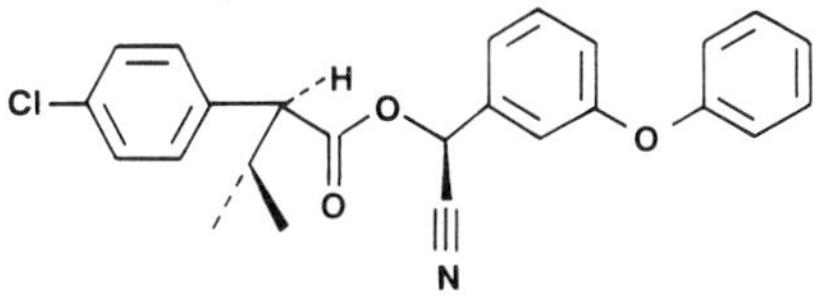

Esfenvalerate (24)

Figure 1-14. Synthetic pyrethroids—evolution of the present range: II.

chain in bioresmethrin had been identified as sites for photochemical degradation (Ueda *et al.*, 1974), the feasibility of a more photostable pyrethroid was indicated. This was realized in practice as an ester of a dichlorovinyl acid with 3-phenoxybenzyl alcohol, an isostere of 5-benzyl-3-furylmethyl alcohol. Fortunately not only was the relatively stable 3-phenoxybenzyl ester (Fig. 1-15) more active to insects than would have been predicted from previous results, but the toxicity to mammals remained low. Permethrin was thus established as the first member of a range of cyclopropane-based pyrethroids stable enough for agricultural applications.

Chemists from the Sumitomo Chemical Company pioneered another class of more stable compounds when they discovered the insecticidal activities of 5-benzyl-3-furylmethyl and 3-phenoxybenzyl α-ethylphenylacetates (Ohno *et al.*, 1976). From these lead compounds they developed the very effective α-isopropyl 4-chlorophenyl esters and, further, discovered the enhanced activity associated with an α-cyano substituent. These advances led to fenvalerate (**28**) and the fully resolved esfenvalerate (**24**) (Fig. 1-17) (Nakayama *et al.*, 1979), now established as important agricultural insecticides.

When an α-cyano group was similarly introduced into 3-phenoxybenzyl dichloro- and dibromovinyldimethylcyclopropanecarboxylates one of the resulting combinations, the crystalline dibromo isomer called deltamethrin (**22**) (Fig. 1-13), was found to have insecticidal activity greater than that of any previous compound (Elliott *et al.*, 1974). A commercial route was developed to deltamethrin (Roussel-Uclaf, 1982), now probably the most widely applied synthetic pyrethroid (Anonymous, 1991).

C. Developments After First Photostable Pyrethroids

The more stable synthetic pyrethroids (Figs. 1-16 and 1-17) permethrin (**34**), cypermethrin (**18**), deltamethrin (**22**), and fenvalerate (**28**) were very successful in controlling many insect pests (e.g. lepidopterous larvae on cotton) at much lower rates of application than were required with other classes of insecticide.

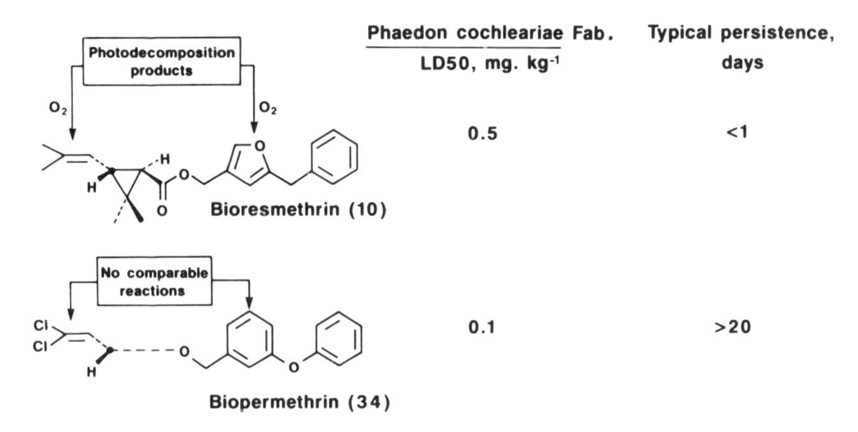

Figure 1-15. Development of a photostable pyrethroid.

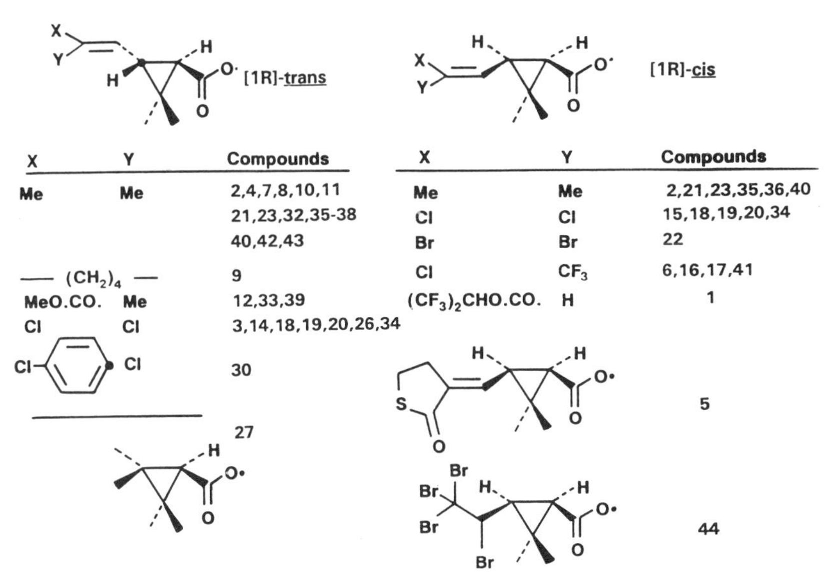

Figure 1-16. Dimethylcyclopropane acid components of pyrethroids.

Permethrin and fenvalerate became available (1977) just as overly-frequent application of organochlorine, carbamate and organophosphorus insecticides to many crops worldwide had led to extensive pest resistance. Such resistance was sometimes so great that control was lost or only possible with such high levels of insecticide that the health of pest control operators was threatened.

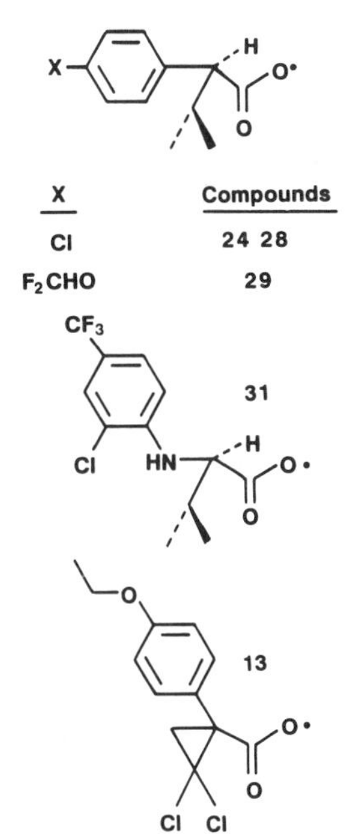

Figure 1-17. Non-dimethylcyclopropane acid components of pyrethroids.

Emergency clearance for application of pyrethroids to cotton was therefore granted and led to rapid commercial promotion of the new class of insecticides.

Although sufficiently stable for efficient action, residues of pyrethroids did not over persist to contaminate the environment but were degraded to more polar, inactive metabolites (Demoute, 1989). These desirable properties stimulated examination of esters from a wide range of related acids and alcohols. Fig. 1-16 indicates the structures of compounds derived from *trans*- and *cis*- dimethylcyclopropanecarboxylic acids and Fig. 1-17 those stemming from Sumitomo's discovery of the activity of α-isopropylphenylacetates. Comparably, Fig. 1-18 indicates structures of the alcohol components of such esters. However, many more compounds (review: Naumann, 1990) have been synthesized and tested in academic and industrial laboratories in the course of developing these products and some have special activities. For example, a 4-fluoro substituent on the benzyl group (Hamman and Fuchs, 1981) in cyfluthrin (**14**) enhances potency generally while replacement of one chlorine in a *trans* side-chain with a 4-chlorophenyl group as in flumethrin (**30**) gives specific activity against cattle ticks (Schnitzerling *et al.*, 1989). Crystalline equiproportion mixtures such as α-cypermethrin (**19**) (Fisher *et al.*, 1983; FMC, 1981) and λ-cyhalothrin (**17**)

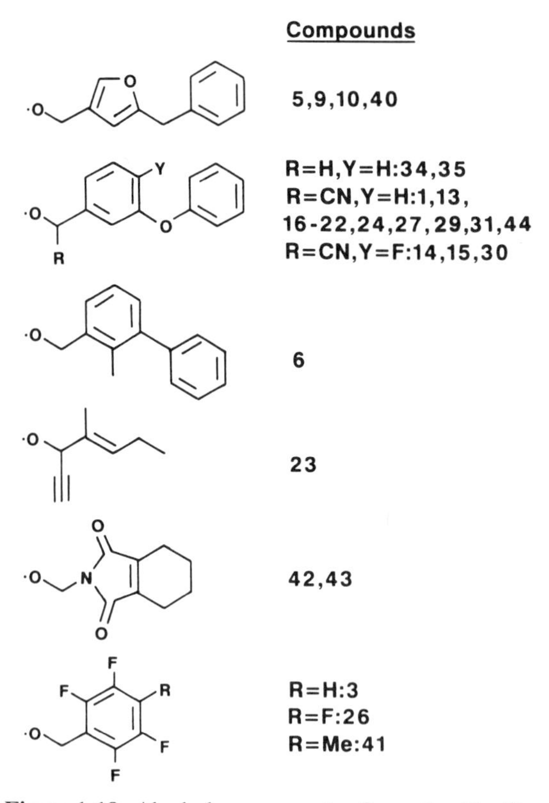

Figure 1-18. Alcohol components of pyrethroids: II.

(Robson *et al.*, 1984) of the most active isomers in *cis*-cypermethrin (**18**) and cyhalothrin (**16,17**), respectively, with their inactive enantiomers are very active insecticides manufactured without specific resolution steps. Bifenthrin (**6**) (Plummer *et al.*, 1983) and acrinathrin (**1**) (Tessier *et al.*, 1983) are active against mites as well as insects. Tralomethrin (**44**) is a two diastereoisomer mixture from addition of bromine to deltamethrin (Roussel-Uclaf, 1978). Cycloprothrin (**13**) (Holan *et al.*, 1978) has diminished fish toxicity and is valuable against pests of rice. The range of structural variations and biological properties of pyrethroids is extended with fluvalinate and tau-fluvalinate (**31**) (Henrick *et al.*, 1980; Anderson *et al.*, 1985) which control spider mites, and are nontoxic to and repel honey bees. Tefluthrin (**41**) (Jutsum *et al.*, 1986) and related compounds (**3**, **26**) have higher vapor pressures than most other pyrethroids and consequently are active against soil pests such as the corn rootworm. The relatively simple compound fenpropathrin (**27**) (Sumitomo Chemical Co., 1973) is an acaricide.

The natural pyrethrins are renowned for their rapid as well as for their powerful action against insects; the pyrethrates (**12**), (**33**) and (**39**) are particularly effective for "knockdown" (Sawicki and Thain, 1962). In general, the more photostable pyrethroids do not act so rapidly as the natural compounds, but the allethrins (**2**, **7** and **8**), tetramethrin (**42**, **43**) (Kato *et al.*, 1964) and kadethrin (**5**) (Lhoste and Rauch, 1976) are fast knockdown agents. Recently prallethrin (**36**) (Matsunaga *et al.*, 1988) has been found to act faster and more effectively than allethrin; empenthrin (**23**) (Fujita, 1984) is valuable in applications that exploit its volatility (e.g. clothes moth control).

D. Nonester Pyrethroids

The structures of cycloprothrin (**13**) and particularly of tau-fluvalinate (**31**) illustrate well the potential for development and refinement of activity once a lead structure such as fenvalerate (**28**) has been established. This is particularly well exemplified by the nonester pyrethroids originated by the chemists of the Mitsui-Toatsu Company from an examination of common features in the structures of permethrin and fenvalerate (Fig. 1-19) (Udagawa, 1988). This stimulated preparation of an ester, showing some activity, in which the 4-chlorophenyl unit had been shifted to the carbon bearing the *gem*-dimethyl group. An ester lacking a methylene group was more active, but most remarkable was the potency of the ether in which methylene replaced the ester carbonyl. These findings led to etofenprox and MTI 800 (Fig. 1-19), effective against a wide range of pests of cereals, rice, top fruit, and vegetables, but having low toxicity to mammals and fish (Udagawa *et al.*, 1985).

Such new compounds suggested further structural variants (Fig. 1-20). Ethers with a single trifluoromethyl group were found active (Tsushima *et al.*, 1988; Bushell, 1990) and flufenprox was developed commercially (Zeneca). Since silicon (Sieburth *et al.*, 1990) and tin (Tsushima *et al.*, 1989) analogs, cyclopropyl substituted hydrocarbons and oxime ethers (Fig. 1-20) are all active, there is now scope for further discovery of active compounds, with either a broad or a narrow spectrum of activity.

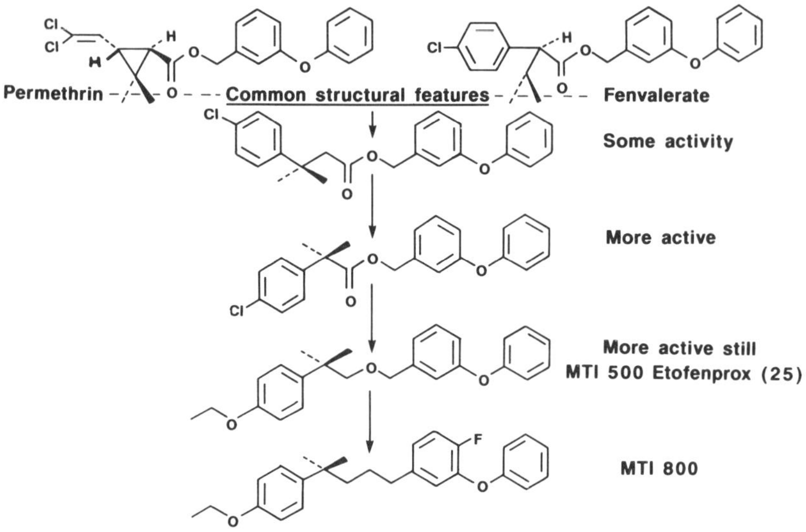

Figure 1-19. Development of nonester pyrethroids.

E. Toxicology of Synthetic Pyrethroids

An important characteristic of pyrethroids is their differential potency between insects and mammals (Casida *et al.*, 1983). These favorable toxicological properties depend on the ability of compounds to penetrate rapidly to, and interact with, sites of action in insects, especially sodium channels of the nerve membrane, where compounds such as deltamethrin are active at concentrations as low as 10^{-12} M (Sattelle and Yamamoto, 1988). In contrast, after external or oral administration to mammals, pyrethroids are largely converted by hydrolytic or oxidative attack to polar metabolites which are then eliminated unchanged or as conjugates in the faeces or urine, before sensitive sites can be reached (Litchfield, 1985). The importance of the route of administration is shown by the very low toxicities of pyrethroids to birds and the contrasting high toxicities to fish, in the bodies of which pyrethroids in water gain access directly into the bloodstream via the gills.

Toxicological and neuroactive properties of pyrethroids are discussed in two comprehensive, authoritative reviews (Aldridge, 1990; Vijverberg and van den Bercken, 1990).

F. Pyrethroids in the Environment

Commercially available photostable pyrethroids are effective in the field at 200 g/ha or less and the most active compounds (deltamethrin, resolved cypermethrin or cyhalothrin) at 10–25 g/ha or even 5 g/ha for special applications. These compounds persist on crops for 7–30 days while infestations are controlled; their physical properties permit resistance to migration in aqueous solution and in wind currents. Residues reaching the soil are

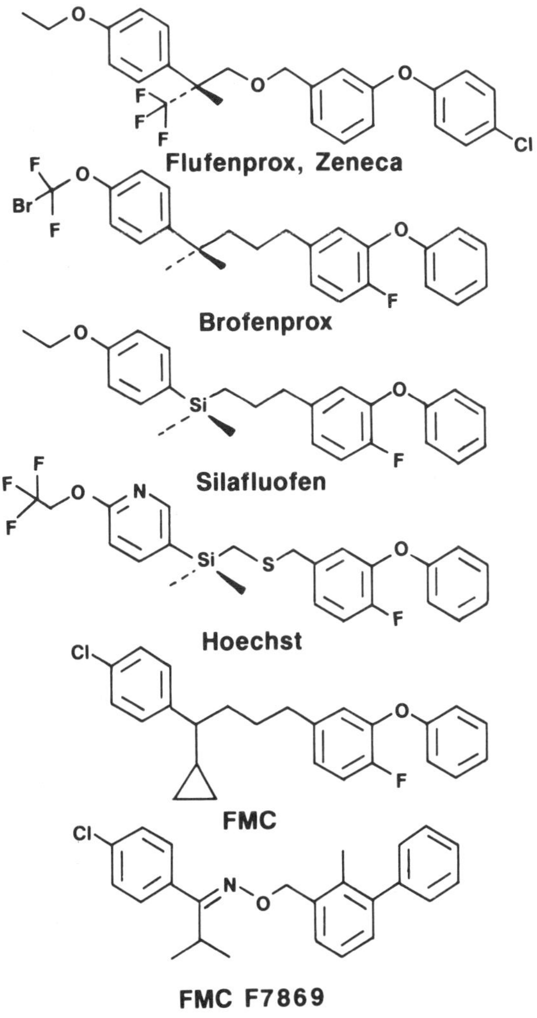

Flufenprox, Zeneca

Brofenprox

Silafluofen

Hoechst

FMC

FMC F7869

Figure 1-20. Developed and proposed nonester pyrethroids.

metabolized to polar products which are bound to soil particles and do not accumulate to contaminate the environment. Impact on fish to which all pyrethroids are more or less active is much diminished because concentrations are lowered by adsorption onto competing lipophilic material in river banks, pond sediments, and organic matter in general (further details and leading references: Demoute, 1989).

IV. RESISTANCE AND MODE OF ACTION OF INSECTICIDES

If insect pests did not develop resistance, the range of insecticides of various classes reviewed here, combined when appropriate with semiochemical attractants and repellents (Pickett, 1988; Pickett *et al.*, 1991), should be adequate to deal with nearly all insect related problems. Unfortunately, successive use of various insecticide groups has led to broad cross-resistance and by 1990, 504 species of insect had been reported to resist insecticides of one or more groups (Georghiou, 1990). For 60–70 species, many affecting economically important crops, control had become extremely difficult, with grave implications for human welfare and health (Denholm and Rowland, 1992). Although the insecticides represented in this review constitute a formidable array of powerful compounds, new effective products are now increasingly rare and expensive to discover and register. Biotechnology and biological control methods can have only limited impact in the immediate future, but fewer treatments delay the onset of resistance and alternative agricultural practices diminish pest populations (Pickett, 1992; Denholm and Rowland, 1992; Croft, 1990; Sawicki and Denholm, 1987). Such approaches are being progressively implemented and are proving successful, for example against pests of cotton (Sawicki and Denholm, 1987, 1989; Forrester, 1990).

The need to overcome resistance has emphasized the importance of understanding the modes of action and metabolism of insecticides (recent authoritative review: Roush and Tabashnik, 1990). Casida (1990) commented upon the possible finite number of biochemical targets useful for pesticide action. Among these the voltage-dependent sodium channel is the site for pyrethroids and DDT, which are cross resistant (Soderlund and Bloomquist, 1990). An important aspect of this relationship must be the structural and biological link between pyrethroids and DDT which Holan *et al.* (1978, 1985) exploited in deriving cycloprothrin (**13**) (Fig. 1-17) and between superficially dissimilar pyrethroids such as pyrethrin I (**37**) and etofenprox (**25**) (Fig. 1-19). However a structural connection between (**37**) and (**25**) can be discerned via the sequences indicated in Figs. 1-13 and 1-19 and by considering related features in the three dimensional structures of dimethyl- and dichlorovinyldimethylcyclopropanecarboxylates (Fig. 1-16), and nondimethyl cyclopropanecarboxylate pyrethroids (Fig. 1-17) esterified with various alcohol components (Fig. 1-18) (Elliott, 1985).

Recognizing the implications of recent molecular biological research on the primary structure of sodium channels, Devonshire *et al.* (1992) anticipate the ability to characterize regions of the protein in house flies which differ in susceptible and resistant strains, giving insight into the topography of the target site for pyrethroids. Recent work has identified linkage between a mutation at the sodium channel locus and the *kdr* phenotype in house flies, providing clear genetic evidence for their association and enabling future detailed structural studies of this gene (Williamson *et al.*, 1993).

Resistance to gamma HCH and the three cyclodienes still used (Fig. 1-1) arises from nerve insensitivity associated with diminished binding of insecticides to the receptor in the GABA-gated chloride channel (Devonshire *et al.*, 1992).

The organophosphorus and carbamate insecticides act on insect acetylcholinesterases which are relatively insensitive in some resistant strains of insects. Besides this target-site related resistance, the other broad group of mechanisms is based on the enhanced ability to detoxify insecticides by glutathione *S*-transferases, monooxygenases, and esterases (Oppenoorth, 1985).

V. COMPARISON OF CLASSES OF INSECTICIDES

Table 1-3 shows mean values for the properties of insecticides relevant to their suitability for various applications in insect control. The insecticidal activities of organochlorine, carbamate, and organophosphorus compounds are only one-hundredth those of the more active pyrethroids such as deltamethrin (**22**), cyhalothrin (**16**), λ-cyhalothrin (**17**), α- and β-cypermethrin (**19** and **20**), and esfenvalerate (**24**). The diminished application rates thus possible, combined with lower toxicities to mammals and in conjunction with limited persistence in the soil and in the environment, markedly lessen the total insecticide burden.

Polarity ranges, expressed as octanol–water partition coefficients, are shown in Table 1-3 for the four main groups of insecticides; from these values the likelihood of compounds moving in the phloem (symplastically) or xylem (apoplastically) of plants may be judged (Bromilow and Chamberlain, 1989). Compounds having log P up to 3-4 are mobile in the xylem and they may move translaminarly even with somewhat higher log P values; however, lipophilic compounds (log $P>5$) will be immobile or nonsystemic. Most carbamates and some organophosphates have the valuable property of systemic action by which plant sucking pests such as aphids are most readily controlled. Organochlorine insecticides and pyrethroids are lipophilic compounds, and only slightly soluble or insoluble in water; they do not act systemically. When detoxified, pyrethroids are converted to more polar products (Section III.F), so structural modifications are unlikely to succeed in generating "systemic" pyrethroids according to present concepts of their structural requirements for activity. However, within the broad range of structures (Table 1-2, compounds **1-44**) of pyrethroids, acrinathrin (**1**), the allethrins (**2**,**7**,**8**), kadethrin (**5**), pyrethrin II (**39**) and the tetramethrins (**42**,**43**), which are somewhat more polar, paralyze ("knockdown") insects faster, possibly because they penetrate more rapidly (Briggs *et al.*, 1974; Elliott and Janes, 1978).

VI. CONCLUSION

Despite the desirability of alternative methods of pest control there is little prospect of feeding the world's increasing population and maintaining health

Table 1-3. Comparison of Classes of Insecticides

	Polarity (log P)[a]	Systemic action	Water solubility	Insects (mg/kg)[b]	Rats (mg/kg)	Selectivity factors[f]	Field rate (kg/ha)	Soil life (years)
Carbamate	−1 to 3	±	±	2.8	45[c]	16	0.7[g]	—
Organophosphorus	1 to 4.5	±	±	2.0	67[d]	33	0.6[h]	—
Organochlorine	4.5 to 7.5	—	—	2.6	230[e]	91	0.5–3	1–10
Pyrethroid	4 to 9	—	—	0.45	2,000[b]	4.500	0.01–0.2	<0.1
Deltamethrin	5.9	—	—	0.01	130	13,000	0.002–0.01	<0.1

[a] P = octanol–water partition coefficient.
[b] Arithmetic means for representative members (LD_{50}).
c Mean for 11 compounds (LD_{50}).
[d] Mean for 83 compounds (LD_{50}).
[e] Mean for 21 compounds (LD_{50}).
[f] Rats/insects.
[g] Means for 12 compounds.
[h] Mean for 45 compounds.

standards without insecticides of the types reviewed here. These compounds can still be applied only relatively inefficiently; therefore, a significant proportion of the dose does not reach the target and is potentially available to contaminate the environment. Such considerations indicate that where possible the most active compounds with the lowest mammalian toxicities and with limited persistence should be used. The data in Table 1-3 indicate, therefore, why the value of pyrethroids, which have these characteristics, has risen from about $10 million per annum in 1972 to some $1,400 million per annum today. This is about one-fifth of the total value (*ca.* $7,500 million) of all insecticides.

The natural pyrethrins have had a particularly beneficial influence on human health and affairs, both as compounds with fascinating structures and exceptional biological activities in their own right, and also by serving as the basis for a group of synthetic insecticides with a very favorable combination of properties.

ACKNOWLEDGMENT

I thank many friends for help and advice in preparing this review, the editors of the Pesticide Manual for permission to quote from it here, the British Technology Group for continued support and McLaughlin Gormley King Co. for inviting me to the symposium on which this book is based.

REFERENCES

Addor, R.W., Babcock, T.J., Black, B.C., Brown, D.G., Diehl, R.E., Furch, J.A., Kameswaran, V., Kamhi, V.M., and Kremer, K.A. (1992). Insecticidal pyrroles. Discovery and overview. *ACS Symp. Ser.* **504**, 283–297.

Aldridge, W.N. (1990). An assessment of the toxicological properties of pyrethroids and their neurotoxicity. *Crit. Rev. Tox.* **21**, 89–104.

Allan, G.G., and Miller, T.A. (1990). Long-acting pyrethrin formulations. *In* "Pesticides and Alternatives: Innovative Chemical and Biological Approaches to Pest Control" (J.E. Casida, ed.), pp. 357–364. Elsevier Science Publishers, Amsterdam.

Anderson, R.J., Adams, K.G., and Henrick, C.A. (1985). Synthesis and insecticidal activity of the stereoisomers of α-cyano-3-phenoxybenzyl 2-[2-chloro-4-(trifluoromethyl)anilino]-3-methylbutanoate (fluvalinate). *J. Agric. Food Chem.* **33**, 508–514.

Anonymous (1969). *Jpn. Pestic. Inf.*, No. 1, 22.

Anonymous (1991). Roussel Uclaf to gain from rising pyrethroidal demand. *Agrow*, No. 140, 3. World Agrochemical News, George Street Publications Ltd.

Baranyovits, F.L.C., and Ghosh, R. (1969). Pirimicarb (PP 062): a new selective carbamate insecticide. *Chem. Ind.* (*London*), 1018–1019.

Barthel, W.F. (1961). Synthetic pyrethroids. *Adv. Pest Contr. Res.* **4**, 33–74.

Berg, W., and Knutti, H.J. (1975). San 155 I—A new insecticide of a novel class of chemicals. *Proc. 8th Br. Insect. Fungic. Conf.* **2**, 683–691.

Black, A.L., Chiu, Y.C., Fahmy, M.A.H., and Fukuto, T.R. (1973). Selective toxicity of *N*-sulfenylated derivatives of insecticidal methylcarbamate esters. *J. Agric. Food Chem.* **21**, 747–751.

Blade, R.J. (1990). Some aspects of synthesis and structure-activity in insecticidal lipid amides. *In* "Recent Advances in the Chemistry of Insect Control II" (L. Crombie, ed.), pp. 151–169. The Royal Society of Chemistry, London.

Briggs, G.G., Elliott, M., Farnham, A.W., and Janes, N.F. (1974). Structural aspects of the knockdown of pyrethroids. *Pestic. Sci.* **5**, 643–649.

Bromilow, R.H., and Chamberlain, K. (1989). Mechanisms and regulation of transport processes: designing molecules for systemicity. *British Plant Growth Regulator Group, Monograph 18*, pp. 113–128.

Bushell, M.J. (1990). Synthesis of fluorinated non-ester pyrethroids. *In* "Recent Advances in the Chemistry of Insect Control II" (L. Crombie, ed.), pp. 125–141. The Royal Society of Chemistry, London.

Busvine, J.R. (1993). "Disease Transmission by Insects: Its Discovery and 90 years of Effort to Prevent it," pp. 1–361 + i–xii. Springer Verlag, Berlin.

Casida, J.E. (1970). Mixed-function oxidase involvement in the biochemistry of insecticide synergists. *J. Agric. Food Chem.* **18**, 753–772.

Casida, J.E. (ed.) (1973). "Pyrethrum, The Natural Insecticide." Academic Press, New York.

Casida, J.E. (1990). Pesticide mode of action: evidence for and implications of a finite number of biochemical targets. *In* "Pesticides and Alternatives: Innovative Chemical and Biological Approaches to Pest Control" (J.E. Casida, ed.), pp. 11–22. Elsevier Science Publishers, Amsterdam.

Casida, J.E., Gammon, D.W., Glickman, A.H., and Lawrence, L.J. (1983). Mechanisms of selective action of pyrethroids. *Ann. Rev. Pharmacol. Toxicol.* **23**, 413–438.

Casida, J.E., Cole, L.M., Hawkinson, J.E., and Palmer, C.J. (1990). Trioxabicyclooctanes: GABA receptor binding site and comparative toxicology. *In* "Recent Advances in the Chemistry of Insect Control II" (L. Crombie, ed.), pp. 212–234. The Royal Society of Chemistry, London.

Colliot, F., Kukorowski, K.A., Hawkins, D.W., and Roberts, D.A. (1992). Fipronil: A new soil and foliar broad spectrum insecticide. *In* "Brighton Crop Protection Conference—Pests and Diseases," Vol. 1, pp. 29–34. British Crop Protection Council, Farnham, UK.

Corbett, J.R., Wright, K., and Baillie, A.C. (1984). "The Biochemical Mode of Action of Pesticides," 2nd Ed., pp. 160–161. Academic Press, London.

Croft, B.A. (1990). Management of pesticide resistance in arthropod pests—research and policy issues. *In* "Managing Resistance to Agrochemicals—from Fundamental Research to Practical Strategies" (M. B. Green, H. M. Le Baron, and W. K. Moberg, eds.), pp. 148–168. *ACS Symp. Ser.*, **421**. Amer. Chem. Soc., Washington, D.C.

Crombie, L. (1994). This volume.

Crombie, L., Josephs, J.L., Cayley, J., Larkin, J., and Weston, J.B. (1992). The rotenoid core structure: modifications to define the requirements of the toxophore. *Bioorg. Med. Chem. Lett.* **2**, 13–16.

Demoute, J.-P. (1989). A brief review of the environmental fate and metabolism of pyrethroids. *Pestic. Sci.* **27**, 375–385.

Denholm, I., and Rowland, M.W. (1992). Tactics for managing pesticide resistance in arthropods: theory and practice. *Annu. Rev. Entomol.* **37**, 91–112.

Devonshire, A.L., Field, L.M., and Williamson, M.S. (1992). Molecular biology of insecticide resistance. *In* "Insect Molecular Science" (J. M. Crampton, and P. Eggleston, eds.), pp. 173–183. Academic Press, New York.

Diehr, H.-J., Gallenkamp, B., Jelich, K., Lantzsch, R., and Shiokawa, K. (1991). Synthesis and chemical-physical properties of the insecticide imidacloprid (NTN 33893). *Pflanzenschutz-Nachrichten Bayer* **44**(2), 107–112 and succeeding papers 113–194.

Dorn, S., Frischknecht, M.L., Martinez, V., Zurflueh, R., and Fischer, U. (1981). A novel non-neurotoxic insecticide with a broad activity spectrum. *Z. Pflanzenkr. Pflanzenschutz* **88**, 269–275.

Dreikorn, B.A., Thompson, G.D., Suhr, R.G., Worden, T.V., and Davis, N.L. (1991). The discovery and development of fenazaquin (EL-436), a new broad spectrum acaricide. *VII Internatl. Cong. Pestic. Chem., Hamburg, 1991.* Abstract 01A 33.

Elbert, A., Overbeck, H., Iwaya, K., and Tsuboi, S. (1990). Imidacloprid, a novel systemic nitromethylene analogue insecticide for crop protection. *In* "Brighton Crop Protection Conference—Pests and Diseases," Vol. 1, pp. 21–28. British Crop Protection Council, Farnham, UK.

Elliott, M. (1958). Isolation and purification of (+)-pyrethrolone from pyrethrum extract: reconstitution of pyrethrins I and II. *Chem. Ind.* (*London*), 685–686.

Elliott, M. (1964a). The pyrethrins and related compounds. Part III Thermal isomerization of *cis*-pyrethrolone and its dervatives. *J. Chem. Soc.*, 888–892.

Elliott, M. (1964b). The pyrethrins and related compounds. Part V. Purification of (+)-pyrethrolone as the monohydrate, and the nature of "Pyrethrolone-C." *J. Chem. Soc.*, 5225–5228.

Elliott, M. (1979). Progress in the design of insecticides. *Chem. Ind.* (*London*), 757–768.

Elliott, M. (1985). Lipophilic insect control agents. *In* "Recent Advances in the Chemistry of Insect Control" (N.F. Janes, ed.), pp. 73–102. The Royal Society of Chemistry, London.

Elliott, M., Farnham, A.W., Janes, N.F., Needham, P.H., and Pulman, D.A. (1974). Synthetic insecticide with a new order of activity. *Nature* **248**, 710–711.

Elliott, M., and Janes, N.F. (1978). Synthetic pyrethroids — a new class of insecticide. *Chem. Soc. Rev.* 7, 473–505.
Farkas, J., Kourim, P., and Sorm, F. (1959). Relation between chemical structure and insecticidal activity in pyrethroid compounds. I. An analogue of chrysanthemic acid containing chlorine in the side chain. *Coll. Czech. Chem. Commun.* **24**, 2230–2236.
Fisher, J.P., Debray, P.H., and Robinson, J. (1983). WL85871 — A new multipurpose insecticide. *In* "Plant Protection for Human Welfare." *Proc. 10th Internat. Plant Prot.*, Vol. 1, pp. 452–459. British Crop Protection Council, Croydon, UK.
Flückiger, C.R., Kristinsson, H., Senn, R., Rindlisbacher, A., Baholzer, H., and Voss, G. (1992). CGA 215'944, a novel agent to control aphids and whiteflies. *In* "Brighton Crop Protection Conference — Pests and Diseases," Vol. 1, pp. 43–50. British Crop Protection Council, Farnham, UK.
FMC (1981). U.S. Patent 4, 261, 921.
Forrester, N.W. (1990). Designing, implementing and servicing an insecticide resistance management strategy. *Pestic. Sci.* **28**, 167–179.
Fujita, Y. (1984). Stages leading to a new pyrethroid — the case of Vapothrin. *Sumitomo World*, No. 3, 2–4.
Georghiou, G.P. (1990). Overview of insecticide resistance. *In* "Managing Resistance to Agrochemicals, from Fundamental Research to Practical Strategies" (M.B. Green, H.M. Le Baron, and W.K. Moberg, eds.), pp. 18–41, *ACS, Symp. Ser.* **421**. Amer. Chem. Soc., Washington, D.C.
Green, M.B., Hartley, G.S., and West, T.F. (1987). "Chemicals for Crop Improvement and Pest Management," 3rd Ed., 59–65. Pergamon Press, Oxford.
Gill, S.G., Cowles, E.A., and Pietrantonio, P.V. (1992). The mode of action of *Bacillus thuringiensis* endotoxins. *Ann. Rev. Entomol.* **37**, 615–636.
Grosscurt, A.C., ter Haar, M., Jongsma, B., and Stoker, A. (1988). PH 70-23: A new acaricide and insecticide interfering with chitin deposition. *Pestic. Sci.* **22**, 51–59.
Hall, R.D., and Fochse, M.C. (1980). Laboratory and field tests of CGA-72662 for control of house fly and face fly in poultry, bovine or swine. *J. Econ. Entomol.* **73**, 564–569.
Hamman, I., and Fuchs, R. (1981). Baythroid, a new insecticide. *Pflanzenschutz-Nachrichten Bayer* **34**, 121–151.
Haynes, H.L., Lambrech, J.A., and Moorefield, H.M. (1957). Insecticidal properties and characteristics of 1-naphthyl N-methylcarbamate. *Contrib. Boyce Thomson Inst.* **18**, 507–13.
Heller, J.J., Mattioda, H., Klein, E., and Sagenmüller, A. (1992). Field evaluation of RH 5992 on lepidopterous pests in Europe. *In* "Brighton Crop Protection Conference — Pests and Diseases," Vol. 1, pp. 45–64. British Crop Protection Council, Farnham, UK.
Henrick, C.A., Staal, G.B., and Siddall, J.B. (1973). Alkyl 3,7,11-trimethyl-2,4-dodecadienoates, a new class of potent insect growth regulators with juvenile hormone activity. *J. Agric. Food Chem.* **21**, 354–359.
Henrick, C.A., Garcia, B.A., Staal, G.B., Cerf, D.C., Anderson, R.J., Gill, K., Chinn, H.R., Labovitz, J.N., Leippe, M.M., Woo, S.L., Carney, R.L., Gordon, D.C., and Kohn, G.K. (1980). 2-Anilino-3-methylbutyrates and 2-(isoindolin-2-yl)-3-methylbutyrates, two novel groups of synthetic pyrethroid esters not containing a cyclopropyl ring. *Pestic. Sci.* **11**, 224–241.
Holan, G., O'Keefe, D.F., Virgona, C., and Walser, R. (1978). Structural and biological link between pyrethroids and DDT in new insecticides. *Nature* **272**, 734–736.
Holan, G., Johnson, W.M.P., O'Keefe, D.F., Quint, G.L., Rihs, K., Sparling, T.H., Walser, R., Virgona, C.T., Frelin, C., Lazdunski, M., Johnston, G.A.R., and Chen Chow, S. (1985). Multidisciplinary studies in the design of new insecticides. *In* "Special Publication No. 53" (N. F. Janes, ed.), pp. 115–131. The Royal Society of Chemistry, London.
Homeyer, B. (1975). Curaterr (carbofuran), a broad spectrum root-systemic insecticide and nematocide. *Pflanzenschutz-Nachrichten Bayer* **28**, 3–54.
Ishaaya, I. (1990). Benzoylphenyl ureas and other selective insect control agents — mechanism and application. *In* "Pesticides and Alternatives — Innovative Chemical and Biological Approaches to Pest Control" (J. E. Casida, ed.), pp. 365–376. Elsevier Science Publishers, Amsterdam.
Jackson, G.J. (1989). The pyrethroid insecticides: a scientific advance for human welfare? *Pestic. Sci.* **27**, 335–467.
Johnson, M.K. (1988). Sensitivity and selectivity of compounds interacting with neuropathy target esterase. Further structure-activity studies. *Biochem. Pharmacol.* **37**, 4095–4104.
Jutsum, A.R., Gordon, R.F.S., and Ruscoe, G.N.E. (1986). Tefluthrin, a novel pyrethroid soil insecticide. *In* "Brighton Crop Protection Conference — Pests and Diseases," pp. 97–106. British Crop Protection Council, Croydon, UK.
Kanno, H., Ikeda, K., Asai, T., and Maekawa, S. (1981). 2-tert-Butylimino-3-isopropyl-5-

phenylperhydro-1,3-5-thiadiazin-4-one (NNI-750). *In* "British Crop Protection Conference—Pests and Diseases," Vol. 1, pp. 59–66. British Crop Protection Council, Farnham, UK.
Kato, T., Ueda, K., and Fujimoto, K. (1964). New insecticidally active chrysanthemates. *Agric. Biol. Chem.* **28**, 914–915.
Lange, W., and Krueger, C.V. (1932) Über Ester der Monofluorphosphorsäure. *Ber. Dtsch. Chem. Ges.* **65**, 1598–1601.
Leahey, J.P. (1985). "The Pyrethroid Insecticides." Taylor and Francis, London.
Leisy, D.J., and van Beek, N. (1992). Baculoviruses: possible alternatives to chemical insecticides. *Chem. Ind.* (*London*), 250–254.
Ley, S. V., Denholm, A.A., and Wood, A. (1993). The chemistry of azadirachtin. *Natural Products Report* **10**, Part 2, 109–157.
Litchfield, M.H. (1985). Toxicity to mammals. *In* "The Pyrethroid Insecticides" (J. P. Leahey, ed.), pp. 99–150. Taylor and Francis, London.
Lhoste, J., and Rauch, F. (1976). RU 15,525, a new pyrethroid with a very strong knockdown effect. *Pestic. Sci.* **7**, 247–250.
Lovell, J.B., Wright, Jr., D.P., Gard, I.E., Miller, T.P., Treacy, M.F., Addor, R.W., and Kamhi, V.M. (1990). AC 303,630—An insecticide/acaricide from a novel class of chemistry. *In* "Brighton Crop Protection Conference—Pests and Diseases," Vol. 1, pp. 37–42. British Crop Protection Council, Farnham, UK.
Maas, W., van Hess, R., Grosscurt, A.C., and Deul, D.H. (1981). Benzoylphenylurea insecticides. *Chem. Pflanzenschutz Schaedlingsbekaempfungsmittel* **6**, 423–470.
Martin, H. (ed.) (1972). "The Pesticide Manual," 3rd Ed. British Crop Protection Council, Croydon, UK.
Matsunaga, T., Makita, M., Higo, A., Nishibe, I., Dohara, K., and Shinjo, G. (1988). Studies on prallethrin, a new synthetic pyrethroid for indoor applications. I. The insecticidal activities of prallethrin isomers. *Sumitomo World*, No. 11, 1921.
Mellanby, K. (1992). "The DDT Story." British Crop Protection Council, Farnham, UK.
Metcalf, R.L., and Metcalf, R.A. (1993). "Destructive and Useful Insects: Their Habits and Control." McGraw-Hill, New York.
Metcalf, R.L., Fukuto, T.R., and Winton, M.Y. (1962). Insecticidal carbamates: position isomerism in relation to activity of substituted phenyl *N*-methylcarbamates. *J. Econ. Entomol.* **55**, 889–894.
Miyamoto, J. (1990). Pesticides in the 21st century. *In* "Recent Developments in the Field of Pesticides and their Application to Pest Control," Proceedings of an International seminar, Shenyang, China, October 8–12, pp. 377–389.
Nakayama, I., Ohno, N., Aketa, K., Suzuki, Y., Kato, T., and Hirosuke, Y. (1979). Chemistry, absolute structures and biological aspect of the most active isomers of fenvalerate and other recent pyrethrins. *Adv. Pestic. Sci.*, Part 2, 174–181.
Naumann, K. (1989). Acetylcholinesterase inhibitors. *In* "Progress and Prospects in Insect Control" (N.R. McFarlane, ed.), Monograph No. 43, pp. 21–41. British Crop Protection Council, Farnham, UK.
Naumann, K. (1990). "Chemistry of Plant Protection," Vol. 4. "Synthetic Pyrethroid Insecticides: Structures and Properties," 241 pp., Vol. 5. "Synthetic Pyrethroid Insecticides: Chemistry and Patents," 350 pp. Springer-Verlag, Berlin.
Ohno, N., Fujimoto, K., Okuno, Y., Mizutani, T., Hirano, M., Itaya, N., Honda, T., and Yoshioka, H. (1976). 2-Alkylalkanoates, a new group of synthetic pyrethroid esters not containing cyclopropanecarboxylates. *Pestic. Sci.* **7**, 241–246.
Oppenoorth, F.J. (1985). Biochemistry and genetics of insecticide resistance. *In* "Comprehensive Insect Physiology, Biochemistry and Pharmacology" (G. S. Kerkut and L. I. Gilbert, eds.), Vol. 12, Insect Control, pp. 731–773. Pergamon Press, Oxford.
O'Shea, M. (1985). Neuropeptides in insects: possible leads to new control methods. *In* "New Approaches in Insect Control" (H. von Keyserlingk, ed.), pp. 133–151. Springer-Verlag, Berlin.
O'Shea, M., and Schaffer, M. (1985). Neuropeptide function: the invertebrate contribution. *Ann. Rev. Neurosci.* **8**, 171–198.
Pickett, J.A. (1988). Chemical pest control—the new philosophy. *Chem. Brit.*, 137–142.
Pickett, J.A. (1992). Potential of novel chemical approaches for overcoming insecticide resistance. *In* "Resistance '91: Achievements and Developments in Combatting Pesticide Resistance" (I. Denholm, L. Devonshire, and D. W. Hollomon, eds.), pp. 354–365. Elsevier Science Publishers, London.
Pickett, J.A., Wadhams, L.J., and Woodcock, C.M. (1991). New approaches to the development of semiochemicals for insect control. *In* "Proceedings of a Conference on Insect Ecology, Tabor, 1990," pp. 333–345. SPB Academic Publishing, The Hague.
Plummer, E.L., Cardis, A.B., Martinez, A.J., van Saun, W.A., Palmere, R.M., Pincus, D.S., and

Steward, R.K. (1983). Pyrethroid insecticide derived from substituted biphenyl-3-yl methanols. *Pestic. Sci.* **14**, 560–567.
Post, L.C., and Mulder, R. (1974). Insecticidal properties and mode of action of 1-(2,6-dihalobenzoyl)-3-phenylureas. *ACS Symp. Ser.* **2**, 136–143.
Robson, M.J., Cheetham, R., Flethes, D.J., and Crosby, J. (1984). Synthesis and biological properties of PP 321, a novel pyrethroid. *In* "British Crop Protection Conference," pp. 853–857. British Crop Protection Council, Croydon, UK.
Roussel-Uclaf (1978). New esters of cyclopropane-carboxylic acids with a polyhalogenated substituent. Preparation and pesticidal compositions. *Ger. Offen* 2,742,546.
Roussel-Uclaf (1982). "Deltamethrin Monograph," 412 pp. Roussel-UCLAF, Paris.
Roush, R.T., and Tabashnik, B.E. (eds.) (1990). "Pesticide Resistance in Arthropods," 303 pp. Chapman and Hall, New York
Sakai, M., Sato, Y., and Kato, M. (1967). Insecticidal activity of 1,3-bis-(carbamoylthio)-2-(N,N-dimethylamino)propane hydrochloride, Cartap with special reference to the effectiveness for controlling the rice stem borer. *Jpn. J. Appl. Entomol. Zool.* **11**, 125.
Sattelle, D.B., and Yamamoto, D. (1988). Molecular targets of pyrethroid insecticides. *Adv. Insect Physiol.* **20**, 147–213.
Sawicki, R.M. (1985). Resistance to pyrethroid insecticides in arthropods. *In* "Insecticides" (D. H. Hutson and T. R. Roberts, eds.), pp. 143–192. John Wiley and Sons Ltd., Chichester.
Sawicki, R.M., and Denholm, I. (1987). Management of resistance to pesticides in cotton pests. *Trop. Pest Manage.* **33**, 262–272.
Sawicki, R.M., and Denholm, I. (1989). Insecticide resistance management revisited. *Brit. Crop Prot. Council Monog.*, No. 43, Progress and Prospects of Insect Control, pp. 193–203.
Sawicki, R.M., and Thain, E.M. (1962). Insecticidal activity of pyrethrum extract and its four insecticidal constituents against houseflies. IV. Knockdown activities of the four constituents. *J. Sci. Food Agric.* **13**, 292–297.
Schechter, M.S., Green, N., and LaForge, F.B. (1949). Constituents of pyrethrum flowers. XXIII. Cinerolone and the synthesis of related cyclopentenolones. *J. Amer. Chem. Soc.* **71**, 3165–3173.
Schnitzerling, H.J., Nolan, J., and Hughes, S. (1989). Toxicology and metabolism of isomers of flumethrin in larvae of pyrethroid-susceptible and resistant strains of the cattle tick *Boophilus microplus* (Acari: Ixodidae). *Exp. Appl. Acarol.* **6**, 47–54.
Schrader, G. (1963). "Die Entwicklung neuer Insektizide auf Grundlage organischer Fluor- und Phosphorverbindungen," 3rd Ed. Verlag Chemie, Weinheim.
Sieburth, S. McN., Lin, S.Y., and Cullen, T.G. (1990). New insecticides by replacement of carbon by other Group IV elements. *Pestic. Sci.* **29**, 215–225.
Smith, C. (1992). "Pyrethroid Pesticides—The Market (DS61)," p. 73. PJB Publications Ltd., Surrey, UK.
Soderlund, D.M., and Bloomquist, J.R. (1990). Molecular mechanisms of insecticide resistance. *In* "Pesticide Resistance in Arthropods" (R.T. Roush and B.E. Tabashnik, eds.), pp. 58–96. Chapman and Hall, New York.
Soloway, S.B., Henry, A.C., Kollmeyer, W.D., Padgett, W.M., Powell, J.E., Roman, S.A., Tieman, C.H., Corey, R.A., and Horne, C.A. (1978). Nitromethylene heterocycles as insecticides. *In* "Pesticides and Venom Neurotoxicity" (D. L. Shankland, R.M. Hollingworth, and T. Smyth, Jr., eds.), pp. 153–158. Plenum Press, New York.
Sousa, A.A., Frazee, J.R., Weiden, M.H.J., and D'Silva, T.D.J. (1977). UC 51762, a new carbamate insecticide. *J. Econ. Entomol.* **70**, 803–807.
Staudinger, H., and Ruzicka, L. (1924). Insektentötende Stoffe I-V and X. *Helv. Chim. Acta* **7**, 177–259 and 448–458.
Staudinger, H., Muntwyler, O., Ruzicka, L., and Seibt, S. (1924). Insektentötende Stoffe VII. *Helv. Chim. Acta* **7**, 390–406.
Sumitomo Chemical Co. (1973). Alpha-cyanobenzylesters of cyclopropane carboxylic acid compounds. Preparation and use as insecticides and acaricides. *Ger. Offen.* 2,231,312.
Takehashi, H., Mitsui, J., Takakusa, N., Matsuda, M., Yoneda, H., Suziki, J., Ishimitsu, K., and Kishimoto, T. (1992). NI-25, a new type of systemic and broad spectrum insecticide. *In* "Brighton Crop Protection Conference—Pests and Diseases," Vol. 1, pp. 89–96. British Crop Protection Council, Farnham, UK.
Tessier, J.R., Tèche, A.P., and Demoute, J.P. (1983). Synthesis and properties of new pyrethroids, diesters of the nor-pyrethric series. *IUPAC Pesticide Chemistry* **1**, 95–100.
Tomizawa, M., and Yamamoto, I. (1992). Binding of nicotinoids and the related compounds to the insect nicotinic acetylcholine receptor. *J. Pestic. Sci.* **17**, 231–236.
Tsushima, K., Yano, T., Takagaki, T., Matsuo, N., Hirano, M., and Ohno, N. (1988). Preparation

and insecticidal activity of optically active 3-phenoxybenzyl 3,3,3-trifluoro-2-phenylpropyl ethers. *Agric. Biol. Chem.* **52**, 1323–1325.

Tsushima, K., Yano, T., Umeda, K., Matsuo, N., Hirano, M., and Ohno, N. (1989). Synthesis, insecticidal activity and pyrethroidal mode of action of new tin ether derivatives. *Pestic. Sci.* **25**, 17–23.

Udagawa, T. (1988). Trebon (etofenprox), a new insecticide. *Jpn. Pestic. Inf.*, No. 53, 9–13.

Udagawa, T., Numara, S., Oda, K., Shivaishi, S., Kodaka, K., and Nakatani, K. (1985). A new type of synthetic pyrethroid. *In* "Recent Advances in the Chemistry of Insect Control" (N. F. Janes, ed.), pp. 192–204. The Royal Society of Chemistry, London.

Ueda, K., Gaughan, L.C., and Casida, J.E. (1974). Photodecomposition of resmethrin and related pyrethroids. *J. Agric. Food Chem.* **22**, 212–220.

Unterstenhöfer, G. (1963). Über zwei neue insektizid wirksame Carbamate. *Meded. Landbouwhogesch. Opzoekingsstn. Staat Gent.* **28**, 758–766.

Umetsu, N. (1984). Modified methylcarbamate insecticides of selective toxicity. *Nippon Noyaku Gakkaishi* **9**, 169–180.

Velluz, L., Martel, J., and Nominé, G. (1969). Syntheses d'analogues de l'acide *trans*-chrysanthémique. *C. R. Acad. Sci., Paris, Ser. C* **268**, 2199–2203.

Vijverberg, H. P. M., and van den Bercken, J. (1990). Neurotoxicological effects and the mode of action of pyrethroid insecticides. *Crit. Rev. Tox.* **21**, 106–126.

Wachs, H. (1947). Synergistic insecticides. *Science* **105**, 530–531.

Weiden, M.H.J., Moorefield, H.H., and Payne, L.K. (1965). O-(Methylcarbamoyl)oximes: a new class of carbamate insecticide—acaricides. *J. Econ. Entomol.* **58**, 154–155.

Williamson, M.S., Denholm, I., Bell, C.A., and Devonshire, A.L. (1993). Knockdown resistance (*kdr*) to DDT and pyrethroid insecticides maps to a sodium channel gene locus in the housefly (*Musca domestica*). *Mol. Gen. Genet.* **240**, 17–22.

Wing, K.D., and Ramsay, J.K. (1989). Other hormonal agents: ecdysone agonists. *Brit. Crop Prot. Council Monog.* No. 43, 107–118.

Wood, H.A., and Granados, R.R. (1991). Genetically engineered baculoviruses as agents for pest control. *Ann. Rev. Microbiol.* **45**, 69–87.

Worthing, C.R., and Hance, R.J. (eds.) (1991). "The Pesticide Manual, A World Compendium," 9th Ed. British Crop Protection Council, Unwin Brothers Ltd., Old Woking, UK.

2

History of Pyrethrum in the 1970s and 1980s

WILLIAM D. GULLICKSON, SR.

I. INTRODUCTION

George McLaughlin, a warm and wonderful man, wrote a brief history of the pyrethrum industry to open a Pyrethrum Symposium in 1972 (McLaughlin, 1973). This update 20 years later covers some of the events in the interim presented from the author's personal point of view with no attempt to be complete but rather to record portions of the history that might otherwise be lost.

The 1970s and 1980s were challenging times for pyrethrum production and marketing. There were major developments in the agronomy sector, particularly in Kenya, where Dutch agronomists J.G. Parlevliet and J. Krull did 5 years work on pyrethrum at the Molo experimental station (Parlevliet, 1970, 1975a,b). Other experts working in Kenya at that time were Brewer (Brewer, 1973), McTaggart (McTaggart *et al.*, 1958), Otieno (Otieno and Pattenden, 1980), Ottaro (Ottaro, 1977) and Tuikong (Tuikong, 1984). The supply of pyrethrum was limited and somewhat erratic. The synthetic pyrethroids became well established. These were in fact interrelated since much of the problem in obtaining sufficient quantities of pyrethrum in the late 1970s and in the 1980s was that several effective synthetic pyrethroids became available. Many household aerosol producers, who had been major pyrethrum users, decided to switch to the more dependable synthetic pyrethroids. However, despite the loss of some major markets, the use of natural pyrethrum continues to grow.

Famous people associated with pyrethrum will play a major role in this compilation. One of them is Denys Glynne Jones, an employee of Endura SpA, Bologna, Italy, and former, long-term employee of the Kenya Pyrethrum Board. Denys wrote a short history of the pyrethrum industry in March, 1989 (Glynne Jones, 1989), in a monograph dedicated to Dr. Joseph Moore who was a good friend to many of the people who read this book and was a very close associate of all of us at McLaughlin Gormley King Co. (MGK). Another of them is Donald Maciver, a major contributor to pyrethrum science and marketing since 1960.[1] A third is Dr. Luis Levy, PhD from Stanford University, an Ecuadorean citizen who teaches chemistry at the University of Quito and owns and operates a botanical extraction plant under the name of INEXA. These "pyrethrum scientists" have provided much of the information on the history of pyrethrum during the past two decades. Emphasis is given to pyrethrum production in different parts of the world, the development of mosquito coils and water-based aerosols, and the impact of synthetic pyrethroids.

II. PYRETHRUM IN KENYA AND TANZANIA

Chapter 3 by Job Wainaina clearly lays out the reasons for the roller coaster production figures with which all of us are too familiar when it comes to pyrethrum. Fig. 2-1 and 2-2 give the relevant data from Kenya and Tanzania, the two largest producers in the 1970s and throughout most of the 80s.

III. PYRETHRUM IN ECUADOR

The pyrethrum industry in Ecuador for the past 40 years is closely linked to Dr. Luis Levy of Quito who provided a brief history for this chronicle (Levy, 1992). In 1941 there was a crisis in the local insecticide market of Ecuador because of the interruption of supplies of pyrethrum flowers, which had become

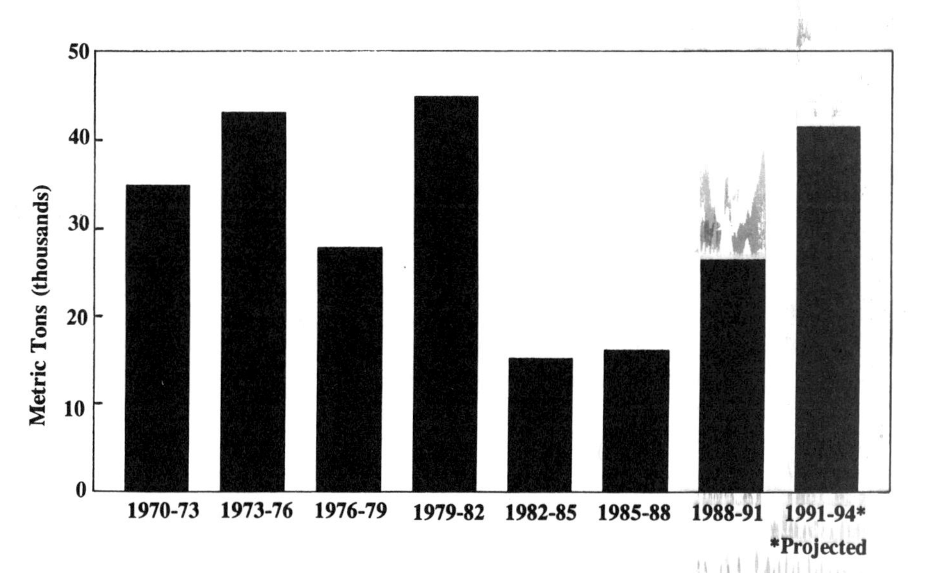

Figure 2-1. Production figures for pyrethrum flowers in Kenya in 1970–94. Data provided by Pyrethrum Board of Kenya.

[1] Don Maciver (Edinburgh University graduate in Chemistry with First Class Honors in special organic chemistry; member of The Royal Society of Chemistry) was a researcher on polymers before joining the Pyrethrum Board of Kenya in 1960 starting a continuing interest and commitment to pyrethrum. He has developed refining methods for pyrethrum and novel synergists (coinventor of Tropital®), and set up or directed the Kenya Pyrethrum Information Centers sequentially in Malaysia, Kenya, and Hong Kong in 1964–80. Since 1980 he has been senior scientist with Fairfield American, Wellcome Environmental Health, and the Roussel Corporation.

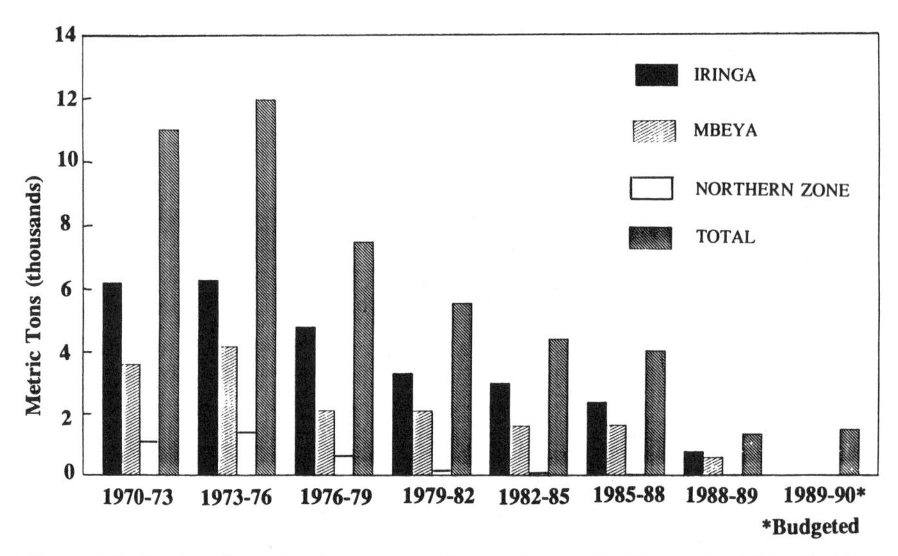

Figure 2-2. Regional production of pyrethrum flowers in Tanzania 1970–90. Data provided by Pyrethrum Board of Tanzania.

a "strategic material" for the war effort. This prompted the idea of perhaps growing pyrethrum in Ecuador as a way to remedy the shortage of insecticides in the country. Kaj Arends, retiring General Manager of the major insecticide producer of Ecuador, bought a farm near Ambato and planted the first pyrethrum with a packet of seeds, probably of Kenyan origin, purchased from a commercial seed firm in England. Samples of the first pyrethrum flowers harvested in 1942 by Arends were sent to a commercial laboratory in England for analysis. The pyrethrins content of 1.5% encouraged expansion of Arends' plantation to 5 hectares. The following year, after hearing about Arends' success, the Economic War Board of the United States requested the US Department of Agriculture to arrange for an independent effort of pyrethrum production in Ecuador. This was carried out on a farm owned by an Ecuadorean Government agency without much success. For the remainder of the 1940s Arends supplied pyrethrum flowers to the local insecticide industry, which was the original purpose for his plantation. In 1951 Arends felt that he was ready to increase his pyrethrum production beyond the requirements of the local industry. He contacted two American firms in search of encouragement and requested an advance purchase order for 5 to 20 metric tons of dried pyrethrum flowers to be dispatched two years later (1953). The firms he approached were US Industrial Chemicals Co. (USI), which operated a pyrethrum extraction plant in Baltimore, and S.B. Penick and Co. of New York. A positive answer from USI, signed by Russell B. Stoddard as manager of Pyrethrum Operations, was received a day before a similar acceptance came from Penick. Arends decided to confirm to USI because of the chronological order the acceptances were received.

Dr. Luis Levy entered the pyrethrum scene in 1952 when he toured Ecuador as a scientific journalist for the American Chemical Society. He met Kaj Arends and decided to publish the interview in *Chemical and Engineering News*. Upon reading the article, Russell B. Stoddard of USI contacted Levy and proposed a meeting in Quito toward the end of the year for a possible cooperation in the future. The following year the first shipment of dried flowers was dispatched by Arends to USI. The decisions taken at the first meeting of Stoddard and Levy were (1) to establish a pyrethrum research program at the Politechnic Institute of Quito, where Levy was Head of the Organic Chemistry Department, and (2) to encourage Levy's father (Dr. Leopoldo Levy) to produce domestic insecticide in Ecuador based on a kerosene percolation of pyrethrum flowers and combinations with USI's piperonyl butoxide, as well as a pyrethrum-based grain protectant. Both decisions were carried through. Levy Sr.'s products were marketed in Ecuador under the trademark "PIX." USI offered to fund the research program at the Politechnic Institute. The contract was signed on January 15, 1954. During the period of 1953-8 Levy and his colleagues published a series of technical and scientific papers on analytical aspects of pyrethrum (Levy and Estrada, 1954; Levy and Geller, 1959, 1960; Levy and Molina, 1957; Levy and Usubillaga, 1956a,b, 1957; Levy *et al.*, 1960). Over the next few years Dr. Levy branched out into the extraction of scopolamine from *Datura*, xanthophyll from marigolds, ginger, and other plant extractives. He did the first tissue culture work with pyrethrum in concert with Professor Emeritus Toshio Murishige of the Botany and Plant Sciences Department, University of California, Riverside. Dr. Levy produced 65 clones of high-content, vigorous pyrethrum plants, but the cost of harvesting large acreage in Ecuador made the clones merely an interesting, scientific curiosity.

In 1958 the insecticide division of USI was acquired by National Distillers Products, Inc. The team of Russell Stoddard, Herman Wachs (discoverer of piperonyl butoxide), John Rodda, George Kirbey, and Harry Iwata remained intact during the transfer. Subsequently, National Distillers Products sold this division to Food Machinery and Chemical Corp. (FMC). In close coordination with Russell Stoddard, now with FMC, Levy established the Ecuadorean firm, INEXA, C.A., for processing pyrethrum flowers into crude extract, based on a 10-year purchase contract for the pyrethrum extract. The contract came together with a financial package to allow Arends to expand his annual pyrethrum production to 400 metric tons. The Arends operation had become the Ecuadorean-American Pyrethrum Company (ECAMPYCO) and was joined by Kaj's brother Poul and then by a third brother, Knud. With the acceptance by FMC, Russell Stoddard became a minority stockholder of INEXA. S.B. Penick and Co., still hurting from the 1951 rebuff by Arends, established their own pyrethrum plantations in Ecuador through a subsidiary, S.B. Penick del Ecuador. In 1961 INEXA and FMC entered into a long-term contract with S.B. Penick whereby the latter's pyrethrum flower production was sold to INEXA, and the equivalent quantity of crude pyrethrum extract was sold by INEXA to FMC. This became the second source of pyrethrum flowers for INEXA/FMC.

On another front, serious disturbances occurred in the Belgian Congo (the Lumumba uprising). Pyrethrum flower production there ceased due to the mass exodus of European farmers. One of them, Leo Hamburger, visited the London office of Mitchell Cotts, his former pyrethrum clients, asking for employment. William Probert of Mitchell Cotts contacted Russell Stoddard of FMC, who in turn contacted Luis Levy of INEXA. Stoddard arranged for a meeting of Probert, Hamburger, and Levy in Quito. Mitchell Cotts and MGK entered into a joint venture for the large-scale production of pyrethrum flowers in Ecuador. The question of whether to have INEXA triple its extraction capacity, or build a second extraction plant was decided in favor of the second alternative. INEXA was placed in charge of building the second plant and of its management for 10 years. The Mitchell Cotts-MGK extraction plant was erected in 1965. In 1968 Leo Hamburger, through leased farms and on one farm north of Quito, produced almost 4,000 metric tons of dried pyrethrum flowers.

At the same time that a truly commercial crop of pyrethrum was harvested, a consortium of Gulf Oil and Texaco made major strikes in test wells located on the east side of the Andes. The promise of oil wealth drove the Ecuadorean politicians into a frenzy of social programs, one of which was a minimum wage law that included farm labor. As soon as that law was passed, the hand picking of pyrethrum by groups of 50–100 women and children was doomed. However, pyrethrum is still grown by small farmers who sell it to INEXA.

One of the fallouts from the impact of the pyrethrum shortage in the late 1970s occurred when the S.C. Johnson Wax people, who strongly supported natural pyrethrum, approached MGK with the idea that we, in consort with Dr. Levy, form a pyrethrum growing and extraction company in Ecuador. We did and in 1979 we incorporated the Pyrethrum Development Company, or PDC. In 1979 and 1980 we tried to grow commercial quantities of pyrethrum flowers on Cochasqui, a large farm about 40 miles north of Quito. Due mostly to weather problems, we were not successful.

IV. PYRETHRUM IN CHINA

In 1981 the S.C. Johnson people were approached by an Austrian firm which claimed that the Chinese in the Province of Yunnan wanted to buy the technology to grow, dry, extract, and refine pyrethrum. Despite all of our Latin American setbacks, PDC became potentially valuable because it possessed all of the information the Yunnan authorities wanted. On June 21, 1981 two members of the Austrian group, two people from S.C. Johnson, and Dr. Luis Levy made an information trip to Yunnan. For 7 days they looked at potential growing areas, and Dr. Levy, who had the most experience in growing pyrethrum, thought that a couple of the large farms they had seen would be suitable. At the end of the trip, Yunnan officials requested that PDC and the Austrians begin negotiations that could produce a contract for the agricultural technology, processing equipment, and marketing of pyrethrum extract. By May of 1982, negotiations with the Austrians had broken down since they insisted on being paid for an extraction plant in advance of any pyrethrum being planted. The key to the operation was obviously not an extraction plant, but rather the transfer of high-content pyrethrum tissue culture technology, leading to commercial plantings. As of June 1982, the Austrians were out of the picture, and Yunnan turned to PDC to furnish the technology they wanted.

A delegation of five from Yunnan visited MGK on December 7 and 8, 1982, going to Racine to visit S.C. Johnson on December 9 and 10, and then traveling to Quito to see the INEXA pyrethrum operation. Since the Chinese were seeking a crop for peasant farmers in the Himalayan foothill country 470 kilometers west of Kunming, Country of the Yi, poorest of the Chinese farmers in Yunnan, they were particularly impressed by the small farm production of pyrethrum in the area between Ambato and Riobamba in the vicinity of the volcano Chimborazo, an area about 100 kilometers south of Quito. They had a chance to observe picking, drying, and transport of the dried flowers to the extraction plant in Quito. Following surveys of the capabilities of the PDC partners, the Chinese returned to Yunnan, and in late spring 1983 an agreement was sent covering the agronomy, technology, and terms of payment, all of which had been discussed with the delegation while in the United States and Ecuador.

A PDC delegation went to Yunnan in August of 1983 to start, and hopefully, to finish negotiations on the technology transfer. We were real neophytes in dealing with the layers of bureaucrats and left Yunnan with nothing accomplished. We were told that road blocks had been removed and returned to Yunnan in December 1983 to try to execute a contract. On that trip we were taken to more promising pyrethrum growing areas. In Kunming the meetings soon turned into shambles as we began to realize that the farmers' interests were at the bottom of the pecking order as compared to University of Yunnan chemistry academics, import/export industries, local industrial development agencies and, perhaps, some jealous folks in Beijing who wanted to use the money we asked for, for their own local projects. At any rate, the PDC project for China was a washout and we accomplished nothing.

V. PYRETHRUM IN NEW GUINEA

As all of us who have had the pleasure of being involved in the pyrethrum industry are aware, we have had more than our share of competent people. One such is Alexander T. Dalgety, now of Craig Farms at Cornhill-on-Tweed, England. Alex was an employee of the Mitchell Cotts organization in East Africa from 1959 to 1967 where he was involved in all aspects of the pyrethrum industry. MGK and Mitchell Cotts had jointly built a pyrethrum extraction plant in Arusha, Tanzania in 1963. In 1965 we built a plant in Quito, and Alex went to Ecuador in 1967 to 1970. From 1971 to 1974 he worked as project manager on a pyrethrum extraction plant in Rwanda. In 1975 he was named project manager for a new pyrethrum extraction plant at Mafinga, Southern Tanzania. This experience proved to be invaluable in New Guinea.

To me, one of his most interesting projects was the replacement of the old 1963 Stafford Allen extraction plant at Kakamuga Natural Products Company, Mt. Hagen, Papua New Guinea. Alex first saw the plant in 1975 and it was in pretty bad shape. It was 12 years old, leaky, inefficient, and potentially dangerous. In 1980 Alex undertook a detailed study of the pyrethrum industry in New Guinea, including crop production, and a study of the extraction plant to determine what, if anything, was usable. The grinding mill was serviceable as were the boiler house and other services. Therefore, Alex proposed that a new, smaller extraction plant be built in the existing factory at Mt. Hagen. The new plant had a nominal capacity of 1 ton of dried flowers per day. The plant was to be built of modular construction in Ireland. On completion this plant would be "dry run" tested, dismantled, shipped to New Guinea, reerected on site, and fully commissioned. The plant was constructed in four modules, plus three base units. Each module was sized to fit inside a standard 20 foot steel shipping container. Handling facilities at the Port of Lae and transport at Mt. Hagen precluded the use of larger containers. Two modules contained the extraction vessels and all supporting structures, walkways, and piping. The third and fourth modules held the solvent evaporators, miscellaneous holding vessels, pumps, vacuum pumps, refrigeration plant, etc. Fabrication got under way in September 1981 and was completed in December for testing. It was then dismantled, packed into five, twenty-foot containers and loaded. The ship arrived at Lae on March 2, 1982, and over the next 10 days all five containers arrived by air at Mt. Hagen. By March 12 the old plant had been decommissioned, all solvent removed, and all main mechanical installation on the new plant completed. New solvent was put into the plant and full commissioning commenced exactly eight weeks after the first container arrived on site.

VI. PYRETHRUM IN AUSTRALIA AND TASMANIA (see also Chapter 4)

The commercial development of pyrethrum in Australia and Tasmania has an interesting history. The earliest report of pyrethrum production in Australia by Baron von Mueller, Director of Melbourne Botanical Gardens, stated that the crop was grown commercially around 1890 in the Lower Latrobe River, Victoria and used in the insecticide preparation "Insectobane" (von Mueller, 1895). In 1931 the Plant Introduction Section of the Division of Plant Industry introduced four strains of pyrethrum from Switzerland, England, Japan, and the United States. The crop was grown in October 1931 at Black Mountain, Canberra. It flowered in the second year and dry flower yield was estimated at 1200 to 1400 kg/ha with pyrethrins content ranging from 0.8 to 1.0%. After the initial experiments, the seed was made available to 170 farmers in New South Wales (NSW) and Victoria. The Domino Chemical Company of Canterbury, NSW took particular interest in this work and processed about 909 kg (1 ton) of locally-produced flowers. However, by 1938 Domino Chemical Company reported that they had failed to establish commercial production because the crop did not bring enough return to the growers. From 1932 to 1935 the NSW Department of Agriculture experimented with the crop at Hawkesbury and Bathurst. The pyrethrins content in the flowers was very low, around 0.22%. However, the crop did not fully survive because of severe *Fusarium* root rot.

Another serious attempt in growing pyrethrum commercially in Australia was made during World War II. In 1944 it was targeted to produce about 182 tonnes (200 tons) of dry flowers at the Prisoners of War campsites at Hay and

Cowra in NSW and at Murchison, Red Cliffs, and Loveday in Victoria. A total of 24 tonnes (26.5 tons) of dried flowers were produced with an average yield of 364 kg/ha (324 lbs/acre) in the first season and 632 kg/ha (563 lbs/acre) in the second season. During the third season, DDT became available and increased supplies of flowers came from Kenya. The need to struggle further with local production ceased, and the plantations were abandoned.

The most serious and meaningful attempt to grow pyrethrum commercially in Australia was made in 1981. In 1979, Dr. B. Krishen Bhat was offered a research fellowship by the University of Tasmania (UT) to explore the possibility of growing "high-value, low-volume crops" in the island state of Tasmania. Dr. Bhat brought with him seeds of several medicinal and aromatic plants, which included pyrethrum, from the Regional Research Laboratory, Srinagar, Kashmir, India. All the plants were first grown in the greenhouse and later in test plots at the Horticultural Research Center (HRC) of UT in Hobart. Further test plots of a few selected plants like digitalis (*Digitalis lanata* L), pyrethrum (*Chrysanthemum cinerariaefolium*), salvia (*Salvia sclerea* L), and a few others were established at the privately-owned farms of C. Alcorso in Berriedale, B. Brain in Ouse, G. Casimaty in Richmond, and Australian Hop Marketers in Bushy Park. Almost all of the crops Dr. Bhat experimented with from September 1979 to June 1981 grew successfully in Tasmania. However, pyrethrum attracted the special attention of the industry.

Two companies, Tasmanian Alkaloids Pty. Ltd. (TA) and Glaxo Australia Pty. Ltd. (Glaxo), approached the UT for release of seed and technical know-how of pyrethrum production in Tasmania. Later, another company, Commonwealth Industrial Gases Ltd. (CIG), also approached the University for the same reasons. TA and Glaxo already had an agricultural base established in Tasmania producing poppies commercially. CIG, on the other hand, was a nonagricultural company dealing mainly with industrial gases and operating from Sydney. Negotiations between the University and the companies ultimately culminated in two of the three companies signing a 5-year research contract with UT and the Department of Industrial Development, Tasmania (presently known as Tasmanian Development Authority — TDA). Glaxo formally signed this three-party agreement in August and CIG in December 1981. TA did not enter into any formal agreement with the University even though they continued to work with the pyrethrum seed initially given them by UT. The formal agreement Glaxo and CIG entered with the UT required the latter to release seed, planting material of the superior clones, and all aspects of production technology — cultivar development, clonal propagation, mechanical harvesting, chemical weed control, disease identification and control, drying, and extraction — to the two companies. TDA was required to provide government assistance, wherever possible, in establishing pyrethrum as a new commercial crop in Tasmania. Also, as part of the agreement, the companies were restricted from cultivating pyrethrum commercially outside the state of Tasmania.

Soon after signing the agreement the companies received the improved seed and clones of pyrethrum developed at UT. Eight clones were released at that stage. This material, both seed and clones, had already established its superiority in multilocation yield trials in Tasmania (Bhat and Menary, 1984a). In the initial stages of this work both Glaxo and CIG laid out extensive test plots, and herbicide and agronomic trials began throughout Tasmania at places like Hobart, Bushy Park, Ulverstone, Ouse, Devonport, Hamilton, Gretna, Scottsdale, Pipers River, Ringarooma, Deloraine, Meander, Latrobe, and many others. A broad-based approach covering most aspects of pyrethrum culture like chemical weed control, macro- and micropropagation, harvester development, drying, extraction, and others were tackled right from the start. Despite the initial success, Glaxo, in restructuring their new product-development strategy globally, decided to discontinue the work on pyrethrum, and they terminated the agreement with the University and TDA in August 1983. The company, though, was generous enough to honor its 5-year commitment to fund the pyrethrum research at UT. From then on CIG was the only company left in the field to develop pyrethrum as a new commercial crop in Tasmania. Over the years they had some ups and downs in this game.

The University, Glaxo, and CIG laid out extensive test plots throughout the state of Tasmania to establish the most suitable areas for commercial production (Bhat, 1991; Bhat and Menary, 1984a). As a consequence of the breeding research, the seed and eight clones released to Glaxo and CIG for commercial evaluation in 1981 were highly superior to any pyrethrum material that had been previously introduced (McTaggart, 1933; Bhat and Menary, 1984a). The mean pyrethrins content in the improved seed population at that time was 1.97%, and in the best clones it was 2.16% as against 1.33% in the unselected base population (Bhat and Menary, 1984b; Bhat *et al.*, 1985). Dry flower yield was also high in the improved seed and clones (Bhat and Menary, 1984a). Previously, commercial production was attempted with crop raised from seed. The inherent problem with such an approach, due to the cross-pollinating nature of the crop, was its high variability in most plant characteristics including pyrethrins content and flower yield. A variable crop is difficult to handle mechanically. In the present case, commercial production was attempted with selected clones from the start. The clones released for cultivation had been selected for high flower yield, high pyrethrins content, lodging resistance, uniform plant canopy characteristics, and synchronous flowering. Hypy, the world's first patented cultivar of pyrethrum, was an outcome of this breeding program (Bhat and Menary, 1987). Hypy was initially selected by Dr. Bhat at the HRC, assigned the experimental number H80014, and was one of the eight clones released to Glaxo and CIG in 1981. This clone was later registered by the Crop Science Society of America in 1984 (Bhat and Menary 1984b) and is the first ever registered clone of pyrethrum in the world. The clone has withstood the test of time and occupies a major portion of the area presently cultivated in Tasmania. It has produced consistently high flower yield and pyrethrins content and is lodging resistant. In addition to the eight clones released in 1981, three more were released to CIG in 1984 and 1985. One of the three clones has found commercial acceptability and is being cultivated presently.

Vegetative propagation techniques through tissue culture and mist propagation of stem cuttings were developed and perfected at UT. Later these techniques were used commercially by nurserymen like Don Ormandy in Victoria and Ian Burrows, John Hill, Greg Fehlberg, B. French, and Rob van der Staay (Westland Nurseries) in Tasmania. The various vegetative propagation techniques discussed here were all employed together to increase the number of plants of any selected genotype in the shortest possible time. The whole propagation process was so well perfected that millions of plants of the selected genotypes were produced each year for field planting.

There is no denying the fact that this recent attempt to establish commercial production of pyrethrum in Australia has finally succeeded, as in 1990-1 CIG harvested around 2,000 ha of the crop. This success came from applying all available information and developing new knowledge as required plus pulling together many different aspects of breeding, propagation, and production.

From the start of the program in 1981, there was strong financial backing by two multinational companies — Glaxo and CIG — to the pyrethrum research at UT. Without industry funds the breeding research initiated by Dr. Bhat at UT could not have been continued for long. CIG involved not only the researchers at UT but a number of consultants to advise them from time to time on various aspects of pyrethrum culture. All parties — industry, researchers, administrators, consultants, farmers, and nurserymen — involved with establishing pyrethrum as a new commercial crop in Tasmania worked on various production aspects of the crop to make it a success. Similarly, chemical weed control, mechanical harvesting, drying, and extraction were looked after by different experts working in the program. CIG showed a strong commitment to helping farmers accept pyrethrum as a new crop. In doing so, they paid a reasonable return to the growers and absorbed all of the financial losses. The approach was unlike previous occasions when researchers went directly to the growers with a new crop without anyone being prepared to share the risk. Hop kilns became readily available for drying the flowers. A primary extraction plant for making crude pyrethrum extract was installed by the Davey Brothers (John and Kenneth) in Tonganah, Tasmania.

VII. PYRETHRUM IN THE UNITED STATES

Pyrethrum production in the United States has two phases initiated almost a century apart. Don Maciver provided information on the first phase obtained from the Haggin Museum, Stockton, California. Pyrethrum was grown as a commercial crop near Stockton in 1877. Seed was obtained from Dalmatia by G.N. Milko and the best results were found in the sandy soils of Merced County. Over 140 acres were planted and pyrethrum powder was produced and sold by J.D. Peters and associates under the trade name BUHACH. The United States government was a large customer since BUHACH gave them control of insects in Panama. A 1,400 acre ranch in Merced was converted to pyrethrum and vineyards. Production of BUHACH continued until the 1930s or 1940s.

Efforts to restart a pyrethrum industry in the United States were not successful until the 1980s. From about 1983, Mark Sims, who had a patent on a carbon dioxide pyrethrum refining process, tried to get farmers interested in pyrethrum, and he had some help from Oregon State University at Corvallis. His company, Botanical Resources, was bought out by the John I. Haas Company in 1987. Haas brought Dr. Krishen Bhat, a world class plant breeder, to the United States to work on newer strains of tissue-cultured pyrethrum clones. Dr. Bhat has developed several strains of high yielding pyrethrum clones. The Haas Company is looking at growing areas in California, Mexico, and elsewhere since they are in the business of growing hops worldwide. Because of the confidential nature of the Haas program to date, we have no specifics of Dr. Bhat's work in the United States.

VIII. MOSQUITO COILS

Mosquito coils based on pyrethrum became a major industry in the Far East and Latin America during the 1970s (Maciver, 1992). They are designed to burn for about 8 hours thereby providing protection from biting mosquitoes throughout the night. They consist of pyrethrum, an organic filler capable of smoldering for a long time, a binder, and auxillary materials such as fungistats and dyes (Maciver, 1963). The developments in India, Southeast Asia, and China were due largely to Don Maciver who in turn credits the late Peter Chadwick for developing data showing the very high effectiveness of pyrethrins as blockers of neurosensing, food-searching mechanisms in adult female mosquitos. Don followed up on the work and showed that pyrethrins were extremely effective and that allethrin was, on a cost basis, much less effective. Don set up his office in Kuala Lumpur and had contacts with coil manufacturers from Kashmir in the west to Korea and Japan in the north and Indonesia in the south. His largest and most challenging market was China. It could not be entered due to the Red Guard, and China and its diplomats were *persona non grata* in Kenya. Since Don was not a Kenyan citizen, he could call on the Chinese Embassy, and he did so issuing an invitation to a dinner at his home in Nakuru to the Trade Secretary. The dinner — all Chinese food — so endeared his guests that shortly afterward they placed an order for 500 tons of pyrethrum powder worth just less than $1 million.

In the 1970s South America and the Caribbean became a major market for pyrethrum powder and marc for mosquito coil production (Keane, 1992).[2] This was the result of improved technology in the handling of pyrethrum flowers. Prior to this time, Kenya had shipped baled flowers to these markets, but the pyrethrins losses in shipment caused both Kenya and Tanzania to abandon this method. However, stabilized, ground pyrethrum flowers at 1.3% from Kenya and 0.9% from Tanzania found a ready market. The potential for Kenyan pyrethrum sales caused the Pyrethrum Board to open a technical liaison office

[2] Paul Keane (graduate of the University of London in England with advanced degrees in Biological Sciences) was involved in marketing pharmaceuticals before joining the Pyrethrum Board of Kenya in 1967. He worked in market development in South and Central America and the Middle East until 1972 when he relocated to Chicago to act as liaison between the Pyrethrum Board of Kenya and its major United States customers with emphasis on regulatory issues. In 1979 he founded his own consulting company and was elected Chairman of the Pyrethrin Steering Committee in 1985.

in Caracas under Paul Keane's direction. From 1972 through 1977 demand increased each successive year reaching well over 1,500 tons of 1.3% powder as well as a considerable amount of 0.6% powder sold to smaller manufacturers. The major markets were Venezuela and Argentina with smaller quantities shipped to Brazil, Guyana, and the Caribbean nations. Brazil had a small pyrethrum industry based at Taquara in the Rio Grande del Sol and had protective tariffs for this business.

When Kenya was supplying pyrethrum powders to the Far East and Latin America, taking 1975-6 as a typical year, the total market for pyrethrins in this form was equivalent to 41,740 kg of 100% pyrethrins or 83,000 kg of 50% extract. Pyrethrum powder is relatively simple to manufacture and ship as compared to extracts, which call for large capital investments in equipment and solvents. However, the short fall in Kenya's pyrethrum flower production in 1977-8 caused the Kenya Pyrethrum Board to concentrate all their supplies on extract production, and the pursuit of the coil market was dropped. Now, practically all mosquito coils are formulated with synthetic pyrethroids.

IX. WATER-BASED AEROSOLS

New developments and rapid growth in the water-based insecticide aerosol market contributed greatly toward the use of pyrethrum in household insecticides. The concept of using water as a carrier for synergized pyrethrins and pyrethroids in insecticidal space sprays became reality in the mid to late 1960s. The transition was rapid for two reasons: the substitution of water for more expensive solvents and the use of hydrocarbon propellents offered substantial savings in cost; the registration process for pyrethrum formulations was not the slow and frustrating procedure that is now experienced 25 years later. The cost savings from using water as a carrier allowed marketers to increase pyrethrins levels, thereby offering greater efficacy characteristics for the aerosol market. During the past 10 years, there has been further development of water-based products with improvements in emulsion systems and new propellent concepts. During the early 1970s many pyrethrum-based aerosols displayed the "pyrethrum" logo with the claim "Safe for Humans and Pets".

Nonpressurized products were developed for use in the household insecticide market establishing another important direction. Even though the non-pressurized products have become more popular recently, the concept of a clear, stable emulsion containing pyrethrum and pyrethroids was under development over 20 years ago. In 1970 a patent was issued for an insecticide product with these characteristics (Baker, 1970). Since then, the microemulsion concept has led to an increase in the use of pyrethrum in pump spray containers and other applications for control of insects. The interest in microemulsions intensified during the late 1970s as the cost of petroleum solvents escalated. The household insecticide industry is presently undergoing still another transition involving water-based products. Regulatory changes and environmental concerns have dictated the reduction of volatile organic compounds in insecticide aerosols. These regulations and concerns will likely magnify the use of both pressurized and nonpressurized water-based products containing pyrethrum and other active ingredients.

X. SYNTHETIC PYRETHROIDS (see also Chapter 1)

No history of pyrethrum in the 1970s and 80s would be complete without reference to the synthetic pyrethroids, their appearance, and impact on the markets for natural pyrethrum. Since the 1960s, Sumitomo Chemical Company Ltd., Japan, manufactured and marketed the first synthetic pyrethroid called allethrin. Sumitomo found a major market for allethrin in mosquito coils, and sold over 100 tons of allethrin per annum into this market in Japan, Southeast Asia, Italy, and Latin America. In 1968, Roussel-Uclaf, a large, French pharmaceutical manufacturer, obtained a patent on preparing the two most active of the eight isomers of allethrin, which they called Bioallethrine® and which is registered in the United States as D-Trans® allethrin. Later they succeeded in commercially producing the most active isomer, which is called *S*-bioallethrin. During the 1970s these active pyrethroids replaced many pyrethrum formulations since the supply of the latter became more and more constricted and supplies of synthetics were reliable.

In the early 1970s, Dr. Michael Elliott and his coworker, Dr. Norman Janes, supported by the National Research and Development Corporation (NRDC) of Great Britain, developed a series of synthetic pyrethroids; starting with NRDC 104, resmethrin; NRDC 107, bioresmethrin; NRDC 143, permethrin; NRDC 149, cypermethrin; and finally, NRDC 161, deltamethrin. These last three compounds were halogenated to give them residual activity and found immediate interest among agricultural marketing companies. Originally, NRDC had licensed these compounds to Mitchell Cotts, UK; Burroughs Wellcome, UK; S.B. Penick, US; and Roussel-Uclaf, France. The importance of the agricultural chemical market caused a significant change in NRDC thinking, and very shortly, licenses to produce and market NRDC 143, 149, and 161 were offered to Royal Dutch Shell, UK and The Netherlands; ICI, UK; FMC Corporation, USA; and Sumitomo Chemical. While Dr. Elliott and collaborators were conjuring up these new pyrethroids, Sumitomo Chemical introduced two new synthetic analogs—Neo-Pynamin® (tetramethrin) and Sumithrin® (phenothrin). Neither of these were halogenated, but Sumithrin had some residual activity and soon found a major market in cockroach control. In the mid-1970s, Sumitomo Chemical produced a halogenated pyrethroid called fenvalerate, which they licensed to Royal Dutch Shell worldwide except Japan, and Shell marketed this product very aggressively, particularly in the United States.

A very interesting facet of the new, stablized pyrethroid esters was that the alcohol moiety in all of them was either *m*-phenoxybenzyl alcohol or derived from the corresponding aldehyde. Sumitomo Chemical's fenvalerate used *m*-phenoxybenzaldehyde, and Shell manufactured the product in the United Kingdom and in the United States. While all these interesting compounds were being discovered, MGK had a sister company in Elgin, South Carolina called

Hardwicke Chemical Company that was started with Dr. James Hardwicke in 1968. Tom Bogaard, a chemist with MGK, knew that Hardwicke was making benzyl alcohol, and noticed that *m*-phenoxybenzyl alcohol, present in most of the new residual pyrethroids, was not patented. He suggested to Dr. Hardwicke that it would be a very interesting and particularly profitable product for our company. Within 3 months of the suggestion, Hardwicke Chemical delivered 10 kilos of *m*-phenoxybenzyl alcohol to FMC, and the race was on. Hardwicke became the leading supplier of alcohol and aldehyde to the major agricultural pyrethroid marketers. In 1976, 6 metric tons of *m*-phenoxybenzaldehyde were shipped to Shell at $30 per pound. Dr. Hardwicke and I had a wonderful time calling on ICI in England and Shell in The Hague, and merrily dropping the price whenever it looked as if a competitor might crop up. While Hardwicke was supplying the large agricultural marketers, it was also the supplier of piperonyl butoxide and diethyltoluamide (DEET) to MGK. In 1978, MGK and Dr. Hardwicke decided that an Ethyl Corporation offer to buy Hardwicke made sense for both companies, and so Hardwicke was sold to Ethyl that year.

XI. PYRETHRIN TASK FORCE

The United States Environmental Protection Agency (EPA) issued a data call-in for natural pyrethrum on September 27, 1985. The producers, manufacturers, and marketers of pyrethrum decided to form a task force to jointly defray the costs of the toxicological work demanded by the agency. The task force members were Chemical Specialties Manufacturers Association (CSMA), Fairfield American Corporation, MGK, OPYRWA, Penick BioUclaf Corporation (subsidiary of Roussel-Uclaf), Prentiss Drug and Chemical, Pyrethrum Board of Kenya, S.C. Johnson Wax, and Tanganyika Pyrethrum Board. This group formed the official task force December 1, 1985. A new member, CIG of Australia, joined the task force November 18, 1986. On June 30, 1990, Roussel left the group. Some of the technical papers in this book are a result of the EPA data call-in 8 years ago.

XII. THIRTY YEARS IN THE PYRETHRUM BUSINESS

As a final note, I have spent more than 30 years at MGK and so have been intimately involved in the pyrethrum business. Because of its worldwide operations, I have been blessed with the fortune to meet many, many interesting people involved in agronomy, production, analysis, formulation, and marketing from the United States to the United Kingdom, Ecuador, East Africa, and Australia. It has been a wonderful experience for me, and I hope pyrethrum will continue to be a fascinating subject for all who read this book.

REFERENCES

Baker, G.J. (1970). U.S. Patent Number 3,492,402.
Bhat, B.K (1991). Personal communication.

Bhat, B.K., and Menary, R.C. (1984a). Pyrethrum production in Australia: Its past and present potential. *Aust. Inst. Agric. Sci.* **50**, 189–192.
Bhat, B.K., and Menary, R.C. (1984b). Registration of Hypy pyrethrum. *Crop Sci.* **24**, 619–620.
Bhat, B.K., and Menary, R.C. (1987). Chrysanthemum plant named Hypy. U.S. Patent Number: Plant 5,848.
Bhat, B.K., Menary, R.C., and Pandita, P.N. (1985). Population improvement in pyrethrum (*Chrysanthemum cinerariaefolium* Vis). *Euphytica* **34**, 613–617.
Brewer, J.G. (1973). Micro-histological examination of the secretory tissue in pyrethrum florets. *Pyrethrum Post* **12**, 17–22.
Glynne Jones, D. (1989). "Important Dates and Famous People Associated with Pyrethrum, the Natural Insecticide, with a History Extending Over 150 Years," March. Unpublished monograph.
Keane, P.A. (1992). Personal communication.
Levy, L.W. (1992). Personal communication.
Levy, L.W., and Estrada, R.E. (1954). Rapid colorimetric determination of total pyrethrins by reaction with sulfur. *J. Agric. Food Chem.* **2**, 629–32.
Levy, L.W., and Geller, D. (1959). Note on the detection of isopyrethrins. *Pyrethrum Post* **5**, 12.
Levy, L.W., and Geller, D. (1960). Speed of formation of isopyrethrins on heating of pyrethrum extract. *Cienca y Naturaleza* **2**, 20.
Levy, L.W., and Molina, H. (1957). A hydrolytic method for the estimation of pyrethrin II in pyrethrum flowers. *Pyrethrum Post* **4**, 22–24.
Levy, L.W., and Usubillaga, A. (1956a). Determination colorimetrica de la aletrina por reaccion con azufre y aplicacion al analisis de piretro. *Boletin Inform. Cientif. Nacls. Ecuador* **9**, 54.
Levy, L.W., and Usubillaga, A. (1956b). Estudio cuantitavito del formoldebido producido por oxydacion de la aletrina con permanganato-periodato. *Boletin Inform. Cientif. Nacls. Ecuador* **9**, 242–252.
Levy, L.W., and Usubillaga, A. (1957). Experiments of biosynthesis of radioactive pyrethrins. *In* "First Inter-American Symposium on Peaceful Application of Atomic Energy." *Proceedings* **2**, 603.
Levy, L.W., Munoz, M.O., Muggia, F., and Jimenez, J.O. (1960). Formation of pyrethrins in detached pyrethrum flowers. A study with radio-isotopes. *Pyrethrum Post* **5**, 3–6.
Maciver, D.R. (1963). Mosquito coils. Part I. General description of coils, their formulations and manufacture. *Pyrethrum Post* **7**, 22–27.
Maciver, D.R. (1992). Personal communication.
McLaughlin, G. (1973). History of pyrethrum. *In* "Pyrethrum, The Natural Insecticide" (J.E. Casida, ed.), pp. 3–15. Academic Press, New York.
McTaggart, A. (1933). Pyrethrum. *J. of the Council for Scientific and Industrial Research* **6**, 204–210.
McTaggart, N.G., Thornton, E., and Harford, A.D. (1958). The determination of pyrethins and other insecticidal compounds by infra-red spectometry. *Pyrethrum Post* **4**, 12–15.
Otieno, D.A., and Pattenden, G. (1980). Degradation of the natural pyrethrins. *Pestic. Sci.* **2**, 270–278.
Ottaro, W.G.M. (1977). The relationship between the ploidy level and certain morphological characteristics of *Chrysanthemum cinerariaefolim* vis. *Pyrethrum Post* **14**, 10–11.
Parlevliet, J.E. (1970). The effect of rainfall and altitude on the yield of pyrethrins from pyrethrum flowers in Kenya. *Pyrethrum Post* **10**, 20–25.
Parlevliet, J.E. (1975a). The genetic variability of the yield components in the Kenyan pyrethrum population. *Pyrethrum Post* **13**, 23–28.
Parlevliet, J.E. (1975b). Breeding pyrethrum in Kenya. *Pyrethrum Post* **13**, 47–54.
Tuikong, A.R. (1984). Pyrethrum breeding in Kenya: A historical account. *Pyrethrum Post* **15**, 113–117.
von Mueller, B.F. (1895). "Select Extra—Tropical Plants," 9th Ed., p. 120. Robert S. Brain, Government Printer, Melbourne.

II

Pyrethrum Flowers

3

Pyrethrum Flowers — Production in Africa

JOB M. G. WAINAINA

I. INTRODUCTION

A. Origin of Pyrethrum

A few plant species that contain active insecticides have been commercially developed. Pyrethrum, which is perhaps the most widely-used botanical insecticide, is derived from the flowers of a plant in the genus *Chrysanthemum* (*Tanacetum*), which belongs to the family Compositae. This genus contains many species of which only a few (e.g. *C. roseum* and *C. cinerariaefolium*) produce insecticidal substances which have been exploited at one time or the other.

The early Persian "insect powders," which were processed and commercialized in Europe in the 1820s, are believed to have been a mixture of *C. roseum* and *C. corneum*. On the other hand, the Dalmatian "insect powders," produced later in the century, were probably based on the flowers of *C. cinerariaefolium*. It is this species that subsequently spread to Africa and the rest of the world and is the basis of the pyrethrum industry in Kenya (Gnadinger, 1936).

B. History of World Production

The pattern of world production of pyrethrum has been greatly influenced by the effects of World Wars. Dalmatia remained the main source of pyrethrum up to World War I, following which Japan took over as the principal producer. The late 1920s saw the introduction of pyrethrum into the highlands of Eastern Africa, including Kenya, Tanzania, Rwanda, and Zaire. Due to the higher flower yields per unit area and superior pyrethrins content, production increased faster in Africa than in any other part of the world. By 1941, Kenya overtook Japan as the main world producer and, following the outbreak of World War II, Japan ceased to be a significant pyrethrum producer (Moore and Levy, 1975).

There have been various attempts over the years to introduce, or expand, the growing of pyrethrum in other parts of the world, including: Bolivia, Brazil, Ecuador and Peru in South America; China, India, Indonesia, Nepal, and Thailand in Asia; Ethiopia, the Republic of South Africa, Zaire, and Zimbabwe in Africa; Hungary and Yugoslavia in Eastern Europe; and California and Oregon in the USA (Anonymous, 1977). Many of those attempts, however, did not prove economically worthwhile and some were discontinued altogether. Consequently, the distribution of world production of pyrethrum flowers has not changed significantly since the end of World War II, with Eastern Africa (Kenya, Tanzania, and Rwanda) remaining the main source of supply. Over the past decade, a major effort has been made to develop commerical pyrethrum production in Tasmania, where significant harvests have been realized in recent years (MacDonald, 1994). Additionally, Papua New Guinea has maintained commercial production of pyrethrum flowers on a limited scale (300–500 tons annually) for several decades. The estimated total production from all sources (Kenya, Tanzania, Rwanda, Tasmania and Papua New Guinea) in 1992 was 18,100 metric tons to which Kenya contributed about 69%.

II. PRODUCTION OF PYRETHRUM FLOWERS IN KENYA

A. Background

In the publication "Pyrethrum, The Natural Insecticide," McLaughlin (1973) relates the early history of pyrethrum production in Dalmatia, the background leading to its subsequent introduction into the rest of the world, and the origin and significance of its development in Kenya. In the same book, Glynne Jones (1973) briefly describes the prevailing cultivation practices in Kenya, the improvements achieved through agronomic research and plant breeding work, and the development and management of large-scale, plant-propagation nurseries. He highlights Kenya's production achievements and challenges, giving comparative world production figures for 1967–72. Glynne Jones also discusses the influences of weather, pests, diseases, and other factors, as well as the role played by agricultural extension staff, on pyrethrum cultivation in Kenya.

B. Recent Production Trends and Achievements in Kenya and Eastern Africa

Over the period of approximately two decades since the publication of "Pyrethrum, The Natural Insecticide," Kenya has consistently maintained a leading position in pyrethrum production (Table 3-1). However, during the same

Table 3-1. African Pyrethrum Production 1971–92 (Sources: Tanganyika Pyrethrum Board, Office du Pyrethre au Rwanda, and Pyrethrum Board of Kenya)

	5-year totals in metric tons					
Country	1971–5	1976–80	1981–5	1986–90	1991[a]	1992[b]
Kenya	63,617	52,511	49,655	32,739	9,943	12,452
Rwanda	7,072	5,281	5,195	3,404	915	1,400[b]
Tanzania	19,041	13,055	8,473	6,887	1,710	2,600
Total	89,730	70,847	63,323	43,030	12,568	16,452

[a] 1-year figures.
[b] Estimated.

period, there has been a significant decrease in pyrethrum usage resulting from more intense competition from new introductions of synthetic pyrethroids. Consequently, pyrethrum production in Africa has declined from an annual average of over 16,000 metric tons in the period 1971–80 to about 11,000 metric tons annually from 1981 onwards (Robins, 1984). Moreover, in the early part of the latter period (1982–5), the industry faced extremely severe economic difficulties arising from a major depression in world demand and the consequent excessive inventory levels and declining prices. Inevitably, reduced flower production occurred for several years subsequent to 1983. By 1987, however, demand had gradually recovered and stocks that had piled up from previous years had been used up, with the result that the market was to experience a further period of short supply.

There has been a recent general shift in consumer preferences towards natural products. This trend has kindled a renewed interest by insecticide manufacturers in the use of natural pyrethrum and is viewed as an important opportunity for sustaining the future of pyrethrum production in Africa. Encouraged by this change in consumer attitudes, and the seemingly strong demand for pyrethrum in the period 1986–90, pyrethrum growers in Africa have invested considerable resources lately with the aim of expanding their crop production and processing capabilities. The effect of these efforts is already evident from the output figures for 1991 and 1992 which reflect an increase of about 30% in 1 year. Accordingly, for the 5-year period from 1992–6, pyrethrum production in eastern Africa is projected to increase to about the same level as the previous (1971–5) high of 90,000 tons.

C. Pyrethrum Growing in Kenya

In Kenya, pyrethrum is cultivated almost entirely by small-scale farmers, currently numbering between 50,000 and 60,000, who depend on the crop as their main source of cash income. The crop is also a major source of export revenue for the country, and this aspect provides gainful employment to over 3,500 additional workers. Pyrethrum is therefore of considerable economic and social benefit to Kenya.

The crop is favored by cool temperatures which occur in the higher altitude (1,800–2,900 m) areas of Kenya. A minimum rainfall of about 100 cm, evenly distributed throughout the year, is essential. Higher temperatures and dry weather have negative effects on flower yields and pyrethrins content. Well-drained soils of moderate organic matter are ideal for pyrethrum flower cultivation. New stands in Kenya are established either from seed or from clonal propagation. A network of large scale nurseries is maintained under the management of the Pyrethrum Board of Kenya to provide a continuous source of suitable planting material. The nursery program is supplemented by initial tissue culture propagation of good performance clones. In commercial pyrethrum stands, the first picking of mature flowers occurs within 3–4 months after transplanting, and thereafter at intervals of 10–14 days (Plate 1). Flower harvesting during the season is continued for up to 10 consecutive months, after which the stand is "cut-back" to remove dead and unproductive plant material. The crop is then allowed to "rest," ready for renewed vegetative growth in the subsequent season. Flower harvesting is selectively done by hand, after which the flowers are suitably dried and delivered to the factory for processing (Beckley, 1938; Kroll, 1962; Machooka, 1980). All deliveries of pyrethrum flowers received at the factory are weighed, sampled and analyzed for pyrethrins content. In Kenya, the rate of payment to the farmers is determined by the pyrethrins content of each harvest on arrival at the factory. In this way, a financial incentive for productivity is systematically provided. Growers' payments are disbursed within 30–40 days following delivery of the produce to Nakuru.

D. Participating Organizations

Kenya Agricultural Research Institute (KARI)

Two organizations in Kenya have a close and continuous involvement in the development of pyrethrum growing in the country. One of these is the National Pyrethrum Research Center at Molo, which is part of the national network of agricultural research under KARI. The Center carries out both exploratory and investigative research on all aspects of pyrethrum agronomy, including the development of new and better clones through plant breeding. It is responsible for many of the improvements that have been realized to date in the quality and productivity of the Kenya pyrethrum crop. The research effort at Molo is geared primarily to finding ways and means of improving pyrethrum yields and minimizing production costs. A part of the program also aims to continue conserving soil fertility and integrity (Kenya Agricultural Research Institute, 1989).

Pyrethrum Board of Kenya

The other organization which is of major relevance to the Kenya pyrethrum scene is the Pyrethrum Board of Kenya—a national marketing board, first established by statute almost 60 years ago and organized on the lines of a farmers' cooperative. The Board is responsible for the overall management of the industry, including promotion, control and coordination of pyrethrum flower growing, in addition to the processing and marketing of Kenya pyrethrum. Financed entirely by loan stock capital subscribed to by Kenya pyrethrum growers on a pro rata basis, the Board operates on purely commercial principles with all profits appropriated to eligible producers. The headquarters of the Board are in Nakuru where plant facilities for pyrethrum extraction and refining are also located.

E. Flower Processing and Manufacture of Products

Processing of Kenya pyrethrum flowers is handled by the Pyrethrum Board of Kenya at its factory in Nakuru. The processing plant comprises facilities for producing crude (oleoresin) extract as well as high-quality dewaxed and decolorized extract concentrates (also called refined or pale extract). Refined

Plate 1. Picking pyrethrum flowers and freshly-picked flowers at a farm in the Molo area, Kenya. Sampling dried flowers with an auger for analysis of pyrethrins content at Pyrethrum Board of Kenya in Nakuru.

extract is the main form in which Kenya pyrethrum is exported. A small quantity of fine pyrethrum powder is also produced for sale to manufacturers of insecticide dusts and mosquito coils. The main buyers of Kenya pyrethrum are located in the USA, Europe, and Southeast Asia, with the local market consuming about 5% of the annual production.

The growing, processing, and marketing of Kenya pyrethrum, including research and development work, entail a great deal of analytical evaluation of products and materials. The Board, therefore, has well-equipped chemical and bioassay laboratory facilities at Nakuru, staffed by a strong team of suitably-qualified scientists and technologists.

III. CONCLUSION

World pyrethrum producers are undoubtedly encouraged by the progress that has been achieved to date in the development of new toxicology data on pyrethrins as demanded by government regulatory agencies in the USA and elsewhere. Hopefully, the imminent completion of this requirement will open the way for greater attention to be given by worldwide insecticide marketers and consumers to the special benefits of natural pyrethrum in the control of nuisance and disease-carrying insect pests.

Pyrethrum producers in Africa remain committed to meeting the future world needs for this valuable natural insecticide. This commitment can be demonstrated by over half-a-century of continuous engagement and investment in pyrethrum cultivation. To maintain their efforts, producers will need the continued support and goodwill of insecticide manufacturers, marketers, and industrial consumers. Speaking with specific reference to Kenya, there is no doubt that pyrethrum will continue to occupy the interest of our rural community for many years. And so long as there is a viable market that offers the farmer fair compensation for his efforts, the Kenyans will continue to play their part in sustaining production. Conversely, pyrethrum producers should find ways of stabilizing production to minimize the occurrence of costly and annoying pyrethrum shortages which have characterized the industry in the past. There is clear need for a pyrethrins buffer-stock to insure against unforeseen production shortfalls (Robins, 1984). Such a buffer-stock would be inevitably expensive to finance, but it is hoped that costs can be widely and equitably shared among both the producers and major users.

REFERENCES

Anonymous (1977). Pyrethrum: its safety sells it. *International Trade Forum* April-June, 15–16, 37.
Beckley, V.A. (1938). Pyrethrum in Kenya. *Bulletin of the Imperial Institute* **36, 31**–44.
Glynne Jones, G.D. (1973). Pyrethrum production. *In* "Pyrethrum, The Natural Insecticide" (J.E. Casida, ed.), pp. 17–22. Academic Press, New York.
Gnadinger, C.B. (1936). "Pyrethrum Flowers," 2nd Ed., pp. 1–4. McLaughlin Gormley King Co., Minneapolis, Minnesota.
Kenya Agricultural Research Institute (1989). "National and Regional Research Mandates," 30 pp. Photo Litho Government Printer, Nairobi.

Kroll, U. (1962). "Pyrethrum, Kenya's Insecticidal Cash Crop," pp. 1–4. World Crops, Technical Press Ltd., London.
MacDonald, W. (1994) This volume.
Machooka, S.M. (1980). "Smallholder Pyrethrum Survey," 117 pp. Kisii District, Egerton University, Kenya.
McLaughlin, G.A. (1973). History of pyrethrum. *In* "Pyrethrum, The Natural Insecticide" (J.E. Casida, ed.), pp. 3–15. Academic Press, New York.
Moore, J.B., and Levey, L.W. (1975). Pyrethrum sources and uses. Part I. Commerical sources of pyrethrum. *In* "Pyrethrum Flowers" (R.H. Nelson, ed.), pp. 1–9. McLaughlin Gormley King Co., Minneapolis, Minnesota.
Robins, S.R.J. (1984). "Pyrethrum: A Review of Market Trends and Prospects in Selected Countries," 73 pp., Tropical Development and Research Institute No. G. 185, United Kingdom.

4

Pyrethrum Flowers — Production in Australia

WILLIAM L. MacDONALD

I. INTRODUCTION

Several attempts have been made since the 1930s to investigate pyrethrum as a potential crop within Australia. These attempts were primarily led by government (State and Federal) research organizations and paralleled similar efforts in other countries such as Kenya and India. In the late 1970s the foundation was laid for the first commercial production of pyrethrum with the establishment of a pyrethrum breeding program at the University of Tasmania, the support of the Tasmanian State Government, and the involvement of a number of significant corporations. A world shortage of the insecticide during 1978 and 1979 and a further shortage during the mid 1980s, combined with encouragement from the major US pyrethrum users, helped to maintain the momentum for this most recent Australian initiative. By 1992, Australia had become the world's second largest producer of pyrethrum, with the first fully mechanized production system.

II. EARLY ATTEMPTS AT PYRETHRUM PRODUCTION IN AUSTRALIA

The earliest report of pyrethrum in Australia was from Baron von Mueller who found the plant growing in the lower Latrobe River valley in Victoria prior to 1895. During the 1930s, the Division of Plant Industry of the Australian Government's Commonwealth Scientific Industrial Research Organisation (CSIRO) introduced plant material derived from the UK, Japan and the USA. Trial cultivation was undertaken at Black Mountain, Canberra (McTaggart, 1933) and also by the Departments of Agriculture in two other Australian States, i.e. New South Wales during 1932–5 and 1952 and Tasmania during 1941–4. The CSIRO conducted further work at five different locations (Hay, Murchison, Cowra, Red Cliffs and Loveday) during 1944. None of this work progressed beyond the research level. A press release made in 1952 by the

CSIRO stated: "Attempts to grow pyrethrum on a commercial scale in Australia over the past 20 years have failed primarily because production costs were too high to allow the products to compete with imported material (East African). CSIRO considers that there is no reason for believing that a further attempt at the present time has any greater prospect of success." However, from 1955 to 1964, a further attempt to commercialize pyrethrum was made by the Australian Pyrethrum Company. This private company established trials in a number of regions across Australia, but once more the venture was unsuccessful and all its activities ceased by 1964.

This historical summary is primarily taken from Bhat and Menary (1984) who noted "While it is difficult to ascertain the reasons for these early failures at commercial cultivation, the introduction of inferior planting material resulting in low flower and pyrethrins yields, diseases and lack of adequate knowledge of the plant, may have been some of the contributing factors." It was their belief that commercial success would depend on the availability of genetic material that could be tailored to meet the needs of the Australian environment and production system.

III. RECENT DEVELOPMENTS

Following a period of low prices for traditional agricultural commodities in the mid 1970s, the Tasmanian State Government was stimulated to search for new crops which would provide higher returns to farmers. High-value, low-volume agricultural products were seen as most appropriate for Tasmania due to the State's export orientation and high transport costs. The University of Tasmania, together with the State Government, identified several essential oil crops including fennel, peppermint, boronia, and pyrethrum as falling into this category.

Research into pyrethrum as a potential commercial crop commenced at the University of Tasmania in 1978. The following year, Dr. R. Menary of the University of Tasmania managed, with University and State funds, to attract Dr. K. Bhat to Tasmania to join the University's program. Dr. Bhat, an experienced plant breeder, came from Kashmir, India. He brought with him two seed lots, one from an unselected base population from Drugs Farms in Kashmir and the other from selected clones maintained at the Council of Industrial and Scientific Research (CSIR) Regional Research Laboratory, Srinagar, Kashmir. This material, containing germplasm from Europe, Kenya, and Japan, provided the foundation for the University of Tasmania's breeding program.

Bhat and Menary's early strategy was to identify clones which exhibited several critical traits: synchronous flowering, even and upright flower presentation and high yields of pyrethrins per plant. It was felt that any successful commercialization of pyrethrum as a crop under Australian conditions would require high yielding, uniform plants that flowered at one time and could thus be harvested mechanically. The initial research established that pyrethrins yield was significantly correlated with three traits in the following order; pyrethrins content (per g of dry flower) > flower yield > flower weight.

By 1980, the University of Tasmania's program had identified a number of superior clones considered suitable for further commercial development. The highest yielding clone produced a mean performance across three sites of 48.8 kg pyrethrins/ha with an average flower pyrethrins content of 2.02%. The dried flower yield was in excess of 2,400 kg/ha. This clone was subsequently patented as "Hypy" by the University of Tasmania.

In 1980 the Commonwealth Industrial Gases Co. Ltd. (CIG) (the Australian subsidiary of the industrial gases multinational, British Oxygen Co. or BOC) became interested in developing a secure supply of pyrethrum for one of its pest control products. At the same time, two other companies, Glaxo (Australia) and Tasmanian Alkaloids, had also become interested in the opportunities for commercially developing pyrethrum. Both Glaxo and Tasmanian Alkaloids were already involved in Tasmania growing poppy crops from which they extracted and supplied alkaloids for the pharmaceutical industry. The growing and extraction of pyrethrum appeared likely to share some of their already developed agrotechnology, crop handling and processing facilities. As such, pyrethrum had the potential to become a profitable second crop.

In 1981 CIG and Glaxo (Australia) signed an agreement with the Tasmanian State Government and the University of Tasmania. Under this arrangement, both companies would foster the development of a pyrethrum industry and fund the University's breeding program through a royalty payment on sales. In return, the University agreed to provide the companies with high yielding pyrethrum clones and seedlines. In 1984, Glaxo withdrew from the agreement leaving CIG as the only commercial party. The Tasmanian Alkaloids company continued from 1980 to 1984 with an independent pyrethrum breeding and development program. During that time they investigated the use of pyrethrum seed lines (not clones) for their commercial plantings, mechanical harvesting and extraction in a carousel extractor. In 1984 they ceased their investigations and concluded that pyrethrum was not financially attractive.

A favorable market study undertaken by CIG in the early 1980s saw the company commence a 5-year feasibility study and research program into the potential for developing a commercial pyrethrum industry in Tasmania. These studies, covering the period 1982 to 1987, were directed from Sydney utilizing the services of two agricultural consultancy groups, Agrisearch and McGowans International (MGI). The studies provided invaluable agronomic data primarily in the areas of clonal comparisons in different regions, basic nutrition, crop irrigation, and weed control. In 1985 the first semi-commercial plantings (35 ha) were made on five properties under the direction of MGI. At this point a rudimentary chemical weed control program had been developed and a prototype harvester was under development at the University. In 1986, CIG appointed Mr. I. Folder as Operations Manager to Tasmania to administer the transition from research trials to early commercialization. With the University advising the use of clonal plant material for commercial plantings and a harvesting system based on removing fresh flowers, CIG engineers were

Table 4-1. Australian Pyrethrum Production

Product	Production by year (tonnes)			
	1989	1990	1991	1992
Dry flowers	350	800	1,300	2,000
100% Pyrethrins	5	10	14	21
20% Pyrethrins (refined product)	25	50	70	105

actively developing tissue culture facilities for mass propagation, and prototype pyrethrum harvesters. In 1987, the first "semi-commercial" harvest yielded a single trailer load of pyrethrum flowers. It was a sobering start to the second phase of the project's development.

With CIG's commercial targets requiring a rapid expansion in pyrethrum production, the company committed significant resources to achieve this end over the following 4 years. These resources were primarily used in developing a number of commercial tissue culture laboratories for the production of clonal plants, in expanding and maintaining the number of commercial hectares under pyrethrum, and in continuing the harvester development. In 1989, 600 ha of new clonal plantings were completed using in excess of 30 million splits derived from tissue culture plants. By that time a management team covering chemistry (Mr. J. Boevink), field operations (Mr. I. Folder), research and processing (Mr. W. MacDonald) and finance (Mr. N. Jones) was in place with a team of field officers working directly with the new pyrethrum farmers. The next 3 years saw pyrethrum plantings rise to 1,200 ha. Table 4-1 summarizes the increase in production.

In comparison to the world's major producer, Kenya, the Australian levels of production are relatively low. However, for the first time in the history of the crop, a fully mechanized and intensively managed pyrethrum production system has emerged.

IV. PRODUCTION SYSTEM

Since 1988, the Australian pyrethrum industry has progressed from trial status to a commercial industry. The emergence of a new world producer, located in a nontropical region and using novel production techniques, is of interest to others involved in pyrethrum. However, full details of the production technologies employed are not available for commercial reasons.

A. Tasmania — Production Environment

1. Location and Climate

Australia is made up of eight States with Tasmania being the southernmost. The State comprises a group of islands lying some 40–43° south off the southeastern corner of the Australian mainland. The largest island covers some 64,409 sq. km., in size somewhat similar to Japan's northern island of Hokkaido or the US State of Michigan. In contrast to mainland Australia, Tasmania is largely mountainous. Its population is some 460,000 with the State capital, Hobart, being the largest center with a population of 172,000 (Fig. 4-1).

The State enjoys a temperate maritime climate with prevailing winds coming from the west. In fact it lies on the edge of the latitude band commonly known as the Roaring 40s. The summer maximum temperatures tend to rise no higher than 26°C, averaging 18–20°C, with average winter maximum temperatures around 10–12°C. Snow and below freezing temperatures do occur on the highlands above 1,000 m but are not generally experienced in the coastal areas of the north and east. Winter lasts for some three months. Average annual rainfall varies from 1,400 mm in the west and northeastern highlands to 500 to 700 mm in the dryer midlands and eastern coastal areas. Rainfall occurs primarily during winter and spring with summer being the driest period.

The temperate Tasmanian conditions have proved ideal for pyrethrum production. The mild winters provide a period of sufficiently low average temperatures to ensure vernalization (essential for flowering) but rarely low enough to develop damaging frosts. The increasing day length during spring promotes the photoperiodic response in the plant (Brown, 1990) and the

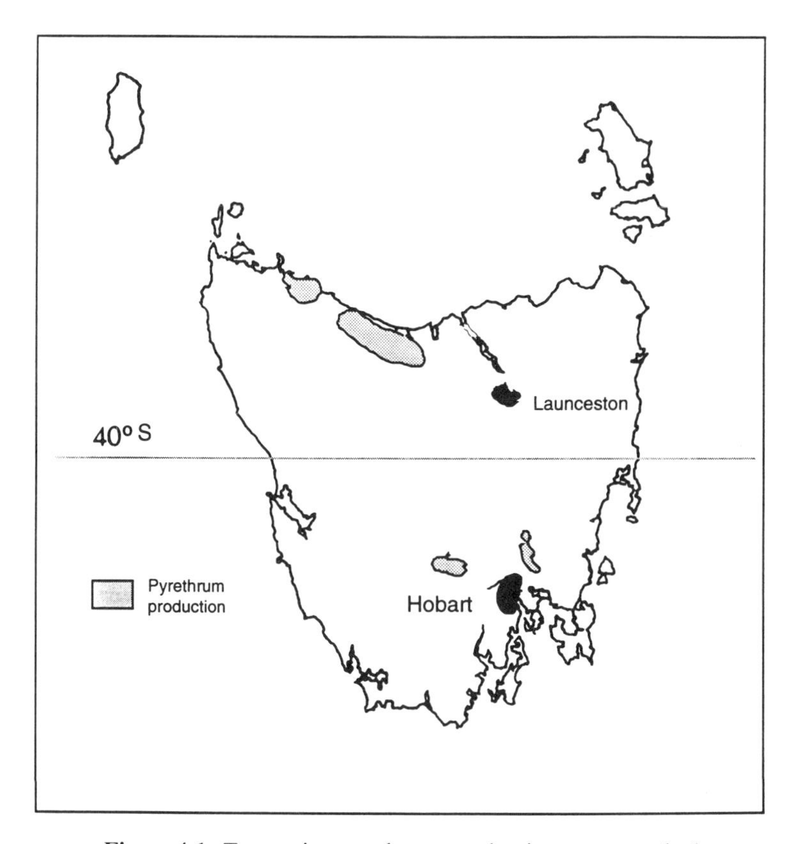

Figure 4-1. Tasmania: pyrethrum production areas marked.

subsequent synchronous flowering in early December. Dependable rainfall during spring and early summer greatly reduces the requirement for supplementary irrigation.

2. Land Use and Infrastructure

The major agricultural products of the State are wool, dairy products, beef, fruit, and vegetables. Areas suitable for crop production are found mainly along the coastal stretches of the northwest, northeast and southeast. These areas produce most of the State's vegetable crops such as potatoes, onions, beans, peas, and the brassicas. Extensive hectarages of poppies are also grown under strict supervision for the pharmaceutical industry. The predominant soil type within these highly productive northern areas is deep, old volcanic, red krasnozems. Most farms in this region are held as freehold and run as family businesses. They operate with a mixture of crops, pastures, and livestock (sheep and cattle). The size would average some 100 to 200 ha.

This highly fertile northern region has also proved to be the most suitable for pyrethrum, not only because of the mild coastal climate and free draining soils, but also because of the already existing high level of farming and management skills developed for the vegetable processors. These skills have been particularly important in the areas of weed and disease control, fertilizer applications, irrigation management and the ability to adopt new practices required for a new crop.

As a State, Tasmania is well serviced with excellent roads, air and sea connections and telecommunications. Agricultural research is provided by the State Department of Primary Industries, the University, and a number of commercial organizations. Fertilizers are manufactured on the island with other inputs being brought in from the mainland of Australia.

3. Corporate Development

To date the pyrethrum industry in Australia has been developed and is controlled by one corporation, CIG. The success of this development can be attributed to several key factors including the company's ongoing financial support, the innovation and vision of the personnel involved, and the support of a number of key farmers, horticulturalists and research workers within the Tasmanian community.

CIG's ability to finance all aspects of the development of the crop and to rapidly develop the necessary technology for its successful production greatly reduced the level of risk to those farmers entering the industry. This ability has been significant. It enabled the company to work with a number of the most progressive and capable farmers and to utilize their skills in developing an understanding of the crop. This package of technical and financial support, essential to ensure the rapid expansion of the industry, included the provision of plant material, full crop planting, nutrition, disease and weed control programs, crop harvesting and a guaranteed return per hectare in the event of poor crop performance. To date there are some 120 farmers contracted to grow pyrethrum. These growers are represented by a Pyrethrum Growers Commodity Group which conducts annual negotiations with CIG on such matters as contract terms and prices.

B. Production — Crop and Processing

There are a number of crop production elements involved prior to eventual extraction and refining of the pyrethrins. The monthly annual farm cycle is shown in Fig. 4-2.

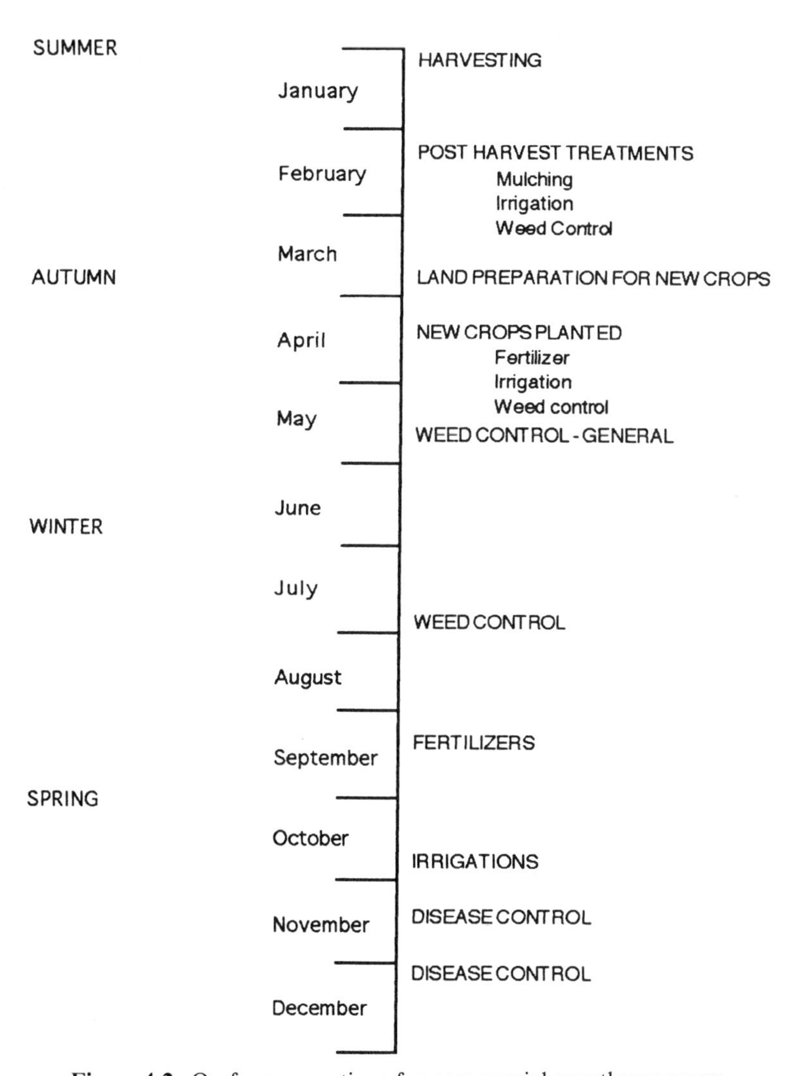

Figure 4-2. On-farm operations for commercial pyrethrum crops.

1. Plant Material—Preparation

The majority of the annual plantings are of two clones. Both clones offer high pyrethrins yields, synchronous flowering, and good upright presentation for harvesting. These clones are bulked up prior to commercial planting through initial tissue culture multiplication and then through outdoor nurseries as splits. The objective of the nurseries is to produce young vegetative splits during March and April for planting. The nurseries are planted in beds for easy mechanical lifting. Plant growth is manipulated through the timely application of irrigation and fertilizers. This is used to produce evenly grown, vegetative material. The material is "hardened off" prior to lifting to reduce plant shock. Once lifted, the plants are rapidly split, packed, drenched, and cooled prior to dispatch for planting.

2. Planting

Considerable attention is paid to selecting suitable sites for planting with pyrethrum splits. The cost of the plant material and the potential of the crop to remain productive in excess of 5 years makes it critical that the selected site is free of difficult weeds, is well drained, has a suitable pH and nutrient status, is irrigable, and is on a well managed farm. Hence over 80% of the pyrethrum grown in Tasmania is located on the productive, deep, red, soils of the northwest coast.

Annual plantings occur in autumn (April/May) using mechanical transplanters. These tractor-drawn transplanters plant beds of three rows at a rate of approximately 3.75 acres (1.5 ha)/day. Three people sit on the planter supplying the plants, while a fourth person walks behind filling in gaps or misplants. The planting density varies between 39,000 and 45,000 plants per hectare. Depending on each site's soil analysis, compound fertilizers are banded in at planting. Irrigation is used to bed in the transplants, with pre-planting and post-planting herbicides applied to reduce weed competition during the winter.

3. Weed Control

Each site's weed problems are assessed separately and an integrated weed control program is developed on this basis. Sheep are occasionally used to graze off weeds prior to land preparation or after harvest. Careful management of this "biological" weed control agent is essential to avoid pyrethrum being grazed flat to the ground! Control of weeds over the growing period is primarily based on the use of chemical herbicides. These are applied by the farmers as directed by the company. Additional hand hoeing and spot-spraying is used, if required.

4. Pests and Diseases

To date, some of pyrethrum's traditional economic pest species, including thrips, aphids, nematodes, and mites, have not proved to be of concern in Tasmania. However, crop damage from slugs and snails in the cooler and wetter areas of

Table 4-2. Primary Diseases and Other Disorders Recorded on Pyrethrum in Tasmania[a]

Pathogen	Disease	Locality	Comments
Sclerotinia minor	Wilt	General	Serious disease
S. sclerotiorum	Wilt	General	Serious disease
Pseudomonas sp. (probably *Ps. syringae*)	Leaf and stem spot/browning	General	May reduce yield
Botrytis cinerea[b]	Flower head browning	General	Occasional disease
Agrobacterium tumefaciens (?)	Crown gall	General (?)	Isolated occurrence

[a]15 June 1989 (3rd revision) — Dr. J. Wong (extract).
[b]Recorded 1992.

the far northwest of the State has occurred during late autumn. Simple and cost effective control has been achieved through the timely application of metaldehyde slug baits. Pyrethrum plants showing poor vigor during spring have been shown to contain levels of the root nematode, *Pratylenchus sp.* The distribution and overall economic damage of this pest across the commercial crops is still to be determined.

With plant diseases, the story is different. As in most parts of the world, the pathogens *Sclerotinia sclerotiorum* and *Sclerotinia minor* (commonly known as Crown Rot) are well established throughout Tasmania and especially within the intensive vegetable growing areas of the northwest. An extensive 4-year research program to establish the life cycle of the diseases within pyrethrum, plus the use of site-by-site risk assessment analysis, has enabled the development of an effective control program. Selection for "resistant" varieties continues as part of the University's overall plant breeding program. A bacterial leaf rot caused by *Pseudomonas sp.* has also caused damage in commercial crops and occasionally in nurseries. This bacterium causes most damage during cold wet periods when the pyrethrum plants are not actively growing. Control is readily maintained through the use of traditional copper-based sprays. Dr. J. Wong, a senior plant pathologist working with the Tasmanian Department of Primary Industry, has identified a number of primary and suspected secondary pathogenic disorders occurring in pyrethrum; the primaries are listed in Table 4-2.

5. *Nutrition and Irrigation*

In 1990 a major research program commenced investigating the role of water stress and added nutrition on pyrethrum yields. Initial results have shown significant yield increases through careful water management within the soil profile (again based on a site-by-site appraisal system using neutron probes) and the timely application of artificial fertilizers. All crops receive some form of fertilizer at planting with subsequent top dressings of nitrogen during spring and further compounds during each autumn period over the life of the crop.

Given good soil structure and the absence of any periodic waterlogging, pyrethrum under Tasmanian conditions will develop deep, active root structures down to two meters. Crops with deep roots have proven to be the most productive. Irrigation tends to be required in early summer when soil moisture deficits may appear, or, after harvest when the crop is restimulated to produce fresh growth prior to becoming dormant as temperatures fall in June and July. Irrigation is either through self-propelled rain guns or solid set pipe systems. The ability to provide supplementary irrigation is a prerequisite for all farmers wishing to obtain a pyrethrum contract from CIG.

6. Soil Conservation

Being a perennial, pyrethrum requires less cultivation than annual crops, an advantage for soil structure. However, during the first year after planting, pyrethrum provides little ground cover as protection against heavy rains or wind. Soil erosion through wash from heavy rainfall is a potential problem. CIG has been active in promoting the use of soil conservation measures and requires farmers to put in place appropriate drains and grassed waterways. For these initiatives, in 1990 and 1991, the company won State awards under the Australian Federal Goverment's Landcare program.

7. Harvesting

Unlike East Africa, under Australian conditions hand harvesting of pyrethrum is not an option. Developing an effective, one-pass, harvesting system has proven to be one of the greatest technical hurdles to overcome. Having established the physiological stage during which the pyrethrum crop contains the maximum yield, CIG investigated several harvesting systems capable of capturing this yield. The first approach was to attempt to remove the flowers and a proportion of stem while still green and sappy (Plate 2). The flowers could contain up to 85% by weight of water. The green material was then transported in ventilated trucks to local hop dryers for drying to 10% moisture levels. The second approach was to field-dry the material prior to harvest. This current approach is still being developed (Plate 2).

Over the last 4 years, monitored field recovery off commercial sites has risen significantly. Total pyrethrins recovery is audited at each processing step from field to final refined product.

8. Farmer Support.

During the farming year, pyrethrum growers receive regular visits from the company's field officers. Recommendations are made on site preparation, crop hygiene, and nutrition. Regular field days are presented to extend new information on a range of research findings or new developments in managing the crop. As CIG is the only company in Australia growing pyrethrum, the total annual hectarage grown is determined by the number of contracts already in place between CIG and the farmers plus the new contracts entered into each year. A grower's commodity group represents the farmers in annual discussions on contract alterations and prices.

Plate 2. Fresh-flower harvesting of pyrethrum — 1989 (top); dry-flower harvesting of pyrethrum — 1992 (bottom).

9. Plant Breeding

Over the last 10 years the University of Tasmania has continued with its pyrethrum breeding program. In 1987, Dr. W. Potts took over from Dr. Bhat. From a system of recurrent selection the program has moved to a system based on index selection and assortive mating. This program has been successful in significantly raising the available pyrethrins content with the mean of the latest generation yielding some 17% higher pyrethrins than the current top commercial clone. With a 3-year generation interval, the need for rapid and cost-effective introduction of improved lines into the commercial program has seen the reemergence of seed as a major source of future plant material.

10. Processing

Unlike the new territory of growing pyrethrum under a mechanized system, considerable information on pyrethrum processing is available. However, as with the other major pyrethrum producers, the problems of post-harvest losses and extraction recoveries remain a constant challenge. Given the natural instability of pyrethrins once harvested, the crop must be processed rapidly. Processing involves crop cleaning, pelletizing, storage, and extraction of the pyrethrins. The potential to lose pyrethrins rapidly during and after harvest has been shown over the last 5 years. During 1990 an expected yield of 10 tonnes of active ingredient was reduced by 50% through harvest and post-harvest losses. Improved harvesting, rapid processing, and the addition of the food antioxidant butylated hydroxytoluene (also known as BHT) have proven to be effective in reducing these losses.

Pyrethrum dust, generated during harvest and processing, has caused mild to severe allergenic reactions to a number of operators. This aspect of the process has required the implementation of rigorous occupational health and safety procedures.

Currently, crude extraction is by a private contractor utilizing a relatively old plant. This plant was originally set up to process hops for the international beer brewing markets. With a series of upgrades and use of computer controls, a fine oleoresin is produced averaging 39% by weight of pyrethrins. Extraction recoveries lie in the range of 91–93%. Refining of the oleoresin is carried out under contract by the McLaughlin Gormley King Company, Minneapolis, USA, with sales of the Australian pyrethrum being primarily within the US market.

V. FUTURE

The next stage in the development of CIG's production capabilities will be the construction of an integrated crop processing, extraction, and refining facility. A major influence on this development will be the successful outcome under the current US Environmental Protection Agency's data call-in for both pyrethrum and its primary synergist, piperonyl butoxide. The ability to produce pyrethrins in a cost-effective manner has largely been demonstrated. The long road to repositioning the product in world markets remains the next challenge.

ACKNOWLEDGMENTS

Helpful suggestions were provided by Professor R.C. Menary of the University of Tasmania and Dr. J. Wong of the Department of Primary Industry and Fisheries in Tasmania.

REFERENCES

Bhat, B.K., and Menary, R.C. (1984). Pyrethrum production in Australia: its past and present potential. *J. Australian Inst. Agric. Sci.* **50**, 189–192.

Brown, P.H. (1990). Morphological and physiological aspects of flower initiation and development in *Tanacetum cinerariaefolium* L. PhD Thesis. University of Tasmania.

McTaggart, A. (1933). Pyrethrum. *J. Counc. Sci. Ind. Res.* (*Aust*) **6**, 204–210.

5

Breeding Methodologies Applicable to Pyrethrum

B. K. BHAT

I. INTRODUCTION

Insecticidal properties of the flowers from two species of pyrethrum have been known for more than a century and a half. The species cultivated earliest for this purpose was *Chrysanthemum coccineum*, also called the Persian pyrethrum, that was grown in its native region of Caucasus (area between the Black Sea and Caspian Sea) extending up to Northwestern Persia (present-day Iran) (Contant, 1976). Many Caucasian tribes used the ground flowers for the control of body lice (Barthel, 1973). "Persian Insect Powder," traded in the first half of the nineteenth century, was made from the dried flowers of *C. coccineum* (Culbertson, 1940; Contant, 1976). Commercial cultivation of this species started in Armenia in 1828 and continued until 1840, when the greater insecticidal activity of *Chrysanthemum cinerariaefolium* became known.

C. cinerariaefolium, also known as Dalmatian pyrethrum, exists naturally along the east coast of the Adriatic Sea extending from Italy to northern Albania and up in the mountainous regions of Croatia, Bosnia and Herzegovina, and Montenegro—areas, until 1992, constituting the republic of Yugoslavia (Culbertson, 1940; Contant, 1976; Parlevliet *et al.*, 1979). It was cultivated in Yugoslavia (Dalmatia) in 1840 because of its high insecticidal activity. The crop was later introduced in many parts of the world. Historically the major pyrethrum producing countries have been Yugoslavia (until World War I), Japan (until World War II), and at present Kenya. This history is described more fully by Gnadinger (1936, 1945), Moore and Levy (1975) and Gullickson (1994).

According to Contant (1976) the taxonomic status of *C. coccineum* is still controversial, partly because of variation in morphology and flower color. The names *C. roseum*, *C. marschallii*, *C. carneum*, *Pyrethrum roseum* and *P. carneum* in the literature all apply to forms of *C. coccineum* (see also Culbertson, 1940). Both species bear a diploid chromosome number of $2n = 2x = 18$ (Contant, 1976).

Even though the plant has been under commercial cultivation in many parts of the world for the past 160 years or so, very little breeding and genetical work has been attempted. Breeding programs were initiated in countries like Australia, India, Japan, Kenya, New Zealand, and the United States, but only a few have been followed seriously. This review considers the morphological and genetical features of the plant that are relevant to breeding and then discusses the breeding work done in the crop so far. An attempt has also been made to suggest appropriate breeding methodologies that may be followed.

II. MORPHOLOGICAL FEATURES OF THE PLANT

Pyrethrum is a perennial herbaceous plant that propagates easily by seed, vegetative splits, stem cuttings rooted under mist, and tissue culture. When the plant flowers, many shoots originate from the crown and grow on average to 75 cm. The shoots branch a few times before terminating into a white, daisy-like flower (a few to several hundred flowers per plant). Sporophytic incompatibility renders the plant totally cross pollinating and makes the crop highly variable in almost all of the characters.

The flower head of pyrethrum, like any other species belonging to the family *Compositae*, is a compound inflorescence with small flowers—the florets—aggregated together on a convex receptacle, the capitulum (Plates 3A, B, Fig. 5-1A). The florets are of two kinds—the disc florets with yellow corollas occupying the center of the receptacle, and the ray florets with white corollas forming the outer rim of the flower head (Plates 3A, B). For more details about flower morphology and seed setting, see Chandler (1951) and Brewer (1968).

The disc florets possess both male and female organs (Figs. 5-1A, B). Five petals join to form the yellow tubular corolla and five stamens, borne on short stalks, are located inside this tubular corolla. The stigma is two-lobed and each lobe is tipped with a minute brush of short hairs (Brewer, 1970; Anonymous, 1972). In the unopened floret, the two lobes are vertical and appressed. As the style elongates and the corolla opens, the stigmatic brush sweeps out of the corolla, the pollen already shed into the staminal tube, and the lobes of the stigma curve open outwards exposing the papillate inner receptive surfaces.

The ray florets are unisexual (female) having no stamens (Figs. 5-1A, C). Strap-shaped, three-toothed white petals form the corolla that becomes partially tubular at the base where it joins the small calyx. The prominent style projects through the folded base of the corolla and its bilobed stigma has the same essential structure as that of the disc florets. The stigmas may be pollinated from the disc florets of the same flower, but selfing is prevented by the sporophytic incompatibility in the plant. The ripened ovule constitutes the true seed which, with the matured ovary wall, forms the fruit or the "achene."

III. CHARACTERS IMPORTANT IN PYRETHRUM BREEDING

A. Pyrethrins Yield and Quality

Pyrethrins yield and quality are determined by pyrethrins content, flower yield, and the ratio of pyrethrins I to pyrethrins II. Pyrethrins content is chemically analyzed in a representative sample of the flowers harvested and expressed as

Plate 3. Pyrethrum flower head. (A) Whole flower showing white-colored ray florets and yellow-colored disc florets. (B) Flower head—vertical section.

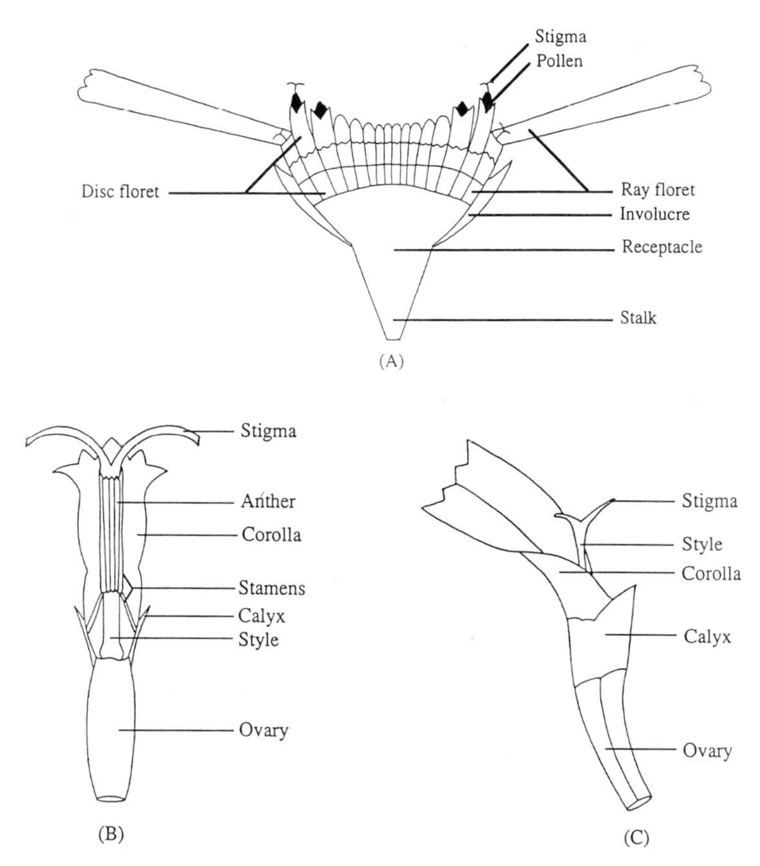

Figure 5-1. Structure of pyrethrum flower head. (A) Inflorescence. (B) Disc floret. (C) Ray floret.

a percentage in the dry matter content. In order to get a true estimation of the pyrethrins content it is important that sample-to-sample variation in flower maturity, drying temperature, and the moisture content of the flowers should be avoided by collecting samples at the same maturity, drying them at the same nondegrading temperature (50°C) (Ngugi and Ikahu, 1990) and reporting the results on a dry-weight basis.

Flower yield is expressed as fresh or dry weight per plant or per unit area; the area could be a plot, an acre, or a hectare. It is important to specify whether the yield expressed is fresh or dry. Further, it is equally important that the moisture content of the flowers be stated because variation will obviously lead to different results. It is most appropriate to express dry flower yield on a moisture-free basis. The moisture content in flowers at harvest is determined by drying the flowers, or the samples thereof, completely and weighing for dry-matter content. As this trait is used in the calculations of the pyrethrins content and the flower yield, it is important that the dry-matter content of the flowers is adjusted to zero moisture. Flower weight is estimated as fresh or dry weight of 100 flowers and is also sometimes referred to as flower size (Parlevliet, 1974). In order to avoid any variation in the expression of this value, the weight should be reported as moisture free.

The ratio of pyrethrins I/pyrethrins II (PI/PII ratio) determines the quality of the pyrethrins extract. In mosquito coils PI is known to have a good knockdown effect while PII has a better kill effect (Winney and Webley, 1969). Pyrethrins quality is also assessed by estimating the relative proportions of the six related esters by high performance liquid chromatography (HPLC) and gas chromatography (GC).

B. Maturity

Flower maturity determines the time for harvesting, which is generally the 1/2 to 3/4 disc-florets-open stage. The harvesting period can be considerably extended if early, medium, and late maturing clones are planted.

In pyrethrum a large number of flowers are borne and their development is variable in different genotypes. In some, the majority of the flowers will develop synchronously while in others the development is asynchronous which results in buds and overblown flowers—and the stages in between—being present at the same time. Asynchronous flowering may not be a serious disadvantage in places like Kenya, Tanzania, and others where selective hand harvesting is practiced while it will be highly undesirable if mechanized harvesting is contemplated.

C. Morphological Characters

Plant height in pyrethrum is taken as the length of the central flowering shoots from the ground to the level of the flowers. Plant canopy has an important bearing on the design and development of appropriate harvesting equipment. Besides, tall plants with weak stems and heavy flowers tend to lodge.

Pyrethrum plants become bushy with age and bush size depends upon genetic and environmental factors like temperature, soil type, and soil moisture. It is generally believed that a larger bush is likely to yield more flowers. Also bush diameter influences the plant spacing that should be followed in field plantings.

Flower shape and size are generally determined by the diameter of the disc florets. These characters are also affected by the length of central and outer whorls of the disc and ray florets. Large flowers with long ray florets may be quite attractive, but their mass may be a disadvantage by causing lodging in weaker plants. Besides, large and heavy flowers take longer to dry, increasing production costs.

D. Miscellaneous

Lodging is determined as the angle at which the flowering shoots stand in relation to the ground level; plants with erect standing shoots (close to 90°) are termed "lodging resistant" while those with more acute angles are "lodging susceptible." A completely lodged plant is one in which the shoots are prostrate.

Different workers have scored this character differently. Chamberlain and Clark (1947) used the terms "compact," "open," or "prostrate" depending upon whether the growth was dense, open, or very open, and whether the flower stems grew erect, inclined to spread, or lay on the ground. On the other hand, Bhat (unpublished) has used a scale of 1 to 5, with 1 being the most resistant type while Parlevliet and Contant (1970) and Parlevliet (1974) have used scales of 1 to 10 and 1 to 5, respectively, with 1 being the most susceptible. Erect (lodging resistant) plants are always easy for picking and weeding. This trait is of extreme importance in places where mechanized harvesting is attempted (Australia and the United States).

Several diseases and pests attack pyrethrum and screening the breeding populations for resistance is not only desirable but also necessary. The pathogens identified so far are: root rot fungi like *Sclerotinia*, *Fusarium*, *Phytophthora*, *Pythium*, *Rhizoctonia*, *Thielaviopsis*, *Phoma*, *Stemphyllium*, *Colletotrichum*, *Cylindrocarpun*; nematodes like *Meloidogyne hapla* and *Pratylenchus neglectus*; flower bud diseases caused by *Ramularia belluensis*, *Coniothyrium*, *Gloeosporium*; and pests like thrips, red spider mite and eelworms.

The ease of vegetative propagation is important and clones must split well and the splits should easily establish after planting. Genotypes should easily propagate by stem cuttings under mist and in tissue culture. These characters are difficult to measure, but large genotypic differences exist. In breeding nurseries the easiest way to score genotypes for vegetative propagation and establishment of propagules after transplanting is to take the percentage strike each time these operations are performed. Commercially the genotypes get automatically evaluated for these two traits because clonal selection involves splitting and establishment of the clones several times before the final selection. Any genotype that does not split and/or establish well gets automatically eliminated (Parlevliet, 1974).

Screening cultivars for adaptation to variable climates, like extremely low winter temperatures (Bhat, unpublished), or extremely high summer temperatures (McDaniel, 1991) is also important in pyrethrum breeding.

IV. AIMS OF THE BREEDING PROGRAM

For pyrethrum, the most important breeding criterion is to increase the yield of pyrethrins per unit area. Pyrethrins yield is a product of two components—pyrethrins content in the dry matter of the flowers and the dry-flower yield per unit area. Dry-flower yield, in turn, is dependent upon the fresh-flower yield and the moisture content. Flower yield is influenced by the number of flowers per unit weight which means that the weight of individual flower heads will be important. Generally, genotypes with high-flower number per plant produce high-flower yield. The relative importance of these characters in increasing the pyrethrins yield is discussed in Section VIII.

The other breeding objectives are to develop varieties which are: of different maturities; nonlodging; tolerant/resistant to diseases and pests; and adapted to widely variable environments. However, whatever the aim, objectives should be well defined before the commencement of the program.

V. BREEDING PROCEDURES EMPLOYED

The first step in starting a breeding program on pyrethrum is to establish genetically heterozygous and heterogeneous populations raised from seed. These populations are screened for the character(s) needing improvement. Variability should be high to maximize improvement potential. Where needed, creation of variability will be required before commencing the selection. Fortunately, the cross-pollinating nature of pyrethrum, due to flower morphology and sporophytic incompatibility, has rendered it a highly variable crop in almost all the characters (Drain and Shuey, 1934; Gnadinger *et al.*, 1933, 1936; Culbertson, 1940; Gnadinger, 1945; Chamberlain and Clark, 1947; Chamberlain and Procter, 1947; Kroll, 1958; Parlevliet, 1969, 1974; Bhat and Pandita, 1982; Pandita and Bhat, 1984; Bhat and Menary, 1984c, 1986b; Singh *et al.*, 1987; Pandita, 1989; Ikahu and Ngugi, 1990; Malley and Mwalukasa, 1991). Individual plants (also referred to as genotypes in this text) showing merit are isolated and evaluated by different methods according to the intended breeding methodology. Age of the plant significantly affects some characters and this should be remembered while evaluating the population (more detail in Section VI).

A few of the subtle differences amongst clones and populations of pyrethrum are often overlooked in breeding programs. Variability occurs not only in flower yield, pyrethrins content, and various morphological traits, but also in the quality of the pyrethrins, as measured by the relative composition of the six esters and PI/PII ratio (Head, 1967; Broadbent and Hagarty, 1969; Winney, 1971). The PI/PII ratio generally varies from 0.8 to 1.0, but a range of 0.47–3.5 has been observed in different breeding lines (Head, 1967; Bhat, unpublished).

Similarly, clone-to-clone differences have been noticed in the flower maturity at which the pyrethrins content maximizes. Generally the pyrethrins content maximizes when 3/4 of the disc florets are open (Head, 1966), but some clones reach the peak a few days prior to this stage while others peak at the overblown stage (Parlevliet, 1970b; Ikahu and Ngugi, 1989).

A. Development of Superior Clones

For superior clones, initial selection of the genotypes is based on the phenotype of the trait under evaluation that is determined by three main factors: (1) genetic make up; (2) the environment; and (3) the interaction of the genotype with the environment. For understanding the key factors contributing to the expression of the trait, the clone is vegetatively propagated and the performance is monitored for several years under diverse environmental conditions. Those clones that perform consistently well across the environments — both years and locations — are regarded as superior with improved quality derived mainly from the genotype. On the other hand, some clones do not perform well under all conditions and their phenotypic expression is determined by the genotype, the environment and environmental interaction.

The success in developing superior clones depends upon precise determination of genetic and environmental factors affecting the trait. Accordingly, Parlevliet

(1975) has proposed that the evaluation of the clones be done in five stages: single plant selection, single line selection, screening trial, replicated yield trial, and adaptability trial.

Single plant selection (*SPS*). Plants in a large heterozygous population, raised from seed, are individually evaluated and desirable genotypes are isolated. In order for the selection to be unbiased the plants should be the same age and grown under identical conditions. If the plants are too young — say only a few months old — the true nature of their production potential may not be revealed (Section VI). Selection criteria here are the production traits: flower yield; resistance to lodging, diseases, and pests; pyrethrins content; and plant canopy shape. At the end of the growing season plants are selected for the next stage of evaluation.

Concerning the assessment of the pyrethrins content, opinions differ. Parlevliet (1975) has suggested that it is unnecessary to evaluate pyrethrins content at this stage. However, it may be desirable to assess pyrethrins content in both agronomically suitable genotypes as well as in a random sample of the population for two purposes. Firstly, the range in the pyrethrins content of the population as a whole is assessed and, secondly, identification of genotypes with high pyrethrins content is facilitated that may also be selected for use as parents in subsequent hybrid breeding programs.

Single line selection (*SLS*). The selected genotypes are removed from the main field and crown split for planting in one-row plots of 10–15 plants. A standard clone (control) is planted every 30–50 rows. Differences amongst genotypes in their splittability and establishment will emerge at this stage. Parlevliet (1975) has suggested discarding all genotypes that do not grow well at this stage. However, it may be desirable to evaluate all the selected genotypes once again. That way it not only allows confirmation of the earlier results at the SPS stage but also helps in identifying genotypes with specific desirable traits even though the growth characteristics may not be adequate. Such genotypes are also selected for use in long-term breeding programs where these specific desirable traits may be utilized. Selection criteria used at this stage are flower yield, pyrethrins content, plant canopy shape, synchronous flower development, lodging resistance, vigor, and disease and pest resistance. In order to eliminate the variation caused by environmental factors like temperature and rainfall that change annually or by location, performance relative to the control clone should be used for selecting the genotypes. Evaluation of the clones commences in the year following planting in single-row plots or when the plants have completed at least 6–8 months of active growth. The final selection is made at the end of the growing season (as in the first year). Only healthy, vigorous, upright, and abundantly flowering clones with high pyrethrins content are selected.

Screening trial (*ST*). Each selected line is split and replanted in single plots of 60 plants with a control clone every 5–8 plots. Only the best clones, surpassing the control in most of the characteristics, including the yield of pyrethrins per hectare, are finally selected.

Replicated yield trial (*RYT*). The clones from the ST are now planted in a RYT. The RYT is conducted at one (Parlevliet, 1975) or more locations (Bhat and Menary, 1984a), depending upon the resources available. Variable numbers of plants per plot and replications per trial have been used. The trial lasts for 2 years during which the flower yield and the pyrethrins content are recorded and after which the best clones are selected for the adaptability trial.

Adaptability trial (*AT*). The adaptability of the selected clones to different areas is tested by planting at 4–5 locations. The best clones are finally selected after 2–3 years of evaluation and then released for large-scale cultivation in the best adapted areas. By these selections, numerous improved clones have been released for commercial cultivation in Australia (Bhat and Menary, 1984b; 1986b); India (Singh *et al*., 1988), and Kenya (Contant, 1962, 1963; Parlevliet and Contant, 1970; Tuikong, 1984). Two clones have also been patented in the US; Bhat and Menary (1987) first patented pyrethrum clone Hypy followed by McDaniel (1991) for the plant named Arizona. Hypy is also the first cultivar of pyrethrum registered by the Crop Science Society of America (Bhat and Menary, 1984b).

It is of interest to review the progress current breeding programs have achieved in the improvement of pyrethrum as a cultivated crop. The results of Parlevliet *et al*. (1979) are worth considering in this regard. Twenty wild populations of pyrethrum were collected from Dalmatia in the summer of 1971 and evaluated at two locations—Molo and Subukia—in Kenya. The average pyrethrins content in these wild collections was 0.89%, ranging from 0.75 to 1.04%. Today it is common to find clones and breeding lines with pyrethrins content of 3.0% or more in programs in Australia, Kenya, and the USA. Other agronomic traits like flower yield, lodging resistance, canopy shape, and synchronous flowering have also been improved in these countries.

Multiplication. The new clones must be multiplied considerably before commercial cultivation can commence. It is important that clonal purity and freedom from diseases and pests are maintained throughout the multiplication stage. Various methods of vegetative propagation can be used. The earliest and the simplest method is crown splitting of the mother plants, but this is too slow to be of significant advantage in establishing commercial plantings rapidly. However, Brown (1965) was able to increase the multiplication by layering the mother plants. Large differences occur among genotypes in the number of splits produced; some give 70–80 or more, and others only 5–10, the average being 15–25. Another method is to root stem cuttings under mist and plant the well established rooted plants out in the field. Bottom heat near 21°C helps rooting that may take between 3–6 weeks.

The more recent method of propagation is tissue culture (Roest and Bokelmann, 1973; Levy 1981; Wambugu and Rangan, 1981; Karki and Rajbhandary, 1984; Bhat, unpublished). In this procedure a genotype can be multiplied considerably and quickly. Explants, generally shoot tips and/or axillary buds, are taken from mother plants, surface sterilized, and then placed on a growth medium for production of adventitious shoots under aseptic conditions. The growth medium generally is that of Murashige and Skoog (1962) supplemented with various auxins and cytokinins. After 4–6 weeks, the adventitious shoots (Plate 4A) are broken off under aseptic conditions, and

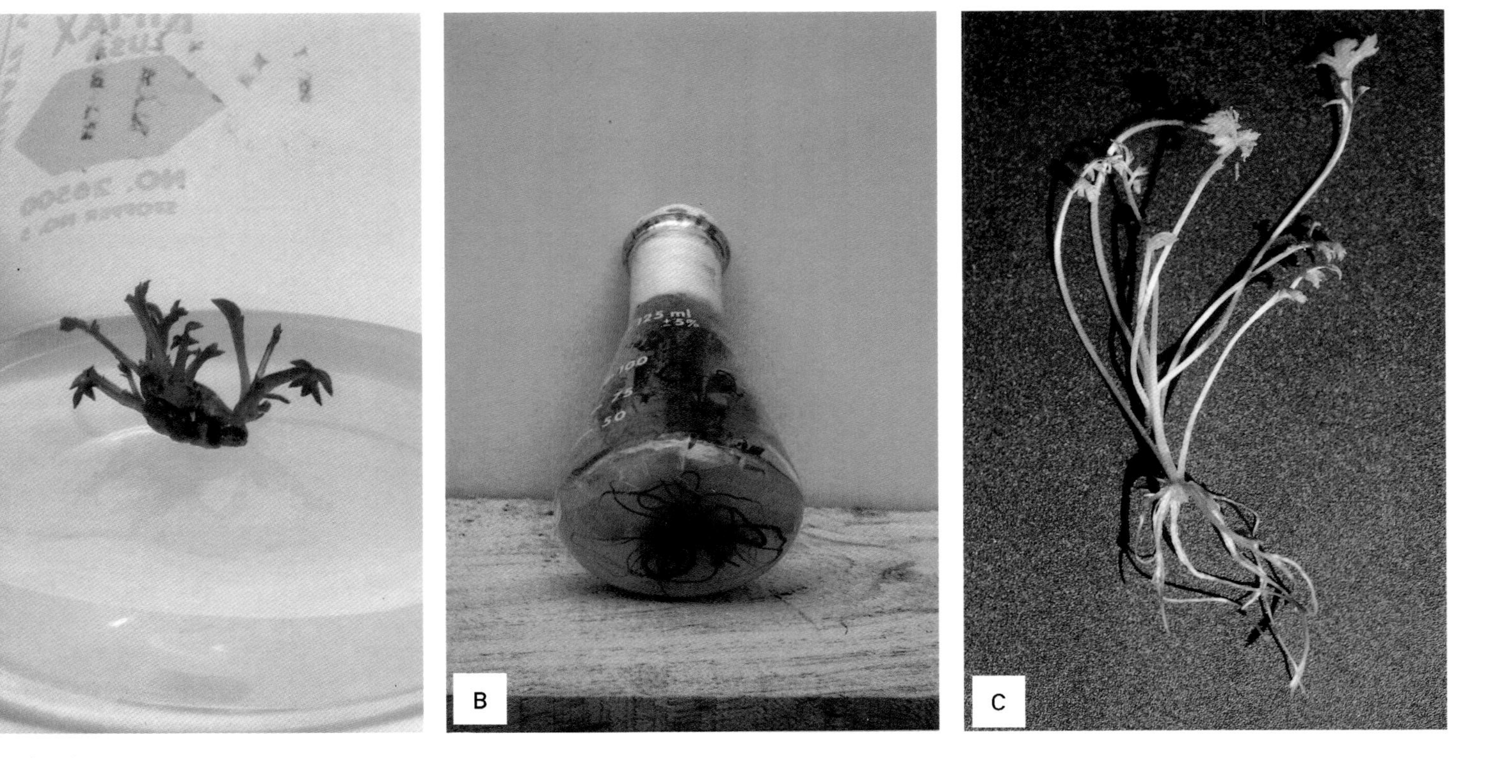

Plate 4. Tissue culture propagation of pyrethrum. (A) Developing adventitious shoots. (B) Plantlets in rooting medium. (C) Plantlet with well-developed shoots and roots.

Plate 4. (*Continued*) Tissue culture propagation of pyrethrum. (D) Rooted plantlets in plugs and under constant mist for hardening off. (E) Fully developed and hardened off plants. (F) Plant with well-developed root ball and ready for transplanting in the field.

recultured on the propagation medium. The cycle is repeated until the required number of plants has been obtained. The adventitious shoots may also be transferred to a rooting medium (Plate 4B) if the propagation medium does not induce rooting. Generally a propagation rate of 3–6 plantlets in 4 weeks has been achieved. The rooted plants from tissue culture (Plate 4C) require 6–8 weeks hardening-off before they can be transplanted into the field, during which frequent misting is necessary (Plate 4D), otherwise the young, delicate plants will desiccate and die. Well established plants with a root ball (Plates 4E, F) are transplanted into the field.

Even though techniques for regeneration of pyrethrum plants from callus cultures have been reported (Pal and Dhar, 1985; Paul *et al.*, 1988) these are not recommended for micropropagation because plants regenerated from such cultures are generally variable in many respects from the mother tissue.

General comments about clonal selection. While scoring the variation in a population it is important to remember that past breeding history will greatly influence the mean and range of the characters. Populations maintained under selection pressure for some characters will be different from those maintained without any selection. For example, the pyrethrum germplasm introduced from Kashmir, India, into Tasmania (Bhat and Menary, 1984a) and the US (Bhat, unpublished) is significantly superior to that from other sources. This germplasm was maintained under selection pressure for high pyrethrins content and it has yielded many superior clones, including the patented cultivar Hypy (Bhat and Menary, 1984b, 1987). Other clones (high pyrethrins content and lodging-resistant) selected from this germplasm are under development in the US (Bhat, unpublished).

B. Development of Hybrid Varieties

In the development of hybrid varieties of pyrethrum, clones are crossed to develop the seed that is released for commercial cultivation, either without or after evaluation, and in the latter case only the best hybrids are recommended for release. Since pyrethrum clones are genetically heterozygous, the progeny will be variable. If need be, this freshly created variation may be utilized in the same manner as explained earlier (Section V.A). The importance of the choice of parents (clones) in developing successful hybrids is paramount. The important considerations will be to identify the characters to be improved, understand their inheritance, and identify the source of germplasm. In this process several clones may be involved in the crossing program.

The procedure in Kenya until 1962 involved evaluating the clones for performance over 3 years, selecting the best clones, and crossing these to produce the hybrid seed for commercial cultivation (Gaddum, 1950; Contant, 1962, 1963; Parlevliet, 1969). Different combinations of clones were crossed that also included backcrossing and sib-mating which were not always successful. Occasionally only one, or sometimes none, of the clones produced viable seed (Gaddum, 1950; Kroll, 1958). It is now well understood that sporophytic incompatibility may have contributed to failure of seed setting with crossing of closely related clones. On the whole, the results were variable resulting in only a few superior hybrids (Glynne Jones, 1968). One of the main reasons for the variable performance of the hybrids was that the parents were selected by their own phenotype and no consideration was given to various genetic and nongenetic components producing the phenotype. In order to understand these factors and use them in future hybrid breeding programs it is necessary to consider the basic principles for improvement of polygenically inherited traits.

Phenotypic and genotypic variance. Phenotypic expression of a polygenically inherited trait is determined by the genes from the parents and the extent to which their expression deviates because of genetic and environmental factors. Phenotypic variance (σ_p) of a metric trait is the sum of the genotypic variance (σ_g) and the environmental deviation (σ_e), $\sigma_p = \sigma_g + \sigma_e$ (Falconer, 1989). Genotypic variance can be partitioned further as follows: additive genetic variance (σ_a), which is also the variation in breeding values among individuals; nonadditive genetic variance (σ_d), including dominance deviation; and additive × additive, dominance × dominance, and additive × dominance interactions, all grouped as σ_i. These components allow estimation of the relative importance of the various determinants of the phenotype. Therefore, $\sigma_p = \sigma_a + \sigma_d + \sigma_i + \sigma_e$.

Breeding value. The breeding value of the individual is judged by the mean value of its progeny as measured by mating the individual to several random individuals from a population, determining the mean performance of the progeny, subtracting this performance from the population mean, and multiplying the deviation by two because half of the genes in the progeny are from the individual and the other half come randomly from the population with a value equal to the population mean (Fehr, 1987; Falconer, 1989).

In pyrethrum, two methods have been used to measure the breeding values of different clones: polycross and topcross (testcross) methods. In polycross (Section V.C), the selected clones cross pollinate in all possible combinations and the seed from each clone is harvested separately. The progeny of each clone is raised separately to assess the breeding values of different characters. In topcross, also known as testcross, each clone crosses with a tester from a broad-based heterozygous population. The tester can be the parent population of the clone or totally unrelated. If the parents are selected by their breeding values, the performance of the hybrids is more predictable and consistent. This is clear from the study of Parlevliet and Contant (1970), when they theoretically calculated the flower yield and pyrethrins content of 18 single crosses from the polycross progeny test of the parent clones and then compared the calculated values with the actual performance of the single crosses. The two values were highly correlated (r value between estimated and actual flower yield was +0.87 and for pyrethrins content, +0.86).

In another polycross of 42 clones, the polycross progeny test was used to select clones producing high flower-yielding hybrids and these were used to develop 14 single crosses whose performances were compared with 37 other single crosses developed from parents selected by their own performance. The flower yield of the first group of 14 single crosses was almost double (129 relative units) compared to the latter group (65 relative units). However, pyrethrins content showed no improvement by this selection method (Parlevliet and Contant, 1970).

Combining ability. In a hybrid breeding program, the objective is to identify new lines that, when crossed with other parents, produce superior hybrids. To this end, it is necessary to make all possible combinations of crosses and evaluate the hybrids to identify superior ones. To produce all possible hybrids, numerous crosses are needed which renders this procedure quite laborious. However, this can be avoided by estimating the combining ability of the parents. There are two types of combining abilities—general (gca) and specific (sca). Gca refers to the average performance of a line or a clone when crossed with other parents. Sca is the performance in a cross with a specific parent (Fehr, 1987). Combining ability estimates of the parents allow predictions of performance of the hybrids. Gca and sca can be estimated from topcrosses, diallele crosses, and a few other mating designs (for details, Fehr, 1987; Hallauer and Miranda, 1988; Falconer, 1989). In a study of 18 single crosses of pyrethrum, Parlevliet and Contant (1970) reported that 76 and 74% of variation in flower yield and pyrethrins content, respectively, was due to gca and 14 and 19% due to sca, the remaining due to error. On the other hand, in a 5×5 diallele cross sca was greater than gca (Pandita, 1989).

Heritability. Partitioning of the variance allows estimation of the relative importance of the various determinants of the phenotype. If these components are used to calculate their ratio to the total phenotypic variance, the heritability of the trait can be obtained. Heritability of a metric character is one of its most important properties used extensively in animal and plant breeding studies and generally two types are considered: broad-sense heritability that is the ratio of genotypic variance to the phenotypic variance (σ_g/σ_p); and narrow-sense heritability, or simply heritability that is the ratio of additive genetic variance to phenotypic variance (σ_a/σ_p). Heritability (h^2) is also defined as the regression of the breeding value (b_A) on the phenotypic value (P) ($h^2 = b_{AP}$). From this definition, the breeding value of an individual ($A_{(expected)}$ can be estimated as a product of its phenotypic value and the heritability ($A_{(expected)} = h^2P$). Heritability has a predictive role in indicating the reliability of using phenotype as a guide in determining the breeding value.

A few considerations, generally overlooked while using heritability estimates in breeding studies, should be emphasized here. First, heritability is a property not only of a character but also of the population and the prevailing environmental conditions. A change in the variance of any of these components will affect the heritability. So, whenever a value is stated for the heritability of a given character, it must refer to a certain population under particular conditions (Falconer, 1989). Second, the heritability given by the formula, $h^2 = \sigma_a/\sigma_p$, refers to the ratio of variances of individual values and therefore will not apply, as such, to situations where individual values are not available such as plot means or plot totals in plants. Relevant theory and the formulae for calculating heritabilities from such data are available (Fehr, 1987; Falconer, 1989). In pyrethrum, heritability of various characters has been estimated by

Table 5-1. Heritability of Different Characters in Pyrethrum

Character	Heritability		Method of estimation[a]	Reference[b]
	Broad sense	Narrow sense		
Pyrethrins content	—	0.70	b_{OF} (polycross	A
	—	0.64	b_{OM} (single cross)	A
	0.92	—	σ_g/σ_p	C
Pyrethrins yield	0.77	—	σ_g/σ_p	C
Flower yield	—	0.53	b_{OF} (polycross)	A
	—	0.56	b_{OM} (single cross)	A
	0.73	—	σ_g/σ_p	C
	0.78	—	σ_g/σ_p	D
Flowers/plant	0.85	—	σ_g/σ_p	C
	0.83	—	σ_g/σ_p	D
Flowers size	—	0.87	b_{OF} (polycross)	A
(100-flowers wt.)	0.79	—	σ_g/σ_p	C
Flower diameter	0.49	—	σ_g/σ_p	D
Plant height	0.63	—	σ_g/σ_p	D
Shoot length	0.70	—	σ_g/σ_p	B
Bush diameter	0.72	—	σ_g/σ_p	D
Lodging resistance	—	0.80	b_{OF} (polycross)	A
Root length	0.71	—	σ_g/σ_p	B
Root dry wt.	0.34	—	σ_g/σ_p	B

[a] b_{OF} = regression of progeny on female parent; b_{OM} = regression of progeny on mid-parent; σ_g = genotypic variance; σ_p = phenotypic variance.
[b] A = Parlevliet and Contant (1970); B = Singh and Rajeswara Rao (1985); C = Bhat and Menary (1986b); D = Singh *et al.* (1987).

different methods (Table 5-1). These data suggest that pyrethrins content has high heritability while flower yield does not. However, these results should not be generalized because heritability estimates, as already pointed out, are specific to the population and prevailing conditions.

C. Population Improvement

Sporophytic incompatibility in pyrethrum will prevent continued mating of closely related individuals that may restrict the extent of hybrid development beyond certain limits. Therefore, other means are used such as population improvement, in which the frequency of favorable alleles within the population increases gradually with each cycle of selection, making the program a dynamic process. Further, these populations have wider adaptation because of the broad genetic base. Three methods for population improvement in pyrethrum are (1) positive mass selection, (2) polycross, and (3) recurrent selection for general combining ability.

Positive mass selection consists of selecting individual plants in a genetically heterogeneous and heterozygous population. The final selection is made after evaluating the plants for about two growing seasons or more. The selected plants are then allowed to interpollinate and the seed from these plants is harvested for raising the next generation. Since the plants are selected by their phenotypic performance, the gain for characters with low heritability will be low. For characters with high heritability this selection will be quite effective. One cycle of positive mass selection will take 3 years—2 years for evaluation and selection of the clones and one for seed production by intermating the selected clones (Parlevliet, 1975).

Polycross is widely used for intercrossing parents by natural hybridization in vegetatively propagated species with mechanisms that prevent or minimize self-pollination. This procedure has been used in pyrethrum and was one of the earliest breeding methods followed in Kenya (Kroll, 1958). The method was not strictly a polycross as understood at present but a combination of polycross and positive mass selection.

The polycross followed in pyrethrum these days is as follows. For the general procedure, 40–50 clones are selected as the mating group to produce the polycross seed. The clones are planted in an isolated field in several replicates of a few plants so that each clone has a chance to cross with every other clone in the field. The seed of each mother clone is harvested and planted separately for the polycross progeny test. Parlevliet (1975) recommends that the polycross progeny test should be replicated at more than two locations for at least 2 years. The progeny of each clone is evaluated for flower yield, pyrethrins content, lodging resistance, and other agronomic characters. The clones giving good progeny (high gca) are planted in isolated fields for the final polycross seed. It will take 4 years to complete this breeding procedure—1 year for establishing the selected clones in an isolated plot for producing initial polycross seed; 1 year for raising the seedlings and establishing them in the field; and 2 years for polycross progeny evaluation to know the breeding values of the parent clones.

In order to assure that all the parents contribute equally to the development of the seed, Fehr (1987) has outlined three criteria for making the polycrosses. First, the flowering of all parents (clones) must synchronize for effective crossing otherwise all possible crosses may not occur. This can be achieved by cutting the early flowering shoots regularly until the flowering coincides. Second, clones planted in adjacent plots are more likely to cross with one another than clones in distant plots. Therefore, to ensure production of all possible crosses, the planting is randomized and replicated to increase the chances of intercrossing, but even so, it cannot be ensured that all clones would have crossed with one another. The other procedure to ensure crossing of all the clones in all combinations involves field planting in a Latin square, where the number of replications equals the number of treatments and the number of plots makes a perfect square. Thus, in a polycross of 8 clones there will be 8 replications (total of 64 plots). The assignment of clones to the plots is such that each clone in one row or column (block) occurs only once. Thus each clone occurs adjacent to every other clone somewhere in the Latin square. This planting arrangement is possible when the number of clones to be crossed is small. With many clones the number of replications will be too large and clones are arranged in a randomized complete block. Third, the method of sampling hybrid seed would influence the genetic makeup of the progeny from the polycross.

Fehr (1987) has suggested the following three methods of bulking hybrid seed and described the merits of these methods: (1) seed from all parents may be bulked without regard to the amount of seed produced by each parent; (2) seed from each plot (replication) of each parent may be harvested separately and then bulked regardless of the amount produced per plot (an equal amount of the bulked seed of each parent mixed to form a single population); and (3) the seed from each plot (replication) of each parent is harvested separately and then bulked equally from each plot (an equal amount of seed from the bulk of each parent mixed to form a single population). Only method (3) ensures equal genetic contribution of each parent to the final population though it requires more labor than the other two procedures.

Recurrent selection for general combining ability. In positive mass selection, clones for producing seed for the next generation are selected by their own performance. However, if these clones are selected by the performance of their progeny, an estimate of their gca can be obtained. Clones with high gca are likely to produce better hybrids than those with untested gca.

Parlevliet (1975) has suggested six steps in completing one cycle of recurrent selection in pyrethrum. (1) The clones are selected from a genetically variable population (Section V.A). The clones evaluated in the ST may be used for this purpose. It is recommended to select not only the best clones from the ST but also those possibly missing the replicated-yield trial because of some undesirable traits. It is possible that this latter group of clones may have high gca for specific traits (Parlevliet and Contant, 1970). (2) About 40–50 clones are selected to form the polycross seed (Section V.C). It takes about 1 year from planting in the isolated crossing block to harvesting of the polycross seed. (3) In the next step, the seed harvested from each mother clone is planted separately to estimate the gca by the polycross progeny test. Gca of the clones can also be tested by allowing the selected clones to cross pollinate with a population having a wide genetic base and then harvesting the seed from each clone separately for the progeny test. It takes nearly 1 year from harvest of polycross seed until establishment of the polycross progeny seedlings in the field. The polycross progeny evaluation will continue for 2 years; thus, it will be 4 years before the breeding values of the clones are known. (4) Clones with high gca are selected and multiplied vegetatively in just enough numbers to plant in pairs to produce hybrids. Hybrid seed is produced in the fifth year of the program. (5) The actual performance of the hybrids is tested by raising seedlings in the 6th year and evaluating in years 7–8. (6) The clones producing the best hybrids are multiplied vegetatively in large numbers and planted in the field for commercial hybrid seed. Even though this breeding method is quite laborious, selection efficiency is greatly improved and the gains are enormous because the clones have been selected on the basis of their gca.

Gains achieved by population improvement. Insufficient published work exists comparing the gains from different population improvement programs in pyrethrum. Parlevliet and Contant (1970) showed that selection for pyrethrins yield by the polycross method gave 80% higher yield than by positive mass selection. Bhat *et al.* (1985) have reported significant increases in the pyrethrins

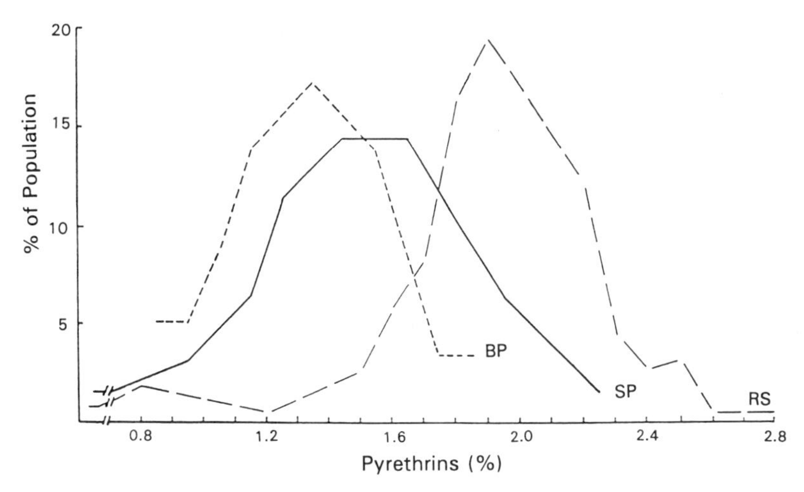

Figure 5-2. Frequency polygon of three pyrethrum populations for pyrethrins content. BP — base populations; SP — selected populations; RS — recurrent selection population. (Bhat *et al.*, 1985).

content following one cycle each of positive mass and recurrent selections. When the three types of populations — the base population (BP), the selected population (SP), and the recurrent selection population (RS) — were compared, the average pyrethrins content increased from 1.33% in BP to 1.46% in SP and 1.97% is RS (Bhat *et al.*, 1985) (Fig. 5-2). Besides the increase in the pyrethrins content, for which conscious selection was made, the pyrethrins yield also increased in each population, partially due to an increase in flower yield attributed to unconscious selection of high-flower-yielding plants.

In essence, the breeding programs discussed so far are interrelated and clones selected from heterozygous populations can be utilized in various ways (Fig. 5-3). The whole program is dynamic as long as enough variation is generated at

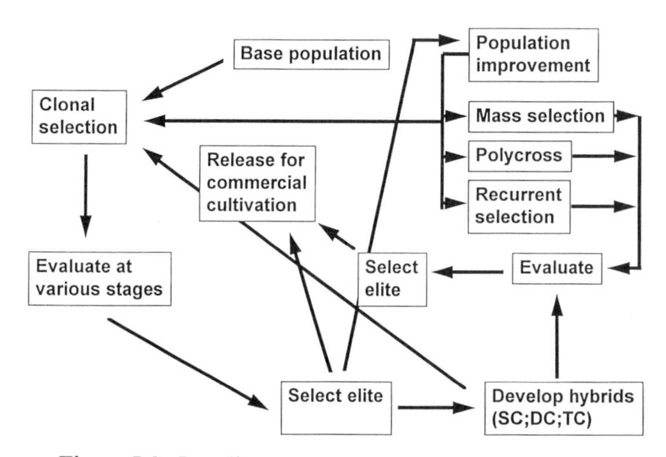

Figure 5-3. Breeding strategy applicable to pyrethrum.

each stage. Once variability is completely exhausted or the gains achieved are not significant, creation of fresh variability is necessary.

D. Miscellaneous Breeding Procedures

Mutation breeding is not in the published work on pyrethrum. This may be because it does not offer advantages over the breeding methods already discussed. The crop is highly variable in all the characters studied so far and this variation needs to be utilized first prior to exploiting mutation breeding. Therefore, this method is not discussed further. However, the general procedures will be the same as followed in other perennial plants (Gottschalk and Wolff, 1983; Fehr, 1987).

Polyploidy breeding has been used in many vegetatively-propagated perennial plants. Chromosome number is increased in multiples of the basic (x) number and this increase in many instances results in increased product yield and vegetative vigor. Generally triploids are more vigorous than diploids, tetraploids, or higher polyploids. In pyrethrum, polyploidy has been used on a limited scale as it provides no significant advantage over other breeding methods.

The greatest advantage in pyrethrum breeding will be from haploids which, when doubled in chromosome number, will give homozygous diploid plants that are otherwise difficult to produce because of sporophytic incompatibility that prevents selfing or mating amongst closely related individuals. Haploids have been widely used in the improvement of a number of crops (Bajaj, 1990a).

Triploids and tetraploids have been produced in pyrethrum. To produce a triploid plant having 27 chromosomes ($2n=3x=27$). first one must produce a tetraploid plant with 36 chromosomes ($2n=4x=36$) and then cross this tetraploid with a diploid parent and select the triploids from this cross. Even though Ottaro (1977) has reported a close relationship between triploidy and an increase in the expression of some morphological characters, it is recommended that triploids be selected only after the actual count of the chromosomes is verified cytologically and not rely merely on the morphology. In pyrethrum, tetraploids have been produced by treating the seed with 0.2% colchicine for 5 days (Contant, 1962; Tuikong, 1984). Vegetative buds can also be treated with colchicine to double the chromosome number. Shoots originating from such treated buds can be rooted *in situ* by air layering or by rooting cuttings of these shoots under mist. The rooted plants can then be established into full-grown tetraploid plants. Triploids in pyrethrum are taller with increased flower diameter and flower weight, compared to diploid and tetraploid plants (Ottaro, 1977).

VI. HOW SOON, AFTER TRANSPLANTING, SHOULD DIFFERENT CHARACTERS BE EVALUATED?

Since pyrethrum is a perennial plant and the testing of the clones continues for 2–3 years, it is important to know when to evaluate and select various characters, and whether the selection after the first year is as reliable as that made after testing for 2–3 years.

Table 5-2. Effect of Planting Data on Pyrethrins Content, Dry Flower Yield, and Pyrethrins Yield[a]

Planting date	Pyrethrins content (%)	Dry flower yield (kg/ha)	Pyrethrins yield (kg/ha)
28 Sep 90	1.82	1,210	22
16 Oct 90	1.98	739	15
28 Nov 90	1.80	550	10
07 Jan 91	1.84	440	8
25 Jan 91	1.67	379	6

[a]Bhat, unpublished.

A factor that should be emphasized is that the age of the plants affects characters like flower yield and pyrethrins yield, while less so the pyrethrins content. A study by Bhat (Table 5-2) showed that different planting dates, which in turn determined the age of the plants at evaluation time in the first year, did not affect the pyrethrins content, but flower yield and pyrethrins yield were considerably reduced in young plants, a difference which disappeared in the second year when plants had fully matured. Therefore, it is suggested that the final evaluation and selection of the genotypes should be deferred until after at least 9–10 months of active growth following transplanting.

Concerning the reliability of first year results, as compared to those obtained after 2–3 years of evaluation, the work of Parlevliet (1969) is noteworthy. In this study pyrethrins content and flower yield of 110 clones were measured in 7 replicated yield trials lasting 3 years. The pyrethrins content and flower yield in the first year were highly correlated over 3 years (Table 5-3). Further, out of the 110 clones evaluated, the 37 highest pyrethrins yielding (kg/ha) clones were selected by their first and third years performance and it was found that 30 of the 37 clones were common to both the groups. Thus, selection for the two traits made in the first year would be as reliable as that after 3 years.

In general, for characters with high heritability, like pyrethrins content, early selection will be as effective as that made in later years. On the other hand, for characters with moderate to low heritability and those which are affected by the age of the plants, like dry-flower yield and pyrethrins yield, selection should be deferred until the plants have fully matured and should be repeated for at

Table 5-3. Percentage Increase in Three Characters of Pyrethrum Clones Selected After One or Three Years of Evaluation (Parlevliet, 1969)

	Percentage increase after evaluation of:		
Character	1 year	3 years	r[a]
Pyrethrins content	2.4	2.6	0.93
Flower yield	31.0	34.1	0.91
Pyrethrins yield	34.3	37.5	—

[a]Correlation coefficient between the results of the two evaluations.

least 2–3 years to get a true picture of the yield potential of the plants. While evaluating the pyrethrins content of various clones from different locations and/or years, one should remember that even though this character is fairly stable, weather conditions affect the expression considerably — warm and dry conditions affect it adversely while cool, moderately wet conditions favor high pyrethrins content (Muturi *et al.*, 1969; Parlevliet, 1970a; Bhat, unpublished).

VII. ADAPTATION OF CLONES

Pyrethrum needs low temperature for flower bud initiation (Glover, 1955), and subsequent flowering — both duration and yield — are determined by the temperature and photoperiod. Genetic differences amongst clones and hybrids make them variable in their chilling requirement for flower bud initiation. Because hybrid populations are heterozygous they have a wider adaptation than the clones.

That different clones have different chilling requirements for flower bud initiation has been shown by Roest (1976) in the only study conducted under controlled conditions. Three clones were grown at constant temperatures of 9, 13, 17, or 21°C and their flowering response and the average number of flowers per plant were different at these four temperatures. This differential growth response of clones to temperature — and possibly to precipitation also — may affect their adaptation to different climates (Parlevliet *et al.*, 1969; Parlevliet, 1970a; Ikahu and Ngugi, 1988; Singh *et al.*, 1988; Wanjala, 1991). That these differences in adaptation are genetic can be observed from Ikahu and Nagugi (1988) who noticed that good performing varieties in all sites had clones 4743 and SB/65/58 as common parents. Therefore, with proper selection of parents it is possible to breed varieties that are adapted to a wide range of climatic conditions. It may be more difficult to breed high-flower-yielding and high-pyrethrins-content cultivars for lower altitudes —consequently for higher temperatures — than for temperate areas. High temperature not only reduces flower yield and the pyrethrins content but also renders many genotypes blind (Muturi *et al.*, 1969; Bhat, unpublished). However, recently some success has been achieved in developing fairly productive clones adapted to high temperature in Arizona (McDaniel, 1991). In the development of new clones and hybrids, testing for adaptability and general performance should occur in yield trials at locations varying in temperature, precipitation, severity of winter, and related climatic factors.

VIII. CORRELATIONS

Many workers have studied relationships amongst agronomically important characters in pyrethrum, more specifically with regard to pyrethrins content. Identifying such a relationship would facilitate selection as quantitative differences in pyrethrins content of different genotypes could be recognized without elaborate chemical analyses. However, pyrethrins content does not appear to be strongly associated with any morphological character (Table 5-4).

Table 5-4. Correlation Coefficient (*r*) Among Different Characters of Pyrethrum

Character	Flowers/ plant	100-Flowers dry weight	Flower diameter	Flower dry matter content	Plant height	Bush diameter	Tillers/ plant	Pyrethrins content	Pyrethrins yield/plant
Dry flower yield/plant	0.945[a] 0.943[b] 0.750[d] 0.979[c] 0.923[e]	0.330[b] 0.430[s] 0.177[e]	0.201[a] −0.171[c]	0.202[a]	0.622[a] 0.186[c]	0.136[c]	0.768[a]	0.023[a] 0.170[b] 0.090[d] 0.201[e]	0.968[b] 0.790[d] 0.948[e]
Flowers/ plant		0.036[b] −0.250[d] −0.191[e]	0.089[a] −0.172[c]	0.288[a]	0.603[a] 0.186[c]	0.136[v]	0.858[a]	0.035[a] 0.148[b] −0.040[d] 0.151[e]	0.913[b] 0.510[d] 0.862[e]
100-Flowers dry weight								0.139[b] 0.160[d] 0.144[e]	0.321[b] 0.430[d] 0.199[e]
Flower diameter				0.165[a]	0.231[a] −0.022[c]	−0.048[c]	−0.227[a]	0.032[a]	
Flower dry matter content					0.146[a]		0.286[a]	0.087[a]	
Plant height						0.201[c]	0.378[a]	0.027[a] −0.210[f]	−0.410[f]
Bush diameter									
Tillers/plant								−0.059[a]	
Pyrethrins content									0.381[b] 0.660[d] 0.330[f] 0.484[e]

[a] Pandita and Bhat (1984).
[b] Bhat and Menary (1986a).
[c] Singh *et al.* (1987).
[d] Bhat and Menary (1986b).
[e] Bhat (unpublished).
[f] Malley and Mwalukasa (1991). Pyrethrins yield reported in this reference is (t/ha).

On the other hand, pyrethrins yield was highly correlated with dry-flower yield and to a lesser extent with pyrethrins content. Correlations of various microscopical observations of floral parts with pyrethrins content have also been reported (Pandita and Bhat, 1984; 1986) but are not included in Table 5-4. In one of these studies, width of disc florets was the only morphological trait significantly correlated with pyrethrins content in the florets ($r=0.416$) while the pyrethrins content in the whole flowers was not ($r=0.133$) (Pandita and Bhat, 1986).

Other correlations of interest revealed that the number of flowers per plant gave a higher r value with dry-flower yield than did the weight of individual flower heads. Two more traits significantly correlated with flower yield were plant height and bush diameter. Bush diameter, in turn, was positively correlated with plant height and the number of flowers per plant. On the contrary, negative relationships were observed between: flower size (diameter) and flower yield; flower size (diameter) and number of flowers per plant (Table 5-4). When Parlevliet (1974) grouped progenies of two single crosses and one polycross of 42 clones by flower weight (actual term used is "flower size," in essence, flower weight in grams) (Table 5-5), it was found that: flower weight was negatively associated with dry-matter content of the flowers and flower yield; pyrethrins content was not associated, in any way, with flower weight or flower yield. From these negative correlations it was concluded that with a decrease in flower size (weight and/or diameter), the number of flowers per plant would increase, resulting in increased flower yield. On the other hand, an increase in flower size (diameter) caused drastic reduction in flower yield (Singh *et al.*, 1987).

The relationship amongst different characters affecting pyrethrins yield is of interest from the breeding point of view and will be discussed in more detail (Section IX). The relationship of great importance in pyrethrum breeding is between flower yield and lodging. Parlevliet (1974) has reported that lodging was markedly affected by flower size and the number of flowers per stem but

Table 5-5. Mean Performance of Four Characters in the Progeny of Two Single Crosses and One Polycross of 41 Clones of Pyrethrum. The Genotypes Were Grouped According to Flower Size (Parlevliet, 1974)

Group	Flower size (g)	Dry matter content (%)	Flower yield (kg/plot)	Pyrethrins content (%)
Single crosses				
1	47	24.0	—	1.58
2	52	23.8	—	1.52
3	56	23.5	—	1.62
4	62	23.2	—	1.53
5	67	23.1	—	1.54
6	78	22.6	—	1.58
Polycross				
1	54	—	7.3	1.74
2	66	—	6.0	1.69
3	79	—	5.4	1.75

minimal relationship was revealed between lodging and total flower yield. This is quite contrary to the general observation in the field—the higher the flower yield, the more lodging (Bhat, unpublished). Parlevliet's conclusion was based on the observation that the number of flowers per stem and flower yield were not related; therefore, total flower yield and lodging were also unrelated. However, the more probable explanation could be that in Kenya pyrethrum flowers continuously through most of the year and the mature flowers are harvested every 2–3 weeks, thus reducing the weight of the flowers on the stems. On the other hand, if the plant produced just one flowering flush a year, which means that all the flower buds matured more or less simultaneously, as happens in places away from the equator, the number of flowers per stem would increase as maturity approached, making stems prone to lodging. This also means that the higher the flower yield, the more susceptible the plants are to lodging, unless inherent lodging resistance is present. This explanation supports Parlevliet's own observation that "lodging was markedly affected by flower size and the number of flowers per stem" which means that the heavier the load of flowers on the stems—due to either heavy flowers or high numbers—the more lodging.

IX. PATH-COEFFICIENT ANALYSIS

Correlation coefficients merely indicate the total relationship between two variables, including all factors making the two variables correlated, but do not necessarily imply any cause-and-effect relationship between the variables. Path-coefficient analysis (Wright, 1921), on the other hand, provides an effective means of sorting out causes of association amongst mutually related traits and has been used in many crops and was extended to pyrethrum also by Bhat and Menary (1986a). They proposed a model wherein pyrethrins yield was affected by four component characters: namely, pyrethrins content, flower yield, flowers per plant and flower weight. However, path-coefficient analysis revealed that even though the correlation coefficients between pyrethrins yield and flowers per plant, and pyrethrins yield and flower weight were positive and very high to moderate ($r = 0.913$ and 0.321, respectively), the direct effects of flowers per plant and flower weight on pyrethrins yield were negative (Bhat and Menary, 1986a). Therefore, this model has now been revised and more logical relationships amongst component characters are proposed. In the revised model, it is assumed that pyrethrins yield is basically determined by pyrethrins content and flower yield, while flowers per plant and flower weight are the components of the flower yield (Fig. 5-4). The correlation coefficients used in the revised model were derived from the data from 579 randomly selected F_1 genotypes studied in 1991 at the Botanical Resources Inc. Experiment Station, Independence, Oregon, USA.

From Table 5-6, the studies of Bhat and Menary (1986a) and Pandita (1989), it may be concluded that pyrethrins yield will be increased by selecting for both high-flower yield and high-pyrethrins content, but the flower yield will increase it more effectively than pyrethrins content because the direct effect of flower yield on pyrethrins yield was higher (0.89) than pyrethrins content (0.31). By

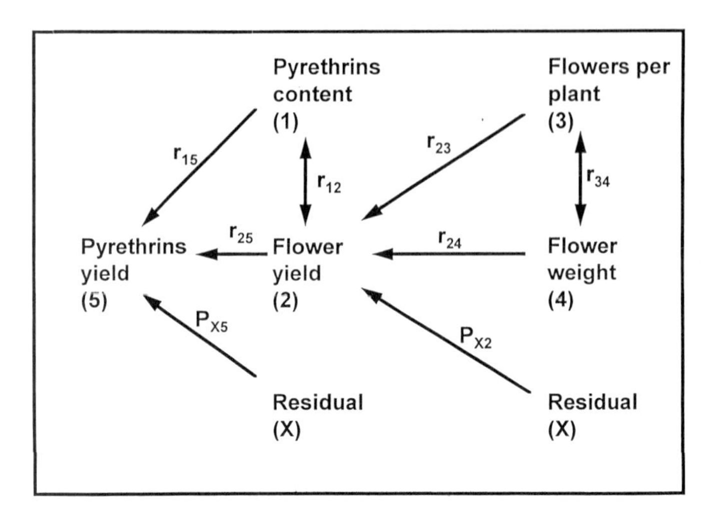

Figure 5-4. Path diagram showing causal relationship of factors influencing pyrethrins yield in pyrethrum.

the same reasoning, flower yield will be more effectively increased by selecting for a high number of flowers per plant rather than selecting for heavy flowers. Increasing the weight of individual flower heads may not be advantageous because in some cases it has decreased the flower yield (Parlevliet, 1974; Singh *et al.*, 1987) and the number of flowers per plant (Singh *et al.*, 1987). Besides,

Table 5-6. Path-coefficient Analysis of Factors Influencing Pyrethrins Yield in Pyrethrum (Bhat, unpublished)

Pathway of association	Path-coefficient		Correlation coefficient (r)
	Direct effect	Indirect effect	
A. Pyrethrins yield (5) and pyrethrins content (1)			
Direct effect, P_{51}	0.3057		
Indirect effect via flower yield (2), $r_{12}P_{52}$		0.1779	
Total			0.4837
B. Pyrethrins yield (5) and flower yield (2)			
Direct effect, P_{52}	0.8862		
Indirect effect via pyrethrins content (1), $r_{12}P_{51}$		0.0614	
Total			0.9476
C. Residual, P_{X5}	0.1113		
$1-P^2_{X5}$			0.9876
D. Flower yield (2) and flowers/plant (3)			
Direct effect, P_{23}	0.9935		
Indirect effect via flower weight (4), $r_{34}P_{24}$		−0.0703	
Total			0.9232
E. Flower yield (2) and flower weight (4)			
Direct effect, P_{24}	0.3675		
Indirect effect via flowers/plant (3), $r_{34}P_{23}$		−0.1901	
Total			0.1774
F. Residual, P_{X2}	0.1327		
$1-P^2_{X2}$			0.9824

heavy flowers take long to dry, increasing the costs. Ideally, the best approach to increase the pyrethrins yield would be to breed for numerous average-sized flowers per plant containing high pyrethrins content.

X. TISSUE CULTURE

Besides micropropagation (Section V.A), the potential of pyrethrins production by cultured cells has been explored by Cashyap *et al.* (1978), Zieg *et al.* (1983), Staba *et al.* (1984), Ravishankar *et al.* (1989), and Rajasekaran *et al.* (1990). The results have been variable, but generally the pyrethrins synthesis and accumulation in the culture were determined by the extent of differentiation of the cells. Undifferentiated callus, or callus differentiated into roots only, did not produce pyrethrins (Cashyap *et al.*, 1978) while cultures showing high differentiation, like the shoot cultures, invariably did (Cashyap *et al.*, 1978; Zieg *et al.*, 1983).

The explant source may affect *in vitro* pyrethrins production and Zieg *et al.* (1983) observed that cultures derived from high pyrethrins selections provided higher yielding lines than those derived from the low yielding selections. This suggests that the genotype of the plant contributing the explant may influence *in vitro* pyrethrins production. Concerning the type of explant tissue, it appeared that the plant part used for the culture had little influence on the pyrethrins production capacity (Zieg *et al.*, 1983), but some plant tissues yielded more genetically and cytologically-variant plants than others (Pal and Dhar, 1985). Cashyap *et al.* (1978) found that the quality of the pyrethrins, as judged by the relative proportion of the six esters, produced by shoot cultures from callus, was similar to that in 4 weeks old seedlings. However, to compare the quality of the pyrethrins in shoot cultures with that in natural flowers, the proportions reported by Cashyap *et al.* (1978) have been converted into ratios by percentage (Table 5-7). The relative proportions of the six esters in the two types of materials—flowers and callus differentiated shoots—were not identical. The cinerins were far too low in callus-differentiated shoots compared to flowers while pyrethrins II was high. That pyrethrins extracted from callus cultures may be somewhat different from the standard pyrethrins extract has also been shown by Rajasekaran *et al.* (1991), comparing the bioefficiency of the pyrethrins

Table 5-7. Relative Composition of Six Esters of Pyrethrins Produced by the Differentiated Shoots and Naturally-Produced Flowers

Type of material	PI	CI	JI	PII	CII	JII
Callus differentiated shoots						
Proportion (Cashyap *et al.*, 1978)[a]	76.0	3.0	14.0	6.0	0.2	0.8
Ratio (%)[b]	81.7	3.2	15.1	85.7	2.9	11.4
Flowers (Head, 1967)						
Ratio—mean (%)[b]	72.9	16.0	11.1	68.0	20.2	11.8
Ratio—range (%)	65.3–84.2	4.5–27.2	7.2–15.8	53.1–79.2	7.9–39.0	7.3–16.2

[a] The six esters total 100%

[b] The pyrethrins I and pyrethrins II each total 100%

from the two sources; pyrethrins extracted from 45 days old callus tissue was only 80% as toxic to *Drosophila melanogaster* as the standard pyrethrins extract. It is possible that these differences may be partly genotypic and partly tissue-culture-induced.

This tissue-culture-induced variation, also termed "somaclonal variation," has been used in the improvement of a number of crops (Bajaj, 1990b; Lal and Lal, 1990). The underlying factor exploited is that a population of cultured cells is not always homogeneous, rendering some of the cells unstable in their gene expression and metabolic activity. Cell lines or plants regenerated from such cultured cells are variable in biochemical and phenotypic characteristics. Techniques specific to each crop and problem have been devised which produce superior crop plants or highly productive cell lines with increased average yield (Bajaj, 1990a).

For pyrethrum Zieg *et al.* (1983) selected high pyrethrins-producing cultures from both callus and shoot cultures using a common explant. One of the cultures produced 114 mg/100 g dry weight (0.11% pyrethrins). Recently Ravishankar *et al.* (1989) and Rajasekaran *et al.* (1990) have reported a pyrethrins content of 0.22% in callus from leaf tissue of an elite pyrethrum line containing about 2.5% pyrethrins. In the latter case, the increase in pyrethrins content of the callus was achieved by altering the CNP (carbon, nitrogen, and phosphorus) ratio of the growth medium (Murashige and Skoog's medium) and supplementing with 2 mg/l of 2,4-D and 5 mg/l of kinetin.

Development of new cultivars through somaclonal variation can be augmented by using explants that yield a relatively high proportion of variable plants. For example, Pal and Dhar (1985) have reported that plants regenerated from petiole explants were genetically and cytologically more variable than plants regenerated from leaf tissue. A wide range of cytotypes — aneuploids, mixoploids, and polyploids — were found in petiole-regenerated plants. Another technique with limited success in pyrethrum so far is the regeneration from protoplast cultures (Malaure *et al.*, 1989)

It may be concluded that cultured cells of pyrethrum can be used commercially for the extraction of pyrethrins provided the yield of pyrethrins per unit dry mass of the tissue is increased considerably. This increase can be achieved by manipulating the composition of the growth medium (Rajasekaran *et al.*, 1990) and selecting high-pyrethrins-producing cell lines (Zieg *et al.*, 1983). Other selection criteria include fast cell growth — in order to reduce the tissue doubling time — and cells not needing high differentiation to produce a high yield of pyrethrins.

XI. APPLICATION OF GENETIC ENGINEERING TO PYRETHRUM

In this era of biotechnology it is becoming routine to isolate genes with known functions from one organism and introduce these into another to produce desired end products. Details of the techniques employed can be found in Lal and Lal (1990), Murray (1991), and many other texts on the subject. Application of this rapidly-expanding technology to pyrethrum is being attempted by a few privately owned research companies in the United States. Broadly speaking, the idea is to identify and isolate pyrethrins-producing genes and gene products (enzymes) and introduce these into microorganisms like yeast or bacteria to mass produce pyrethrins in culture. While theoretically it is possible to do so, practically it has not been achieved so far because of associated technological problems. Pyrethrins is a mixture of six esters and so far it is not known how many genes are involved in the synthesis of these compounds, even though pyrethrins content as such shows polygenic inheritance. The other problem is that precise metabolic steps in the synthesis of these compounds have not been completely determined.

One of the possibilities of genetic engineering application to pyrethrum warrants special mention. It is well-known that plant tissues accumulate specific mRNAs, indicating differential transcription of the genome in different tissues. Furthermore some organ-specific transcripts may be restricted to certain cell types within an organ (Lindsey and Jones, 1990). It is well established that pyrethrins synthesis and accumulation are both tissue-specific and developmentally regulated as the maximum concentration of the pyrethrins is found only in the achenes and that too only when the flower maturity reaches about 3/4-disc-florets-open stage. This means that the genes encoding pyrethrins synthesis and accumulation in the achenes are being actively transcribed at this specific stage of flower maturity. Therefore, developmentally regulated, as well as tissue-specific mRNAs, can be isolated from the cells synthesizing pyrethrins and cDNA copies used to pick out specific genes from a genomic library and these genes can be identified as pyrethrins-synthesizing genes. After identifying the genes, it should be possible to isolate and sequence them so that the regulatory code for tissue-specificity and developmental regulation can be altered and reintroduced into the plant, or simply into cultured cells. If properly expressed, it may be possible to produce a green bioreactor that accumulates enhanced yields of pyrethrins in all tissues at all the developmental stages of the plant or in undifferentiated cultured cells.

Some other strategies for increasing the yield of secondary metabolites, like pyrethrins, by genetic engineering have been discussed by Lindsey and Jones (1990) who concluded that the major hurdle in the genetic engineering of secondary metabolites is the lack of basic biochemical information, like identification and isolation of key regulatory enzymes involved in the formation of secondary metabolites, and not the lack of novel genetic manipulation techniques.

REFERENCES

Anonymous (1972). Unusual electron-microscope "3D" pictures of pyrethrum. *Pyrethrum Post* **11**, 118–119.

Bajaj, Y.P.S. (1990a). "Haploids in Crop Improvement I." Springer Verlag, Berlin.

Bajaj, Y.P.S. (1990b). "Somaclonal Variation In Crop Improvement." Springer Verlag, Berlin.

Barthel, W.F. (1973). Toxicity of pyrethrum and its constituents to mammals. *In* "Pyrethrum, The Natural Insecticide" (J.E. Casida, ed.), pp. 123–142. Academic Press, New York.

Bhat, B.K., and Menary, R.C. (1984a). Pyrethrum production in Australia: Its past and present potential. *Aust. Inst. Agric. Sci.* **50**, 189–192.

Bhat, B.K., and Menary, R.C. (1984b). Registration of Hypy pyrethrum. *Crop Sci.* **24**, 619620.
Bhat, B.K., and Menary, R.C. (1984c). Genotypic and phenotypic variation in floral development of different clones of pyrethrum (*Chrysanthemum cinerariaefolium* Vis). *Pyrethrum Post* **15**, 99–103.
Bhat, B.K., and Menary, R.C. (1986a). Path-coefficient analysis of pyrethrins yield in pyrethrum (*Chrysanthemum cinerariaefolium* Vis). *Acta Horticulturae* **188**, 111–116.
Bhat, B.K., and Menary, R.C. (1986b). Genotypic and phenotypic correlation in pyrethrum (*Chrysanthemum cinerariaefolium* Vis) and their implication in selection. *Pyrethrum Post* **16**, 61–65.
Bhat, B.K., and Menary, R.C. (1987). Chrysanthemum plant named Hypy. U.S. Patent Number: Plant 5,848.
Bhat, B.K., and Pandita, P.N. (1982). Variation in components of yield of pyrethrins in pyrethrum (*Chrysanthemum cinerariaefolium* Vis). *In* "Cultivation and Utilization of Medicinal Plants" (C.K. Atal and B.M. Kapoor, eds.), pp. 704–707. Regional Research Laboratory. Council of Scientific and Industrial Research, Jammu-Tawi.
Bhat, B.K., Menary, R.C., and Pandita, P.N. (1985). Population improvement in pyrethrum (*Chrysanthemum cinerariaefolium* Vis). *Euphytica* **34**, 613–617.
Brewer, J.G. (1968). Flowering and seed setting in pyrethrum (*Chrysanthemum cinerariaefolium* Vis). A review. *Pyrethrum Post* **9**(4), 18–21.
Brewer, J.G. (1970). The fine structure of pyrethrum pollen (*Chrysanthemum cinerariaefolium* Vis.). *Pyrethrum Post* **10**(3), 3–6.
Broadbent, D. J., and Hagarty, J.D. (1969). The insecticidal properties of various pyrethrum clones compared to the synthetic pyrethroids: P I, P II, and Cinerin I. *Pyrethrum Post* **10**(1), 17–20.
Brown, A.F. (1965). A pyrethrum improvement programme. *Pyrethrum Post* **8**(1), 8–10.
Cashyap, M.M., Kueh, J.S.H., Mackenzie, I.A., and Pattenden, G. (1978). *In vitro* synthesis of pyrethrins from tissue cultures of *Tanacetum cinerariifolium*. *Phytochemistry* **17**, 544–545.
Chamberlain, E.E., and Clark, P.J. (1947). Investigations on growing pyrethrum in New Zealand. II. Selection of clones: their variation in habit of growth, susceptibility to root-rot, yield, and pyrethrins content of flowers. *New Zealand J. Sci. and Tech.* **29**, 215–222.
Chamberlain, E.E., and Procter, C.H., (1947). Investigations on growing of pyrethrum in New Zealand. I. Methods of propagation, cultivation, harvesting, and drying. *New Zealand J. Sci. and Tech.* **28**, 353–361.
Chandler, S.E. (1951). Botanical aspects of pyrethrum. General considerations: the seat of the active principles. *Pyrethrum Post* **2**(3), 1–9.
Contant, R.B. (1962). Annual report of the plant breeder (pyrethrum). Kenya Dept. of Agric., Annual Report 2, pp. 124–128.
Contant, R.B. (1963). The current position of pyrethrum breeding in Kenya. *Proc. East Africa Acad.*, pp. 93–96.
Contant, R.B. (1976). Pyrethrum, *Chrysanthemum spp.* (*Compositae*). *In* "Evolution of Crop Plants" (N.W. Simmonds, ed.), pp. 33–36. Longman, London, New York.
Culbertson, R.E. (1940). An ecological, pathological, and genetical study of pyrethrum (*Chrysanthemum cinerariaefolium* Vis) as related to possible commercial production in the United States. Unpublished PhD Thesis, 272 pp. The Pennsylvania State College, Department of Horticulture.
Drain, B.D., and Shuey, G.A. (1934). The isolation and propagation of high pyrethrin strains of pyrethrum. *Proc. Am. Soc. Hort. Sci.* **32**, 190–191.
Falconer, D.S. (1989). "Introduction to Quantitative Genetics," 3rd Ed. Longman Scientific and Technical, Essex, England.
Fehr, W.R. (1987). "Principles of Cultivar Development." Vol. 1. "Theory and Technique." Macmillan Publishing Company, New York.
Gaddum, E.W. (1950). "Pyrethrum Breeding in Kenya." Booklet, Pyrethrum Board of Kenya.
Glover, J. (1955). Chilling and flower bud stimulation in pyrethrum (*Chrysanthemum cinerariaefolium*). *Ann. Bot.* (*London*) **19**, 138–148.
Glynne Jones, G.D. (1968). The pyrethrins content of pyrethrum clones and hybrids. *Pyrethrum Post* **9**(3), 28–29.
Gnadinger, C.B. (1936). "Pyrethrum Flowers," 2nd Ed. McLaughlin Gormley King Co., Minneapolis, Minnesota.
Gnadinger, C.B. (1945). "Pyrethrum Flowers." Supplement to 2nd Ed. 1936–1945. McLaughlin Gormley King Co., Minneapolis, Minnesota.
Gnadinger, C.B., Evans, L.E., and Corl, C.S. (1933). "Pyrethrum Investigations in Colorado. I. Preliminary Report on Factors Affecting Pyrethrin Content." Bulletin 401, Colorado Agricultural College, Colorado Experiment Station, Fort Collins.
Gnadinger, C.B., Evans, L.E., and Corl, C.S. (1936). "Pyrethrum Plant Investigations in Colorado.

II. A Review of the Progress Since 1932." Bulletin 428, Colorado State College, Colorado Experiment Station, Fort Collins.
Gottschalk, W., and Wolff, G. (1983). "Induced Mutations in Plant Breeding." Springer Verlag, New York.
Gullickson, W.D. (1994). This volume.
Hallauer, A.R., and Miranda Fo., J.B., (1988). "Quantitative Genetics in Maize Breeding," 2nd Ed. Iowa State University Press, Ames, Iowa.
Head, S.W. (1966). A study of the insecticidal constituents in *Chrysanthemum cinerariaefolium.* (1) Their development in the flower head. (2) Their distribution in the plant. *Pyrethrum Post* **8**(4), 32–37.
Head, S.W. (1967). A study of the insecticidal constituents of *Chrysanthemum cinerariaefolium.* (3) Their composition in different pyrethrum clones. *Pyrethrum Post* **9**(2), 3–7.
Ikahu, J.M.K., and Ngugi, C.W. (1988). Yield assessment of newly developed pyrethrum varieties in different ecological zones of Kenya. *Pyrethrum Post* **17**, 21–23.
Ikahu, J.M., and Ngugi, C.W. (1989). Investigations into yield losses in some pyrethrum clones through picking of flowers at improper stage of development. *Pyrethrum Post* **17**, 56–59.
Ikahu, J.M., and Ngugi, C.W. (1990). Floral development in some pyrethrum clones and its implication in picking. *Pyrethrum Post* **18**, 11–14.
Karki, A., and Rajbhandary, S.B. (1984). Clonal propagation of *Chrysanthemum cinerariaefolium* Vis (Pyrethrum) through tissue culture. *Pyrethrum Post* **15**, 118–121.
Kroll, W. (1958). The breeding of improved pyrethrum varieties. *Pyrethrum Post* **4**(4), 16–19.
Lal, R., and Lal, S. (1990). "Crop Improvement Utilizing Biotechnology." CRC Press, Boca Raton.
Levy, L.W. (1981). A large-scale application of tissue culture: The mass propagation of pyrethrum clones in Ecuador. *Environ. Exper. Botany* **21**, 389–395.
Lindsey, K., and Jones, M.G.K. (1990). "Plant Biotechnology in Agriculture.' Prentice Hall, Englewood Cliffs, New Jersey.
Malaure, R.S., Davey, M.R., and Power, J.B. (1989). Isolation and culture of protoplasts of *Chrysanthemum cinerariaefolium* Vis. *Pyrethrum Post* **17**, 90–94.
Malley, Z.J.U., and Mwalukasa, E.H. (1991). Association amongst some phenotypic traits and their importance in pyrethrum improvement. *Pyrethrum Post* **18**, 47–51.
McDaniel, R.G. (1991). Pyrethrum plant named Arizona. U.S. Patent Number: Plant 7,495.
Moore, J.B., and Levy, L.W. (1975). Pyrethrum sources and uses. Part 1. Commercial sources of pyrethrum. *In* "Pyrethrum Flowers" (R. H. Nelson, ed.), 3rd Ed., pp. 1–9. McLaughlin Gormley King Co., Minneapolis, Minnesota.
Murashige, T., and Skoog, F. (1962). A revised medium for rapid growth and bioassays with tobacco tissue cultures. *Physiol. Plant* **15**, 473–497.
Murray, D.R. (1991). "Advanced Methods in Plant Breeding and Biotechnology." CAB International, Wallingford, Oxon.
Muturi, S.N., Parlevliet, J.E., and Brewer, J.G. (1969). Ecological requirements of pyrethrum. I. A general review. *Pyrethrum Post* **10**(1), 24–28.
Ngugi, C.W., and Ikahu, J.M.K. (1990). The effect of drying temperatures on pyrethrins content in some pyrethrum clones. *Pyrethrum Post* **18**, 18–21.
Ottaro, W.G.M. (1977). The relationship between the ploidy level and certain morphological characteristics of *Chrysanthemum cinerariaefolium* Vis. *Pyrethrum Post* **14**,10–14.
Pal, A., and Dhar, K. (1985). Callus and organ development of pyrethrum (*Chrysanthemum cinerariaefolium* Vis.) and analysis of their cytological status. *Pyrethrum Post* **16**, 3–11.
Pandita, P.N. (1989). Summary of a Ph.D. thesis entitled "Cytomorphological and Genetical Studies on *Chrysanthemum cinerariaefolium* Vis." *Pyrethrum Post* **17**, 111–113.
Pandita, P.N., and Bhat, B.K. (1984). Variation and correlation in pyrethrum (*Chrysanthemum cinerariaefolium* Vis). *Indian Drugs* **22**, 113–117.
Pandita, P.N., and Bhat, B.K. (1986). Correlations in phenotypic traits of pyrethrum (*Chrysanthemum cinerariaefolium* Vis). *Pyrethrum Post* **16**, 93–94.
Parlevliet, J.E. (1969). Clonal selection for yield in pyrethrum, *Chrysanthemum cinerariaefolium* Vis. *Euphytica* **18**, 21–26.
Parlevliet, J.E., (1970a). The effect of rainfall and altitude on the yield of pyrethrins from pyrethrum flowers in Kenya. *Pyrethrum Post* **10**(3), 20–25.
Parlevliet, J.E. (1970b). The effect of picking interval and flower head development on the pyrethrins content of different pyrethrum clones. *Pyrethrum Post* **10**(4), 10–14.
Parlevliet, J.E. (1974). The genetic variability of the yield components in the Kenyan pyrethrum population. *Euphytica* **23**, 377–389.
Parlevliet, J.E. (1975). Breeding pyrethrum in Kenya. *Pyrethrum Post* **13**, 47–54.

Parlevliet, J.E., and Contant, R.B. (1970). Selection for combining ability in pyrethrum, *Chrysanthemum cinerariaefolium* Vis. *Euphytica* **19**, 4–11.

Parlevliet, J.E., Muturi, S.N., and Brewer, J.G. (1969). Ecological requirements of pyrethrum. II. Regional adaptation of clones. *Pyrethrum Post* **10**(1), 28–29.

Parlevliet, J.E., Brewer, J.G., and Ottaro, W.G.M. (1979). Collecting pyrethrum, *Chrysanthemum cinerariaefolium* Vis. in Yugoslavia for Kenya. *Proc. Conf. Broadening Genet. Base Crops*, Wageningen, 1978. pp. 91–96, Pudoc, Wageningen.

Paul, A., Dhar, K., and Pal, A. (1988). Organogenesis from selected culture lines of pyrethrum, *Chrysanthemum cinerariaefolium* Vis. Clone HSL 801. *Pyrethrum Post* **17**, 17–20.

Rajasekaran, T., Rajendran, L., Ravishankar, G.A., and Venkataraman, L.V. (1990). Influence of medium constituents on growth and pyrethrins production in callus tissues of pyrethrum, *Chrysanthemum cinerariaefolium* Vis. *Pyrethrum Post* **17**, 121–124.

Rajasekaran, T., Ravishankar, G.A., Rajendran, L., and Venkataraman, L.V. (1991). Bioefficacy of pyrethrins extracted from callus tissues of *Chrysanthemum cinerariaefolium. Pyrethrum Post* **18**, 52–54.

Ravishankar, G.A., Rajasekaran, T., Sarma, K.S., and Venkataraman, L.V. (1989). Production of pyrethrins in cultured tissues of pyrethrum (*Chrysanthemum cinerariaefolium* Vis). *Pyrethrum Post* **17**, 66–69.

Roest, S. (1976). Flowering and vegetative propagation of pyrethrum (*Chrysanthemum cinerariaefolium* Vis) *in vivo* and *in vitro*. *Agric. Res. Rep.* 860. Centre for Agricultural Publishing and Documentation, Wageningen.

Roest, S., and Bokelmann, G.S. (1973). Vegetative propagation of *Chrysanthemum cinerariaefolium in vitro*. *Scientia Horticulturae* **1**, 120–122.

Singh, S.P., and Rajeswara Rao, B.R. (1985). Root and shoot development in pyrethrum (*Chrysanthemum cinerariaefolium* Vis). *Pyrethrum Post* **16**, 12–13.

Singh, S.P., Rao, B.R.R., Sharma, J.R., and Sharma, S. (1987). Genetic improvement of pyrethrum. I. Assessment of genetic variability and clonal selection. *Pyrethrum Post* **16**, 120–124.

Singh, S.P., Sharma, J.R., Rajeswara Rao, B.R., and Sharma, S.K. (1988). Genetic improvement of pyrethrum. III. Choice of improved varieties and suitable ecological niches. *Pyrethrum Post* **17**, 12–16.

Staba, E.J., Nygaard, B.G., and Zito, S.W. (1984). Light effects on pyrethrum shoot cultures. *Plant Cell Tissue Organ Culture* **3**, 211–214.

Tuikong, A.R. (1984). Pyrethrum breeding in Kenya: A historical account. *Pyrethrum Post* **15**, 113–117.

Wambugu, F.M., and Rangan, T.S. (1981). *In vitro* clonal multiplication of pyrethrum (*Chrysanthemum cinerariaefolium* Vis.) by micropropagation. *Plant Sci. Lett.* **22**, 219–226.

Wanjala, B.W. (1991). Performance of low-altitude, unimproved pyrethrum clones grown in Kisii, Kenya. *Pyrethrum Post* **18**, 61–64.

Winney, R. (1971). The biological activity of mosquito coils with a high pyrethrin content. *Pyrethrum Post* **11**, 55–57, and 71.

Winney, R., and Webley, D.J. (1969). The biological activity of mosquito coils of different 'pyrethrins' composition. *Pyrethrum Post* **10**(1), 44–48.

Wright, S. (1921). Correlation and causation. *J. Agric. Res.* **20**, 557–585.

Zieg, R.G., Zito, S.W., and Staba, E.J. (1983). Selection of high pyrethrin producing tissue cultures. *Planta Medica* **48**, 88–91.

III

Pyrethrins Chemistry

6

Pyrethrum Extraction, Refining, and Analysis

DAVID J. CARLSON

I. INTRODUCTION

Pyrethrum has been commercially produced for more than 50 years, yet even today it continues to pose challenges to successfully extract, refine, and analyze. As a natural product, some difficulties would be expected in processing and analyzing pyrethrum and indeed we continue to learn the effects of variations between growing regions and conditions, time of harvest, and other parameters. This compilation describes developments in the extraction, refining, and analysis of pyrethrins since reviews around 1972 (Balbaa *et al.*, 1972; Head, 1973; Moore, 1973, 1975b; Stevenson, 1972). In some instances, the details of new developments are proprietary and the comments presented are based on the information that has been made generally available to the industry.

II. EXTRACTION

Pyrethrins are extracted from chrysanthemum flowers which are grown as an agricultural crop in many parts of the world (Glynne Jones, 1973; Moore, 1975a). These regions include Kenya, Tanzania, and Rwanda in East Africa, New Guinea, and recently Tasmania. The plots in Tanzania, Rwanda, and New Guinea have typically been smaller plantings while Tasmanian and Kenyan operations have been more substantial. Most growers and suppliers of crude pyrethrum extract use extraction procedures that have not changed since shortly after World War II (Bozhanov *et al.*, 1972; Moore, 1975c).

Pyrethrum flowers are harvested by hand at optimum maturity and full bloom to obtain the maximum pyrethrins content. Flowers are dried in the open air shortly after picking to avoid fermentation and pyrethrins losses. The dried flowers are transported to the nearest processing facility where they are ground and extracted with hexanes or another suitable solvent. Undissolved plant matter from the ground pyrethrum flowers is filtered out and the solvent is flashed off to produce a crude oleoresin that typically contains about 30%

pyrethrins. The crude oleoresin, produced in a batch process, is a black, viscous product consisting of vegetable matter, waxes, and pyrethrins. This oleoresin is then ready for refining by one of the two primary refiners of pyrethrum extract in the world—the Pyrethrum Board of Kenya or McLaughlin Gormley King Company in Minneapolis, Minnesota.

At least one grower and supplier of crude oleoresin has introduced numerous innovations to the extraction process. Producers in Tasmania have genetically selected plants to fully bloom in a rather narrow window of time (about 2 weeks). The pyrethrum flowers are mechanically harvested rather than using labor-intensive, hand-picking techniques found in other regions. The flowers are allowed to dry in the field, cut from the plant and separated from stems. The dried flowers are then ground, pelletized, and extracted using hexane. The solvent is flashed off leaving a crude oleoresin that consistently has had an increased pyrethrins content of typically greater than 35%.

III. REFINING

Crude oleoresin requires further processing to remove pigments and extraneous plant matter, including vegetable waxes and resins (Anand *et al.*, 1977). This refining step results in a clear, amber solution of pyrethrins normally diluted in a refined kerosene to a standard concentration for marketing. An acceptable refined extract will have low staining properties and a minimum level of inert insolubles. Although a light colored extract has always been an objective of refining operations, a high recovery of the pyrethrins actives continues to be the most important goal. A recovery of 95 to 97% is achievable and typically can be met in concert with the low staining, light colored properties that are being sought.

The two primary refiners of pyrethrum extract in the world use processes that have remained relatively unchanged in the last 20 years. They both operate batch processes based upon proprietary solvents for extraction to separate the pyrethrins from the unwanted vegetable waxes and resins. Although solvent extraction operations are fairly simple in concept, the two refiners have accumulated experiences over the years with the variations found in the different sources of extract or even in different lots which must be resolved to achieve a high quality product with good pyrethrins recovery.

The refined pyrethrins are typically diluted to a standard concentration based upon the titrimetric analysis method from the Association of Official Analytical Chemists (AOAC) (Anonymous, 1990) and butylated hydroxytoluene (BHT) is added as an antioxidant (Head and Rebello, 1971). In the United States the standard concentration has historically been 20% pyrethrins which has been marketed as a technical grade material. The 20% material will eventually be blended with synergist, emulsifiers, and solvents to produce insecticide concentrates and formulations primarily for consumer product applications.

In the middle 1970s, Kenya began producing and marketing a refined pyrethrum concentrate which contained 50 to 60% pyrethrins. One impact of this highly concentrated product was to reduce transportation costs when compared to those of a more dilute material. In addition, the concentrated extract appealed to pyrethrum users who preferred to formulate technical grade material with a higher purity. This refined pyrethrum concentrate continues to be marketed by Kenya today.

Work has been conducted to develop a pyrethrum refining process using carbon dioxide in a super-critical-fluid extraction (SFE) procedure. An advantage of this process would be to avoid exposing the extract to heat when flashing off the solvents used in the conventional procedures operated by the two primary pyrethrum refiners. Another attractive aspect of SFE is that it can be operated as a continuous process. Early samples were high in pyrethrins with a low color, but they exhibited the unwanted characteristics of solidification and polymerization of the pyrethrins. More recent productions from the SFE process have been of a high quality with no indications of the unwanted characteristics seen in the early samples. At this time, details of recovery efficiencies, process capacity, and costs are not available. All of these parameters must be considered and will have to be favorable when compared to the conventional solvent extraction procedures before SFE becomes a viable, commercial process.

Pyrethrum refining is an art form to some degree. Each source and batch of crude oleoresin can be different. Pyrethrum refiners must be able to recognize the differences and respond adequately to process the material properly to achieve acceptable recoveries and a high quality product.

IV. ANALYSIS

The primary quantification methodology for world trade of crude oleoresin and refined extract today is the AOAC/titrimetric procedure. This method was established after collaborative studies by the AOAC in conjunction with the Chemical Specialty Manufacturers Association (CSMA) and has been accepted as reproducible by both industry and government regulators. The AOAC method quantitates pyrethrum as pyrethrins I and II and standards for other analytical methods are typically based on this procedure. This method involves extensive manipulations that require the use of meticulous technique to achieve reliable, reproducible results. The methodology is not suitable for the analysis of residue levels of pyrethrins, mixtures of pyrethrins and synergists, or mixtures of pyrethrins with other active ingredients. The AOAC method involves saponification to chrysanthemum mono- and dicarboxylic acids that can be quantitated via titrimetric procedures. The pyrethrins I group consists of pyrethrin I, jasmolin I, and cinerin I. In the AOAC method, saponification of pyrethrins I with alcoholic caustic produces chrysanthemum monocarboxylic acid from each of these components which are quantitated together via the mercury-reduction procedure. The pyrethrins II group similarly consists of pyrethrin II, jasmolin II, and cinerin II. The saponification step in the AOAC procedure converts these compounds to chrysanthemum dicarboxylic acid which is quantitated by titrating against a standardized base solution.

The major areas of analytical method development for pyrethrins in the last 20 years have paralleled the improved quality and increased availability of gas

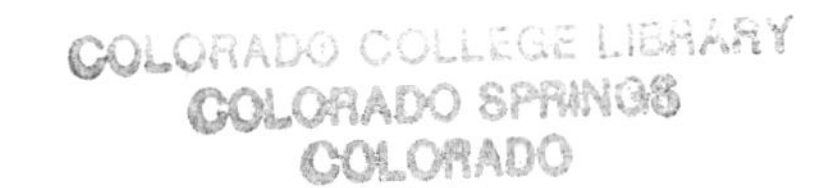

and liquid chromatographs. However, these techniques are comparative procedures requiring a known standard to quantitate an unknown sample. The pyrethrins content of the standard typically has been quantitated using the AOAC/titrimetric methodology as above. The pyrethrins content of flowers, crude oleoresin, or refined extract is usually shown as the percent pyrethrins I, pyrethrins II, and total pyrethrins. Since pyrethrins II are more polar than pyrethrins I, they are readily separated in titrimetric or chromatographic procedures. The three components of both pyrethrins I and pyrethrins II are differentiated by the alkyl chain on the alcohol portion of the molecule. Gas liquid chromatography (GLC) and high pressure liquid chromatography (HPLC) procedures have been developed that separate all six constituents. The general pyrethrins structure lends itself to detection by a number of methods. The conjugation of the ketone to the double bond on the cyclopentyl ring results in good detection by GLC using an electron capture (EC) detector. This methodology has been used extensively in pyrethrins residue analysis work. The conjugated double bonds on the alkyl side chain of the alcohol portion of the molecule give greater UV sensitivity for pyrethrin I and pyrethrin II which has proven helpful in developing HPLC and spectrophotometric procedures.

The most active area for method development in the last 20 years is GLC (Anonymous, 1979; Class, 1991; Class and Kintrup, 1991; Meinen and Kassera, 1982; Sherma, 1976) (Fig. 6-1). Almost half the methods in the literature have used this procedure in some form. This has involved analysis of formulations (Bevenue *et al.*, 1971; Birdie *et al.*, 1986; Horiba *et al.*, 1975; Kawano and

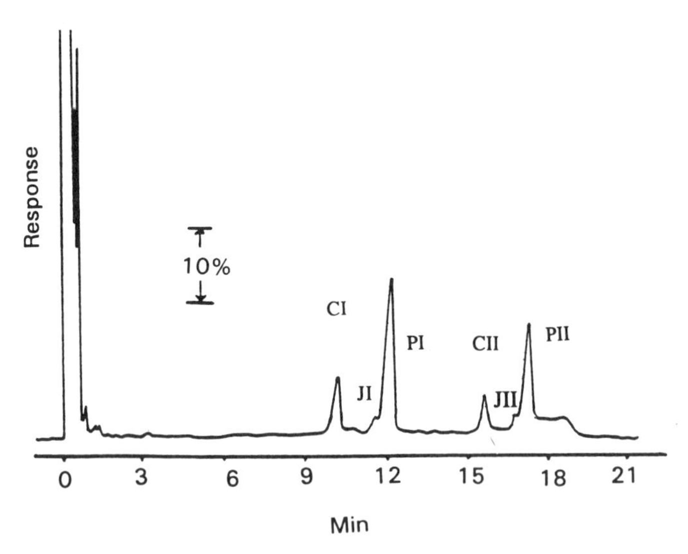

Figure 6-1. Typical GLC analysis of pyrethrums. Conditions: column 2% SE-30, 6 ft, glass; column temperature 150 (5 minutes) to 210°C at 5°C/minute; sensitivity 1×10^{-10} AFS, attenuation −32; sample 40 μg/ml; volume injected 2.0 μl. Designations: PI and PII, pyrethrin I and pyrethrin II; CI and CII, cinerins I and II; JI and JII, jasmolins I and II.

Bevenue, 1972, 1975; Kawano *et al.*, 1974; Latif *et al.*, 1984; Rickett and Chadwick, 1972), mosquito coils (Abe and Fujita, 1976; Sakaue *et al.*, 1985; Takiura *et al.*, 1973), water (Anonymous, 1982; Koehle and Haberer, 1990), soils (Siltanen *et al.*, 1978), urine (Zhang, 1991), air (MacLeod and Lewis, 1982; Thomas and Nishioka, 1985), poisoned bees (Ebing, 1987), fruit (Ryan *et al.*, 1982), grain (Desmarchelier, 1980; Dicke *et al.*, 1988; Okada *et al.*, 1983), flour (Scheidi *et al.*, 1980), and other agricultural products (Mestres *et al.*, 1979; Nakamura *et al.*, 1990). GLC has the advantages of speed and reduced manipulations. Usually quantitation by GLC is based only on pyrethrins I since the pyrethrins II degrade at the high temperatures involved. This can create problems if the standard and sample are not from similar sources with similar pyrethrins ratios. GLC procedures for analyzing pyrethrins have been developed using both packed and capillary columns (Nakamura *et al.*, 1990; Sedea *et al.*, 1983; Stringham and Schulz, 1985) with an assortment of detectors including EC (Tetenyi *et al.*, 1971), flame ionization, thermoconductivity, mass spectrometry (Etemad-Moghadam and Salajegheh, 1975; Holmstead and Soderlund, 1978; Lidgard *et al.*, 1986; Nikiforov, 1988; Nikiforov and Kohlmann, 1983), and Fourier-transform infrared. Each detector has its advantages and weaknesses and a universal detector for all analyses involving pyrethrins has not emerged.

The second most active area of methodology development for pyrethrins involves HPLC (Ando *et al.*, 1986; Baker *et al.*, 1973; Bushway, 1985; Bushway *et al.*, 1985; Debon and Segalen, 1989; McEldowney and Menary, 1988; Mourot *et al.*, 1978; Nijhuis *et al.*, 1985; Otieno *et al.*, 1982; Perez, 1983; Rickett, 1972; Wagner-Loeffler, 1985; Westwood *et al.*, 1981) (Fig. 6-2). This technique also has

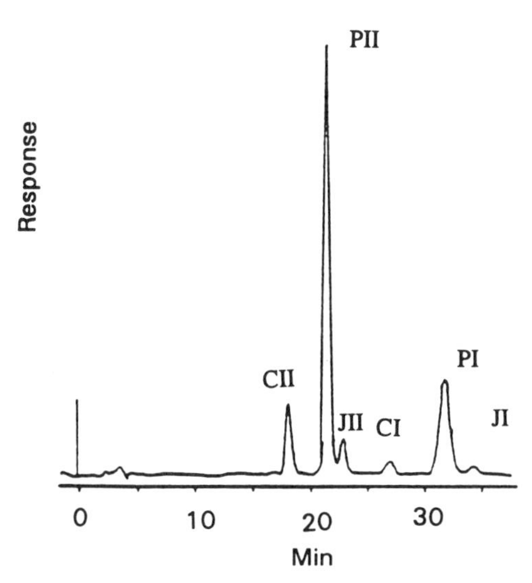

Figure 6-2. Typical HPLC analysis of pyrethrins. Conditions: column μ Bondapak CN; eluant 40% acetonitrile, 60% water, 1 ml/min; detector UV at 254 nm. Designations as in Figure 6-1.

the advantages of speed and reduced manipulation similar to GLC, but HPLC has the additional attribute that pyrethrins II do not degrade since the procedure does not involve high temperatures. Both the pyrethrins I and pyrethrins II can be quantitated by HPLC. This methodology depends on an AOAC quantitated standard that can give variable results if the source and pyrethrins ratios are not similar between the sample and standard. UV has emerged as the most popular HPLC detection although refractive index and fluorescence have also been used. Due to the increased availability of HPLC in the modern analytical chemistry laboratory, this methodology has become as popular as GLC for quantitation of pyrethrins. The choice between GLC and HPLC often is based on other actives, synergists, and which methodologies are best suited to all ingredients in the product.

There are a few methods developed in the last 20 years using colorimetric or spectrophotometric techniques (Baba *et al.*, 1972a,b; Bhavnagary and Ahmed, 1973; Bhavnagary *et al.*, 1973; DePrins, 1973; Donegan *et al.*, 1971; Marshall, 1971; Ohno *et al.*, 1973). In general, these methods focused towards field assays of pyrethrum flowers. Colorimetric and spectrophotometric methods generally have had limited applications due to interferences, lack of specificity, and rather narrow limits of detection. In addition, colorimetric techniques have frequently dealt with unstable colors.

Analysis can also be accomplished by thin layer chromatography (Jork *et al.*, 1981; Olive, 1973; Scharf, 1979; Volkov and Starkov, 1970), but this technique is much slower than other chromatographic methods previously discussed. The universal availability of GLC and HPLC has significantly limited work being conducted with thin layer chromatography.

A recent technique is an adaptation of chromatography using super critical fluids (SCF) as the carrier phase (Lubke, 1991; Wieboldt and Smith, 1988; Wieboldt *et al.*, 1989). This technology is not widely available at this time and the instruments are expensive. The few papers that are available using SCF chromatography have been published by instrument manufacturers. This method still requires a standard based upon the AOAC quantitation, but pyrethrins II degradation is avoided, similar to HPLC methods. Time will tell as to whether this methodology will attain the popularity of GLC and HPLC.

Two newer methods for analysis rely on selective hydrolysis (Macharia and Kamau, 1986) and specific antibody activity (Wing *et al.*, 1978). The radioimmunoassay with specific antibody used *S*-bioallethrin but, in principal, pyrethrins should be equally as good.

Although the standards used in most methods mentioned in this discussion are based upon an AOAC quantitation, there are variations between sources of extract and the constituent ratios which can give erratic results. As examples, Fig. 6-3 and Fig. 6-4 illustrate GLC and HPLC chromatograms, respectively, of four sources of pyrethrum at the same pyrethrins concentration with variation in responses for each source which would result in potential quantitation problems. These variations might be minimized by using a standard with all six pyrethrins independently quantitated. A possible route to this goal might be to separate the six constituents via a preparative HPLC column, quantitate

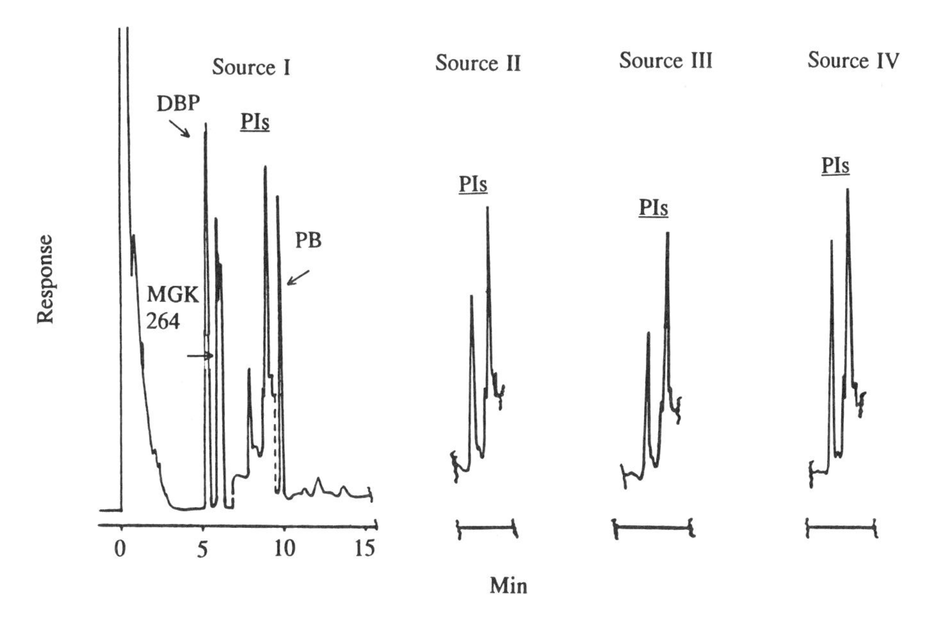

Figure 6-3. GLC analysis showing variations in pyrethrins I and II ratio for four sources of pyrethrum at the same pyrethrins concentration. Standard sample was MGK intermediate 5192. Designation: PIs, pyrethrins I; MGK 264 and piperonyl butoxide (PB) synergists; DBP, dibutyl phthalate internal standard.

Source	Ratio PIs/PIIs[a]	Pyrethrins (% of declared[b])
I	1.06	standard
II	1.17	99.2
III	1.60	128
IV	0.86	103

[a] Ratio calculated from additional data not shown. Note that only PIs are quantitated by GLC since PIIs degrade at the temperatures used.
[b] Declared as 9.00%.

the individual constituents by the AOAC method, and then compare the individual results against the traditional results from an AOAC analysis of the combined constituents. This project would require a significant, dedicated effort, but would provide valuable information if a "six pyrethrins standard" could account for the variability previously presented.

The pyrethrum industry is constantly learning about the analysis of pyrethrins and is continuously challenged to solve problems in trying to quantitate these ingredients in products ranging from crude oleoresin, to refined technical material, to finished consumer products. Current methods based upon the standards quantitated via the AOAC/titrimetric method have shortcomings. Hopefully future method development will result in a suitable analytical standard for pyrethrins where all six constituents have been quantitated.

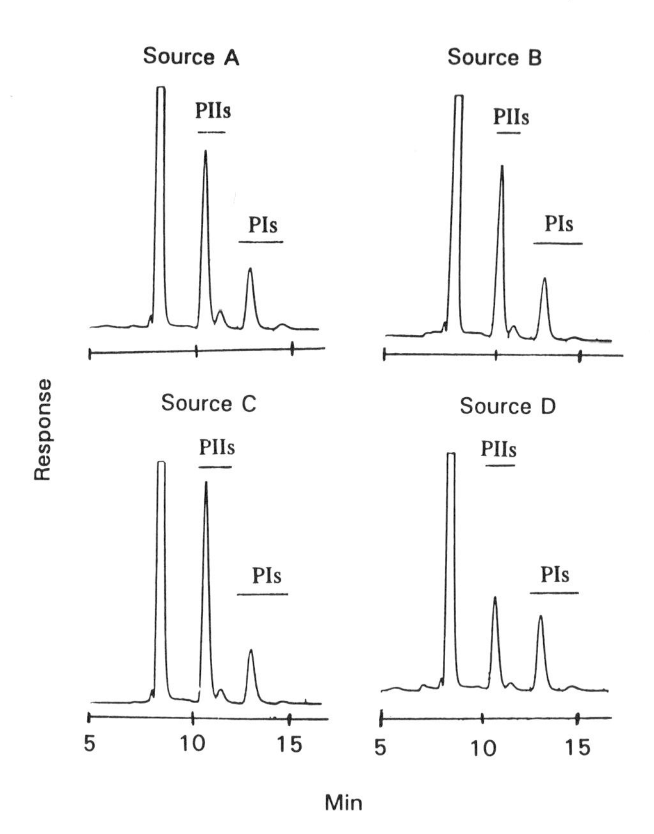

Figure 6-4. HPLC analysis showing variations in pyrethrins I and II for four sources of pyrethrum at the same pyrethrins concentration. Designations: PIs, pyrethrins I; PIIs, pyrethrins II.

Source[a]	Pyrethrins (%) Found	Pyrethrins (%) of Declared
A	standard (20%)	—
B	17.83	89.2
C	19.71	98.6
D	16.81	84.1
A	standard (19.95%)	99.8

[a] Samples injected in the sequence tabulated.

REFERENCES

Abe, Y., and Fujita, Y. (1976). Studies on pyrethroidal compounds. VI. Vaporization ratio of pyrethroids from burning mosquito coils. *Bochu-Kagaku* **41**, 22-28.

Anand, G. K., Sen, T., and Nigam, M. C. (1977). Production of pale dewaxed oleoresin of pyrethrum. *Res. Ind.* **22**, 10-11.

Ando, T., Kurotsu, Y., and Uchiyama, M. (1986). High performance liquid chromatographic separation of the stereoisomers of natural pyrethrins and related compounds. *Agric. Biol. Chem.* **50**, 491-493.

Anonymous (1979). IUPAC reports on pesticide. 9. Recommended methods for the determination of residues of pyrethrins and piperonyl butoxide. *Pure Appl. Chem.* **51**, 1615-1623.
Anonymous (1982). Pyrethrins and permethrin in potable waters by electron-capture gas chromatography 1981. *Methods Exam. Waters Assoc. Mater.* 1981, 15 pp.; *Chem. Abstr.* **98**, 132003m, 1982.
Anonymous (1990). 936.05: Pyrethrins in pesticide formulations—mercury reduction method. *In* "Official Methods of Analysis of the Association of Official Analytical Chemists" (K. Helrich, ed.), pp. 170–171. Official Association of Analytical Chemists, Arlington, VA.
Baba, N., Kirihata, M., Ohno, M., and Takano, T. (1972a). Orthophosphoric acid method for the pyrethrum assay. *Bochu-Kagaku* **37**, 155-161.
Baba, N., Kirihata, M., Takano, T., and Ohno, M. (1972b). Color reaction of pyrethrins and eugenol with orthophosphoric, sulfuric, and hydrochloric acids. *Bull. Inst. Chem. Res., Kyoto Univ.* **50**, 150-159.
Baker, D.R., Henry, R.A., Williams, R.C., Hudson, D.R., and Parris, N.A. (1973). Preparative columns in high-speed liquid chromatography. *J. Chromatogr.* **83**, 233-243.
Balbaa, S.I., Abdel-Kader, E.M., Abdel-Wahab, S.M., Zaki, A.Y., and El-Shamy, A.M. (1972). Comparative study between different methods used for estimation of pyrethrins and a new suggested method. *Planta Medica* **21**, 347-352.
Bevenue, A., Kawano, Y., and DeLano, F. (1971). Analytical studies of pyrethrin formulations by gas chromatography. *Pyrethrum Post* **11**, 41-47.
Bhavnagary, H.M., and Ahmed, S.M. (1973). Spectrophotometric method for the micro-determination of piperonyl butoxide in the presence of pyrethrins. *Analyst (London)* **98**, 792-796.
Bhavnagary, H.M., Ahmed, S.M., and Gupta, M.R. (1973). New colorimetric method for the microdetermination of pyrethrins. *Res. Ind.* **18**, 21-23.
Birdie, N.S., Banerji, R.K., and Chauhan, A.K. (1986). Gas liquid chromatographic separation of pyrethrins from some synthetic pyrethroids in formulations. *Pyrethrum Post* **16**, 77-80.
Bozhanov, B., Kamedulski, V., and Tonev, I. (1972). Obtaining pyrethrins for insecticide preparations. *Farmatsiya (Sofia)* **22**, 27-30.
Bushway, R.J. (1985). Normal phase liquid chromatographic determination of pyrethrins in formulations. *J. Assoc. Off. Anal. Chem.* **68**, 1134-1136.
Bushway, R.J., Johnson, H., and Scott, D.W. (1985). Simultaneous liquid chromatographic determination of rotenone and pyrethrins in formulations. *J. Assoc. Off. Anal. Chem.* **68**, 580-582.
Class, T.J. (1991). Optimized gas chromatographic analysis of natural pyrethrins and pyrethroids. *J. High Resol. Chromatogr.* **14**, 48-51.
Class, T.J., and Kintrup, J. (1991). Pyrethroids as household insecticides: analysis, indoor exposure, and persistence. *Fresenius J. Anal. Chem.* **340**, 446-453.
Debon, A., and Segalen, J.L. (1989). Trace analysis of pyrethrins and piperonyl butoxide in water by high performance liquid chromatography. *Pyrethrum Post* **17**, 43-46.
DePrins, H.C. (1973). Colorimetric estimation of pyrethrins in pyrethrum flowers. *Pyrethrum Post* **12**, 22, 33.
Desmarchelier, J.M. (1980). Comparative study of analytical methods for bioresmethrin, fenothion, d-fenothrin, pyrethrum I, carbaryl, fenitrothion, methacrifos, pirimiphos-methyl, and dichlorvos on various grains. *Nippon Noyaku Gakkaishi* **5**, 521-532.
Dicke, W., Ocker, H.D., and Thier, H.P. (1988). Residue analysis of pyrethroid insecticides in cereal grains, milled fractions, and bread. *Z-Lebensm.-Unters. Forsch.* **186**, 125-129.
Donegan, L., Morrison, J.N., and Webley, D.J. (1971). Rapid field assay for pyrethrum flowers. *Pyrethrum Post* **11**, 36-40.
Ebing, W. (1987). Determination of pesticide residues in poisoned honey bees. II. Pyrethrin and pyrethroid insecticides. *Fresenius Z. Anal. Chem.* **327**, 539-543.
Etemad-Moghadam, P., and Salajegheh, D. (1975). Mass spectrometric evaluation of pyrethrin type insecticides. *Bull. Iran. Pet. Inst.* **59**, 8-14.
Glynne Jones, G.D. (1973). Pyrethrum production. *In* "Pyrethrum, The Natural Insecticide" (J.E. Casida, ed.), pp. 17–22. Academic Press, New York.
Head, S.W. (1973). Composition of pyrethrum extract and analysis of pyrethrins. *In* "Pyrethrum, The Natural Insecticide" (J.E. Casida, ed.), pp. 25–53. Academic Press, New York.
Head, S.W., and Rebello, C. (1971). Butylated hydroxy toluene as an antioxidant for refined pyrethrum extract. *Pyrethrum Post* **11**, 24-28.
Holmstead, R.L., and Soderlund, D.M. (1978). Separation and analysis of the pyrethrins by combined gas-liquid chromatography-chemical ionization mass spectrometry. *Pyrethrum Post* **14**, 79-82.
Horiba, M., Kitahara, H., Kobayashi, A., and Murano, A. (1975). Gas chromatgraphic determination of pyrethroidal insecticides in aerosol formulations. *Bochu-Kagaku* **40**, 123-132.

Jork, H., Reh, E., and Wimmer, H. (1981). To what extent is thin-layer chromatography effective as a pilot techique for high-performance liquid chromatgraphy? *GIT Fachz. Lab.* **25**, 566, 568-570, 572-573.
Kawano, Y., and Bevenue, A. (1972). Analytical studies of pyrethrin formulations by gas chromatography. II. Isolation of the pyrethrins from water-based formulations. *J. Chromatogr.* **72**, 51-59.
Kawano, Y., and Bevenue, A. (1975). Analytical studies of pyrethrin formulations by gas chromatography. Isolation of the pyrethrins from water-based formulations. *Pyrethrum Post* **13**, 71-77.
Kawano, Y., Yanagihara, K.H., and Bevenue, A. (1974). Analytical studies of pyrethrin formulations by gas chromatography. III. Analytical results on insecticidally active components of pyrethrins from various world sources. *J. Chromatogr.* **90**, 119-128.
Koehle, H., and Haberer, K. (1990). Multimcthod for gas chromatographic trace determination of pyrethrins and pyrethroids in water. *Vom Wasser* **75**, 75-82.
Latif, S., Haken, J.K., and Wainwright, M.S. (1984). Gas chromatographic analysis of insecticidal preparations using carbon dioxide propellents. *J. Chromatogr.* **287**, 77-84.
Lidgard, R.O., Duffield, A.M., and Wells, R.J. (1986). Positive and negative ion chemical ionization mass spectrometry of pyrethrin pesticides. *Biomed. Environ. Mass Spectrom.* **13**, 677-680.
Lubke, M. (1991). The advantages of supercritical fluid chromatography for analyzing natural products. *Analusis* **19**, 323-343.
Macharia, B.W., and Kamau, J.N. (1986). A new chemical method of analysis of pyrethrins based on selective hydrolysis. *Pyrethrum Post* **16**, 48-51.
MacLeod, K.E., and Lewis, R.G. (1982). Portable sampler for pesticides and semivolatile industrial organic chemicals in air. *Anal. Chem.* **54**, 310-315.
Marshall, R.A.G. (1971). Colorimetric method for the determination of iron in pyrethrum extract. *Analyst* (*London*) **96**, 675-678.
McEldowney, A.M., and Menary, R.C. (1988). Analysis of pyrethrins in pyrethrum extracts by high-performance liquid chromatography. *J. Chromatogr.* **447**, 239-243.
Meinen, V.J., and Kassera, D.C. (1982). Gas-liquid chromatographic determinations of pyrethrins and piperonyl butoxide: collaborative study. *J. Assoc. Off. Anal. Chem.* **65**, 249-255.
Mestres, R., Atmawijaya, S., and Chevallier, C. (1979). Methods for the study and determination of pesticide residues in cereal products. XXXIV. 1. Organochlorine, organophosphorous, pyrethrin, and pyrethroid pesticides. *Ann. Falsif. Expert. Chim.* **72**, 577-589.
Moore, J.B. (1973). Residue and tolerance considerations with pyrethrum, piperonyl butoxide, and MGK 264. *In* "Pyrethrum, The Natural Insecticide" (J.E. Casida, ed.), pp. 293–306. Academic Press, New York.
Moore, J.B. (1975a). Pyrethrum sources and uses. *In* "Pyrethrum Flowers" (R.H. Nelson, ed.), 3rd Ed., pp. 1–9. McLaughlin Gormley King Co., Minneapolis, Minnesota.
Moore, J.B. (1975b). Pyrethrum evaluation. Part I. Chemical analysis of pyrethrum. *In* "Pyrethrum Flowers" (R.H. Nelson, ed.), 3rd Ed., pp. 41-56, 131-145. McLaughlin Gormley King Co., Minneapolis, Minnesota.
Moore, J.B. (1975c). Manufacture of pyrethrum extract. *In* "Pyrethrum Flowers" (R.H. Nelson, ed.), 3rd Ed., pp. 61–67. McLaughlin Gormley King Co., Minneapolis, Minnesota.
Mourot, D., Boisseau, J., and Gayot, G. (1978). Separation of pyrethrins by high-performance liquid chromatography. *Anal. Chim. Acta* **97**, 191-193.
Nakamura, Y., Hasegawa, Y., Tonogai, Y., and Ito, Y. (1990). Simultaneous determination of seven pyrethroids in agricultural products by capillary ECD-GC. *Eisei Kagaku* **36**, 525-537.
Nijhuis, H., Heeschen, W., and Hahne, K.H. (1985). Determination of pyrethrum and piperonyl butoxide in milk by high performance liquid chromatography. *Pyrethrum Post* **16**, 14-17.
Nikiforov, A. (1988). Characterization of complex organic mixtures by computer averaged integrated mass spectra (CAI) of soft ionization methods. *Spectroscopy* (*Ottawa*) **6**, 47-61.
Nikiforov, A., and Kohlmann, H. (1983). Characterization of natural and synthetic pyrethrins by EI, CI, FI/DI mass spectra and computer averaged integrated FD mass spectrometry (CAI-FD). *Int. J. Mass Spectrom. Ion Phys.* **48**, 141–144.
Ohno, M., Takano, T., Baba, N., Kirihata, M., and Kawano, K. (1973). New assessment of the phosphoric acid method for pyrethrum assay. *Pyrethrum Post* **12**, 91-92.
Okada, T., Uno, M., Nozawa, M., and Tanigawa, K. (1983). Studies on pyrethroid insecticides. II. ECD-gas chromatographic determination of pyrethroid insecticides in grain. *Skokuhin Eiseigaku Zasshi* **24**, 147-154.
Olive, B.M. (1973). Color specific reagent for identification and semiquantification of pyrethrins and piperonyl butoxide by thin-layer chromatography. *J. Assoc. Off. Anal. Chem.* **56**, 915-918.

Otieno, D.A., Jondiko, I.J., McDowell, P.G., and Kezdy, F.J. (1982). Quantitative analysis of the pyrethrins by HPLC. *J. Chromatogr. Sci.* **20**, 566-570.

Perez, R.L. (1983). Simultaneous determination of folpet, piperonyl butoxide, and pyrethrins in aerosol formulations by high-pressure liquid chromatography. *J. Assoc. Off. Anal. Chem.* **66**, 789-792.

Rickett, F.E. (1972). Preparative-scale separation of pyrethrins by liquid-liquid partition chromatography. *J. Chromatogr.* **66**, 356-360.

Rickett, F.E., and Chadwick, P.R. (1972). Measurements of temperature and degradation of pyrethroids in two thermal fogging machines, the Swingfog and Tifa. *Pestic. Sci.* **3**, 263-269.

Ryan, J.J., Pilon, J.C., and Leduc, R. (1982). Composite sampling in the determination of pyrethrins in fruit samples. *J. Assoc. Off. Anal. Chem.* **65**, 904-908.

Sakaue, S., Doi, T., and Doi, T. (1985). Determination of allethrin and other pesticides in mosquito coils by the shaking extraction method. *Agric. Biol. Chem.* **49**, 921-924.

Scharf, K.H. (1979). Isolation and characterization of pyrethrins from the petals of *Chrysanthemum cinerariaefolium* and insect sprays. *Prax. Naturwiss. Biol.* **28**, 309-315.

Scheidi, I., Pfannhauser, W., and Woidich, H. (1980). Gas chromatographic determination of pyrethrin residues in flour samples. *Dtsch. Lebensm.-Rundsch.* **76**, 309-311.

Sedea, L., Toninelli, G., and Sartorel, B. (1983). Gas chromatographic analysis of pyrethrum extracts using glass capillary columns. *Riv. Ital, Sostanze Grasse* **60**, 133-137.

Sherma, J. (1976). Pyrethrum. *Anal. Methods Pestic. Plant Growth Regul.* **8**, 225-238.

Siltanen, H., Rosenberg, C., and Tiittanen, K. (1978). Pyrethrin residues in the soils. *Pyrethrum Post* **14**, 65-67.

Stevenson, D.S. (1972). Application of liquid-gel chromatography to the analytical characterization of pyrethrum extract. *Pyrethrum Post* **11**, 90-93.

Stringham, R.W., and Schulz, R.P. (1985). Capillary gas chromatographic determination of pyrethrins in low level formulations. *J. Assoc. Off. Anal. Chem.* **68**, 1137-1139.

Takiura, K., Yamaji, A., Oe, M., and Yuki, H. (1973). Analysis of smoke from commercial mosquito smoke coils by gas chromatography. *Bochu-Kagaku* **38**, 26-29.

Tetenyi, P., Hethelyi, E., Okuda, T., and Szilagyi, I. (1971). Use of programmed temperature in analytical determination of pyrethrins by electron capture detector. *Pyrethrum Post* **11**, 29-31, 47.

Thomas, T.C., and Nishioka, Y.A. (1985). Sampling of airborne pesticides using Chromosorb 102. *Bull. Environ. Contam. Toxicol.* **35**, 460-465.

Volkov, Y.P., and Starkov, A.V. (1970). Detection of the components of pyrethrins in pyrethrum extracts using thin-layer chromatography and their 2,4-dinitrophenylhydrazones. *Tr. Tsent. Nauch.-Issled. Dezinfek. Inst.* No. 19, 257-259.

Wagner-Loeffler, M. (1985). Determination of pyrethrins in pharmaceutical formulations. *GIT Fachz. Lab.* **29**, 982-984.

Westwood, S.A., Games, D.E., and Sheen, L. (1981). Use of circular dichroism as a high-pressure liquid chromatography detector. *J. Chromatogr.* **204**, 103-107.

Wieboldt, R.C., and Smith, J.A. (1988). Supercritical fluid chromatography with Fourier-transform infrared. *ACS Symp. Ser.* **366**, 229-242.

Wieboldt, R.C., Kempfert, K.D., Later, D.W., and Campbell, E.R. (1989). Analysis for pyrethrins using capillary supercritical fluid chromatography and capillary gas chromatography with Fourier-transform infrared detection. *J. High Resol. Chromatogr.* **12**, 106-111.

Wing, K.D., Hammock, B.D., and Wustner, D.A. (1978). Development of an *S*-bioallethrin specific antibody. *J. Agric. Food Chem.* **26**, 1328-1333.

Zhang, H. (1991). Use of X-5 resin miniature cartridges in the determination of trace pyrethrin insecticide in urine by gas chromatography. *Lizi Jiaohuan Yu Xifu* **7**, 38-41.

7
Constituents of Pyrethrum Extract

DONALD R. MACIVER

I. INTRODUCTION

In response to a data call-in by the US Environmental Protection Agency (EPA), an industry task force, The Pyrethrum Joint Venture (PJV), was formed to develop data to satisfy current requirements of the Federal Insecticide, Fungicide, and Rodenticide Act (FIFRA). An extensive study was undertaken to define the chemistry of pyrethrum extract by identifying all significant components. This was accomplished by analysis of a current blend of typical products of refined extract from various producing locations. This review describes the analytical profile of the composite blend which led to the definition of pyrethrum extract. Since the only significant item of commerce in the USA is refined or "pale" pyrethrum extract, this was the material which the committee decided to analyze.

The PJV started by preparing general specifications for the percentage of actives and ratio range for pyrethrins I: pyrethrins II together with the amount of added stabilizer and diluents for EPA approval. These being accepted by the Agency, a blend of production batches was prepared which would serve as the test material. To satisfy the EPA requirements for identification of components, both already present and intentionally added as defined in the Pesticide Assessment Guidelines (EPA, 1982), a full analytical study was initiated of the constituents of the extract composite. The composition of various crude, oleoresins and other Kenyan extracts was described about 20 years ago by Head (1973). The present contribution details the typical composition of modern refined pyrethrum extracts from a variety of sources as described to the EPA, together with all the declared additives used during processing.

Pyrethrum flowers are grown in many diverse locations from the foothills of the Himalayas (Malla *et al*,. 1977) to the deserts of the Southwestern United States (McDaniel, 1991). Pyrethrum is commercially produced for local consumption in countries such as Ecuador, China, India, Nepal, Indonesia, and Brazil (Wilson, 1973). Imports of extracts to the USA are from the major commercial producers in East Africa, Papua New Guinea and in recent years from Australia. Experimentally, pyrethrum has been produced in the United

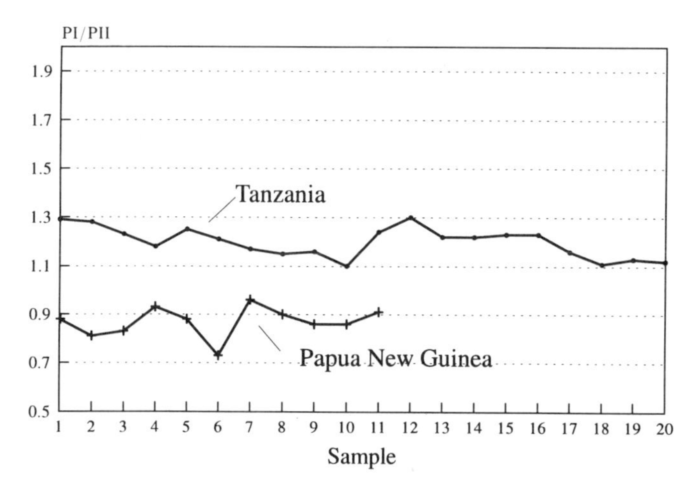

Figure 7-1. Pyrethrins ratio (PI/PII) for dried pyrethrum flowers from Tasmania (20 samples 1963–83; average 1.22, median 1.21; standard deviation, 0.058) and Papua New Guinea (11 samples, January to April, 1986; average 0.87; median, 0.88; standard deviation, 0.045).

States in locations such as New Jersey, Oregon, and Arizona. The great variety of pyrethrum plant clones and the different growing conditions contribute to the variations in composition. However, within a growing locale, there is always a significant conformity in the composition of the six insecticidal pyrethrins. For example, the ratio of pyrethrins I to pyrethrins II over a period of years is fairly constant and is uniform for multiple samples (Fig. 7-1). The importance of the ratio of pyrethrins I to pyrethrins II has been shown to be the different neuroactive properties of the pyrethrins, with pyrethrins II having greater speed of knockdown while pyrethrins I kill faster (Elliott, 1971; Sawicki and Thain, 1962a,b). Many plant constituents are present in pyrethrum extract. The formidable task of identifying and quantifying all components is fortunately simplified. EPA requires normally only the identification and quantification of compounds at 0.1% and above, together with classification and a total "tally" of classes (EPA, 1982).

Extracts are prepared by the following general procedure. Crude or "brut" extract is produced as the starting material which is refined further. Crude production involves coarse grinding of the dried flower heads and extracting the grist by percolation with low boiling isoalkanes. Since most of the pyrethrins are located in the achenes or small dry indehiscent fruits of the flower heads, it is important that these are exposed to the extracting alkanes. The filtered solution of extractives, known as the micella, is evaporated to leave a viscous dark greenish-brown liquid called "brut" or, simply, oleoresin. Refining the crude extract involves a number of means (Hopkins, 1964). The crude extract is usually partitioned into a solvent such as nitromethane, acetonitrile (Matsumoto, 1976; Moore and Kassera, 1975), ethylene glycol monomethyl ether (Prasad, 1969), methanol (Alexander and Forster, 1973; Anand *et al.*,

1977) or methanol/acetone (Lloyd, 1970) and then decolorized. The partitioning solvent is distilled and the residue is dissolved in the final diluent or higher boiling isoalkanes. At the final stage, the concentration of pyrethrins may be adjusted conveniently to a standard level by adding higher boiling isoalkanes (Hopkins,1962). Addition of butylated hydroxytoluene (BHT), a US approved food preservative, is made during the processing to prevent loss of pyrethrins by oxidation.

This review describes the first formal composition of the final refined "pale" extract and identifies some other components of pyrethrum flowers which do not appear in refined extract.

II. COMPONENTS

A. General Information

An account of the composition of an assortment of extracts, including crude "pale" produced in Kenya around 1970 by the Pyrethrum Board of Kenya (PBK), was presented by Head (1973). In addition to the active esters of chrysanthemic and pyrethric acids, other plant extractives found in crude "oleoresin" and refined "Kenya Pale Extract" were sterols, triterpenols, alkanes, fatty acids from triglycerides and carotenoids.

The PJV consortium's blend of extracts (FEK-99) was analyzed by gas chromatography coupled with mass spectrometry (GC/MS) to identify and broadly quantify all components of the blend (Fig. 7-2). Nonelutable components such as triglycerides were studied as methyl esters of their fatty acids after saponification. It is accepted that, if they are present, certain high-molecular-weight compounds such as triglycerides will not elute and that some compounds such as the pyrethrins will undergo thermal isomerization on the gas chromatographic column.

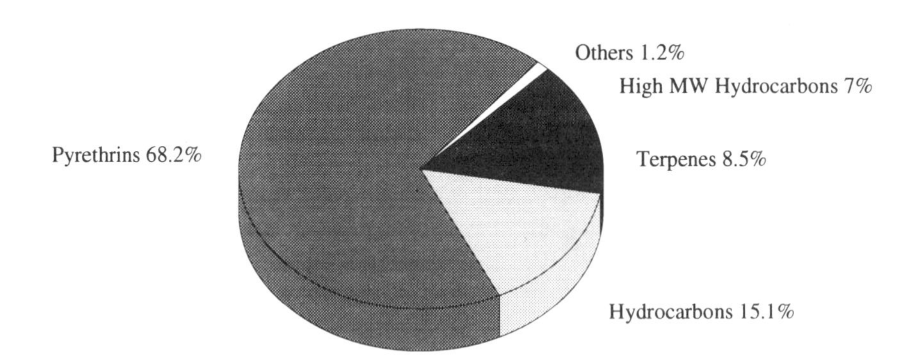

Figure 7-2. Constituent groups for pyrethrum blend FEK-99 (GC/MS determination). Data from Shrader Laboratories (Detroit, MI).

B. Pyrethrins

These actives have been studied progressively since the early years of the century and their structures are well established (Crombie and Elliott, 1961; Elliott and Janes, 1973). Full elucidation of their relatively complex stereochemical features is a tribute to the many years of work from some of the World's most distinguished organic chemists, among them Professors Hermann Staudinger, Leopold Ruzicka, Leslie Crombie and Michael Elliott. This work has also paved the way for the modern synthetic pyrethroid industry. Chemical Abstracts Service (CAS) numbers for individual pyrethrins are given on p. xvii.

The ratio of pyrethrins I to pyrethrins II, commonly known as the PI/PII ratio, varies from different geographical areas. In East Africa where the pyrethrum plant is in asynchronous flowering stages from September through March, flowers are hand selected and harvested usually at the stage when the ray florets are horizontal (FMI—Flower Maturity Index = 4 to 5). The PI/PII ratio in extracts from Kenya and Tanzania is 1.2 to 1.5 over a period of years (Fig. 7-1). Some extracts from Rwanda have been known to reach a PI/PII ratio of over 2.0. In contrast, extracts from Papua New Guinea have a greater relative amount of pyrethrins II and consequently lower PI/PII ratios in the region 0.7 to 1.0 (Fig. 7-1). Thus, in defining pyrethrum extract we were compelled to consider the regional ratio variations and submit appropriate proposals to the EPA. In tests of varying ratios of pyrethrins, the PJV found little or no difference in the activity of extracts of differing ratios. The Agency currently allows PI/PII in a range from 0.8 to 2.8 (EPA, 1986). The difference in ratio may be due to genetic variation in sub-species of the plant or possibly to the stage of maturity of the plant. Late maturity flowers appear to form higher levels of pyrethrins II (Fig. 7-3). An explanation may be that the rise of total pyrethrins up to FMI stage 5–6 and simultaneous drop in PI/PII ratio indicates that pyrethrins I are being converted biochemically to pyrethrins II as the flower approaches full maturity (Boevink, 1990). This idea has arisen from time to time from a number of observers (Pattenden, 1970).

The concentration of total pyrethrins in extracts was set at $50 \pm 5\%$ (w/w) by the AOAC current or 13th Edition chemical assay. It is convenient to reduce this concentration with a diluent to 20%. Since final formulation concentrates are usually only several percent total pyrethrins, 20% pyrethrins forms a practical base for virtually all formulations.

Pyrethrins are altered by heat and light. Heat may induce rearrangement and formation of less active isopyrethrins (Brown *et al.*, 1956, 1957; Levy and Geller, 1959) (Fig. 7-4), while light acting on dilute solutions will induce severe degradative changes starting with *Z* to *E* isomerization in the (*Z*)-pentadienyl side-chain (Kawano *et al.*, 1980). There are many other light-induced degradative processes outside the scope of this review. Our GC/MS study (Fig. 2) showed isopyrethrins which are included with the general category of "pyrethrins." These thermally-altered pyrethrins probably are produced largely by heating on the column. Both isopyrethrin I (CAS number 5768-81-1) and isopyrethrin II (CAS number 64655-46-5) are thought to be thermal artifacts and not natural products, having been described by a number of workers (Brown *et al.*, 1956, 1957; Holmstead and Soderlund, 1978; Sedea *et al.*,1983).

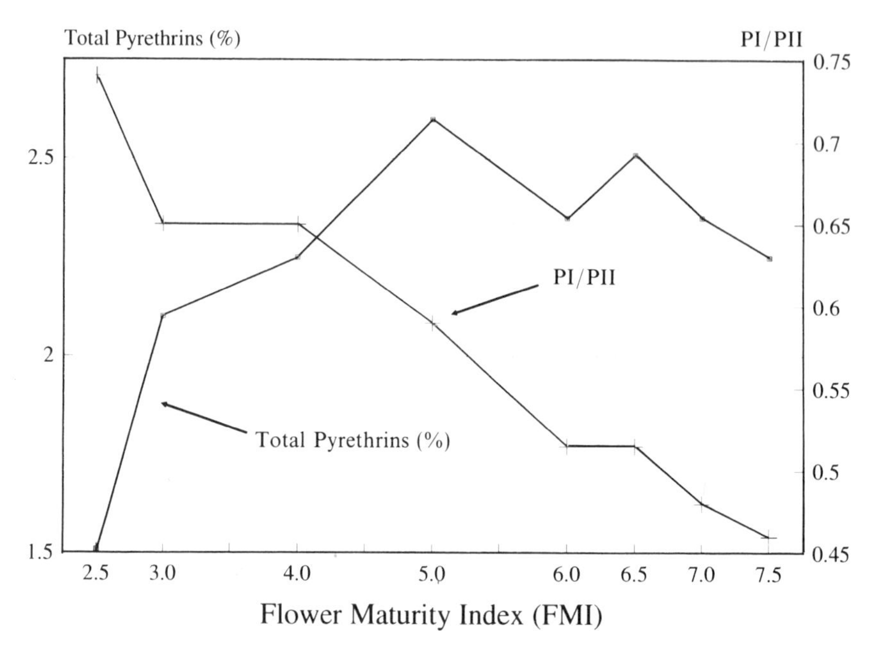

Figure 7-3. Pyrethrins content and ratio (PI/PII) as a function of flower maturity index. Data from CIG Pyrethrum.

C. Hydrocarbons

Hydrocarbons in the extract (Fig. 7-5) result from the deliberate addition of alkanes and isoalkanes during and after processing and are distinguished from the higher-molecular-weight phytoalkane "waxes" previously reported by Head (1973). The alkanes are mainly 5,6-dimethyldecane, 4-methylundecane and *n*-dodecane. They are often referred to as solvents but in fact possess poor solvency properties and are more strictly diluents which enable extracts to be standardized for commercial purposes. Typical of this group are the odorless isoparaffins (Fiero, 1964) commercially known as "Isopars®" (Exxon Corp.), "Shellsols®" (Shell Chemicals) or "Vistas®" (Vista Corp.), which may be produced by altering and highly refining oils or by structuring from natural gas and are available in various boiling ranges.

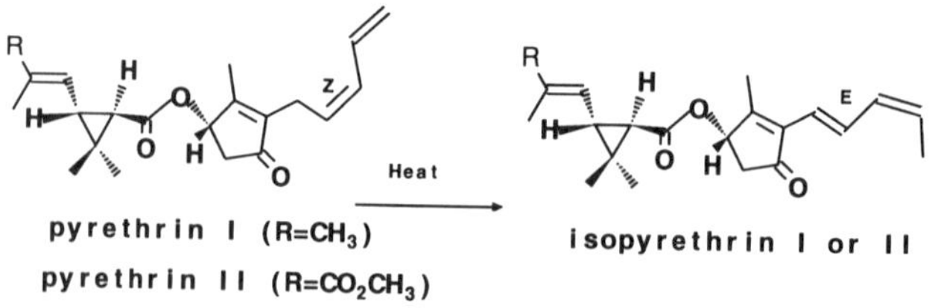

Heat

pyrethrin I ($R=CH_3$)
pyrethrin II ($R=CO_2CH_3$)

isopyrethrin I or II

Figure 7-4. Thermal isomers (isopyrethrins).

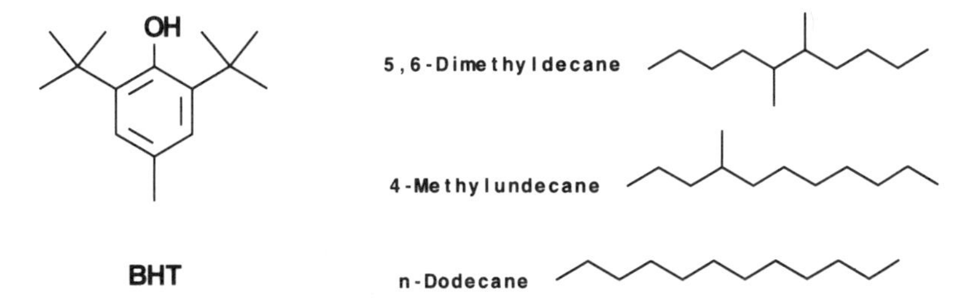

Figure 7-5. Stabilizer and isoparaffins added intentionally to pyrethrum extract.

Natural high-molecular-weight alkanes were discussed by Head (1973) with reference to crude extract "oleoresin". They ranged from C_{23} to C_{31} with a molecular-weight-range of 325 to 437. Our GC/MS study suggested the presence of hydrocarbons similar to oleanene, ursene or amyrin, but these were not specifically identified.

D. Terpenoids

A variety of phytoterpenes are retained in the most highly refined extracts and certainly contribute to the "flowery" fragrance of pyrethrum extract. The lower-molecular-weight terpenes aid the dissolution of pyrethrins in isoalkanes. However, certain of the pentacyclic triterpenols are waxy and can give precipitates or gels if present in high enough concentrations. While these triterpenol materials are inert, they can cause formulations to be cloudy or dull in appearance. Some may even cause plugging of fine spray nozzles.

(*E*)-*β*-Farnesene (CAS number 18794-84-8) (commonly known as *β*-farnesene) (Fig. 7-6) has not been previously reported in commercial pyrethrum extracts. It is one of the most common lower-molecular-weight aliphatic triterpenes identified, is quite fragrant and boils at 80–82°C, and may be easily

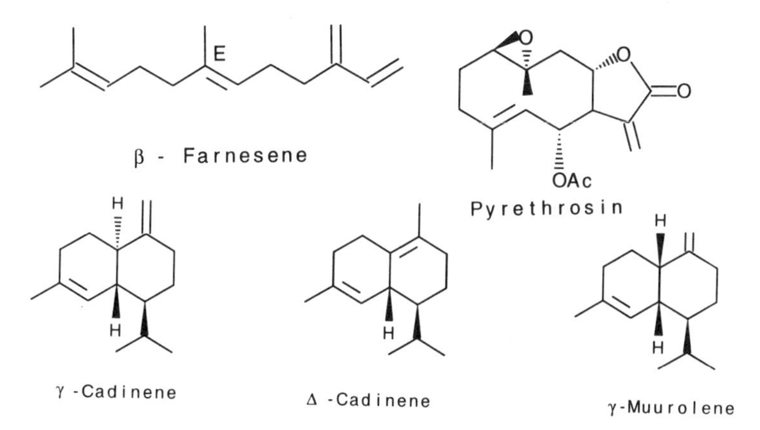

Figure 7-6. Sesquiterpenoids in pyrethrum extract.

prepared by dehydration of farnesol from citronella, neroli, cyclamen, rose, and other essential oils. It occurs naturally in *Pinus sylvestris*, spruce, other conifers, and in oils of the starthistle, walnut and fig leaf. *E*-β-Farnesene also appears in a lesser, non-commercial strain of pyrethrum species (*Chrysanthemum balsamita L.*).

Several sesquiterpenoids occur in pyrethrum extract. In the original GC/MS examination pyrethrosin (CAS number 28272-18-6) was characterized by the reporting laboratory as "pyrethrin isomer," but the structural assignment was corrected after reexamination. This crystalline compound (Fig. 7-6) has been observed frequently in pyrethrum extracts (Rose and Haller, 1937) and the structure was established by Barton *et al.* (1960). It may also be called chrysanthin or chrysanthene. Sesquiterpenes identified in the present analysis but not previously recorded in the commerical extract are: γ-muurolene (CAS number 24268-39-1); γ-cadinene (CAS number 483-74-9), an epimer of γ-muurolene, which is also called cadina-4,10(15)-diene, *d*,γ-cadinene, (+)-γ-cadinene, or even (+)-*epi*-muurolene; Δ-cadinene (CAS number 483-76-1) called alternatively (+)-γ-cadinene or cadina-1(10),4-diene (Fig. 7-6).

The pentacyclic triterpenols, taraxasterol (CAS number 1059-14-9) and β-amyrin (CAS number 559-70-6) (Fig. 7-7), are waxy or crystalline solids which have poor solubility in isoalkanes and partitioning solvents used in the refining of pyrethrum. It is important to ensure that partitioning solvents are chilled during refining so as to reduce the solubility of these compounds. If the temperature is permitted to rise, larger quantities of these materials are carried through to the final extract where they can cause later formulation problems. A typical profile of the relative amounts of these substances and pyrethrosin which can be isolated from pyrethrum extract is presented in Fig. 7-8. Taraxasterol remains the predominant triterpenol in pyrethrum extract. The gelling of pyrethrum extract appears to be a phenomenon closely linked to the presence of taraxasterol. The gel is usually reversible to the sol state by agitation and reappears only after the gel is undisturbed for some hours. Taraxasterol will not cause gelling if dissolved in isoalkanes which are the usual diluents; pyrethrins must also be present. We speculate that the phenomenon may be caused by weak hydrogen bonding between the carbonyl group of the pyrethrins and the hydroxyl group of the pentacyclic triterpenols, respectively (Fig. 7-9).

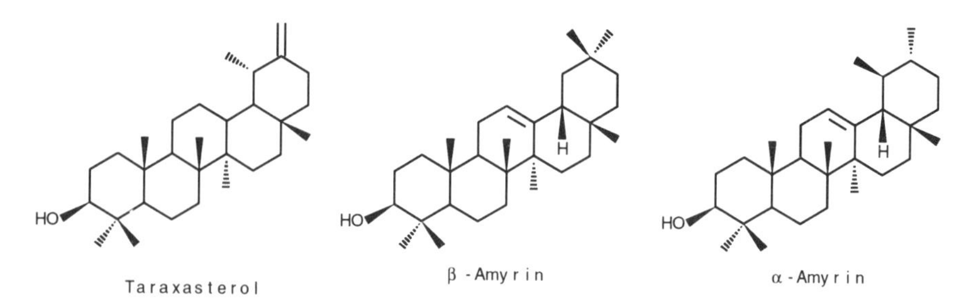

Figure 7-7. Pentacyclic triterpenols in pyrethrum extract.

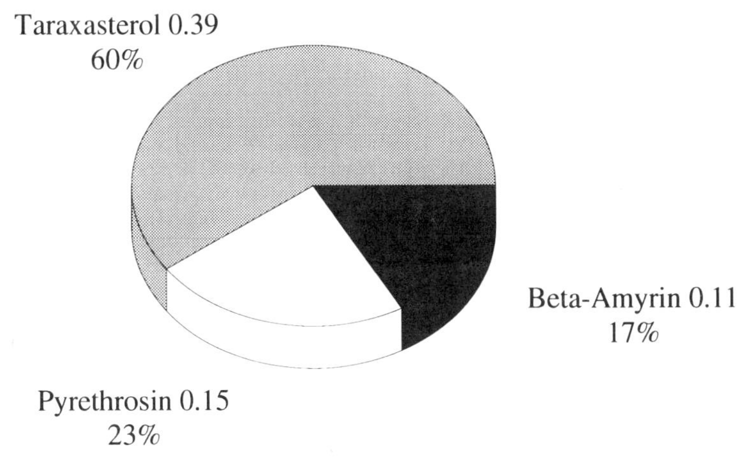

Figure 7-8. Relative distribution of pyrethrosin, taraxasterol, and β-amyrin in pyrethrum blend FEK-99 (GC/MS determination). Data from Shrader Laboratories (Detroit, MI).

Popularly known as "dandelion wax," taraxasterol is also derived from the roots of *Taraxicum officinale*, the common hedgerow dandelion, charmingly known in French as "pissenlit." Being a member of the Compositae plant family to which pyrethrum belongs, a taxonomic similarity of terpenoids may exist. Taraxasterol has been recorded previously in pyrethrum extract (Fujitani, 1909; Staudinger and Ruzicka, 1924) as "pyrethrol" and its full identification was unequivocally established by Hertz and Mirrington (1966) and has been reviewed (Elliott and Stephenson, 1966). It is also called ursenol or alphactucerol. β-Amyrin was reported previously in pyrethrum extract by Fukushi (1952) and Stephenson (1972) and it is documented thoroughly. Other names are olean-12-en-3-β-amyrin, amyrin, oleanenol, 12-oleanen-3-β-ol or simply β-amyrenol. β-Amyrin is widely distributed in nature and occurs in the latex and resins of many plants. Miscellaneous terpenoids reported in the literature include α-amyrin (CAS number 474-40-8) (Fig. 7-8), also known as α-amyrenol, which is isomeric with α-amyrin. It is known to occur widely in nature, especially as the acetate in latex of rubber trees (*Hevea spp*), in latexes from *Ficus variegata*

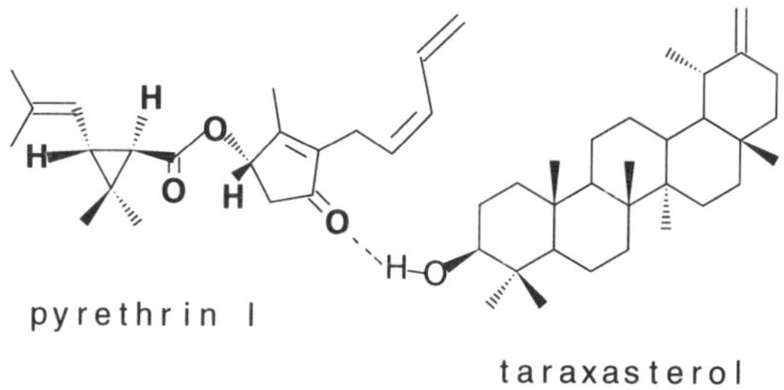

pyrethrin I

taraxasterol

Figure 7-9. Hypothetical explanation of gelling in pyrethrum extract involving hydrogen bonding between pyrethrin I and taxasterol.

Table 7-1. Fatty Acids From Hydrolysis of Pyrethrum Glycerides[a]

Type		Percentage of total fatty acids	Composition (g/100g)
Saturated			
Myristic	$C_{14:0}$	2.05	0.003
Palmitic	$C_{16:0}$	19.4	0.033
Stearic	$C_{18:0}$	12.8	0.022
Total		34.3	0.058
Monounsaturated			
Myristoleic	$C_{14:1}$	2.73	0.005
Palmitoleic	$C_{16:1}$	6.37	0.011
Oleic	$C_{18:1}$	41.7	0.071
Total		50.8	0.087
Polyunsaturated			
Linoleic	$C_{18:2}$	11.9	0.020
TOTAL		97	0.165

[a]Source: Nutrition International Inc. (East Brunswick, NJ).

Blume (Moraceae), in *Balanophora elongata* Blume (Balanophoraceae), and may be isolated from *Manila elemi* Vesterberg. β-Sitosterol (CAS number 83-46-5) has been reported in extracts analyzed by column chromatography (Stephenson, 1972).

E. Glycerides

The plant glycerides known to be present in pyrethrum extract (Head, 1973) cannot be analyzed on GC/MS columns. The combined fatty acids (as glycerides or fixed oils) had been reported in Kenyan commercial extracts. Major fatty acids separated from the glycerides were palmitic, linoleic, linolenic, and oleic (Table 7-1, Fig. 7-10), while much smaller quantities of a large variety of fatty

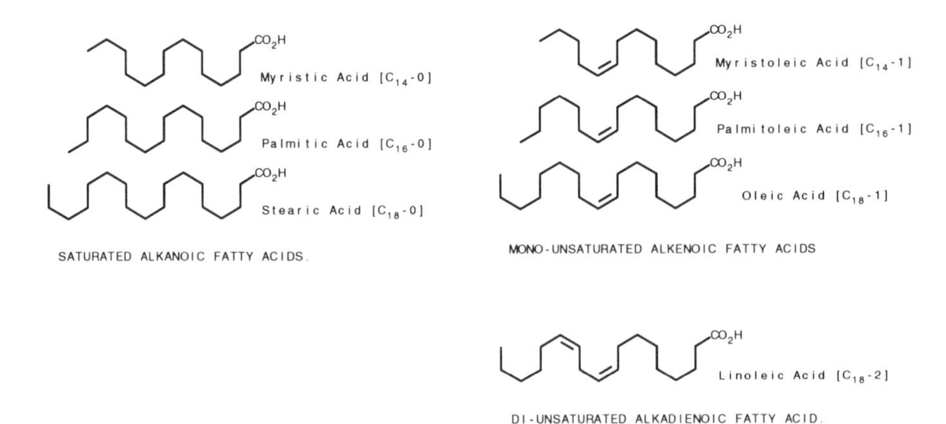

Figure 7-10. Fatty acids present in pyrethrum extract (as triglycerides).

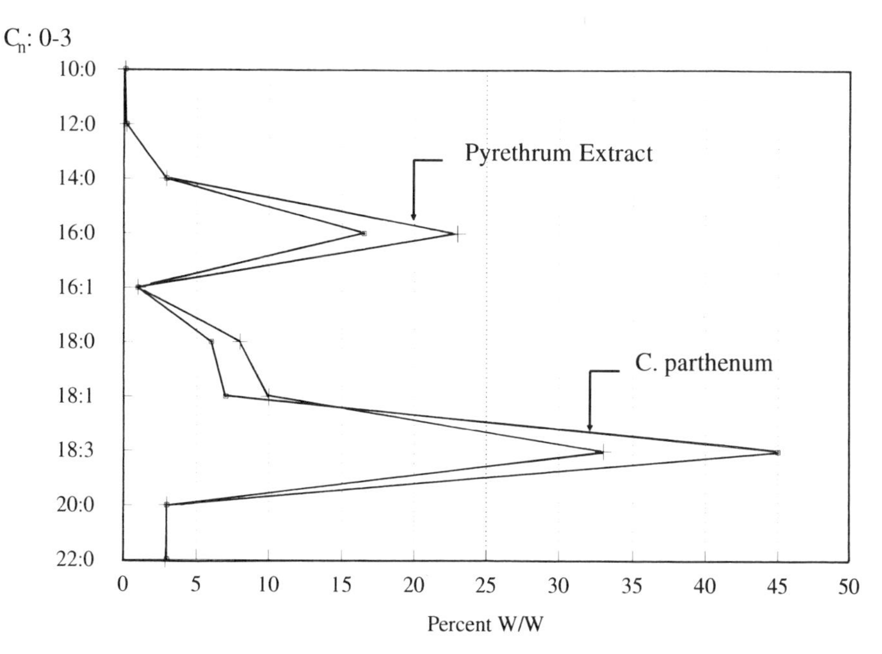

Figure 7-11. Identity of C_{10} to C_{22} fatty acids in pyrethrum extract triglycerides comparing composite blend FEK-99 with *Chrysanthemum parthenum.*

acids as their glycerides were found also. There was no evidence for free fatty acids in pyrethrum extract, as reported in the early work of Ripert (1934).

We separated the pyrethrins in the PJV composite as their semicarbazones and submitted the residue for normal study of the fatty acids and glycerides. The procedure involved a standard conversion of the hydrolyzed glyceride fatty acids to methyl esters analyzed by gas chromatography (Table 7-1). The report concluded that 73.8% of the residues (after pyrethrins removal) were triglycerides. These results compare favorably with the determinations of Kenyan extracts (Head, 1973).

We also compared the fatty acid/glyceride profile of the composite blend FEK-99 pyrethrum extract with an extract prepared from another common species of noninsecticidal pyrethrum, *Chrysanthemum parthenum*, which is a roadside wildflower of New Jersey (Fig. 7-11). The profile of fatty acids from glycerides in this related plant closely parallels that from pyrethrum-producing plants.

F. High-Mass Substances

High-molecular-weight polymers of pyrethrins, which would not elute by GC, have been known for some years under the names "green oils" (Otieno 1983) or "yellow oils" or "false" pyrethrins (Brown *et al.*, 1954). They have been identified as polymers with a molecular mass around 100 kilodaltons and are composed mostly of units of pyrethrins II with some pyrethrins I and additional oxygen (Boevink, 1990). It is not known at what stage in the plant or in the subsequent processing that these polymers appear. They are sparingly soluble in alkane solvents used in process refining and are relatively easily removed. One explanation for their formation may be an epoxide production followed by ring opening probably in the plant tissue. Much of the early work in chemical analysis by the Pyrethrum Board of Kenya or AOAC methods involved discussions of "false" pyrethrins. It may be that these polymers could contribute to errors in chemical analysis by mercury-reduction methods but would not appear in HPLC assays. For practical purposes of reducing losses in the pyrethrum industry, efforts to minimize formation of polymers have varied from the strategic addition of the antioxidant (BHT) to conducting the entire extraction and refining under an inert gas blanket.

G. Trace Metals

From time to time metals, and particularly iron (Marshall, 1971), have been indicated as causing problems in the extract. Chiu *et al.* (1974) studied the effects of various metals on the stability of pyrethrum extract and concluded that no effects were observable after deliberate additions of metals. We examined two extracts for trace metals by emission spectrographic analysis (Table 7-2). Additional testing demonstrated the absence of other metals at the following levels: <0.1 ppm (Hg), <0.01 ppm (As, Te, W, P, Tl, Nb, Li), <0.001 ppm (Mn, Sb, Pb, Cr, Sn, Ni, Bi, V, In, Cd, Zn, Ti, Zr, Co) and <0.0001 ppm (Be, B, Ge, Ga, Ag).

H. Miscellaneous

For the sake of completeness and to answer questions on the components which are discarded or refined out of crude "brut" extract, I have assembled a list of the most significant substances which have been reported. The purpose of refining the extract largely is to remove triglycerides, high-mass plant waxes, plant pigments, and polymers, which can cause blocking of spray nozzles and staining of light-colored fabrics or surfaces. Pigments which were extensively recorded for Kenyan extracts by Head (1973) are at such low levels that they

Table 7-2. Metals in Pyrethrum Extracts Determined by Semi-Quantitative Emission Spectrographic Analysis[a]

		Pyrethrum extract	
Abundance	ppm	No. 539	No. 557
Principal	10–100	nil	nil
Major	1–10	nil	nil
Medium	0.01–0.1	Si	Si
Weak	0.001–0.01	Mg, Fe, Al	Mg, Fe, Al
Trace	0.0001–0.001	Ca	Ca, Cu
Faint Trace	<0.0001	Cu	nil

[a]Source: PTL Testing Laboratory, Inc. (Princeton, NJ).

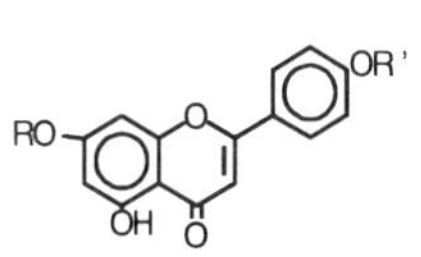

Apigenin 7-glucoside (R=Glc.;R'=H)
Apigenin 7-glucuronide (R=Gluc.;R'=H)
Apigenin 4'-glucuronide (R=H;R'=Gluc.)

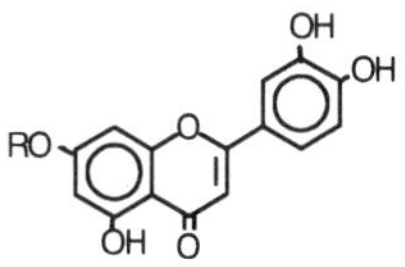

Luteolin 7-glucoside (R=Glc.)
Luteolin 7-glucuronide (R=Gluc.)

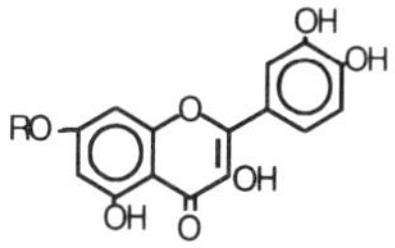

Quercetin 7-glucoside (R=Glc.)
Quercetin 7-glucuronide (R=Gluc.)

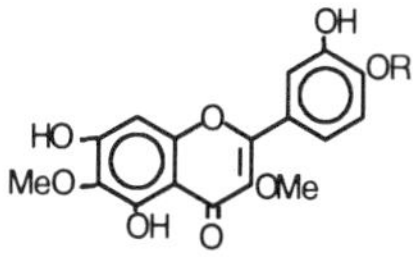

Quercetagetin 3,6-dimethyl ether (R=H)
Quercetagetin 3,6,4'-trimethyl ether (R=Me)

Figure 7-12. Pyrethrum flavonoids.

are difficult to quantify and need not be reported to the EPA. These materials form the bulk of the dark-colored paste residues which are discarded for use as boiler fuel during the refining of the crude extract.

Certain components of pyrethrum flowers are not found in extracts. These materials remain either in the spent ground pyrethrum flowers (marc) or in the residues (sludges) left after the refining process. In general, they are either high-molecular-weight, polar, polymeric materials or are highly polar, lower-molecular-weight substances which are insoluble in extracting alkanes and/or partitioning solvents or both. Certain constituents, particularly the flavonoids (Glennie and Harborne, 1972; Rao *et al.*, 1973) (Fig. 7-12), were *formerly* associated with the skin allergies from unrefined or powdered pyrethrum flowers containing pollen. These allergies were reported in subjects who were highly sensitive to ragweed. Modern refined extracts, which have been marketed widely since 1957, have not produced any skin allergies when tested on sensitive subjects (Rickett *et al.*, 1972). These results should be noted by abstracting services and reviewers since much of the existing literature still refers to crude pyrethrum extracts or pyrethrum flowers.

Other components in pyrethrum extract include sesamin and β-cyclopyrethrosin (Fig. 7-13) (Doskotch and El-Faraly, 1969; Doskotch *et al.*, 1971). Sesamin, also known as fagarol and pseudocubebin, occurs widely in nature, e.g. in *Sesame indicum* oil and fruits of *Piper lowong* (Piperaceae).

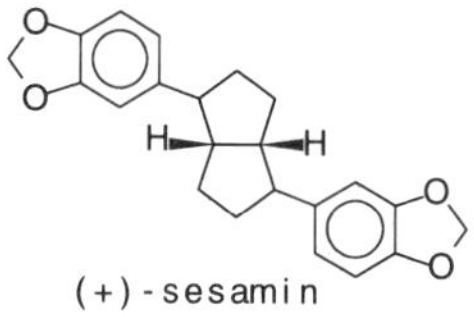

(+)-sesamin

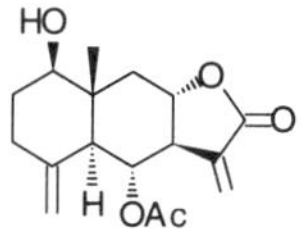

β-cyclopyrethrosin

Figure 7-13. Structures of sesamin and β-cyclopyrethrosin in pyrethrum extract.

For processing stability and preservation of the pyrethrins, the antitoxidant BHT (Fig. 7-5) is a deliberate addition during the processing. BHT and isoalkane diluent are the only materials added to pyrethrum extract. BHT may also be called 2,6-di-*tert*-butyl-4-methylphenol or 2,6-ditertiary-butyl-*para*-cresol or butylated hydroxytoluene. Additions may be up to 5% of the pyrethrins. BHT use is important for stabilization of pyrethrins during storage and processing (Head and Rebello, 1971; Glynne Jones and Head, 1965).

III. CONCLUSION

The description above gives the chemical identifications recorded for the composite blend of pyrethrum extract FEK-99. The composition of this standard extract can be summarized as follows (Anonymous, 1992):

Pyrethrum Extract

'—Pyrethrum Extract is a mixture of three naturally occurring, closely related insecticidal esters of chrysanthemic acid (pyrethrins I) and three closely related esters of pyrethric acid (pyrethrins II). It contains not less than 45.0 percent and not more than 55.0 percent of the sum of pyrethrins I and pyrethrins II in a mixture consisting of approximately 20 to 25 percent (w/w) light isoparaffins. The ratio of pyrethrins I to pyrethrins II in the extract is not less than 0.8 and not greater than 2.8. It may also contain 3 to 5 percent butylated hydroxytoluene as an antioxidant and 23 to 25 percent phytochemical extracts containing triglyceride oils, terpenoids, and carotenoid plant colors. It contains no other added substances. ...'

I believe that this definition will cover all commercial extracts for some time to come. While new means of preparing extracts are currently being investigated, such as refining with liquid carbon dioxide as a partitioning solvent (Stahl and Schuetz, 1980; Sims, 1982; Bunzenberger *et al.*, 1983), such methods are not yet commercially available. The advantages new methods offer would be in the economies of refining and possibly more concentrated extracts without alkane diluents. Some problems remain to be solved, but the future of pyrethrum extracts seems to lie in this direction.

The literature referring to pyrethrum extract often needs to be updated to reflect the improvements in refining techniques. This is especially needed in references to allergic skin reactions which are no longer witnessed. The current studies encouraged by the EPA to develop modern toxicological data will certainly bring natural pyrethrum, with its long track record of safety in use, into the full ranks of modern insect control.

ACKNOWLEDGEMENTS

I wish to acknowledge permission to use proprietary data developed by the Pyrethrum Joint Venture for EPA data call-in. I also thank Mr. John Boevink of Commonwealth Industrial Gases for certain data, Dr. Krishen Bhat for supplying flowers at various stages of maturity, Mr. David Carlson of MGK company for various data and Mr. Mike Grunauer of FMC for the GC chart of pyrethrum extract.

REFERENCES

Alexander, D.G., and Forster, A. (1973). Great Britain Patent 1,332,962.

Anand, G.K., Sen, T., and Nigam, M.C. (1977). Production of pale dewaxed oleoresin of pyrethrum. *Res. Ind.* **22**, 10–11.

Anonymous (1992). Pyrethrum extract. *Pharm. Rev.* **18**, 3046.

Barton, D.H.R., Bockman, O.C., and de Mayo, P. (1960). Sesquiterpenoids, Part XII. Further investigations on the chemistry of pyrethrosin. *J. Chem. Soc.*, 2263.

Boevink, J. (1990). Private Communication.

Brown, N.C., Phipers, R.F., and Wood, M.C. (1954). The estimation of "false" pyrethrins in pyrethrum extract. *Pyrethrum Post* **3**(3), 3–5.

Brown, N.C., Phipers, R.F., and Wood, M.C. (1956). The analysis of pyrethrins. Part III. A correlation of various analytical methods with bioassays. *Pyrethrum Post* **4**(1), 24–29.

Brown, N.C., Hollinshead, D.T., Phipers, R.F., and Wood, M.C. (1957). New isomers of the pyrethrins formed by the action of heat. *Pyrethrum Post* **4**(2), 13–19.

Bunzenberger, G., Lack, E., and Marr, R. (1983). Wissenschaftliche Forschungsarbeit CO_2 Extraktion-Vergleich der Überkritschen Fluid-Extraktion mit der unterkritischen Betriebsweise anhand von Problemen der Naturstoff-Extraktion. *Chem-Ing-Tech* **55**, 320–321.

Chiu, F.-T., and Wu, N.-C. (1974). The interaction of certain metals with diluted refined pyrethrum extract. *Pyrethrum Post* **12**(3), 7.

Crombie, L., and Elliott, M. (1961). Chemistry of the natural pyrethrins. *Fortsch. Chem. Org. Naturst.* **19**, 120.

Doskotch, R.W., and El-Faraly, F.S. (1969). Isolation and characterization of (+)-sesamin and β-cyclopyrethrosin from pyrethrum flowers. *Can. J. Chem.* **49**, 2103.

Doskotch, R.W., El-Farly, F.S., and Hafford, C.D. (1971). Sesquiterpene lactones from pyrethrum flowers. *Can. J. Chem.* **49**, 1142–1146.

Elliott, M. (1971). The relationship between the structure and the activity of pyrethroids. *Bull. W.H.O.* **44**, 315.

Elliott, M., and Janes, N.F. (1973). Chemistry of the natural pyrethrins. *In* "Pyrethrum, The Natural Insecticide" (J.E. Casida, ed.), pp. 55–100. Academic Press, New York.

Elliott M., and Stephenson, H. (1966). "Pyrethrol"—taraxasterol—A note. *Pyrethrum Post* **8**(3), 27–28.

EPA (1982). "Pesticide Assessment Guidleines—Subdivision D, Product Chemistry." US Environmental Protection Agency, Office of Pesticides and Toxic Substances, Washington, D.C. Publication PB83-153890/EPA 540/9-82-018, October 1.

EPA (1986). PJV Archives, Correspondence (JWA-RE) with Pyrethrum Task Force from Registrations Division dated October 1, 1966.

Fiero, G.W. (1964). Isoparaffinic solvents as bases for pyrethrum insecticides. *Pyrethrum Post* **7**(4), 3.

Fujitani, J. (1909). Chemistry and pharmacology of insect powder. *Arch. Exp. Pathol. Pharmakol.* **61**, 47–75.

Fukushi, S. (1952). Studies on the components of the unsaponifiable matter of the wax of *Chrysanthemum cinerariaefolium*. *J. Agr. Chem. Soc. Jap.* **26**, 1–2.

Glennie, J.B., and Harborne, J.B. (1972). Flavonoids of pyrethrum. *Pyrethrum Post* **11**, 82–84.

Glynne Jones, G.D., and Head, S.W. (1965). The manufacture of pyrethrum powder from pyrethrum flowers, Part II: Stabilization of the active constituents. *Pyrethrum Post* **8**(1), 14.

Head, S.W. (1973). Composition of pyrethrum extract and analysis of pyrethrins. *In* "Pyrethrum, The Natural Insecticide" (J.E. Casida, ed.) pp. 25–53. Academic Press, New York.

Head, S.W., and Rebello, C. (1971). Butylated hydroxytoluene as an antioxidant for refined pyrethrum extract. *Pyrethrum Post* **11**, 24.

Hertz, W., and Mirrington, R.N. (1966). Identification of pyrethrol with taraxasterol. *J. Pharm. Sciences* **55**, 104.

Holmstead, R.L., and Soderlund, D.M. (1978). Separation and analysis of pyrethrins by combined gas-liquid chromatography-chemical ionization mass spectrometry. *Pyrethrum Post* **14**, 79–82.

Hopkins, J. (1962). A note on sampling pyrethrum extract. *Pyrethrum Post* **6**, 22.

Hopkins, L.O. (1964). Processes of preparing refined pyrethrum extract: A review. *Pyrethrum Post* 7(3), 41–48.
Kawano, Y., Yanagahara, K., Miyamoto, T., and Yamamoto, I. (1980). Examination of the conversion products of pyrethrins and allethrin formulations exposed to sunlight by gas chromatography and mass spectrometry. *J. Chromatogr.* **198**, 317–328.
Levy, L.W., and Geller, D. (1959). A note on the detection of the "isopyrethrins". *Pyrethrum Post* **5**(1), 12.
Lloyd, A.M. (1970). Extraction of pyrethrin from pyrethrum flowers. *S. African*, Oct. 14, 8 pp.
Malla, S.B., Upadkya, A.V., Bhattari, D.D., and Singh, M.T. (1977). Revival of pyrethrum cultivation in Nepal. *Pyrethrum Post* **8**(4), 39.
Marshall, R.A.G. (1971). Colorimetric method for the determination of iron in pyrethrum extracts. *Analyst* (*London*) **96**, 675–678.
Matsumoto, M. (1976). Highly concentrated extracts of Dalmatian pyrethrum. *Japan. Kokai*, 2 pp.; *Chem. Abst.* **85**(7):43836z.
McDaniel, R.G. (1991). Pyrethrum plant named Arizona. U.S. Patent: Plant 7,495.
Moore, J., and Kassera, D.C. (1975). U.S. Patent 74-432,656.
Otieno, D.A. (1983). Quantitative analysis of the pyrethrins by HPLC. *Pyrethrum Post* **15**, 71.
Pattenden, G. (1970). Biosynthesis of pyrethrins. *Pyrethrum Post* **10**(4), 2–5.
Prasad, S. (1969). Production of concentrated pyrethrum extract. *Chem. Ind.* (*London*), 756–757.
Rao, P.R., Seshadri, T.R., and Sharma, P. (1973). Polyphenolic constituents of pyrethrum flowers (*Chrysanthemum cinerariaefolium*). *Current Science* **42**, 811–812.
Rickett, F.E., Tyszkiewicz, K., and Brown, N.C. (1972). Pyrethrum dermatitis, Part 1: The allergenic properties of various extracts of pyrethrum flowers. *Pyrethrum Post* **11**, 85.
Ripert, F.E. (1934). New method of analysis for products containing extracts of pyrethrum. *Ann. Fals. Fraudes.* **27**, 580–595.
Rose, W.G., and Haller, H.L. (1937). Constituents of pyrethrum flowers XI. *J.Org.Chem.* **2**, 484–488.
Sawicki, R.M., and Thain, E.M. (1962a). Insecticidal activity of pyrethrum extract and its four insecticidal constituents against houseflies. I. Preparation and relative toxicity of the pure constituents and statistical analysis of the action of mixtures of these components. *J. Sci. Food Agric.* **13**, 172.
Sawicki, R.M., and Thain, E.M. (1962b). Insecticidal activity of pyrethrum extract and its four insecticidal constituents against houseflies. IV. Knock-down activities of the four constituents. *J. Sci. Food Agric.* **13**, 292.
Sedea, L., Toninelli, G., and Sartorel, B. (1983). Gas chromatographic analysis of pyrethrum extracts using glass capillary columns. *Riv. Ital. Sostanze Grasse* **60**, 133–137.
Sims, M. (1982). Process uses liquid carbon dioxide for botanical extracts. *Chem. Eng.* (*New York*) **89**, 50–51.
Stahl, E., and Schuetz, E. (1980). Extraction of natural compounds with supercritical gases. 3. Pyrethrum extracts with liquified and supercritical carbon dioxide. *Pharm. Anal. Phytochem. Planta Medica* **40**, 12–21.
Staudinger H., and Ruzicka, L. (1924). Insektentötende Stoffe. X. Über die Synthese von Pyrethrinen. *Helv. Chim. Acta.* **7**, 448–458.
Stephenson, H. (1972). Application of liquid-gel chromatography to the analytical characterization of pyrethrum extract. *Pyrethrum Post* **11**, 90.
Wilson R.J. (1973). A note on the market for pyrethrum. Foreign & Commonwealth Office (ODA), Economics Planning Department, July 1973. Tropical Product Institute Publication, pp. 1–75.

8

Chemistry of Pyrethrins

LESLIE CROMBIE

I. INTRODUCTION

The insecticidal components of pyrethrum, the important botanical insecticide obtained by solvent extraction of pyrethrum flowers (*Chrysanthemum cinerariaefolium*), (syn. *Tanacetum cinerariifolium*), are six esters (**1**)–(**6**) known collectively as "the pyrethrins" (Fig. 8-1). Commercially, the dried flowers are extracted with light petroleum, and the extract is then evaporated to give an oleoresin concentrate containing about 30% of pyrethrins. For commercial use this has to be decolourised, and for chemical purposes the oleoresin may be partitioned between light petroleum and nitromethane (Barthel *et al.*, 1944, Barthel and Haller, 1945) which leaves most of the inactive material in the less polar phase, giving a pyrethrins mixture of about 90% purity. Initial extraction of the flowers with liquefied and supercritical carbon dioxide, an extremely mild method, has also been recommended, producing increased extractives (Stahl and Schuetz, 1980; Sims, 1981).

A number of earlier reviews of pyrethrum insecticide and its chemistry are available (Gnadinger, 1936, 1936–1945; Nelson, 1945–1972; Crombie and Elliott, 1961; Jacobson and Crosby, 1971; Elliott and Janes, 1973, and Crombie, 1980, 1988).

Early in the history of pyrethrum chemistry the components were recognized as esters (Fujitani, 1909; Yamamoto 1919, 1923), and in 1924 an extensive investigation was published in *Helvetica Chimica Acta* (Staudinger and Ruzicka, 1924; Staudinger *et al.*, 1924). Although only two pyrethrin esters were recognized, and the resulting structures were incorrect in detail, enormous progress was made, particularly when the inadequate techniques of the time are taken into consideration. Staudinger and Ruzicka's investigation stands as one of the great chemical classics of its age. Later, the cinerins were recognized as components, and later still, the jasmolins. Continued progress was made on the structural and stereochemical side leading to recognition of the complete structures and stereochemistry of the six natural pyrethrin esters: pyrethrin I (**1**), cinerin I (**2**), jasmolin I (**3**), and pyrethrin II (**4**), cinerin II (**5**), jasmolin II (**6**) (collective name "rethrins").

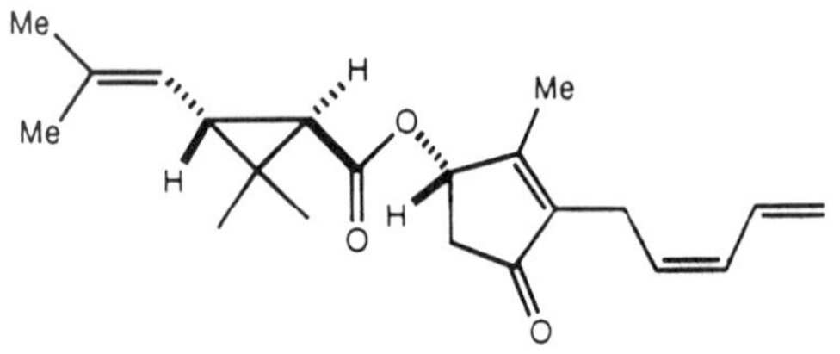

(1) Pyrethrin I

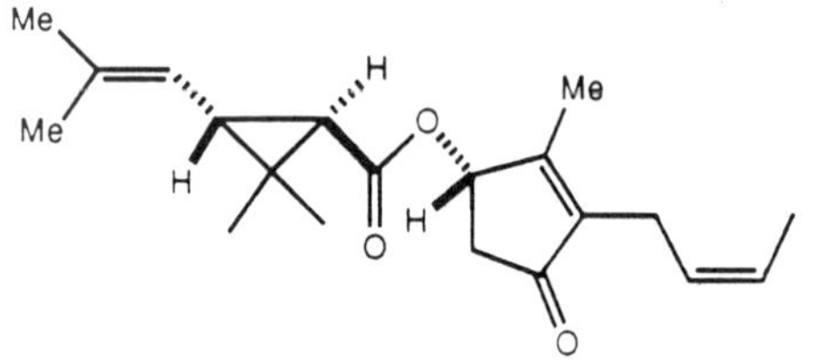

(2) Cinerin I

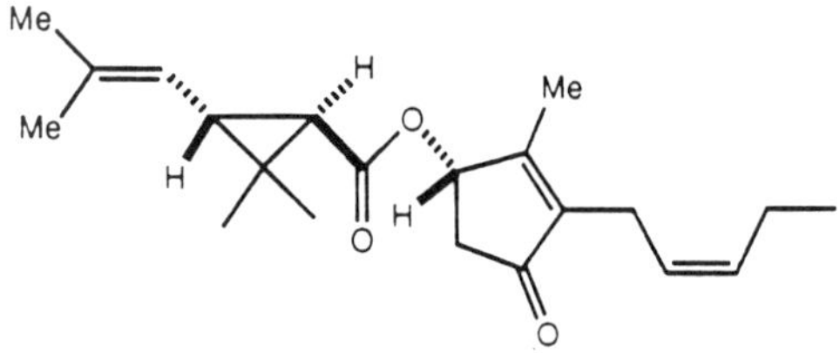

(3) Jasmolin I

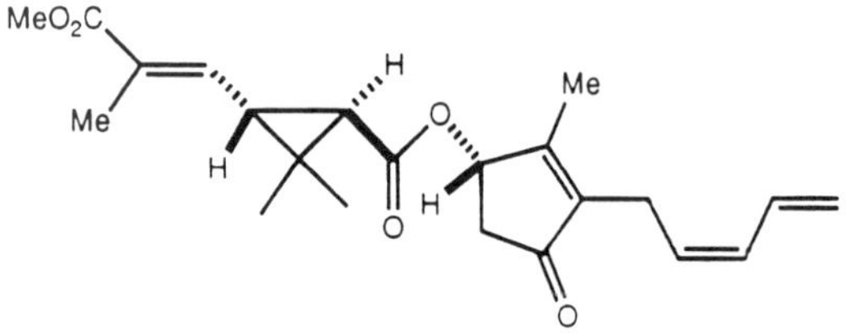

(4) Pyrethrin II

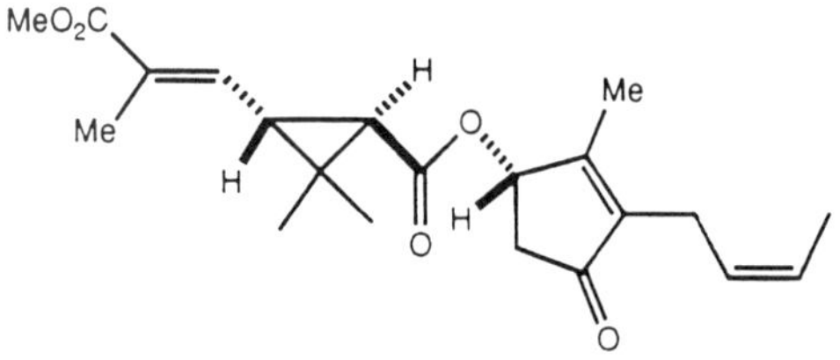

(5) Cinerin II

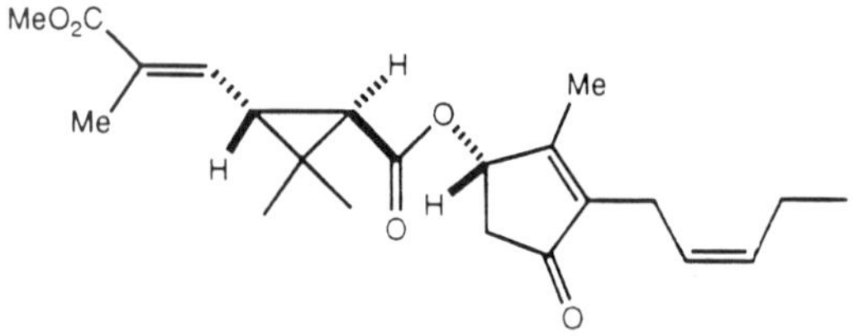

(6) Jasmolin II

Figure 8-1. Structures of natural pyrethrins.

There is considerable variation in the relative proportions of the six esters (**1**)–(**6**) depending on the particular plant type, geographical source, and time of harvest, but a typical extract (Casida, 1973) contains: pyrethrin I (38.0%), cinerin I (7.3%), jasmolin I (4.0%), pyrethrin II (35.0%), cinerin II (11.7%), and jasmolin II (4.0%). Pyrethrins I and II are the dominant, and usually the most valued, components from an insecticide point of view.

Being esters, it might be expected to be easy to hydrolyze the mixture to form the three component ketols pyrethrolone (**10**), cinerolone (**11**) and jasmolone (**12**) ("rethrolones") along with the two component acids chrysanthemum monocarboxylic (chrysanthemic) acid (**7**) and chrysanthemum dicarboxylic acid (**9**) (Fig. 8-2) (Chemical Abstracts names and registry numbers in Section VI). This is so, as far as the latter two acids are concerned, and indeed they are easily separated from each other since the monocarboxylic acid is volatile in steam whilst the dicarboxylic acid is not (Staudinger and Ruzicka, 1924). However, yields of the component ketols can be very poor as elimination (**13**) (Fig. 8-3) competes effectively with hydrolysis leading to cyclopentadienones which readily dimerize and undergo side reactions (LaForge *et al.*, 1952; Elliott *et al.*, 1967). For this reason isolation of the rethrolones usually follows the original Staudinger and Ruzicka (1924) procedure whereby the crude pyrethrins mixture is first converted into semicarbazones and then methanolized with alcoholic sodium methoxide to form rethrolone semicarbazones. The latter are then hydrolyzed using dilute sulfuric acid, potassium hydrogen sulphate, or pyruvic acid. For a long time the only ketol identified was pyrethrolone, cinerolone being recognized by LaForge and Barthel (1944), and jasmolone (after it had been made synthetically (Crombie *et al.*, 1951b)) by Godin *et al.*

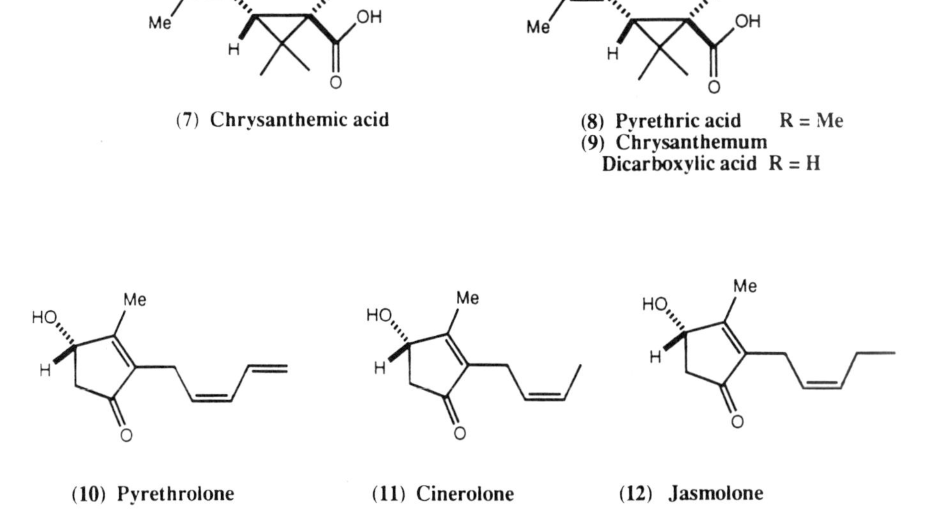

(7) **Chrysanthemic acid**

(8) **Pyrethric acid** R = Me
(9) **Chrysanthemum Dicarboxylic acid** R = H

(10) **Pyrethrolone** (11) **Cinerolone** (12) **Jasmolone**

Figure 8-2. Component acids and ketols of pyrethrins.

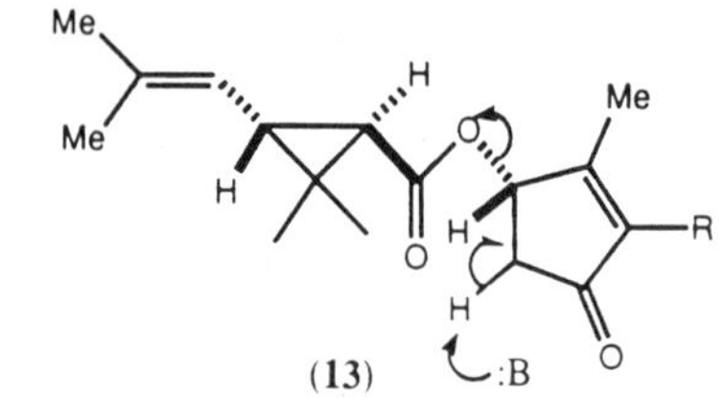

(13)

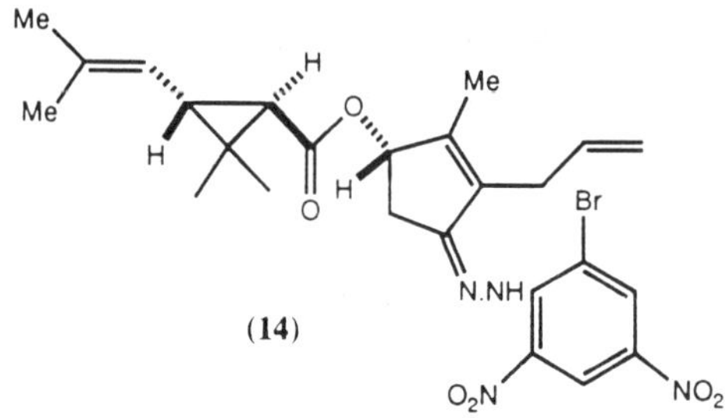

(14)

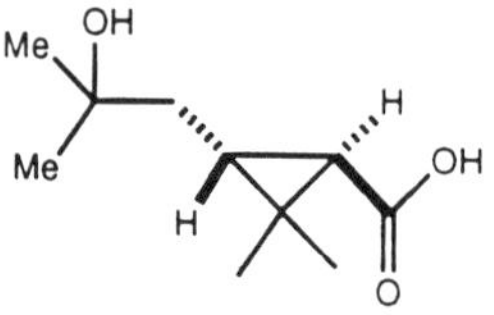

(15) 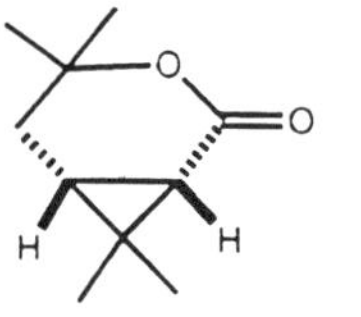 (16) 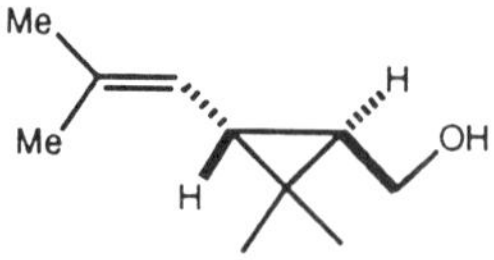 (20)

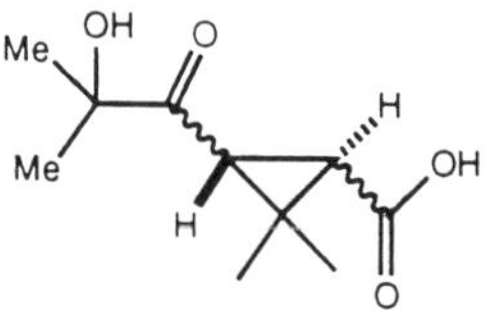

(21) 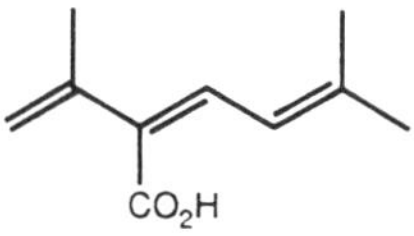 (22)

Figure 8-3. Miscellaneous structures.

(1966) as a result of gas chromatographic studies. From a preparative point of view, it is fortunate that of the three ketols, only the important pyrethrolone forms a stable crystalline hydrate (Elliott, 1964c) enabling it to be separated in pure, and optically pure, (+)-(*S*)-form.

As a group, the rethrins I are easily separated chromatographically from the corresponding rethrins II, but separation of the three components within each group requires more refined methods. Partition chromatography was early used to separate all six natural pyrethrins, employing nitromethane on Celite, eluting with carbon tetrachloride/hexane (1:3) (Rickett, 1972), and since that time high performance liquid chromatography (HPLC) has been used extensively both for analytical and isolation purposes (e.g., Mourot *et al.*, 1978; Otieno *et al.*,

1982). Bushway *et al.*, (1985) used Zorbax ODS, eluting with acetonitrile and water, whilst Ando *et al.* (1986) employed Nucleosil 5-CN or 5-NO_2 as column materials, eluting with tetrahydrofuran in hexane. Since they are α-unsaturated ketones, ultraviolet (UV) methods are very suitable for detection and quantitation, with refractive index suitable for use on a larger scale. In recent years supercritical fluid chromatography employing liquid carbon dioxide as solvent, with Fourier transform infrared detection, has been developed (Nieass *et al.* 1984; Wieboldt and Smith, 1988; Wieboldt *et al.*, 1989). Various systems for thin-layer chromatography (TLC) have been recommended, along with visualization reagents (Stahl and Pfeifle, 1966; Balbaa *et al.*, 1972; Olive, 1973; Scharf, 1979), and can be used to monitor other chromatographic processes.

Gas-liquid chromatography (GLC) provides a suitable analytical and monitoring tool, though it is less satisfactory for preparing small rethrin specimens than HPLC (e.g., Donegan *et al.*, 1962; Head, 1964, 1966a, 1967, 1968b; Beevor *et al.*, 1965; Bevenue *et al.*, 1970; Murano, 1972; Class and Fresinius, 1992; Seda *et al.*, 1983). Suitable stationary phases are OV 1 or OV 101 (Meinen and Kassera, 1982), and Holmstead and Soderlund (1977), using an OV 25 column and GLC mass spectral methods, separated and detected the six natural rethrins using 114 ng of total extract. Apart from the usual flame ionization detection (FID), the α-unsaturated ketonic structural feature in pyrethrins makes possible the use of the more highly sensitive electron capture detector (ECD) (Donegan *et al.*, 1962; Head, 1966a; Kawano *et al.*, 1974). Thus Class (1991) using DB5 and DB1701 in silica capillary columns detected 0.1 ng/μL using FID but 0.01 ng/μL for ECD.

All six natural pyrethrins are high boiling viscous liquids, sensitive to aerial oxidation and difficult to store for extended periods. This, and the difficulties involved in isolating them in pure form, makes the recorded spectral data of more than usual importance. Ultraviolet spectra (West, 1944; Crombie and Elliott, 1961; Elliott, 1964b) are recorded in Table 8-1 and were important in early work on the structure of pyrethrolone, when the nature of the side-chain unsaturation was in doubt. Infrared (IR) spectra are useful in deciding the cyclopropane geometry of a rethrin (Crombie and Harper, 1958) and reference IR spectra for pyrethrins and their relatives are available (Elliott, 1961).

Table 8-1. UV Absorption and Physical Data for Natural Pyrethrins (Crombie and Elliott, 1961)

	λ_{max}(nm)[a]	ε	n^{20}D	[α]D
Pyrethrin I	223	38,500	1.5242	−14°[b]
Pyrethrin II	228	47,000	1.5355	+14.7°[c]
Cinerin I	221	21,000	1.5064	−22°[d]
Cinerin II	229	28,500	1.5183	+16°[e]

[a]Octane or isooctane.
[b]Isooctane, 20°C.
[c]Isooctane-ether, 19°C.
[d]Hexane, 20°C.
[e]Isooctane, 16°C.

Table 8-2. ^1H NMR Data for the Natural Rethrins: Chrysanthemyl Segments[a] (see Fig. 4 for Structure)

	^{a}Me	^{b}Me	^{c}Me	^{d}Me	^{e}H	^{f}H	^{g}H
Pyrethrin I **1**	1.24	1.12	1.71	1.71	4.93	1.38	~2.1
Cinerin I **2**	1.27	1.14	1.73	1.73	4.93	1.41	~2.1
Jasmolin I **3**	1.27	1.15	1.72	1.73	4.94	1.41	~2.1
Methyl chrysanthemate	1.25	1.11	1.69	1.69	4.91	1.37	~2.1

[a]Coupling constants (Hz): $J_{d,e}$ 1.5; $J_{e,g}$ 8; $J_{f,g}$ 5.5.

Table 8-3. ^1H NMR Data for the Natural Rethrins: Pyrethryl Segments[a] (see Fig. 8-4 for Structure)

	^{a}Me	^{b}Me	$^{c'}$Me	^{d}Me	^{e}H	^{f}H	^{g}H
Pyrethrin II **4**	1.30	1.23	3.75	1.93	6.48	1.73	2.21
Cinerin II **5**	1.30	1.23	3.74	1.93	6.53	1.74	2.21
Jasmolin II **6**	1.35	1.26	3.75	1.95	6.50	1.75	2.20
Methyl pyrethrate	1.31	1.23	3.71	1.93	6.47	1.73	2.20

[a]Coupling constants (Hz): $J_{d,e}$ 1.5; $J_{e,g}$ 9; $J_{f,g}$ 5.5.

Proton magnetic resonance data (^1H NMR) for the natural pyrethrins (Bramwell *et al.*, 1969) are given in Tables 8-2 to 8-4 (structures in Fig. 8-4). Full diagrams of the six spectra are to be found in the original paper where there is further detail on the hydrolysis products and related materials. The spectra were determined at lower field strength than would be commonly used today, but they have been of particular use to synthetic chemists who wish to compare their products with natural materials without becoming involved in lengthy reisolations, as well as in confirming certain structural and stereochemical conclusions. Tables 8-5 to 8-7 (structures in Fig. 8-4) list carbon magnetic resonance data (^{13}C NMR) for the natural rethrins, together with full assignments, and again the original publication (Crombie *et al.*, 1975a, 1975b) should be consulted for further data.

The mass spectrometer, with its high sensitivity, is now widely used as a detector giving much structural information in gas chromatography (GC-MS)

Table 8-4. ^1H NMR Data for the Natural Rethrins: Rethrolone Segments[a] (see Fig. 8-4 for Structure)

	^{k}H	^{l}H	$^{l'}$H	^{m}Me	^{n}CH$_2$
Pyrethrin I **1**	5.69	2.88	2.19	2.03	3.12
Cinerin I **2**	5.68	2.89	2.2	2.05	2.98
Jasmolin I **3**	5.72	2.87	2.19	2.05	3.00
Pyrethrin II **4**	5.69	2.90	2.19	2.03	3.12
Cinerin II **5**	5.71	2.90	2.20	2.03	2.98
Jasmolin II **6**	5.74	2.91	2.2	2.05	2.99

[a]Coupling constants (Hz): $J_{k,l}$ 6; $J_{k,l'}$ 2; $J_{l,l'}$ 18; $J_{n,o}$ 6.

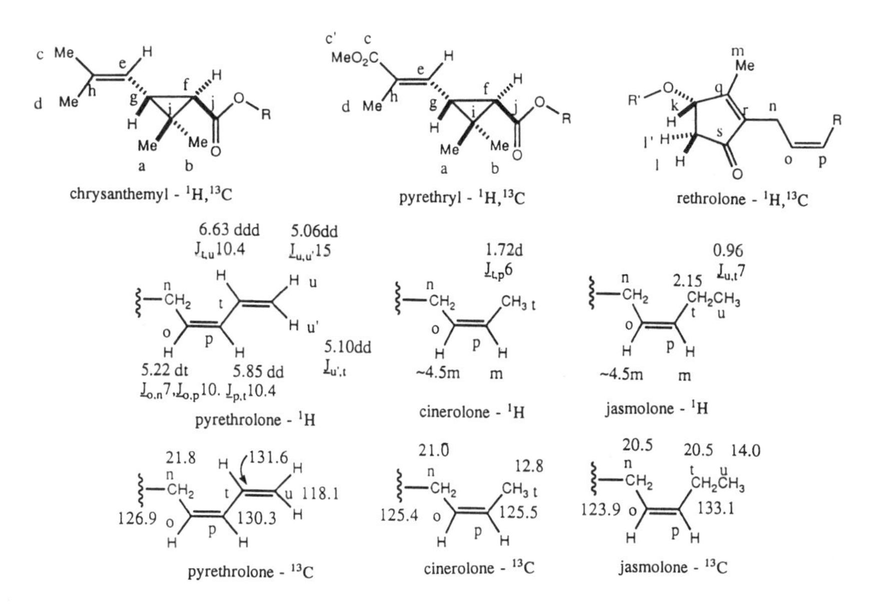

Figure 8-4. Positional designations of chrysanthemyl, pyrethryl, rethrolone, and alkenyl segments of the natural rethrins relative to ^{1}H and ^{13}C NMR data.

Table 8-5. ^{13}C NMR Data for the Natural Rethrins: Chrysanthemyl Segments (see Fig. 8-4 for Structure)

	^{a}Me	^{b}Me	^{c}Me	^{d}Me	^{e}CH	^{f}CH	^{g}CH	^{h}C	^{i}C	^{j}C
Pyrethrin I **1**	22.4	22.1	25.5	18.5	120.9	32.9	34.5	135.7	28.9	172.0
Cinerin I **2**	20.4	22.1	25.6	18.5	120.8	34.6	32.9	135.9	29.1	172.3
Jasmolin I **3**	20.4	22.1	25.6	18.5	120.8	34.6	32.9	135.9	29.1	172.3
Methyl chrysanthemate	20.4	22.2	25.3	18.4	121.2	34.5	32.7	135.2	29.4	172.6

Table 8-6. ^{13}C NMR Data for the Natural Rethrins: Pyrethryl Segments (see Fig. 8-4 for Structure)

	^{a}Me	^{b}Me	^{c}C	$^{c'}$Me	^{e}CH	^{f}CH	^{g}CH	^{h}C	^{i}C	^{j}C
Pyrethrin II **4**	20.5	22.3	168.0	51.8	138.9	35.8	33.0	129.3	30.6	171.1
Jasmolin II **6**	20.6	22.3	168.0	51.8	138.9	35.8	32.9	129.8	30.5	171.2
Pyrethric acid **8**	20.4	22.5	168.2	51.8	139.2	35.8	33.4	129.6	31.0	177.1

Table 8-7. ^{13}C NMR Data for the Natural Rethrins: Rethrolone Segments (see Fig. 8-4 for Structure)

	^{k}CH	$^{l}CH_2$	^{m}Me	^{q}C	^{r}C	^{s}C
Pyrethrin I **1**	73.0	42.0	14.0	165.1	141.9	203.3
Cinerin I **2**	72.7	41.8	14.0	165.0	142.7	203.9
Jasmolin I **3**	72.8	41.8	14.0	164.8	142.7	203.9
Pyrethrin II **4**	73.2	41.7	14.2	165.3	141.7	203.4
Jasmolin II **6**	73.2	41.7	14.1	164.3	142.9	203.6

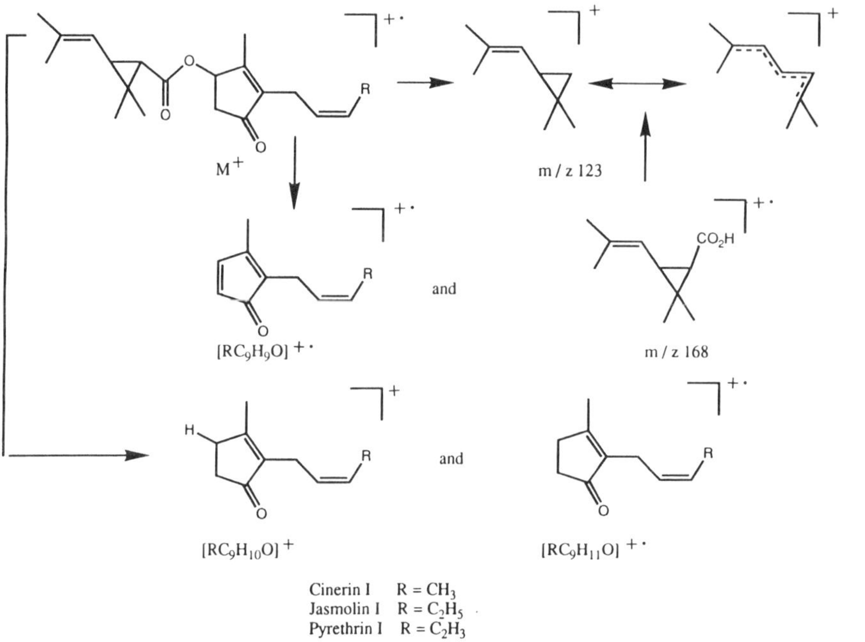

Figure 8-5. EI mass spectral fragmentation of the pyrethrins I: important fragmentations.

(e.g., Kawano *et al.*, 1980; Groneman *et al.*, 1984), and this has underlined interest in the mass spectrometry of the pyrethrins group (King and Paisley, 1969). Fig. 8-5 records the fragmentation patterns deduced for the pyrethrins I, whilst Fig. 8-6 deals with the pyrethrins II (Pattenden *et al.*, 1973). Full line

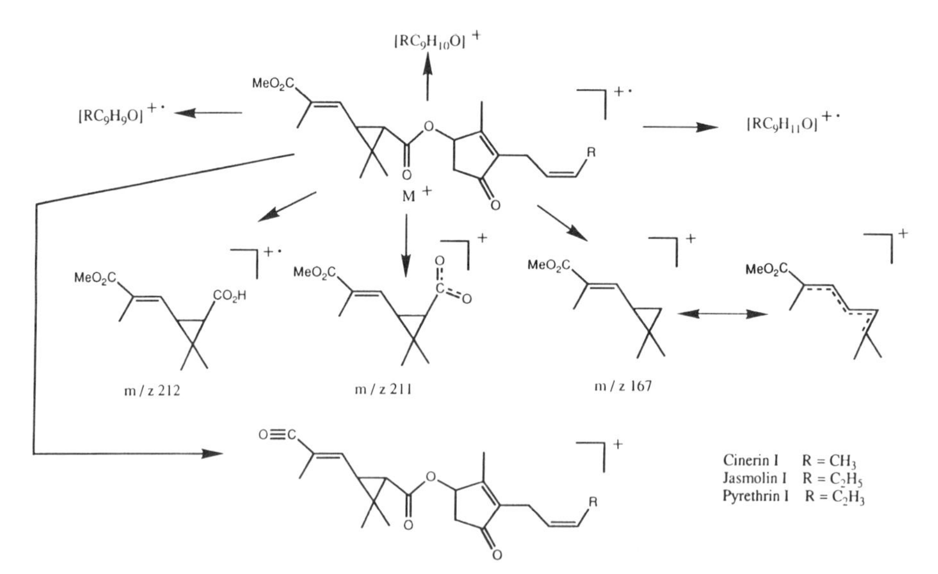

Figure 8-6. EI mass spectral fragmentation of the pyrethrins II: important fragmentations.

diagrams for the spectra are available in the latter reference, as well as a detailed discussion of the fragmentations. Holmstead and Soderlund (1977) give information on the chemical ionization (isobutane) mass spectra of the natural pyrethrins. Various ionization techniques have been applied to the pyrethrins by Nikiforov and Kohlmann (1983).

Fig. 8-7 shows the circular dichroism (CD) spectra for the six natural pyrethrins, whilst Fig. 8-8 shows the corresponding optical rotatory dispersions (ORD) (Begley *et al.*, 1974). The Cotton effects are dominated by the chiral environment around the α-unsaturated carbonyl of the cyclopentenone, i.e., the absolute configuration at C-4, and the familial nature of the sets of curves shows that all the natural esters have the same configuration at C-4, as does the synthetic ester (+)-allethronyl (+)-*trans*-chrysanthemate, (*S*)-bioallethrin, also shown on the figures. Determination of the C-4 absolute configuration for any one of the seven compounds thus applies to the remaining six.

Although the formation of a crystalline derivative suitable for single crystal X-ray analysis from any of the natural rethrins was unsuccessful, a suitable

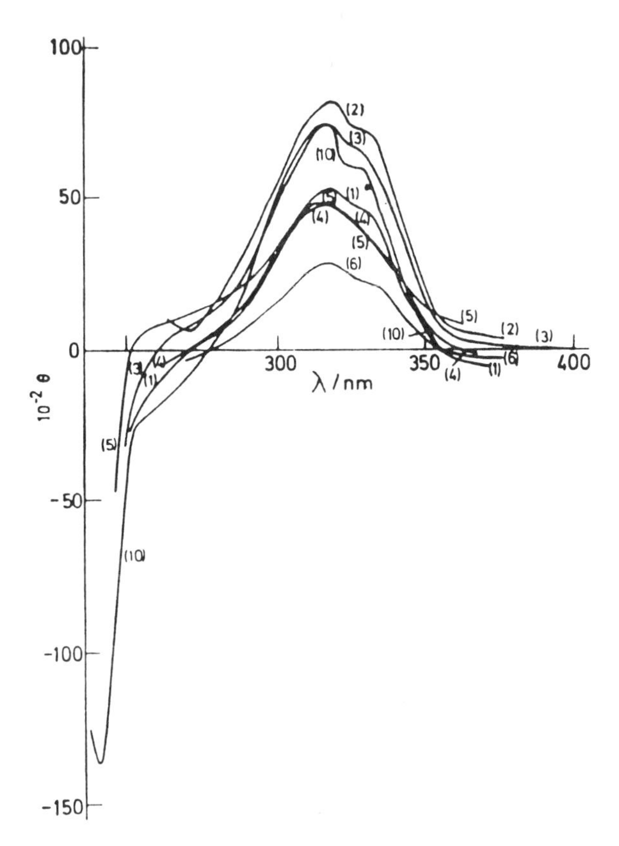

(1) Pyrethrin I. (2) Pyrethrin II. (3) Cinerin I. (4) Cinerin II
(5) Jasmolin I. (6) Jasmolin II. (10) S-Bioallethrin.

Figure 8-7. Circular dichroism spectra for pyrethrins.

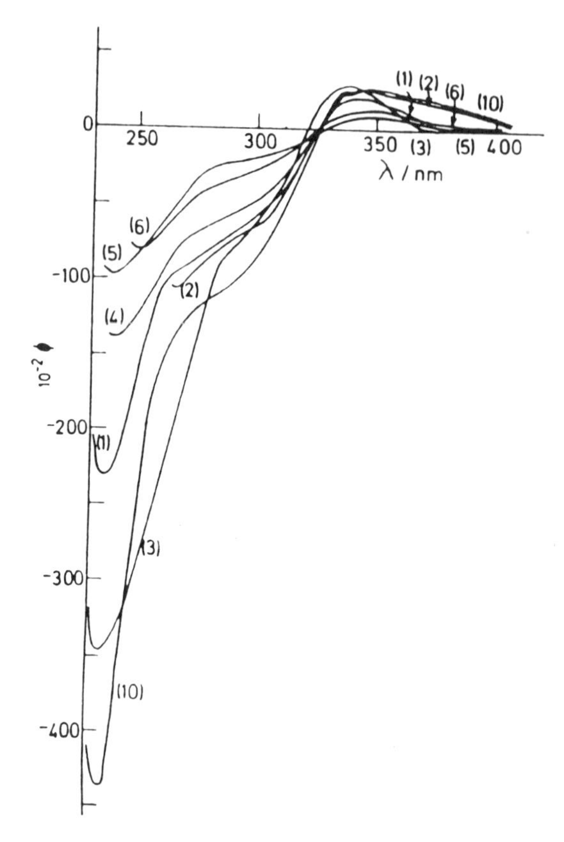

(1) Pyrethrin I. (2) Pyrethrin II. (3) Cinerin I. (4) Cinerin II.
(5) Jasmolin I. (6) Jasmolin II. (10) S-Bioallethrin.

Figure 8-8. Optical rotatory dispersion spectra for pyrethrins.

6-bromo-2,4-dinitrophenylhydrazone (**14**) (Fig. 8-3) was obtained from the chiral allethronyl ester and Figs. 8-9 and 8-10 show two views of the molecule (Begley *et al.*, 1972, 1974). Knowing the absolute configuration of the chrysanthemate segment (see later) it is apparent that it, and all the natural rethrins, have the 4-(*S*)-configuration. Furthermore, by comparison of Bijvoet pairs of reflections (possible because of the anomalous dispersion of the heavy bromine atom), the 4-(*S*)-configuration could be shown without recourse to the stereochemistry of the chrysanthemate segment: the determination in fact also checked the correctness of the earlier chemical determination of the chrysanthemate stereochemistry (see later). The 4-(*S*)-configuration has also been demonstrated by chemical correlation to a prostanoid of known absolute configuration (Miyano and Dorn, 1973) and (though there are certain inconsistencies in the reports) an earlier chemical degradation (Inouye and Ohno, 1958; Katsuda *et al.*, 1958, 1959). Structural and physical data for the pyrethrins have been reviewed (Crombie *et al.*, 1976a).

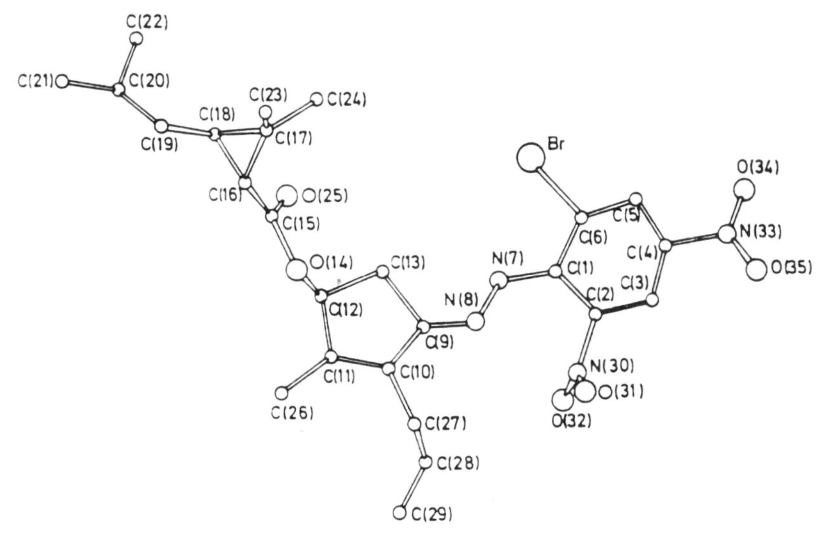

Figure 8-9. X-Ray structure of (+)-allethronyl (+)-*trans*-chrysanthemate 6-bromo-2,4-dinitrophenylhydrazone showing (1*R*, 3*R*, 4*S*) configuration.

The conformation of pyrethrins has been studied by both theoretical (Castellani *et al.*, 1980; Tosi *et al.*, 1982) and NMR means (Pierre *et al.*, 1970), whilst Elliott and Janes (1977) have discussed the relationship of possible conformation to insecticidal activity.

Apart from their occurrence in *Chrysanthemum cinerariaefolium*, all six pyrethrins are present in *C. coccineum* Wild., a species in which pyrethrin II is apparently dominant in the flowers (pyrethrin I, 0.018%; cinerin I, 0.0007%; jasmolin I, 0.001%; pyrethrin II, 0.15%; cinerin II, 0.01%; jasmolin II, 0.02%) (Hogstad *et al.*, 1984). Claims that pyrethrins occur in the peony flower have been refuted (Godin *et al.*, 1967). Various papers claiming that pyrethrins occur in *Tagetes erecta* and its tissue in culture (Khanna *et al.*, 1975, Khanna and Khanna, 1976; Jain, 1977; Kamal, 1977) have not been supported by the Hogstad group (1984) who suspect confusion with natural thiophenes. We too were unable to find pyrethrins in this plant source (Crombie, 1979). There is

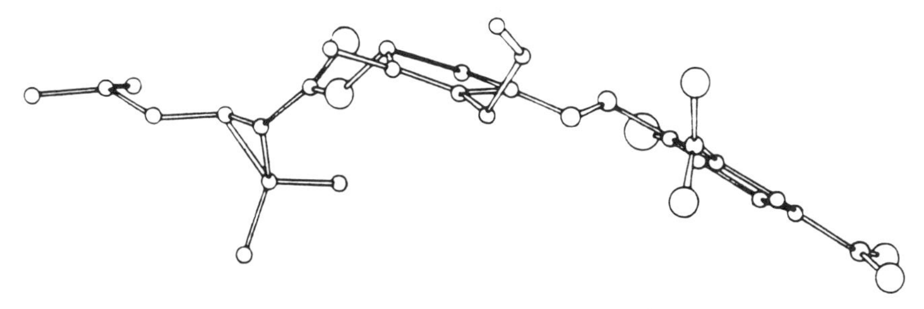

Figure 8-10. X-Ray structure of (+)-allethronyl (+)-*trans*-chrysanthemate 6-bromo-2,4-dinitrophenylhydrazone: side view.

an unconfirmed report that pyrethrins are present in *Tanacetum odessanum* (tansy) (Bohlmann and Knoll, 1978).

To a large extent the chemistry of the six individual natural rethrins is a summation of the chemistry of the appropriate rethrolone with that of the appropriate chrysanthemate. Much of the chemistry has therefore been investigated on the separated ketols and acids and this approach is developed in the following sections.

II. THE CHRYSANTHEMIC ACIDS

As mentioned earlier, chrysanthemum monocarboxylic acid and chrysanthemum dicarboxylic acid are readily available from hydrolysis of a natural pyrethrins mixture, and can be easily separated by steam distillation. In the natural rethrins II the latter acid occurs as a side-chain methyl ester, and the corresponding half methyl ester (**8**) (Fig. 8-2) is known as pyrethric acid; some of the latter is obtained when a pyrethrins mixture is hydrolyzed with less than the theoretical quantity of alkali.

The side-chain double bond of chrysanthemic acid or ester can be catalytically hydrogenated without affecting the cyclopropane ring (Harper, 1954), hydroxylated with osmium tetroxide (Crombie *et al.*, 1970), or epoxidized with peracid (Sasaki *et al.*, 1968a; Ando *et al.*, 1991a,b). Treatment with potassium permanganate gives a diol and a ketol (Matsui *et al.*, 1963b). Under acid conditions the side chain double bond of (±)-*trans*-chrysanthemic acid is hydrated (**15**) (Fig. 8-3), whilst the (±)-*cis*-isomer forms a crystalline δ-lactone (**16**) (Crombie *et al.*, 1951a; Harper and Thompson, 1952).

A. Stereochemical Features

Staudinger and Ruzicka (1924) formulated natural (+)-chrysanthemum monocarboxylic (chrysanthemic or chrysanthemumic) acid correctly and appreciated that it was the *trans*-isomer since it gave an optically active *trans*-caronic acid on ozonolysis whereas the *cis*-isomer would have given optically inactive (*meso*) *cis*-caronic acid. The work did not however provide a solution to the problem of its absolute configuration and this was first solved as indicated in Fig. 8-11. Since all four possible stereoisomers of chrysanthemic acid, (+)-*trans*-, (−)-*trans*-, (+)-*cis*- and (−)-*cis*-, were known from synthesis and optical resolution, the proof was applied to all four.

Pyrolysis of (+)-*trans*-chrysanthemic acid gives a lactone (−)-pyrocin (**17**) (Fig. 8-11) (originally incorrectly formulated (Matsui, 1950)) in which the A center of the former is destroyed leaving the B center (Crombie and Harper, 1954). Since the absolute configuration of (−)-pyrocin is already known through chemical links via (+)-terebic acid with the chiral glyceraldehydes, it follows that natural (+)-*trans*-chrysanthemic acid has the (*R*)-configuration at C-3. Homologation of the latter acid gives the homo-(+)-*trans* acid (**18**) and pyrolysis of this leads to a homolactone (**19**) in which this time the B center is destroyed leaving the A center (Crombie *et al.*, 1963). Degradation of the homolactone

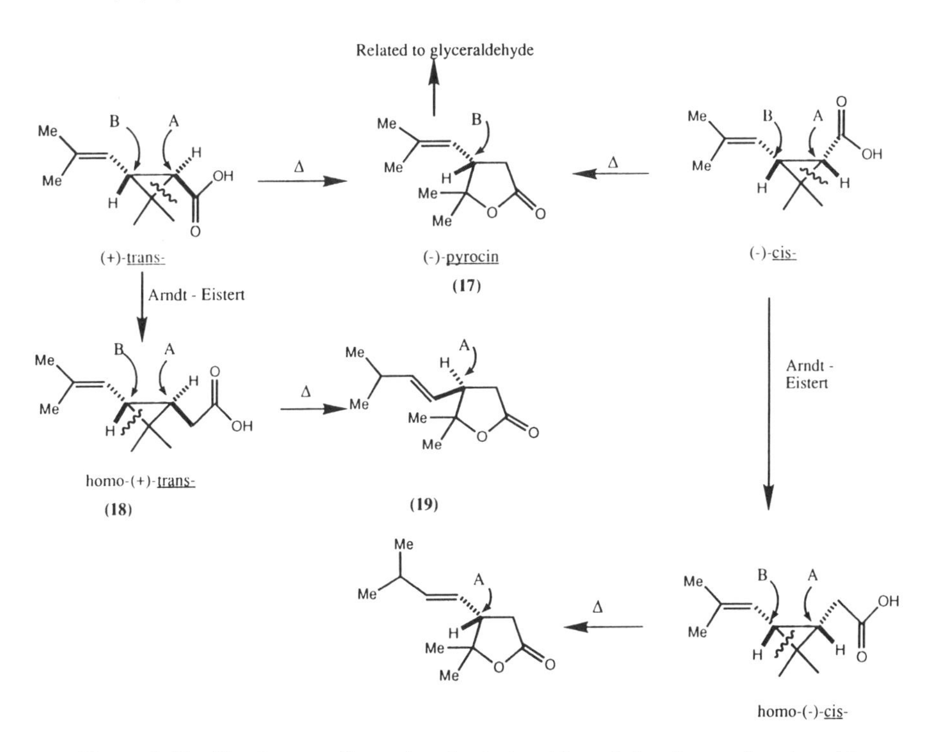

Figure 8-11. Absolute configurational relationships of the chrysanthemic acids.

to (−)-terebic acid shows that (+)-*trans*-chrysanthemic acid has the (*R*)-configuration at C-1. Further, since (−)-*cis*-chrysanthemic acid gives (−)-pyrocin on pyrolysis, its absolute configuration at C-3 is the same as in the (+)-*trans*- acid, i.e., (*R*) (Crombie and Harper, 1954), with the configuration at C-1 being (*S*), as confirmed by pyrolysis of the homo-(−)-*cis*-acid.

The formation of the same chiral (−)-*trans*-caronic acid by ozonolysis of natural chrysanthemum dicarboxylic acid as is obtained from the natural monocarboxylic acid (Staudinger and Ruzicka 1924) indicates the (1*R*, 3*R*)-configuration for the dicarboxylic acid, and this has been confirmed by subsequent syntheses. As a final stereochemical feature, the geometry of the side chain of natural chrysanthemum dicarboxylic acid or methyl pyrethrate was shown to be (*E*) from the deshielded position of the olefinic CH signal in the ^{1}H NMR spectrum (Crombie *et al.* 1963). All the stereochemical conclusions derived chemically and by NMR are confirmed by the X-ray single crystal data of Begley *et al.* (1972,1974) using a complete rethrin mentioned above. In addition, an X-ray structure of (+)-*trans*-chrysanthemic acid as its *p*-bromoanilide, confirming its absolute configuration, is available (Cameron *et al.*, 1975). For analytical purposes the four stereochemical forms of chrysanthemic acid can be separated and estimated quantitatively by GLC as their (+)-α-methylbenzylamides, or after esterification with (*d*)- or (*l*)-octan-2-ol, or menthol (Murano, 1972; Rickett, 1973). A study of ^{13}C-NMR shifts for compounds of the chrysanthemic and presqualene type is available (Crombie *et al.*, 1975b) and lanthanide ^1H NMR shift data have been published (Crombie *et al.*, 1972a).

Chrysanthemic acid or ester is reduced by lithium aluminium hydride to chrysanthemyl alcohol (also known as chrysanthemol), shown in *trans*-form (**20**) (Fig. 8-3). The latter is readily oxidized to the aldehyde by manganese dioxide (Crombie and Crossley, 1963) and a variety of other reagents such as dimethyl sulfoxide/oxalyl chloride (Mancuso *et al.*, 1979), chromic acid supported on silica gel (Singh *et al.*, 1979), and tetrabutylammonium per-ruthenate (Green *et al.*, 1984; Griffith *et al.*, 1987). Chemical oxidation to the carboxylic acid level, as demanded in some syntheses, is less easy, and the microbiological oxidation of chrysanthemol to chrysanthemic acid (Miski and Davis, 1988a; Davis and Miski, 1988) is therefore of interest. Suitable GLC systems (10% QF1) for handling analyses in the latter work have been described (Miski and Davis, 1988b).

Sasaki *et al.* (1968b) have studied the synthesis and reactions of isocyanates derived from chrysanthemic acid and in an interesting study this group (Sasaki *et al.*, 1973) has explored the Bamford-Stevens rearrangement of the tosylhydrazone of chrysanthemic aldehyde. Decomposition in aprotic medium gives a carbene, forming tetramethylbutadiene and a substituted cyclobutene among the products of its decomposition.

Methyl chrysanthemate is deprotonated by lithium diisopropylamide at the C-1 of the ring and can be substituted there (Me, MeS, $CH_2CH=CH_2$) by suitable electrophiles (Reichelt and Reissig, 1985). The decomposition of a nitrosourea derived from chrysanthemic acid has been shown to give an allene-olefin (Holm and Skattebol, 1985)

B. Stereochemical Interconversions

Chrysanthemic acid derivatives carry a center at C-1 which is formally epimerizable through enolization. Thus treatment with base under rather severe conditions (e.g., sodium ethoxide at 190°C) (Julia *et al.* 1959; Matsui and Yoshioka, 1965) permits conversion of methyl (−)-*cis*-chrysanthemate (1*S*,3*R*) into methyl (+)-*trans*-chrysanthemate (1*R*,3*R*). At higher temperatures (240–260°C) epimerization of the ester can be achieved thermally (Hanafusa *et al.*, 1970). When the ester group is replaced by more electron withdrawing groups such as the nitrile or acid chloride, epimerization proceeds at lower temperatures (Suzuki *et al.*, 1970). Interestingly both ethyl chrysanthemate and chrysanthemic acid undergo *cis*- to *trans*-isomerization at room temperarure in solution in the presence of an homogeneous catalyst L_2PdCl_2 where L = MeCN, EtCN or PhCN. Unfortunately the early "stopping" of the reaction is a disadvantage (Williams and Rettig, 1981); apparently the epimerization involves C-3 (Suzukamo *et al.*, 1984). The latter authors find that treatment of ethyl chrysanthemate with a Lewis acid causes cleavage of the 2,3-cyclopropane bond, thereby allowing conversion of the *cis*- to *trans*-ester.

An additional electron withdrawing group placed adjacent to the C-3 center permits epimerization at this center also. Thus, either the (+)-*trans*- or

(−)-*trans*-ketol (**21**) (Fig. 3) gives the (±)-ketol on treatment with potassium *t*-butoxide (Matsui *et al.*, 1963c). The lactonization of *cis*-chrysanthemic acid to a δ-lactone (**16**) (Fig. 3) under acidic conditions, whilst the *trans*-acid merely hydrates (**15**), was mentioned earlier and allows the conversion of (+)-*trans*-acid into the (−)-*cis*-form. Gradual epimerization at C-1 allows the *cis*-form to accumulate as its lactone; the lactone can then be converted into (−)-*cis*-chrysanthemic acid by treatment with magnesium bromide hydrate in pyridine (Martel and Buendia, 1970a).

Less specifically, chrysanthemates of any chirality can be converted into a 34%:64% mixture of (±)-*cis*-/(±)-*trans*-forms photochemically via a presumed diradical which allows epimerization at both C-1 and C-3 (see below).

C. Cleavage Reactions of the Cyclopropane Ring

Chrysanthemic acid is clearly a monoterpene, though of an unusual type. It carries within itself a structural relationship to four acyclic monoterpene skeleta: santolinyl, lavandulyl, and artemisyl, along with a relationship to the C_{10} lower prenylogue of squalene, as shown in Fig. 8-12. All of the first three types of cleavage can be demonstrated by appropriate functionalization of the chrysanthemate framework as indicated (Fig. 8-13) (Bates and Paknikar, 1965; Crombie *et al.*, 1967, 1972c). The fourth relationship of Fig. 8-12 is less obvious, and forms only a very minor pathway *in vitro*, but it parallels the incompletely understood enzyme mediated rearrangement of presqualene pyrophosphate into squalene and has been intensively studied by Poulter and his co-workers (Poulter, 1972; Poulter and Hughes, 1977a,b; Poulter and Rilling, 1981; Poulter *et al.*, 1971, 1977). There is also considerable interest in the biogenetic relationships between chrysanthemyl alcohol and other "irregular" terpenes

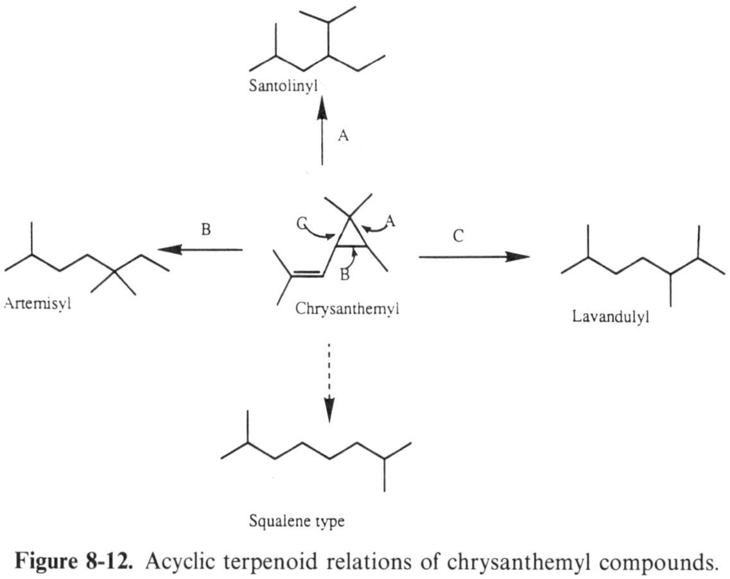

Figure 8-12. Acyclic terpenoid relations of chrysanthemyl compounds.

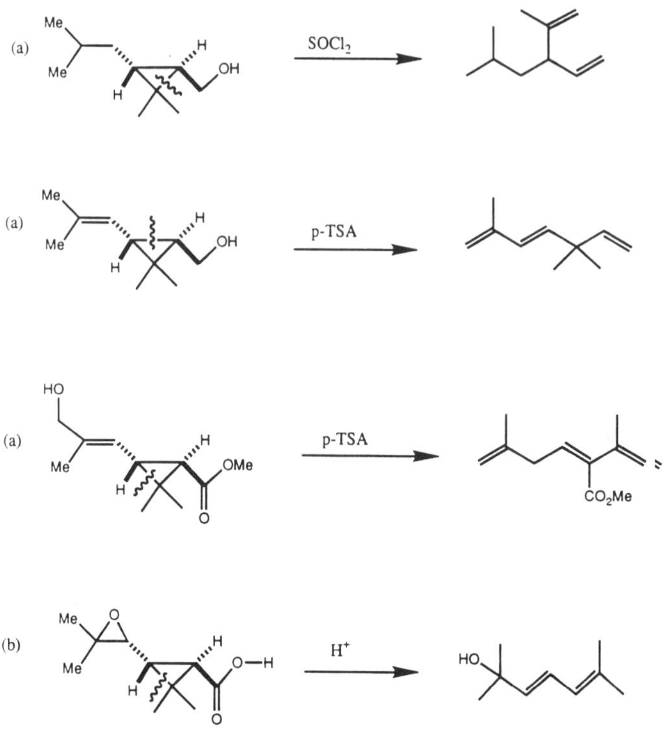

Figure 8-13. Examples of chrysanthemyl/acyclic terpenoid cleavages. (a) Crombie *et al.* (1967, 1972c); (b) Smith and Casida (1981).

(Epstein and Poulter, 1973; Epstein *et al.*, 1991). It is interesting that epoxychrysanthemic acid undergoes rapid acid catalyzed decarboxylation to form an acyclic diene (Fig. 8-13), and this and related processes are part of the metabolic detoxification mechanism of the pyrethrins (Smith and Casida, 1981; Ando *et al.*, 1991a,b).

Matsui *et al.* (1963a,b) have shown that 2,6-dimethylhepta-2,4,6-triene-5-carboxylic acid (**22**) (Fig. 8-3) can be obtained by oxidation and rearrangement of chrysanthemic acid and the Wittig rearrangement of allyl ethers of chrysanthemol e.g., (**23**) (Fig. 8-14), have been studied (Crombie *et al.*, 1976b). Treatment with *n*-butyllithium gives (**24**) by a process which may be written as a homo-[4,5]-sigmatropic rearrangement of the anion but which may in fact be radical in mechanism. Chrysanthemic acid and its derivatives, obtainable in all stereo-forms, have considerable potential as synthetic starting materials. Thus Ohno *et al.* (1986) have studied reactions between chrysanthemol silyl ether and allylic and enolic silanes in the presence of titanium tetrachloride, obtaining products which have 1,3-cyclopropane cleavage and are of the yomogi type. (1*R*,3*R*)-Chrysanthemol has been used as a precursor for synthesis of the natural product (*S*)-lyratol (**25**) (Fig. 8-15) (Gaughan and Poulter, 1979) and (1*R*,3*S*)-(+)-*cis*-chrysanthemic acid has been used in the stereospecific synthesis

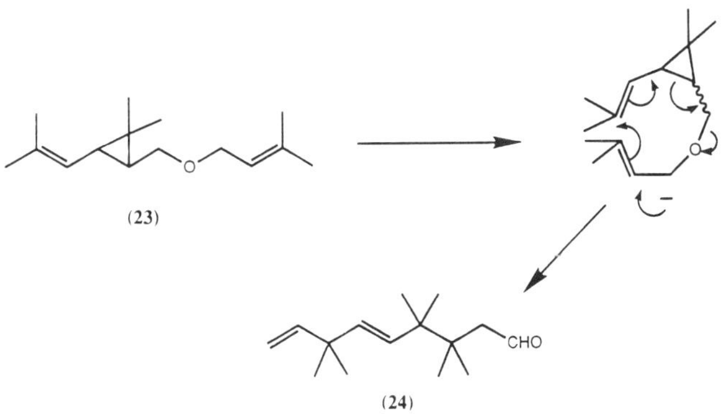

(23)

(24)

Figure 8-14. Wittig rearrangement of *cis*- or *trans*-chrysanthemol dimethylallyl ether.

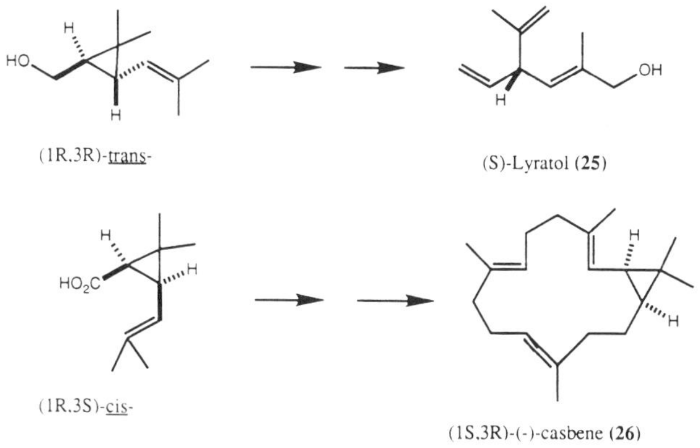

(1R,3R)-trans-

(S)-Lyratol (**25**)

(1R,3S)-cis-

(1S,3R)-(-)-casbene (**26**)

Figure 8-15. Chrysanthemyl precursors in natural product synthesis.

of (1*S*,3*R*)-casbene (**26**), a biogenetically important terpene from *Ricinus communis* (Crombie *et al.*, 1980).

Cleavage reactions of chrysanthemol in superacid (fluorosulphuric acid) have been studied by Baig *et al.* (1990) and implicate the ion (**27**) (Fig. 8-16), though the products from its quenching are complex and include dihydrofurans and tetrahydrofurans. It is reported (Suzukamo *et al.*, 1984) that treatment of an alkyl chrysanthemate with a Lewis acid leads to C-3 epimerization, but protonic acid gives C-2–C-3 bond cleavage leading to a synthesis of tetrahydrolavandulol.

D. Pyrolytic Reactions of the Chrysanthemic Acids

As outlined earlier, pyrolysis of the stereoisomers of chrysanthemic acid leads to the formation of stereoisomers of pyrocin. *trans*-Chrysanthemic esters are more stable, however, but pyrolyzed at 500°C, the cyclopropane bond ruptures with hydrogen transfer to form the branched ester (**28**) (Fig. 8-16) having a lavandulyl carbon skeleton (Ohloff, 1965).

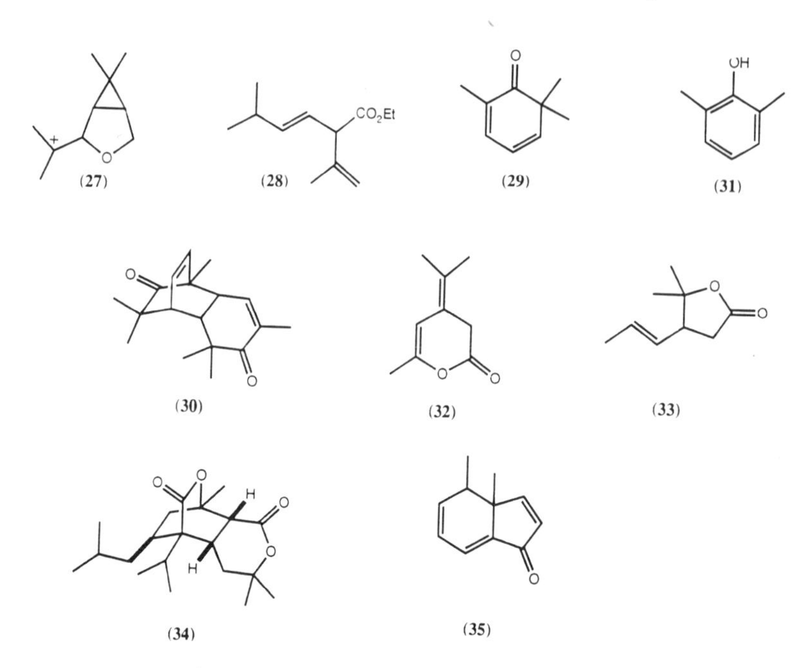

Figure 8-16. Acid and thermal cleavage products of chrysanthemol, chrysanthemic acid, and pyrethrin I.

Pyrolysis of chrysanthemum dicarboxylic acid at 260–320°C gives a mixture of neutral and acidic compounds and the composition of the acid fraction is summarized in Fig. 8-17 (Crombie *et al.*, 1971b). Three of the products involve cyclopropane ring rupture and the original paper should be consulted for mechanistic discussion. The composition of the neutral fraction, which contains a cyclopentadienone (**29**) (Fig. 8-16), its dimer (**30**), 2,6-dimethylphenol (**31**), and two lactones (**32**) and (**33**) has also been studied (Otieno *et al.*, 1977). The pyrolysis of chrysanthemum monocarboxylic acid, at 210°C and in the presence of pyridine hydrochloride, has been examined by Otieno *et al.* (1977) and the main products are summarized in Fig. 8-18. Related work has been continued by others. Elmore *et al.* (1985) have shown that acid cleavage of methyl chrysanthemate (rather than the acid) gives the methyl ester of the diene carboxylic acid of Fig. 8-18 as the major product, and this acid-catalyzed lavandulyl cleavage has been the subject of an informative and detailed mechanistic study by Goldschmidt *et al.* (1984). An interesting bridged dilactone (**34**) (Fig. 8-16) has been found in the products of a thermal acid catalyzed reaction of chrysanthemic acid by Goldschmidt *et al.* (1987) and its structure is verified by X-ray means.

The pyrolysis of pyrethrin I itself has been studied (Nakada *et al.*, 1972) and amongst products originating from the chrysanthemate segment were pyrocin and an unusual trienone (**35**) (Fig. 8-16).

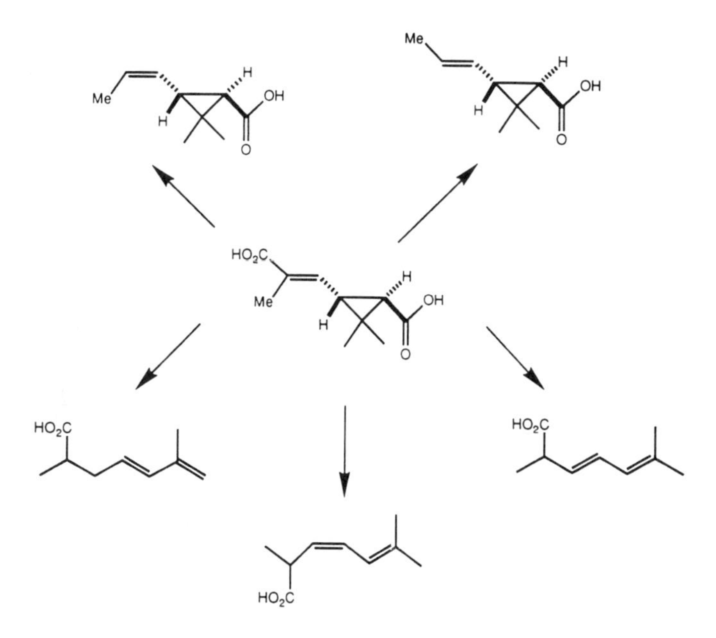

Figure 8-17. Main acidic pyrolysis products from chrysanthemum dicarboxylic acid at 260–320°C (Crombie *et al.*, 1971b).

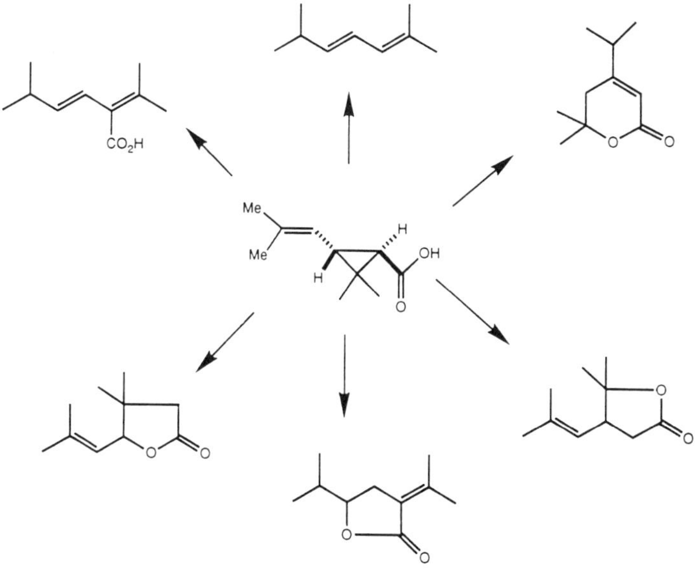

Figure 8-18. Products from pyrolysis (210°C) of chrysanthemic acid under acidic conditions. (Otieno *et al.*, 1977).

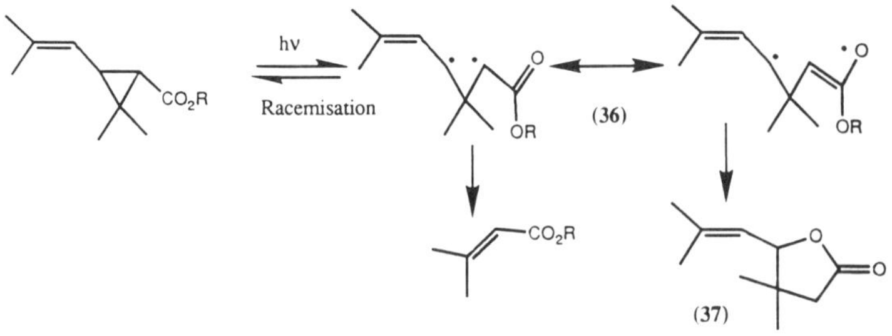

Figure 8-19. Photochemistry of chrysanthemic acid.

E. Photochemistry

The photochemistry of chrysanthemic acid has been studied by various authors (Sasaki *et al.*, 1970; Bullivant and Pattenden, 1971; Ueda and Matsui, 1971) and involves initial formation of a biradical (**36**) (Fig. 8-19). As mentioned above, one fate for this involves reclosure of the cyclopropane with racemization at both the C-1 and the C-3 centers. In addition however, recyclization of the diradical as shown on the second resonance contributor leads to the pyrocin isomer (**37**) whilst radical cleavage forms the methacrylate.

The photochemistry of a chrysanthemate ester (2,4-dimethylbenzyl) has been studied in the presence of oxygen (Chen and Casida, 1969) (Fig. 8-20) and undergoes oxidative attack at the (*E*)-methyl as represented by an alcohol, aldehyde, acid sequence. In addition, the ketone (**38**) and caronic half ester (**39**) were formed. In the presence of benzophenone, chrysanthemates form an oxetane (**40**) under UV irradiation (Sasaki *et al.*, 1968a).

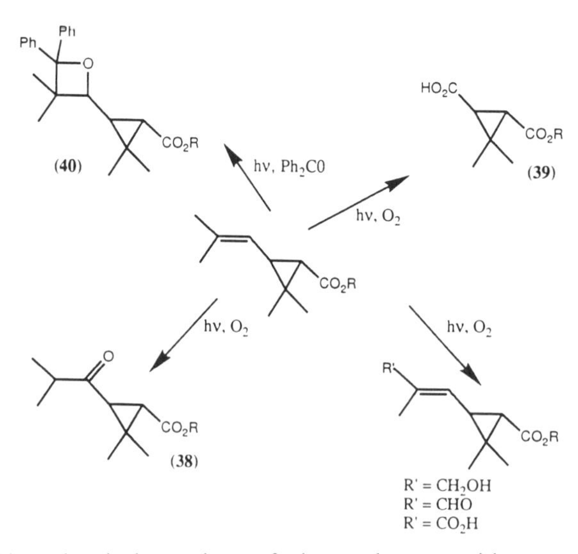

Figure 8-20. Photochemical reactions of chrysanthemate with oxygen and benzophenone.

F. Synthesis of Chrysanthemic Acids

Because of commercial interest in the synthetic pyrethroid area, many synthetic routes to the chrysanthemic acids have been explored in the literature (for specific reviews see Thomas, 1973; Arlt *et al.*, 1981; Krief *et al.*, 1990). The cyclopropane ring can be formed by final closure of the 1-2, 1-3 or 2-3 linkages and may be classified on this basis. In some cases, operationally, if not mechanistically, two of the linkages may be formed together in a reaction and in other cases the synthesis may start from a preformed cyclopropane derivative, e.g., a natural terpene. In more recent years effort has moved to the synthesis of the chrysanthemic acids in homochiral form.

Synthesis of chrysanthemum monocarboxylic (chrysanthemic) acid. Formally the whole chrysanthemate molecule can be made by union of two segments C_8+C_2, C_7+C_3, or C_5+C_5, making two bonds of the cyclopropane either synchronously or sequentially. Thus the classical synthesis, the addition of ethyl diazoacetate to 2,5-dimethylhexadiene in the presence of copper catalyst, follows the first pattern, giving a mixture (64:36) of ($\pm$)-*trans*- and ($\pm$)-*cis*-esters (Fig. 8-21) (Staudinger and Ruzica, 1924; Campbell and Harper, 1945; Harper *et al.*, 1951). Hexadecacarbonylhexarhodium also catalyzes the addition (Doyle *et al.*, 1981). Use of the *tert*-butyl ester of diazoacetic acid is reported to give entirely ($\pm$)-*trans*-ester (Matsumoto *et al.*, 1963). The presumed C_2 carbenoid addend can also be formed from diiodoacetic ester and zinc/copper couple (Büchel and Korte, 1962).

Dimethylallene carbene, readily generated from 3-chloro-3-methylbutyne and base, adds to dimethylallyl alcohol (but not 3-methylbutenoic ester) to give an allenic alcohol (**41**) (C_5+C_5) (Fig. 8-22) which on reduction with sodium and alcohol is stereospecifically reduced to *trans*-chrysanthemyl alcohol (Mills *et al.*, 1973). The latter can be oxidized to chrysanthemic acid, though this process has proved less easy than expected.

Two syntheses involve ylides and are formally C_7+C_3. In the first of these diphenylsulfonium isopropylide is added to methyl 5-methylhexa-2,4-dienoate to give ($\pm$)-*trans*-chrysanthemate (Fig. 8-23) (Corey and Jautelat, 1967), and in the second, methyl (*E*)-4-oxobut-2-enoate is treated with isopropylidenetriphenylphosphorane (**44**) (Fig. 8-24) (De Vos *et al.*, 1976). The first formed betaine (**42**) adds a second mole of the phosphorane to give ($\pm$)-*trans*-

CO_2Et

trans- 64%
cis- 36%

(a) N_2CHCO_2Et / Cu [$:CHCO_2Et$]

Figure 8-21. Ethyl chrysanthemates by diazoacetic ester addition (Staudinger and Ruzicka, 1924; Campbell and Harper, 1945; Harper *et al.*, 1951).

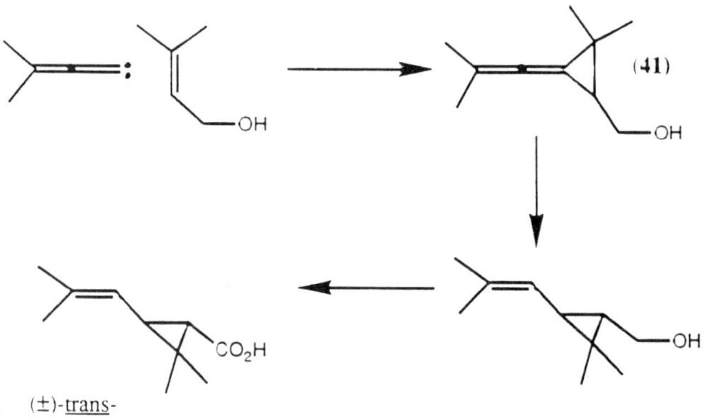

Figure 8-22. The dimethylallenecarbene approach (Mills *et al.*, 1973).

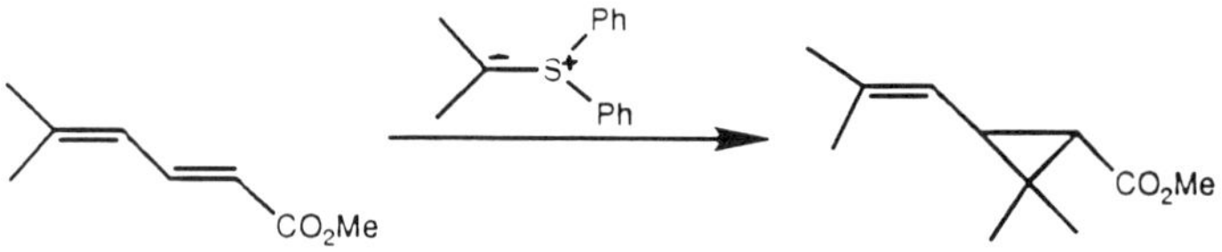

Figure 8-23. Diphenylsulfonium isopropylide addition (Corey and Jautelat, 1967).

chrysanthemate. Curiously, the phosphorane is reported not to add to methyl 5-methyl-2,4-hexadienoate (**43**) itself (De Vos and Krief, 1979a). The addition of isopropylidenediphenylsulfurane to but-2-enolide allows entry to the *cis*-chrysanthemate series (Sevrin *et al.*, 1976). A "one pot" synthesis based on phosphorus ylides has been effected (De Vos and Krief, 1979b); the Wittig reagent forms two sections of the molecule consecutively. Other developments

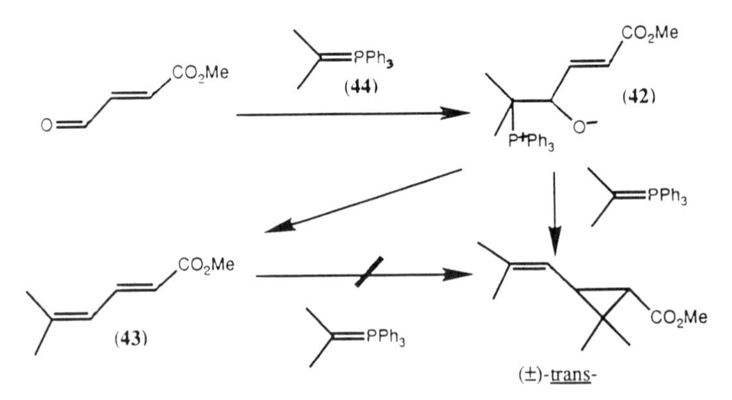

Figure 8-24. Synthesis of (±)-*trans*-chrysanthemic ester via phosphonium ylide method (De Vos *et al.*, 1976; De Vos and Krief, 1979a).

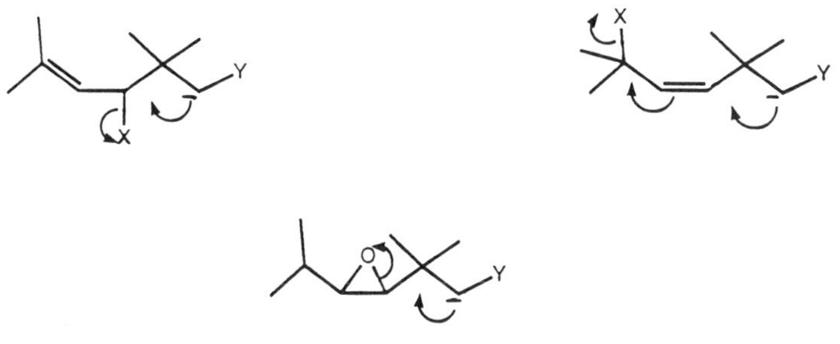

Y = Electron withdrawing group. X = Leaving group

Figure 8-25. C-1 to C-3 cyclizations to form chrysanthemate cyclopropane.

of the reaction have been described (De Vos *et al.*, 1978, De Vos and Krief, 1979c). More recently, 2-metallo-2-nitropropanes have been used as three carbon isopropylidene transfer agents in place of the phosphorane for dimethylcyclopropanation (Krief *et al.*, 1985), and Babler and Spina (1985) use 2-nitropropane in a tandem Michael reaction leading to chrysanthemic acid intermediates.

Formation of the C-1 to C-3 cyclopropane bond by anionic elimination is characteristic of a number of methods. It can take one of the three forms of Fig. 8-25, i.e., S_N2 elimination, S_N2' elimination, or epoxide opening. One of the most successful is sulfone elimination following a C_5+C_5 construction. Thus treatment of 3,3-dimethylacrylic ester with the prenyl sulfone (**45**) gives an isolable intermediate, which on treatment with sodium methoxide in dimethylformamide cyclizes to (±)-*trans*-chrysanthemate (Fig. 8-26) (Martel *et al.*, 1967; Martel and Huynh, 1967; Martel and Goffinet 1968). The prenyl sulfone can also be added to isopropylidene malonate in the presence of a cuprate, and the resulting 1,1-dimethoxycarbonylcyclopropane can be half hydrolyzed and decarboxylated to give a mixture of *cis*- and *trans*-chrysanthemates (Fig. 8-27) (Julia and Guy-Roualt, 1967). The Julia group has

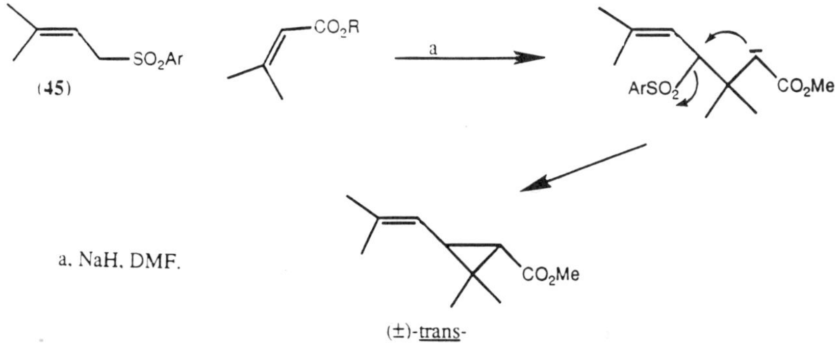

SO_2Ar (45) CO_2R a $ArSO_2$ CO_2Me

a. NaH, DMF. CO_2Me

(±)-trans-

Figure 8-26. Arylsulphone method for (±)-*trans*-chrysanthemic ester (Martel *et al.*, 1967; Martel and Huynh, 1967; Martel and Goffinet, 1968).

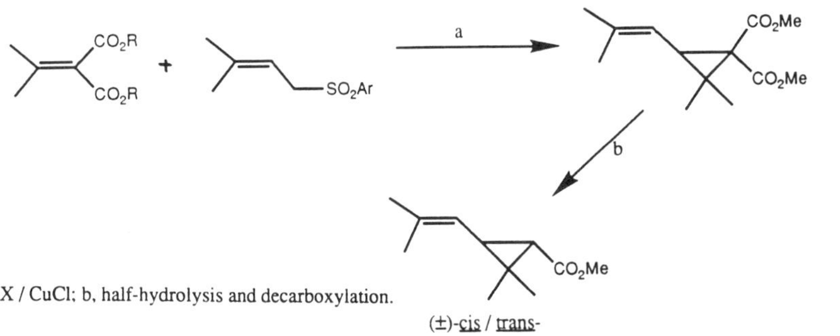

a, RMgX / CuCl; b, half-hydrolysis and decarboxylation.

(±)-cis / trans-

Figure 8-27. Synthesis using addition to isopropylidene malonate (Julia and Guy-Roualt, 1967).

made other important and pioneering contributions to the chrysanthemate synthesis area (e.g., Julia *et al.*, 1964a,c, 1965a,b).

Two syntheses which involve 1,3-anionic elimination set up their final reaction intermediates by means of Claisen rearrangements. In the first of these (Babler and Tortorello, 1976) (Fig. 8-28), the acetal (**46**) is formed from dimethylallyl alcohol and 1,1,1-triethoxyethane under acidic conditions. Claisen rearrangement of this product provides an olefin which is epoxidized by peracid and then treated with lithium diisopropylamide to form the cyclopropane ring with eliminative opening of the epoxide. The alcohol so formed is now oxidized to the aldehyde and synthesis is completed by a Wittig reaction. The pattern is

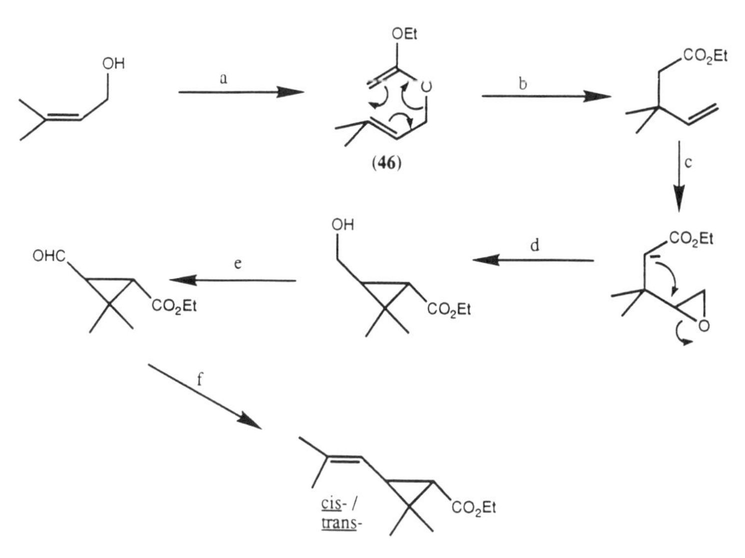

a, $MeCH(OEt)_3$ / H^+; b, Δ; c, m-chloroperbenzoic acid; d, lithium diisopropylamide; e, CrO_3 / pyridine; f, $Me_2C=PPh_3$.

Figure 8-28. Claisen rearrangement approach 1 (Babler and Tortorello, 1976).

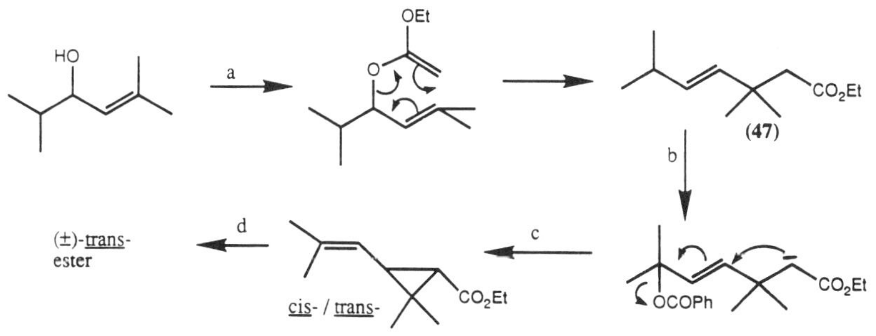

a, $MeC(OEt)_3$ / H^+; b, $PhCO_3Bu^t$ / Cu^+; c, Pr^i_2NLi; d, EtO^-.

Figure 8-29. Claisen rearrangement approach 2 (Ficini and d'Angelo, 1976; Ficini *et al.*, 1983).

$C_5+C_2+C_3$. The pattern of the second synthesis (Fig. 8-29) involving Claisen rearrangement is C_8+C_2 (Ficini and d'Angelo, 1976; Ficini *et al.*, 1983). The Claisen rearrangement is set up to form the ester (**47**) which is allylically substituted using *tert*-butyl perbenzoate. 1,3-Benzoyloxy elimination (S_N2'), base catalyzed by lithium diisopropylamide, yields (±)-*trans*-chrysanthemate.

Other ways of building up a suitable final intermediate for cyclopropane formation by 1,3-elimination have been studied. The approach of Garbers *et al.* (1978) is summarized in Fig. 8-30 and the attractive synthesis of lactone (**16**) from 2,2,5,5-tetramethylcyclohexane-1,4-dione (d'Angelo and Revial, 1983) is of a similar mechanistic type. The syntheses of (±)-*cis*-chrysanthemic acid by Genêt *et al.* (1986) (Fig. 8-31) and (±)-*cis*-/*trans*-chrysanthemic acid by Genêt *et al.* (1980) (Fig. 8-32) both involve the formation of the cyclopropane ring by S_N2' elimination. Majewski and Snieckus (1984) employ the epoxide opening variant of S_N2 attack (Fig. 8-33) and a second synthesis based on conjugate addition of 2-lithio-1,3-dithiane is also described by them. Babler and Invergo (1981) make a useful synthon for *cis*-chrysanthemic acid via the addition of cyanoacetic ester anion to 2-bromo-3,3-dimethylacrylic ester.

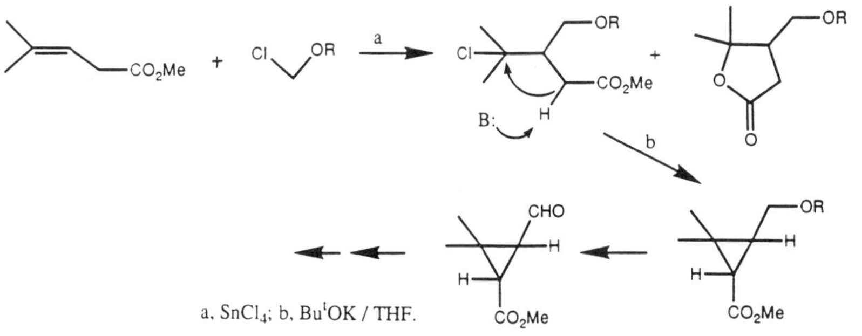

a, $SnCl_4$; b, Bu^tOK / THF.

Figure 8-30. (±)-*trans*-Chrysanthemic acid via chloromethyl ester synthesis (Garbers *et al.*, 1978).

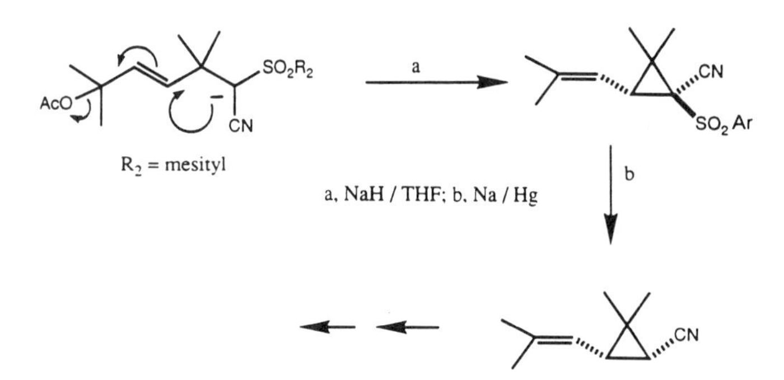

Figure 8-31. Synthesis of *cis*-chrysanthemates (Genêt *et al.*, 1986).

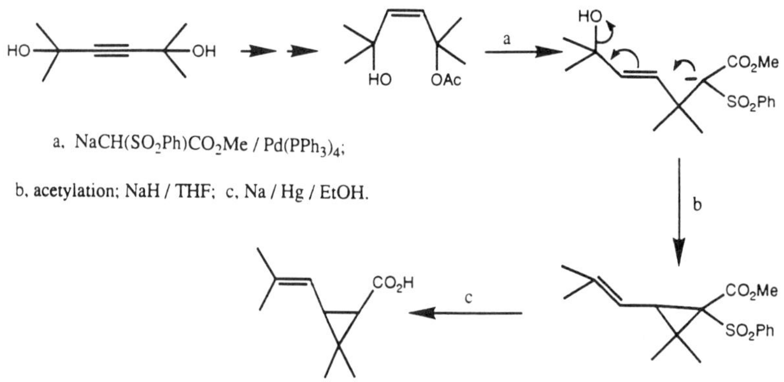

Figure 8-32. Synthesis of ($\pm$)-*cis*-/*trans*-chrysanthemic acid from tetramethylbutyndiol (Genêt, 1980).

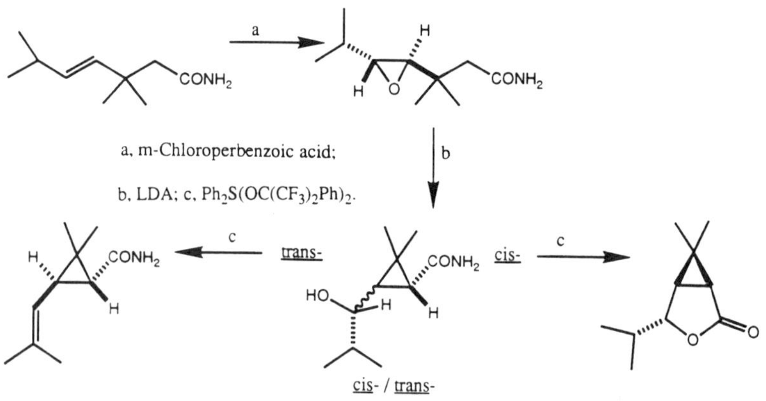

Figure 8-33. Epoxy-cyclization methods for chrysanthemates (Majewski and Snieckus, 1984).

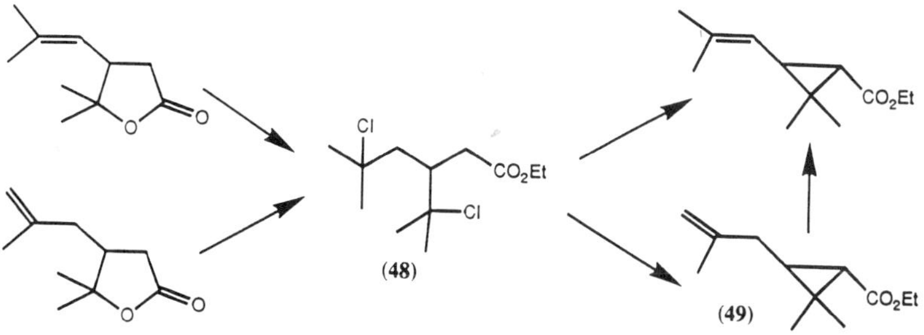

(48)

(49)

Figure 8-34. Pyrocin and isopyrocin routes to chrysanthemates (Matsui and Uchiyama, 1962; Julia *et al.*, 1964b, 1965b).

Several methods have been devised for the synthesis of pyrocin and its isomer (Julia *et al.*, 1962, 1964b; Ueda and Matsui, 1970a; Torii *et al.*, 1977; Takeda *et al.*, 1977; De Vos and Krief, 1978; Scott and Nkwelo, 1985) and these lactones can be converted into the dichloro compound (**48**) (Fig. 8-34) by treatment with thionyl chloride followed by ethanolic hydrogen chloride. The dichloride can be cyclized and dehydrohalogenated by sodium *tert*-amylate or sodium hydride to give (±)-*trans*-chrysanthemate and its double-bond isomer (**49**) (Matsui and Uchiyama, 1962; Julia *et al.*, 1962, 1964b). Saponification and isomerization of the latter by treatment with *p*-toluenesulfonic acid provides further (±)-*trans*-chrysanthemic acid.

Ring contraction of a cyclobutane ring to a cyclopropane by the Favorskii rearrangement, a variant of the 1,3-elimination theme, has been used to make chrysanthemic acid or its relatives (Greuter *et al.*, 1981; Martin *et al.*, 1981). Fig. 8-35 illustrates the approach of Brady *et al.* (1983) while that of Martin *et al.* (1981) involves cine-rearrangement (Fig. 8-36).

An interesting photochemical approach (Fig. 8-37) is provided by the di-π-methane rearrangement of the diene ester which leads to a 1:2 mixture of

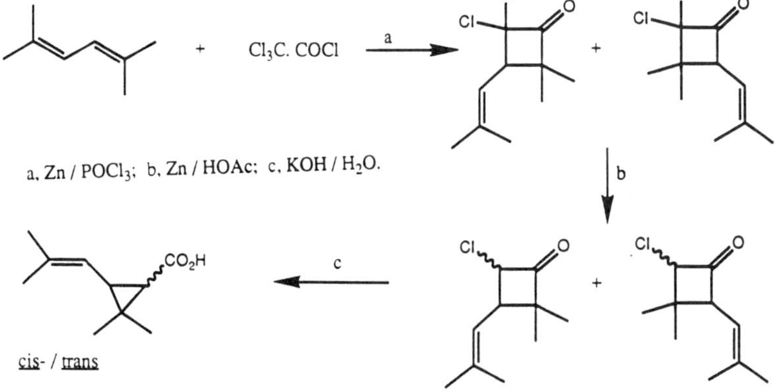

a. Zn / $POCl_3$; b. Zn / HOAc; c. KOH / H_2O.

cis- / trans

Figure 8-35. Chrysanthemic acids by Favorskii reactions (Brady *et al.*, 1983).

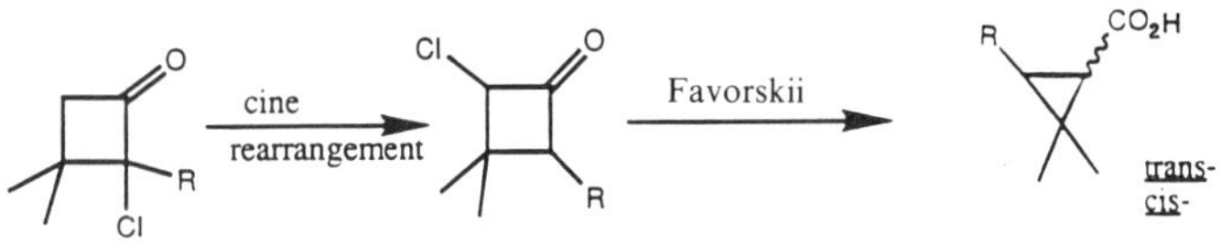

Figure 8-36. cine-Rearrangement in Favorskii approach (Martin *et al.*, 1981).

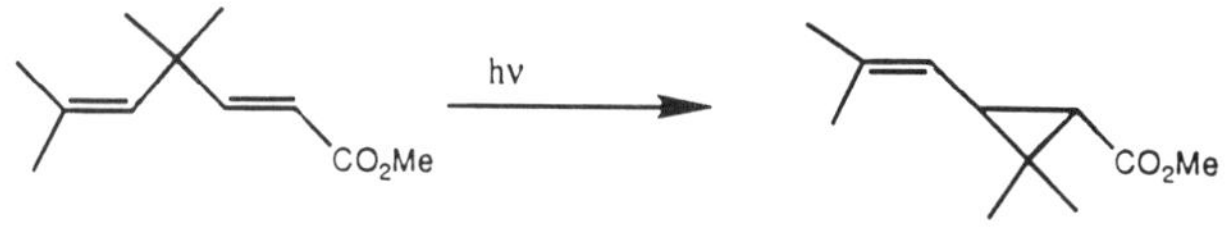

Figure 8-37. Application of the di-π-methane photochemical reaction.

methyl *cis*- and *trans*-chrysanthemates (Bullivant and Pattenden, 1976c; Baeckström, 1978), though yields are poor. The synthesis of Ohkata *et al.* (1978) (Fig. 8-38) also involves a photochemical step. Both the cyclohexanedione (**51**) (Krief *et al.*, 1988b) and the lactone (**52**) (Schmidt, 1985) have been employed as starting compounds for (±)-chrysanthemate synthesis (Fig. 8-39). De Vos and Krief (1979c) have described a useful synthesis of caronic half aldehyde (Fig. 8-40) whilst Cameron and Knight (1985), and Funk and Munger (1985), have applied the Claisen rearrangement of *O*-silyl enolates to the synthesis of *cis*-chrysanthemic acid (Fig. 8-41).

Synthesis of chrysanthemum dicarboxylic acid and its half methyl ester, pyrethric acid. The first synthesis of chrysanthemum dicarboxylic acid employed copper catalyzed addition of diazoacetic ester to 2,6-dimethylsorbic ester (Harper *et al.*, 1954; Crombie *et al.*, 1957; Matsui *et al.*, 1957). The product is a mixture of (±)-(*E*)-*trans*- and (±)-(*E*)-*cis*-isomers which can be separated and optically resolved. A detailed revision of the claims of Takei *et al.* (1958), who added diazopropane to geometrical isomers of the muconic esters, has been made by Scharf and Mattay (1978). Pyrethric acid can be formed by half hydrolysis of the dimethyl ester of chrysanthemum dicarboxylic acid (Crombie *et al.*, 1957), and can be purified as the quinine salt (Matsui and Meguro, 1963). A satisfactory method for obtaining pyrethric acid chloride from natural sources has been described (Elliott and Janes, 1969) and is convenient for making rethrins II by resynthesis.

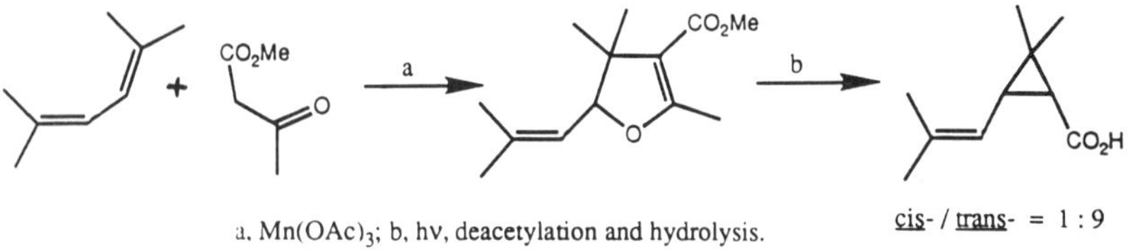

a. $Mn(OAc)_3$; b. hv, deacetylation and hydrolysis.

cis- / trans- = 1 : 9

Figure 8-38. Chrysanthemic acid by dihydrofuran photochemical reaction (Ohkata *et al.*, 1978).

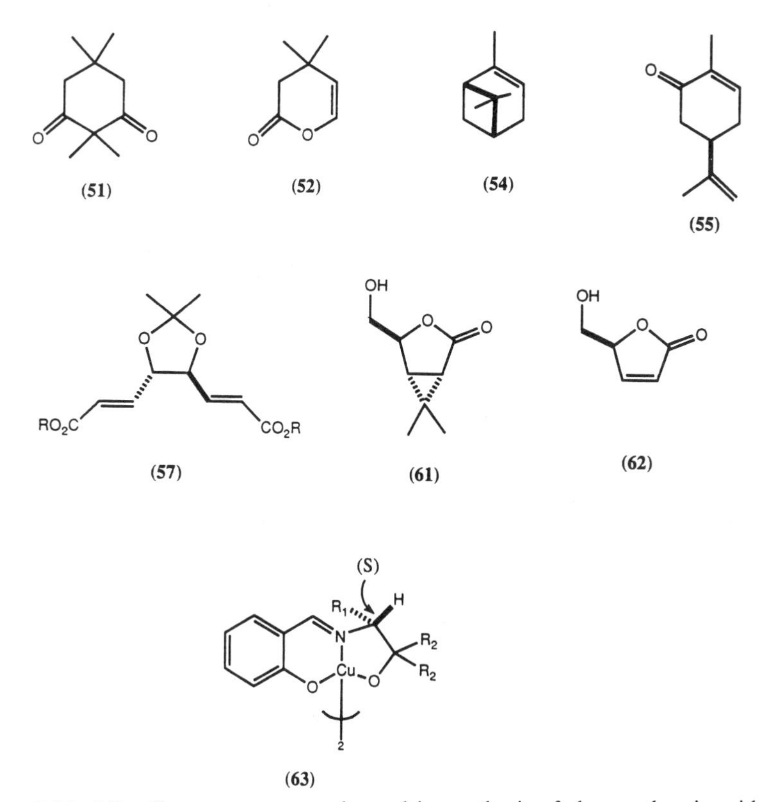

Figure 8-39. Miscellaneous compounds used in synthesis of chrysanthemic acids.

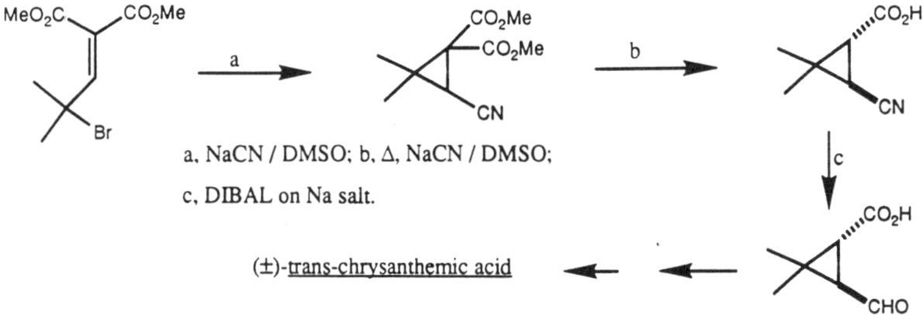

MeO_2C CO_2Me Br → a → CO_2Me CO_2Me CN → b → CO_2H CN → c → CO_2H CHO → → (±)-trans-chrysanthemic acid

a, NaCN / DMSO; b, Δ, NaCN / DMSO;
c, DIBAL on Na salt.

Figure 8-40. Synthesis of caronic hemialdehyde (De Vos and Krief, 1979c).

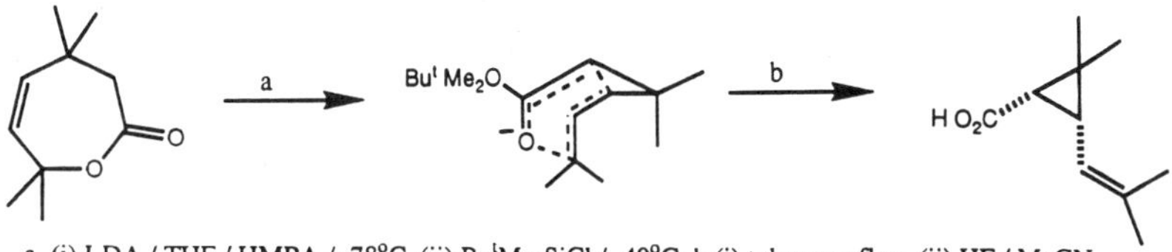

→ a → Bu^tMe_2O → b → HO_2C

a, (i) LDA / THF / HMPA / -78°C, (ii) Bu^tMe_2SiCl / -40°C; b (i) toluene reflux, (ii) HF / MeCN.

Figure 8-41. (±)-*cis*-Chrysanthemic acid by Claisen enolate rearrangement (Cameron and Knight, 1985; Funk and Munger, 1985).

With the increased availability of chrysanthemic acid in its various stereochemical forms it has become convenient to use this as a precursor. Oxidation of a chrysanthemic ester with selenium dioxide introduces an (*E*)-aldehyde function, oxidizable with silver oxide to a carboxylic acid. The latter can be converted into its methyl ester and if the original chrysanthemate was the *tert*-butyl ester, the *tert*-butyl group can be eliminated by pyrolysis to give pyrethric acid (Matsui and Yamada, 1963, 1965). A milder procedure consists in removal of the side-chain of the chrysanthemate either by osmium tetroxide/periodate (Crombie *et al.*, 1970) or by ozonolysis (Martel, 1969; Martel and Buendia, 1970b) to give caronic half-aldehyde. The latter undergoes Wittig type reactions and either chrysanthemum dicarboxylic acid or pyrethric acid are easily accessible in this way (Fig. 8-42). Synthesis of four geometric isomers of (±)-pyrethric acid has been reported by Ueda and Matsui (1970b).

Franck-Neumann *et al.* (1987a,b) describe an unusual synthesis of *cis*-pyrethric acid using the photodegradation of a gem-dimethylpyrazolenine (**50**) to a cyclopropene, followed by *cis*-hydrogenation (Fig. 8-43). Dimethylcyclopropenes have also attracted other investigators. Addition of isobutenyl magnesium bromide to dimethylcyclopropene itself gives an anion which can be directly carboxylated by carbon dioxide to give *cis*-chrysanthemic acid (Lehmkuhl and Mehler, 1978, 1982; Nesmeyanova *et al.*, 1982).

Optical resolution and syntheses of chiral chrysanthemic acids. The original optical resolution of the chrysanthemic acids was carried out by classical crystallization of diastereoisomers, forming salts with quinine and α-phenylethylamine (Campbell and Harper, 1952); since that time a number of other amines have been employed (Goffinet and Locatelli, 1968; Matsui and Horiuchi, 1971; Ueda and Suzuki, 1971; Horiuchi and Matsui, 1973; Nohira *et al.*, 1988). Control of optical resolutions can be effected gas chromatographically [using esters with (+)- or (−)-octanol] (Murano, 1972), or by NMR.

Synthesis of the chrysanthemic acids in optically active form may be by the direct utilization of the pool of natural chirality, or by the chemical utilization of a chiral auxiliary substance, or by enzymic means. Thus (+)-car-3-ene (**53**) (Fig. 8-44) is a cheap and readily available terpene containing a chiral cyclopropane residue, so its use as a chrysanthemate precursor has been much

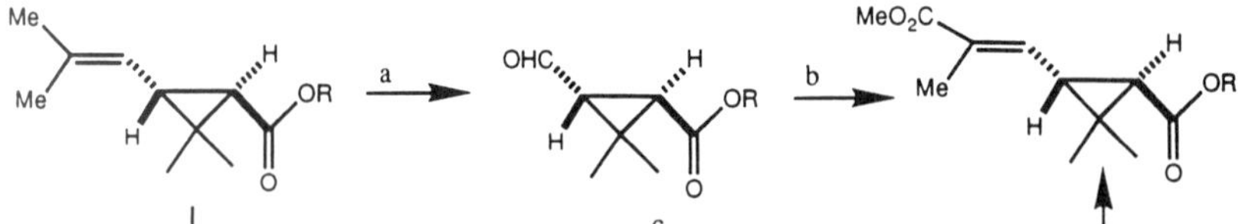

a, O_3 or OsO_4 / IO_4^-; b, $Ph_3P{=}C(Me)CO_2Me$; c, on t-butyl ester (i) SeO_2, (ii) Ag_2O, (iii) CH_2N_2, then Δ for pyrethric acid, R = H.

Figure 8-42. Chrysanthemum dicarboxylic acid derivatives from chrysanthemum monocarboxylic acid derivatives (Crombie *et al.*, 1970; Martel and Buendia, 1970; Matsui and Yamada, 1963, 1965).

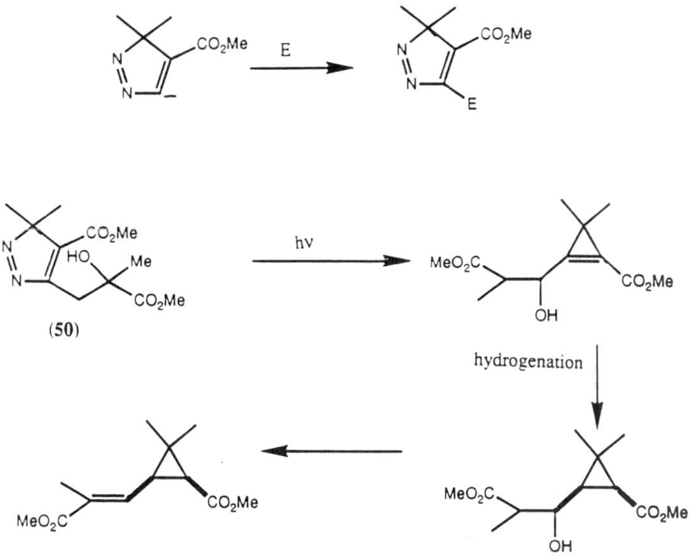

Figure 8-43. *cis*-Pyrethric acid from gem-pyrazolenine synthesis (Franck-Neumann *et al.*, 1987a, b).

studied. One such reconstructive synthesis leading to (+)-*trans*-chrysanthemic acid is summarized in Fig. 8-44 (Sobti and Sukh Dev, 1974). There have been many other schemes directed to the utilization of this terpene for chrysanthemate synthesis (*inter alia*, Matsui *et al.*, 1965, 1967; Cocker *et al.*, 1975; Gopichand *et al.*, 1975; Khanra and Mitra, 1976; Bhat *et al.*, 1981; Ho, 1983; Naik and

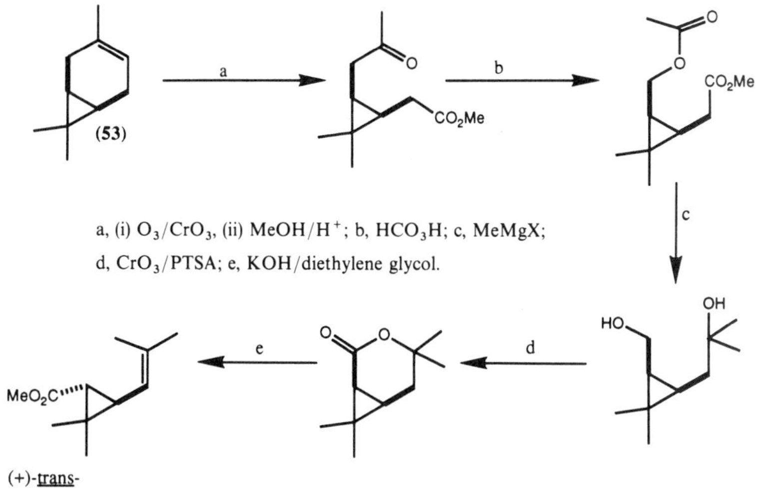

a, (i) O_3/CrO_3, (ii) $MeOH/H^+$; b, HCO_3H; c, MeMgX; d, CrO_3/PTSA; e, KOH/diethylene glycol.

Figure 8-44. Synthesis of (+)-*trans*-chrysanthemic acid from (+)-car-3-ene (Sobti and Sukh Dev, 1974).

Kulkarni, 1983; Bhosale *et al.*, 1985, 1988; Mitra *et al.*, 1987; Joshi and Kulkarni, 1988; Lochynski *et al.*,1988; Bakshi *et al.*, 1989; Rukavishnikov *et al.*, 1989; Shaha *et al.*, 1989; Dhillon *et al.*, 1991, etc.). Unusually, Inokuchi *et al.* (1988) have used electrochemical processes involving selenation-deselenation and anodic decarboxylation to convert (+)-car-3-ene into methyl (+)-*cis*-chrysanthemate.

Other natural terpenes employed to make chiral chrysanthemic acids are (+)-α-pinene (**54**) (Fig. 8-39) (Mitra and Khanra, 1977) and the natural terpene carvone (**55**) (Ho and Din, 1982; Torii *et al.*, 1983). Eucarvone (**56**) is a terpenoid readily accessible from carvone; it does not contain a cyclopropane ring but can be induced to form one and this has led to a synthesis of (±)-*trans*-chrysanthemic acid (Fig. 8-45) (Welch and Valdes, 1977).

Sources of chirality other than natural terpenes have been used. Yadav *et al.* (1989) used (*R*,*R*)-tartaric acid as the chiral source in their synthesis of (1*R*)-(+)-*cis*-chrysanthemic acid (Fig. 8-46), whilst carbohydrate chirality is incorporated by "transcription" into (+)-*trans*- and (−)-*trans*-pyrethric acids using an ingenious, if circuitous, route (Fig. 8-47) (Fitzsimmons and Fraser-Reid, 1979, 1984). Stereoselective syntheses of (1*R*)-*trans*- and *cis*-hemicaronaldehydes, available for chrysanthemate syntheses, originate via (**57**) (Fig. 8-39) from natural tartaric acid as the chiral source (Krief *et al.*, 1989, 1988a)

(2*R*)-(−)-Pantolactone (**58**) is readily available and its chirality is built into (1*S*,3*S*)-(−)-*trans*-chrysanthemic acid via a chiral epoxynitrile (**59**); the (2*S*)-(+)-lactone gives (1*R*,3*R*)-(+)-*trans*-chrysanthemic acid (Matsuo *et al.*, 1976) (Fig. 8-48). Franck-Neumann *et al.* (1985) use the chiral lactone (**60**) to synthesize the two *cis*-methyl chrysanthemates in high optical purity (Fig. 8-49). Mann and Thomas (1986) have described the use of the lactone (**61**) (Fig. 8-39) in *cis*-chrysanthemic acid synthesis; it was made by addition of diazopropane

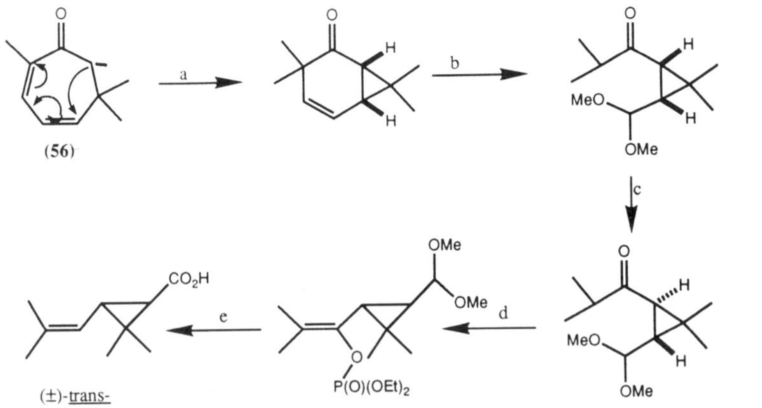

a, (i) $NaNH_2$ / MeI, (ii) OsO_4 / $NaClO_3$ / t-BuOH (to decompose an unwanted isomer); b, O_3 / MeOH / H^+; c, $KOBu^t$; d, lithium diisopropylamide / diethyl phosphorochloridate ; e, (i) lithium / diethylamine, (ii) CrO_3 / H^+.

Figure 8-45. Synthesis of (±)-*trans*-chrysanthemic acid from eucarvone (Welch and Valdes, 1977).

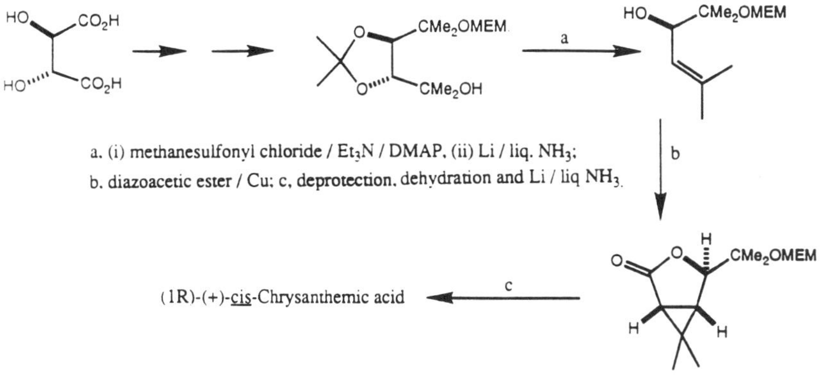

HO
CO_2H
HO
CO_2H
O
CMe_2OMEM
O
CMe_2OH
a
HO
CMe_2OMEM
a. (i) methanesulfonyl chloride / Et_3N / DMAP, (ii) Li / liq. NH_3;
b. diazoacetic ester / Cu; c, deprotection, dehydration and Li / liq NH_3.
b
H
O
O
CMe_2OMEM
H
H
(1R)-(+)-cis-Chrysanthemic acid
c

Figure 8-46. Synthesis of (1*R*)-(+)-*cis*-chrysanthemic acid from (*R*, *R*)-tartaric acid (Yadav *et al.*, 1989).

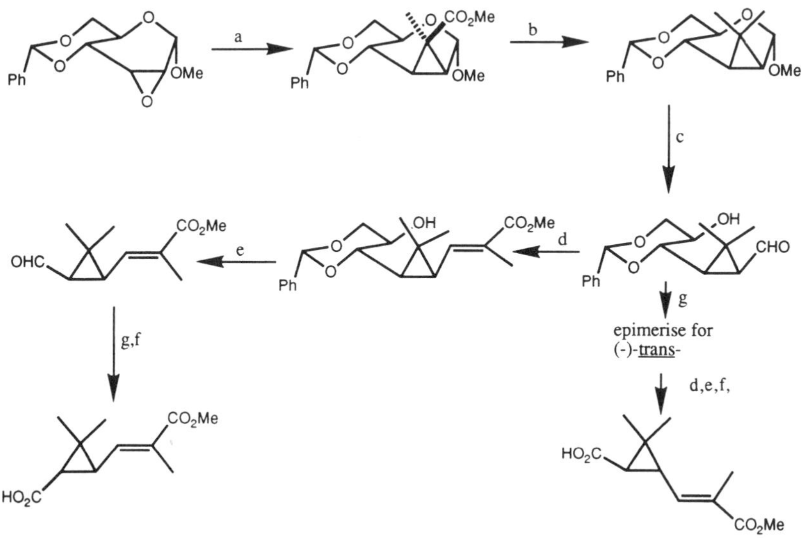

a
b
c
d
e
g
g,f
epimerise for
(-)-trans-
d,e,f,
Ph
OMe
CO_2Me
OH
CHO
OHC
HO_2C

a, $(EtO)_2P(O).CH(Me)CO_2Et$; b, (i) $LiAlH_4$, (ii) $MeSO_2Cl$/DMF, (iii) $LiAlH_4$; c, H_2O/dioxan
d, $Ph_3P{=}C(Me)CO_2Me$; e, (i) MeOH/TsOH, (ii) IO_4^-; f, Ag_2O/NaOH; g, NaOMe.

Figure 8-47. Synthesis of (+)-*trans*- and (−)-*trans*-pyrethric acids using annulated pyranosides (Fitzsimmons and Fraser-Reid, 1984).

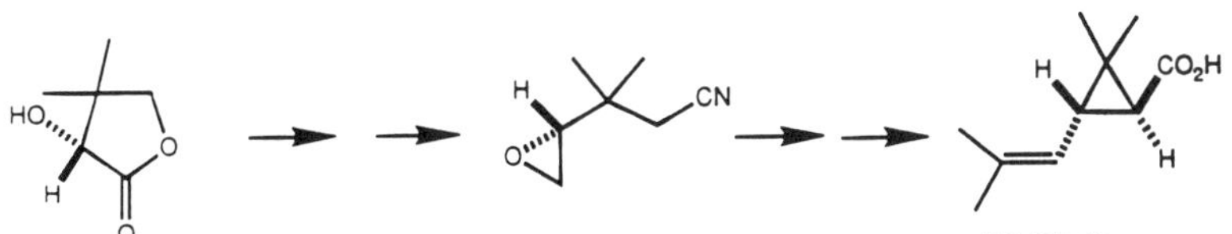

HO
H
O
O
H
O
CN
H
CO_2H
H

(58) **(59)** **(1S,3S)-(-)-trans-chrysanthemic acid**

Figure 8-48. Optically active *trans*-chrysanthemic acid from optically active pantolactone (Matsuo *et al.*, 1976).

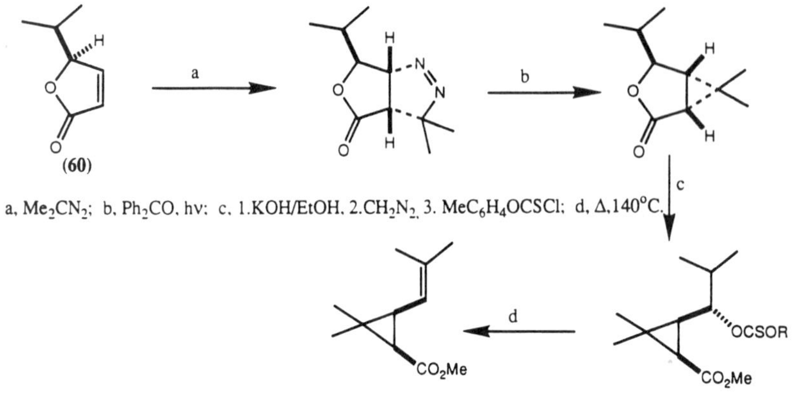

(60)

a, Me_2CN_2; b, Ph_2CO, hv; c, 1.KOH/EtOH, 2.CH_2N_2, 3. MeC_6H_4OCSCl; d, Δ,140°C.

Figure 8-49. Synthesis of optically active *cis*-chrysanthemate (Franck-Neumann *et al.*, 1985).

to the butenolide (**62**) (Fig 8-39), followed by photochemical extrusion of nitrogen from the resulting pyrazole (see also Mann and Weymouth-Wilson, 1991). Takano *et al.* (1985) use 1,3-elimination to make (1*S*,3*R*)-*cis*-chrysanthemol, employing a lactone whose chirality originates from (*S*)-*O*-benzylglycidol.

Asymmetric induction has received attention. Thus l-menthyl diazoacetate in the presence of a chiral copper complex (**63**) (Fig. 8-39) adds to 2,5-dimethylhexadiene to produce *trans-/cis*-chrysanthemate (89:11). The *trans*-material had a (1*R*)-enantiomeric excess of 87%, the *cis*- of 25% (Aratani *et al.*, 1975, 1977). The enantiomeric excess appears not to be affected by the chirality of the alkyl diazoacetate, but the bulkier the alkyl group, the higher the *trans*-content and the higher the enantiomeric excess. A copper carbenoid is thought to be involved. The use of another mixed ligand (copper) complex

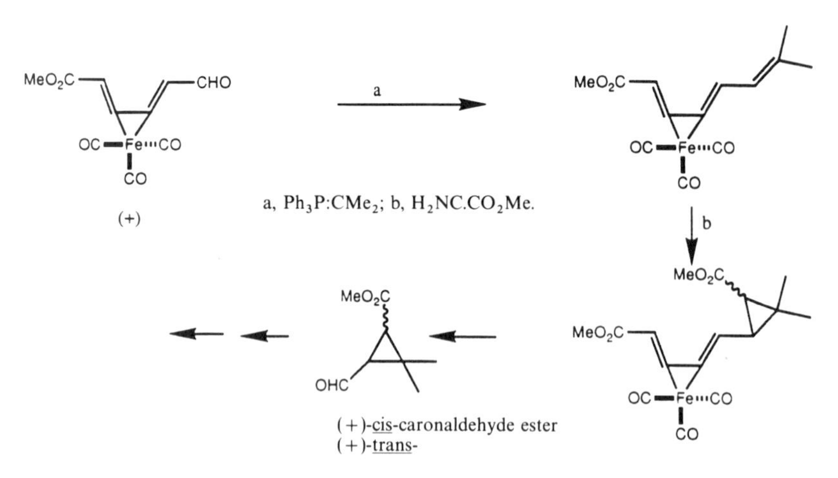

Figure 8-50. Optically active hemicaronaldehydes using a chiral butadiene-iron tricarbonyl complex (Monpert *et al.*, 1983).

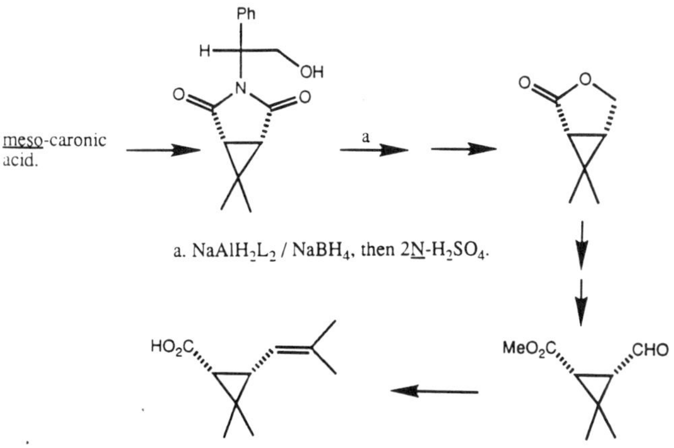

Figure 8-51. Asymmetric synthesis of (1*R*, 3*S*)-(+)-*cis*-chrysanthemic acid (Mukaiyama *et al.*, 1983).

has been reported (Zang and Chow, 1991), whilst Lowenthal and Masamune (1991) have developed chiral bis-oxazoline copper catalysts giving excellent enantioselectivity for *trans*-chrysanthemic ester (94% e.e.). Monpert *et al.* (1983) use a chiral butadiene-iron tricarbonyl complex to make optically active caronaldehydes (Fig. 8-50) and chiral iron complexes have also been employed by Franck-Neumann *et al.*, (1982); Mukaiyama *et al.* (1983) make (1*R*,3*S*)-(+)-*cis*-chrysanthemic acid via the half aldehyde of caronic ester using a chiral amino-alcohol as a chiral auxillary (Fig. 8-51).

De Vos and Krief (1983) have examined the preparation of chiral caronic esters, convertible into chrysanthemic acid, by addition of isopropylidene triphenylphosphorane to di-(*d*)- and di-(*l*)-menthyl fumarate. d'Angelo *et al.* (1986) have turned their synthetic approach (mentioned above) into a synthesis of (1*R*,3*S*)-*cis*-chrysanthemic acid through microbiological reduction of 2,2,5,5-tetramethylcyclohexanedione. A biochemical preparation of optically active chrysanthemic acid has been reported by Mitsuta *et al.* (1985).

III. THE NATURAL RETHROLONES

The three natural rethrolones pyrethrolone (**10**), cinerolone (**11**) and jasmolone (**12**) all belong to the 4(*S*)-configurational series (Begley *et al.*, 1972, 1974) and all have (*Z*)-side chain geometries, as shown originally by IR data and by synthesis (Crombie and Harper 1949, 1950, 1952; Crombie *et al.*,1951b, 1956b). Attention was drawn earlier to their tendency to undergo elimination to form cyclopentadienones which then dimerize (Elliott *et al.*, 1967). All have the usual reactions expected of cyclopentenones, but of the three, pyrethrolone is much the least stable, and has an interesting chemistry. Treated with sodium methoxide in hot methanol it undergoes oxidation and reduction reactions and two cyclopentenediones (**64**) and (**65**) and a cyclopentanedione (**66**) have been identified (Elliott, 1965) (Fig. 8-52). Synthetic allethrolone, often used as a rethrolone model, under similar conditions gives only a cyclopentanedione

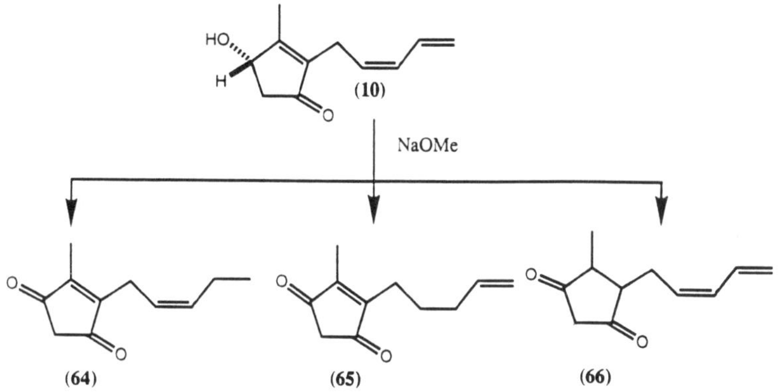

(10)

NaOMe

(64) (65) (66)

Figure 8-52. Formation of the pyrethrolone enols (Elliott, 1965).

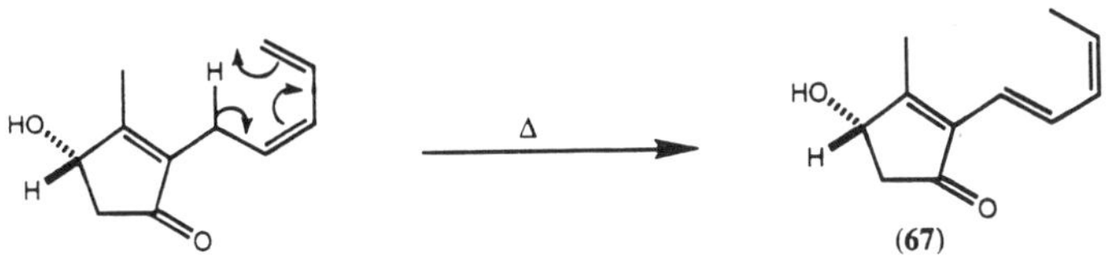

Δ

(67)

Figure 8-53. Thermal sigmatropic rearrangement of pyrethrolone (Elliott, 1964a).

through double bond isomerization with ketonization and there is no hydrogen transfer. Under acid conditions allethrolone is isomerized to 4-methyl-5-(prop-2-enyl)cyclopent-4-ene-1,3-dione (Pattenden and Storer, 1973a, 1974b).

On thermal treatment pyrethrolone undergoes a sigmatropic shift (Fig. 8-53) giving the cross-conjugated triene ketone **(67)** (Elliott, 1964a). The side-chain methylene of pyrethrolone, flanked as it is by double bonds, is a very reactive center. When treated with manganese dioxide, besides the usual oxidation of an allylic alcohol to its ketone, a radical forms at this center (Fig. 8-54) and, being delocalized over the diene, two molecules couple in the least sterically crowded position to give **(68)** as the reaction product (Elliott *et al.*, 1969; Crombie *et al.*, 1971a).

MnO_2

(68)

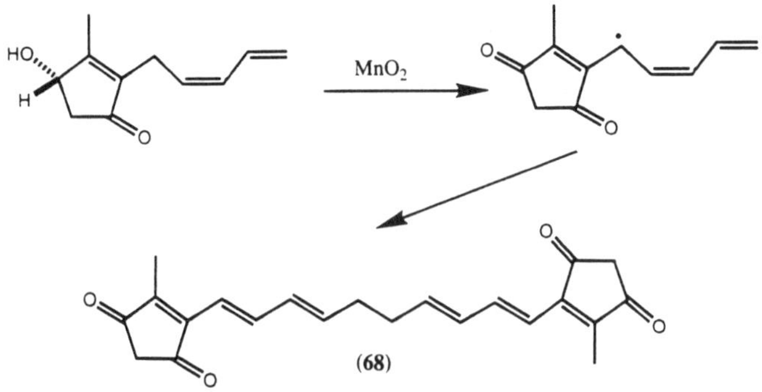

Figure 8-54. Oxidation and coupling of pyrethrolone by manganese dioxide (Elliott *et al.*, 1969; Crombie *et al.*, 1971a).

Epoxides and hydroxy derivatives of the pyrethrin ketols are of interest in metabolic studies. The epoxidation of pyrethrin II with *m*-chloroperoxybenzoic acid yields a mixture of ketol side-chain epoxides which can be hydrolyzed to the corresponding diols. The 4′,5′-diol is the major metabolite of both pyrethrin I and II (Ando *et al.*, 1991a, 1991b).

A. Photochemistry

The pyrethrins themselves, particularly pyrethrin I, undergo photochemical oxidation, their pyrethrolone component being particularly affected (Chen and Casida, 1969; Bullivant and Pattenden, 1976a). Ultraviolet irradiation of jasmolone causes stereomutation of the side-chain double bond leading to a steady-state mixture of 90% (*E*)-olefin:10% (*Z*)-isomer. Allethrolone and (*S*)-bioallethrin however behave differently, isomerizing to a cyclopropyl side-chain (**69**) (Fig. 8-55) (Bullivant and Pattenden 1972, 1976b; Ando *et al.*, 1983).

B. Synthesis of the Rethrolones

Synthesis involves construction of a 4-hydroxycyclopentenone ring system (**70**) and the critical cyclization could involve one of five bonds A to E (Fig. 8-56). The most used method involves formation of bond A, but other interesting approaches have been employed. Introduction of the R group in (**70**), when R is alkyl, or alk-2-enyl as in cinerolone, jasmolone or the nonnatural allethrolone, causes no unusual problems, but introduction of the unstable penta-2′(*Z*),4′-dienyl side-chain presents more difficulty and will be considered separately.

The diastereoisomeric pairs of esters from synthetic (*RS*)-rethrolones with (−)-3,3,3-trifluoro-2-methoxy-2-phenylpropionic acid are separable by GLC (Rickett and Henry, 1974) and are suitable for the assessment of chiral purity.

Hunsdiecker (1947) studied in some detail the base catalyzed cyclization of 1,4-diketones (a specific variant of the aldol condensation) to form cyclopentenones, and this, and the condensation of substituted acetoacetic acids with pyruvaldehyde (Schechter *et al.*, 1949) to give hydroxy 1,4-diketones,

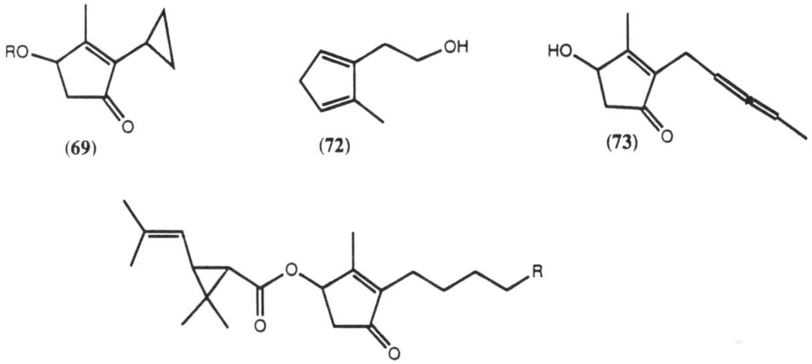

(69) **(72)** **(73)**

(74) R = H
(75) R = Me

Figure 8-55. Miscellaneous compounds.

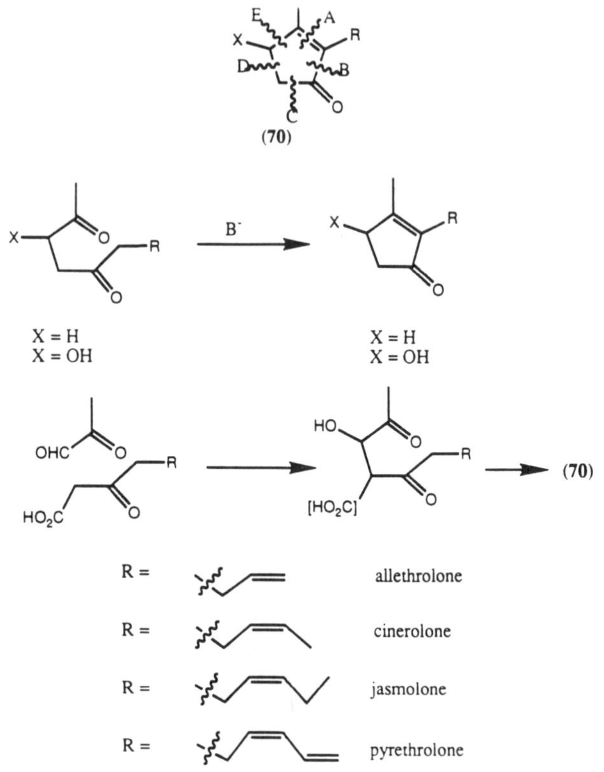

(70)

X = H
X = OH

X = H
X = OH

R = allethrolone
R = cinerolone
R = jasmolone
R = pyrethrolone

Figure 8-56. Synthesis of cyclopentenones and cyclopentenolones.

formed a foundation for the synthesis of rethrones and rethrolones. It was in this way that (±)-cinerolone (Crombie and Harper, 1949, 1950; Crombie *et al.*, 1950a, 1951b; LaForge and Green, 1952; Schechter *et al.*, 1952), (±)-jasmolone (Crombie *et al.*, 1951b), and (±)-pyrethrolone (Crombie *et al.* 1951c, 1956a,b) were synthesized for the first time (Fig. 8-56), ring construction involving closing of bond A (Fig. 8-56). Various analogues have also been synthesized by this method and Sato *et al.* (1981) have used a similar scheme starting from (*Z*)-hept-4-enoic acid.

Ficini's synthesis of cinerolone and jasmolone (Ficini and Genêt, 1975a, 1975b) (Fig. 8-57) uses the same methodology for cyclopentenone formation, as does that of Woessner and Ellison (1972) (Fig. 8-58), though very different methods are employed to form the 1,4-diketone. Ficini ozonizes a cyclobutene (**71**), using the much greater resistance of an acetylene linkage to ozonolysis as compared to an olefin. Woessner and Ellison apply dithiane chemistry to the problem. Curran (1983) makes the hydroxydiketone for dihydrocinerolone synthesis via isoxazoline synthesis (Fig. 8-59).

These methods involve final formation of bond A (Fig. 8-56), but a modification of the pyruvaldehyde method introduced by Martel *et al.* (1966) ensures formation of the final bond D by blocking the reactive aldehyde function

a, O_2; b, (i) Pd / H_2, (ii) NaOH.

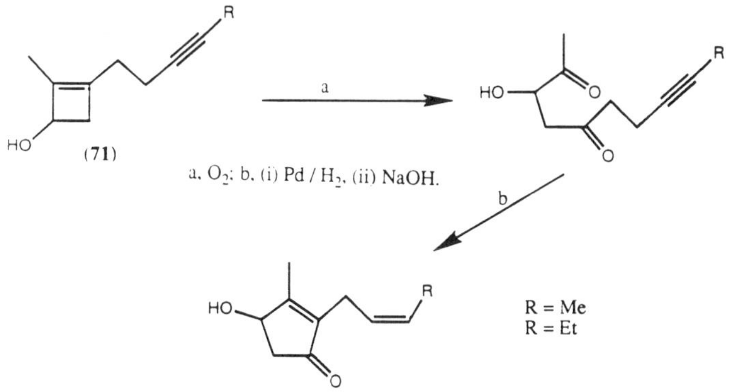

R = Me
R = Et

Figure 8-57. Cinerolone and jasmolone by cyclobutene synthesis (Ficini and Genêt, 1975a, b).

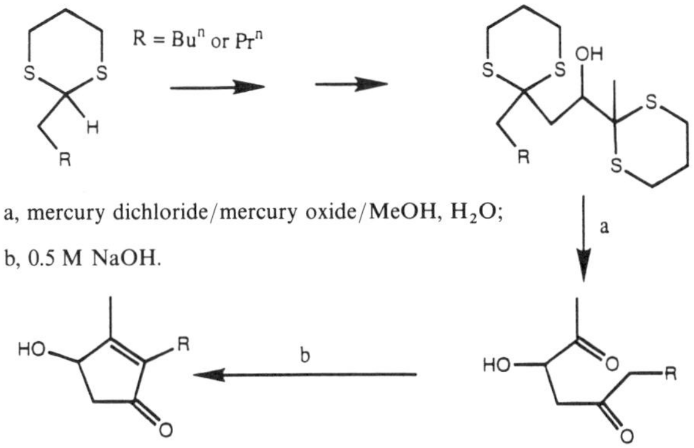

R = Bu^n or Pr^n

a, mercury dichloride/mercury oxide/MeOH, H_2O;

b, 0.5 M NaOH.

Figure 8-58. Rethrolones by dithiane chemistry (Woessner and Ellison, 1972).

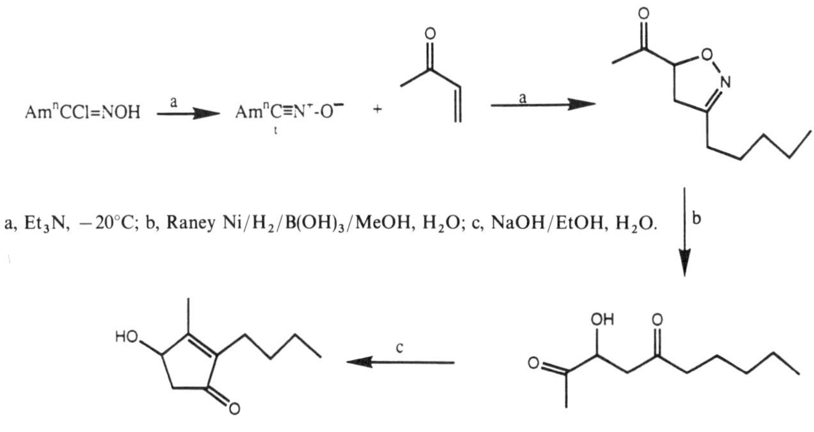

$Am^nCCl{=}NOH \xrightarrow{a} Am^nC{\equiv}N^+{-}O^-$

a, Et_3N, $-20°C$; b, Raney Ni/H_2/$B(OH)_3$/MeOH, H_2O; c, NaOH/EtOH, H_2O.

Figure 8-59. Dihydrocinerolone by Δ^2-isoxazoline synthesis (Curran, 1983).

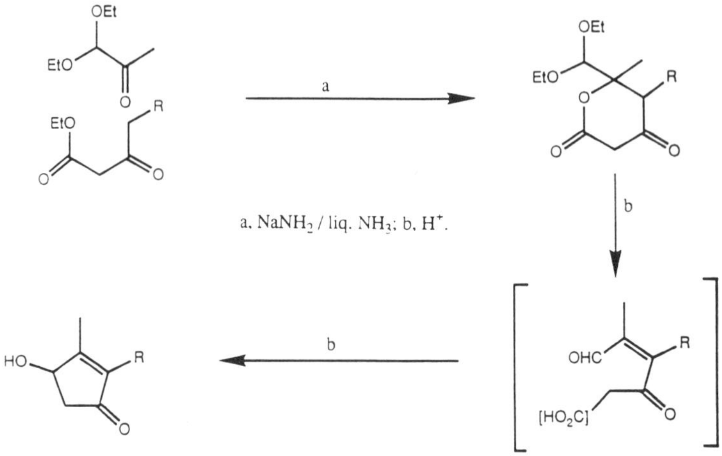

a, $NaNH_2$ / liq. NH_3; b, H^+.

Figure 8-60. Modified pyruvaldehyde approach (Martel *et al.*, 1966).

of the pyruvaldehyde as an acetal (Fig. 8-60). Büchi's synthesis (Büchi *et al.*, 1971) contains a similar bond D closure for the concluding step, but the aldehyde is first formed protected as an unusual chloroepoxide (Fig. 8-61). Romanet and Schlessinger's synthesis (1974), an essay in the use of sulphur chemistry, involves a similar final approach, with the aldehyde appearing protected as a thioacetal monoxide (Fig. 8-62). A synthesis of dihydrojasmolone employing a sulphur-containing vinyl phosphonium compound (Cameron *et al.*, 1984) is shown in Fig. 8-63 and bears some similarity to Kawamoto's synthesis.

The Kawamoto synthesis (1975) (Fig. 8-64) uses the Wittig reaction in a ring closure context forming the final bond as bond E, the synthesis involving an allylic transposition of the hydroxyl and the double bond. Vandewalle and

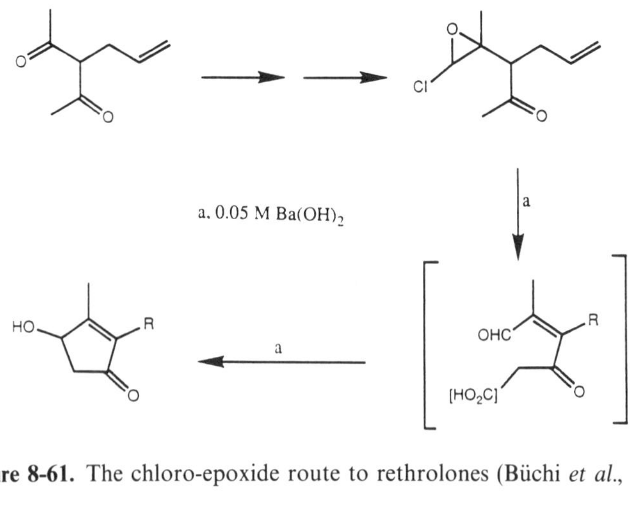

a, 0.05 M $Ba(OH)_2$

Figure 8-61. The chloro-epoxide route to rethrolones (Büchi *et al.*, 1971).

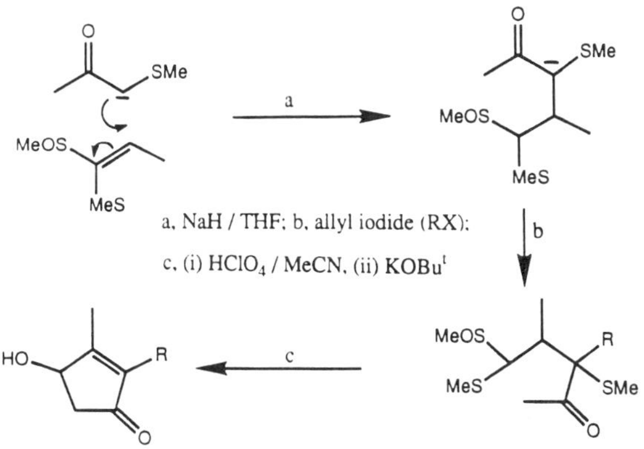

a, NaH / THF; b, allyl iodide (RX); c, (i) $HClO_4$ / MeCN, (ii) $KOBu^t$

Figure 8-62. Rethrolone (allethrolone) synthesis using sulfur activation (Romanet and Schlessinger, 1974).

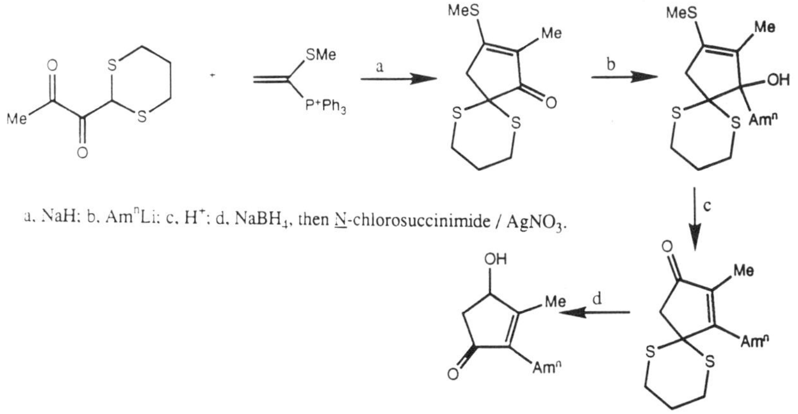

a, NaH; b, Am^nLi; c, H^+; d, $NaBH_4$, then N-chlorosuccinimide / $AgNO_3$.

Figure 8-63. Synthesis of dihydrojasmolone using a vinylphosphonium salt (Cameron *et* 11*al.*, 1984).

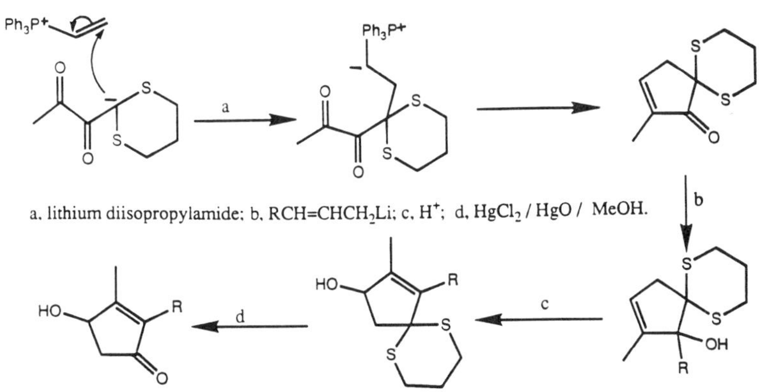

a, lithium diisopropylamide; b, $RCH{=}CHCH_2Li$; c, H^+; d, $HgCl_2$ / HgO / MeOH.

Figure 8-64. Rethrolone (allethrolone) cyclization using Wittig reaction (Kawamoto *et al.*, 1975).

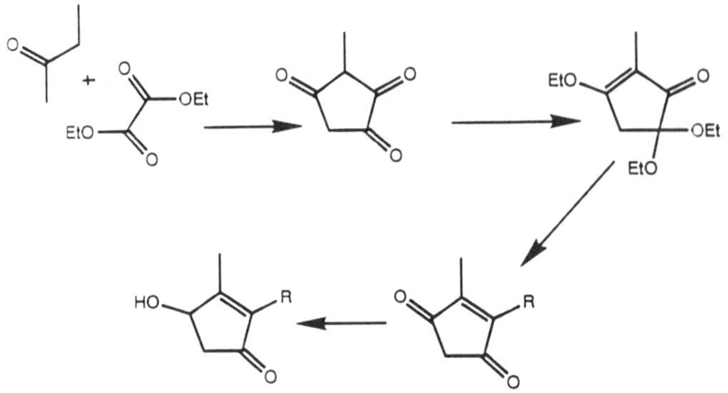

Figure 8-65. Rethrolone synthesis (e.g., allethrolone, dihydrojasmolone) from methyl cyclopentanetrione (Vandewalle and Madeleyn, 1970).

Madeleyn (1970) and Madeleyn and Vandewalle (1973) construct a five-membered ring, a trione, as a first step, making bonds A and C together (Fig. 8-65). The remainder of the synthesis now consists of modification to achieve the rethrolone structure. The approach of Szekléy *et al.* (1980), which leads to chiral syntheses since it uses an optically resolved prostaglandin precursor, also involves a preformed cyclopentane (Fig. 8-66), as does that of Fujisawa *et al.* (1976) who start out from a dialkylated cyclopentadiene (**72**) (Fig. 8-55). Shono *et al.* (1976) have described a general synthesis of 4-hydroxycyclopentenones from furan derivatives which extends earlier work on rethrone synthesis (Crombie *et al.*, 1969b) to rethrolones by an ingenious electrolytic oxidation step. Takahashi *et al.* (1981) report a synthesis of (±)-jasmolone; this employs dialkenyl cuprate addition to a 2-methylenecyclopentanone to insert the side-chain.

Introduction of the side-chains. Natural cinerolone, jasmolone, and pyrethrolone require *cis*-(*Z*)-side-chains and two main methods have been employed. The first is the semihydrogenation of an acetylene linkage, usually by catalytic hydrogenation over a supported palladium catalyst (sometimes with poisoning, e.g., Lindlar catalyst). This is very satisfactory in the case of an isolated acetylene linkage (cinerolone, jasmolone) (Crombie and Harper, 1950. 1952), but in the case of an acetylene conjugated to a terminal ene, as

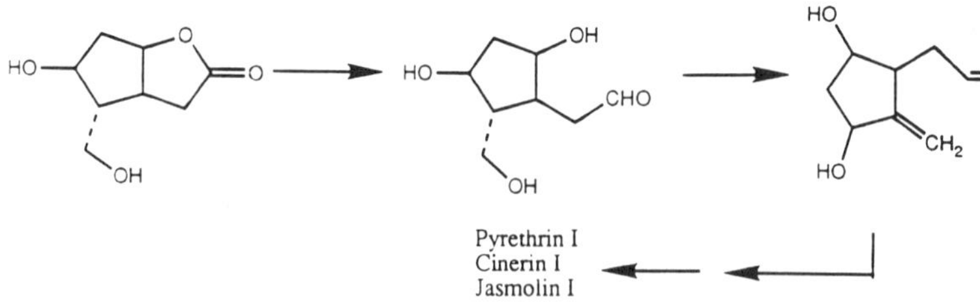

Pyrethrin I
Cinerin I
Jasmolin I

Figure 8-66. Synthesis of chiral pyrethrins from a prostaglandin intermediate (Szekléy *et al.*, 1980).

required for pyrethrolone, the tendency for 1,2-reduction of the exposed olefin linkage, along with 1,4-reduction, gives impure *cis*-product. Nonetheless, the first synthesis of (±)-pyrethrolone was successfully effected in this way (Crombie *et al.*, 1956a,b). More recently, Matsuo *et al.* (1993) reported the first practical synthesis of (*S*)-pyrethrolone from (*S*)-4-hydroxy-3-methyl-2-(2-propynyl)-2-cyclopenten-1-one which is readily available from a commercial pyrethroid. The (*S*)-propynyl ketone was coupled with vinyl bromide in the presence of $Pd(Ph_3P)_4$, CuI, and Et_3N to give an enyne alcohol in 64% yield (Takagaki and Matsuo, 1991). A clean semihydrogenation of the triple bond of the enyne alcohol occurred with activated Zn in 1-propanol at 100°C for 30 hours giving (*S*)-pyrethrolone in 85% yield. (±)-*trans*-(*E*)-Pyrethrolone, a less difficult proposition, had been synthesized at an earlier date (Crombie *et al.*, 1951c). The second procedure is employment of the Wittig reaction using an unstabilized phosphorane to obtain a high (*Z*)-content. (±)-Cinerolone, (±)-jasmolone, and (±)-pyrethrolone have been made in this way (Crombie *et al.*, 1968, 1969a,b), and in the last case small amounts of the (*E*)-isomer formed were removed by the Diels-Alder reaction with benzoquinone (the (*Z*)-isomer does not react). Tsuji *et al.* (1979) have used telomerization of butadiene and palladium catalysts to make the side-chain for (±)-pyrethrolone as a (*Z*)-/(*E*)-mixture (Fig. 8-67). The unnatural (±)-(*E*)-forms of cinerolone and jasmolone as well as pyrethrolone have all been prepared, as too has the allene (**73**) (Fig. 8-55), a structure at one time incorrectly believed to be the structure of pyrethrolone (Crombie *et al.*, 1951c,1968, 1969a).

Since allethrolone stereoisomers are now available, a convenient approach to pyrethrolone and its relatives is excision of the methylene of the side-chain by ozonolysis or OsO_4/IO_4^- and reconstruction of the side-chain from the resulting aldehyde by Wittig synthesis (Fig. 8-68) (Pattenden and Storer, 1974a; Okada *et al.*, 1977; Sasaki *et al.*, 1979). However, suitable protection for the ring substituents is required during these reactions. Thus Sasaki *et al.* (1979), using ozonolysis, block the keto function as a methyldioxalan and the hydroxyl as its acetate. The protection used by Ando and Casida (1983) is mentioned later.

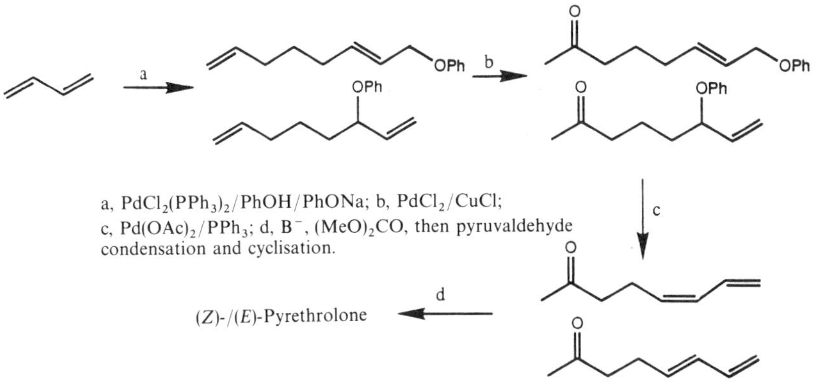

a, $PdCl_2(PPh_3)_2$/PhOH/PhONa; b, $PdCl_2$/CuCl; c, $Pd(OAc)_2/PPh_3$; d, B^-, $(MeO)_2CO$, then pyruvaldehyde condensation and cyclisation.

(*Z*)-/(*E*)-Pyrethrolone

Figure 8-67. Intermediates for (±)-pyrethrolone via telomerization of butadiene (Tsuji *et al.*, 1979).

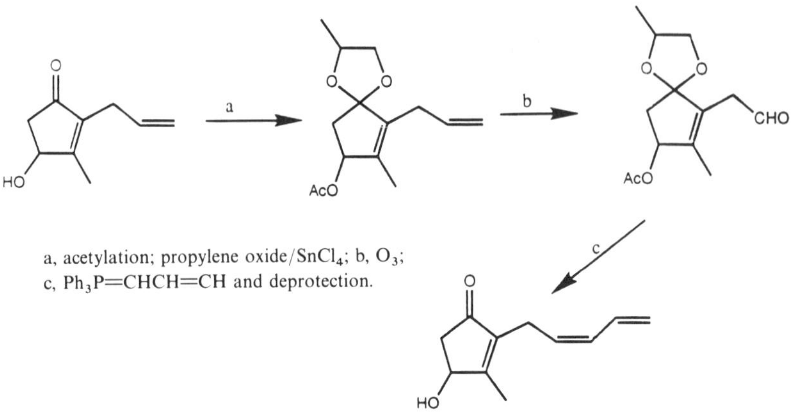
O
a
b
CHO
HO
AcO
AcO
c
a, acetylation; propylene oxide/SnCl4; b, O3;
c, Ph3P=CHCH=CH and deprotection.
O
HO

Figure 8-68. Reconstructive synthesis of pyrethrolone from allethrolone (Sasaki *et al.*, 1979; Pattenden and Storer, 1974a; Okada *et al.*, 1977; Ando and Casida, 1983).

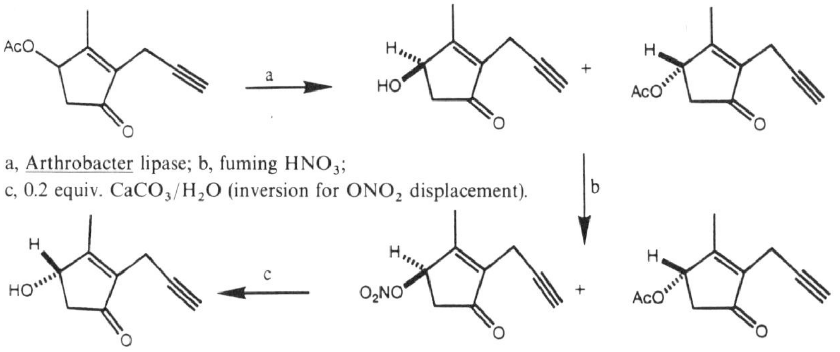
AcO
O
a
H
HO
O
+
H
AcO
O
a, Arthrobacter lipase; b, fuming HNO3;
c, 0.2 equiv. CaCO3/H2O (inversion for ONO2 displacement).
b
H
HO
O
c
H
O2NO
O
+
H
AcO
O

Figure 8-69. Conversion of an (*R*/*S*)-rethrolone acetate into an (*S*) rethrolone (Danda *et al.*, 1991).

Danda *et al.* (1991) have presented a neat method for converting a synthetic (*R*,*S*)-rethrolone acetate into the (*S*)-rethrolone in 74% yield and 82% enantiomeric excess (Fig. 8-69). Replacing the nitrate by mesylate gives even better results (82% yield, 90% e.e).

IV. THE PYRETHRINS

The first pyrethrins to receive total synthesis were dihydrocinerin I (**74**) (Fig. 8-55) and tetrahydropyrethrin I (**75**). Dihydrocinerone and tetrahydropyrethrone (synthesized by 1,4-diketone cyclization) were brominated in the 4-position by *N*-bromosuccinimide (Dauben and Wenkert, 1947) and then treated with the silver salts of (±)-*trans*- and (±)-*cis*-chrysanthemic acids (Crombie *et al.*, 1948, 1950b). The products were of course diastereoisomeric mixtures. Unfortunately, although the rethrones are readily prepared, examples with unsaturated side-chains undergo complex substitution with *N*-

bromosuccinimide and the method cannot be extended to these cases. In this connection, LeMahieu *et al.* (1968) have described how cinerone can be converted by reduction and dehydration into a cyclopentadiene which can be photo-oxygenated in the presence of eosin to give cinerolone and its isomer.

Specimens of pure resynthesized pyrethrins (from natural materials) are obtained by esterification of the appropriate rethrolone with chrysanthemic acid chloride or pyrethric acid chloride (Elliott, 1964c; Elliott and Janes 1969). The latter two acid chlorides are available from natural pyrethrins, or in their various stereochemical forms by synthesis. The synthetic rethrolones are frequently available only in (±)-forms; the totally synthetic rethrins are then a pair of diastereoisomers. Sometimes, as in the case of (±)-cinerolone esterified with (+)-*trans*-chrysanthemic acid the diastereoisomeric pair is separable by crystallization of the semicarbazones (LaForge and Green, 1952; LaForge *et al.*, 1954). Because of crystalline hydrate formation, pyrethrolone from natural sources is readily obtainable freed from other rethrolones (Elliott, 1964c) and natural pyrethrins I and II are thus accessible in pure form by resynthesis.

Diastereoisomeric rethrins have distinguishable chemical shifts in their NMR spectra for protons around the asymmetric ester centres (Bramwell *et al.*, 1969), and such shifts can be emphasized by lanthanide shift reagents to form the basis of an analytical method for rethrin diastereoisomers (Rickett and Henry, 1974).

Isotopically labelled specimens of the pyrethrins esters and their components are required for study of reaction mechanism, biosynthesis, and metabolism, and rethrin molecules provide some good opportunities for positionally specific labelling. "Uniformly" labelled pyrethrins of low specific activity can be obtained by growing *C. cinerariaefolium* in $^{14}CO_2$ (Pellegrini *et al.*, 1952). (±)-*cis-trans*-Chrysanthemic acid having ^{14}C labelling on the carboxy-grouping (**76**) (Fig. 8-70), or C-1 of the ring (**77**) is obtainable by diazoacetic ester synthesis using [1-^{14}C]- or [2-^{14}C]-material, and can be further resolved (Nishizawa and Casida, 1965); it has been used to make [^{14}C]-pyrethrin I (Yamamoto and Casida, 1968). Chrysanthemic acids are readily degraded to caronic half aldehyde and by using [1,3-^{14}C]-isopropyl bromide in a Wittig reaction the methyl labelled chrysanthemic acid (**78**) is available (Crombie *et al.*, 1970). In a similar way using [1-^{14}C]-2-bromopropionic ester, pyrethric acid or chrysanthemum dicarboxylic acid labelled in the side chain methyl can be made (**79**). A sample of pyrethrin II labelled in the side chain ester carboxyl (**80**) has been made by first treating chrysanthemum dicarboxylic acid dichloride with (+)-pyrethrolone and then with [^{14}C]-methanol (Elliott *et al.* 1972). Tritium can be introduced from T_2O into (−)-*cis*-(1*S*,3*R*)-chrysanthemic ester at the C-1 position by exchange and epimerization to give (+)-(1*R*,3*R*)-chrysanthemic acid as shown in (**81**) (Pattenden and Storer, 1976). Direct synthesis (Krief type using isopropylidene triphenylphosphorane) has been employed to make *trans*-chrysanthemic ester having both ring and side-chain substituted with deuterium (**82**) (Saljoughian, 1985).

Exchange of the base-labile protons of (+)-pyrethrolone, under conditions which do not cause racemization, produces tritium or deuterium labelling at C-5 and in the side-chain methyl (**83**) (Elliott and Casida, 1972). Since labelled

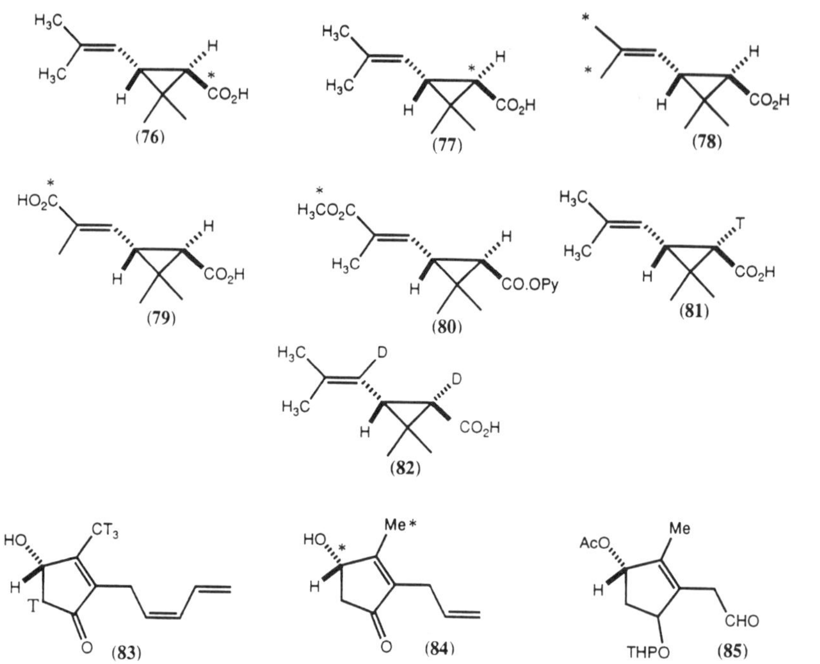

Figure 8-70. Availability of isotopically labeled chrysanthemic acids and rethrolones.

pyruvaldehyde can be made by selenium dioxide oxidation of [1,3-^{14}C]-acetone (Winteringham *et al.*, 1955; Yamamoto and Casida, 1968), this can be employed in synthesis of rethrolones and the reaction has been demonstrated for allethrolone, giving labelling at C-4 and the methyl group (**84**). The protected aldehyde (**85**) is available by degradation from allethrolone, and by suitable reactions with labelled Wittig reagents, and deprotection and reoxidation, all the side-chain labelled rethrolones are potentially available (Ando and Casida, 1983). Both ^{13}C and ^{14}C have been introduced at the side-chain 3′-position of allethrolone in this way.

V. BIOSYNTHETIC ASPECTS

Relative to the dry weight of the flower head, pyrethrum produces ~1–2% pyrethrins (i.e., 2–4 mg per flower head), so it is a very efficient producer (Head, 1966a, 1967). About 93–94% of the pyrethrins are localized in the achenes of the flower (Head,1966b, 1968a). Nonetheless there has been considerable interest in tissue culture for various reasons.

A. Tissue Culture

Tissue culture of high-yielding pyrethrins strains of *Chrysanthemum cinerariaefolium*, followed by redifferentiation into plantlets, is an important way of propagating and improving stock (Wambugu and Rangan, 1981; Staba *et al.*, 1984). Because of the very seasonal supply of pyrethrum flowers, tissue cultures which produced adequate quantities of pyrethrins would provide a more adequate medium for biosynthetic study. It has been claimed that *C. cinerariaefolium* callus produces 0.22% pyrethrins over 3 weeks (Ravishankar *et al.*, 1980) and there is a patent claim that callus (as well as differentiated plantlets and green callus) produces pyrethrins (Aoki *et al.*, 1978). On the other hand, there are reports of less success. Both Cashyap *et al.* (1978) and Kueh *et al.*(1985) found no pyrethrins in nondifferentiated callus, nor in callus that had differentiated only roots. Shoots which had differentiated from the callus contained all six rethrins with pyrethrins dominant over jasmolins and cinerins The callus culture did however produce chrysanthemic acid. Zieg *et al.* (1983) reported that differentiated tissue culture produced more pyrethrins than callus culture and in a later paper it is stated that callus produces chrysanthemyl alcohol, chrysanthemum mono- and dicarboxylic acids but no pyrethrins (Zito and Tio, 1990). Indeed the conversion of isopentenyl pyrophosphate or mevalonic acid into chrysanthemyl alcohol by a cell-free homogenate from callus culture has been reported (Zito and Staba, 1984; Zito *et al.*, 1991). Warszawki *et al.* (1988) have reviewed the possibility of pyrethrins biosynthesis in tissue culture.

It seems generally agreed that callus produces chrysanthemyl alcohol and the chrysanthemum acids, but when the tissue culture redifferentiates into plantlet form, all six natural pyrethrins are formed.

B. Biosynthesis of the Chrysanthemic Acids

Both in connection with the biosynthesis of these acids, and of the rethrolone segments, there is only limited hard scientific information available at the present time. Enzymes have not been isolated and much of the following discussion must proceed on the basis of analogy and informed speculation. Apart from information derived from tissue culture, experiments have usually been carried out on flower heads or achene preparations (Pattenden, 1970), though a cell-free homogenate has been made from ground seedlings (Zito and Staba, 1984) which converts mevalonic acid and isopentenyl pyrophosphate into chrysanthemyl alcohol and the pyrethrins.

There is firm experimental evidence that chrysanthemic acid is a monoterpene formed from mevalonic lactone (Crowley *et al.*, 1961,1962; Godin *et al.*, 1963), the first formed cyclopropane being chrysanthemyl alcohol. Chrysanthemic acid is derived by further oxidation of the latter via the aldehyde and lavandulyl and artemisyl alcohols are not precursors, as demonstrated by radiolabelled feeding experiments using achene and seedling preparations (Pattenden *et al.*, 1975). It is of interest to note that chrysanthemyl alcohol can be oxidized microbiologically to chrysanthemic acid (Miski and Davis, 1988a; Davis and Miski, 1988). (±)-*cis*-Chrysanthemyl alcohol, when oxidized by *Aspergillus ochraceus*, gives (+)-*cis*-chrysanthemic acid having an optical purity of 78.6%; the (±)-*trans*-alcohol gives (+)-*trans*-chrysanthemic acid of 83.4% optical purity. Importantly, the *trans*-acid is oxidized before the *cis*- and can provide a means of separation.

Chrysanthemum dicarboxylic acid is derived by enzymic oxidation at the (*E*)-methyl and it has been demonstrated by labelling that pyrethrin II's are derived from pyrethrin I's (Abou Donia *et al.*, 1973). Such enzyme mediated oxidations are well established in detoxification processes, following the oxidative sequence $\cdot CH_3$ to $\cdot CH_2$–OH to ·CHO to $\cdot CO_2H$ (Casida, 1973; Casida and Ruzo, 1980). Further, Allen *et al.* (1977) have shown that *C. cinerariaefolium* incorporates [Me-^{14}C]-chrysanthemyl alcohol into pyrethrins I and II to the extent of some 2%. The source of the ester methyl in the rethrins II is, not unexpectedly, methionine (Godin *et al.*, 1963; Abou Donia *et al.*, 1973). Chrysanthemyl alcohol (chrysanthemol) is not unique to pyrethrins biosynthesis. It is found for example in *Artemisia ludoviciana* (Alexander and Epstein, 1975) and *Santolina chamaecyparissus* (Banthorpe *et al.*,1977; Baig *et al.*, 1989; Brunke *et al.*, 1992) and as its acetate in *Tanacetum boreale* (Dembitskii *et al.*, 1984). Major interest focuses on the way in which the cyclopropane ring is constructed in nature, and here analogy with an important primary metabolic process seems very relevant.

Squalene (**86**) is formed by head to head union of two molecules of squalene pyrophosphate (**87**) in liver, yeast, *Gibberella fujikuroi* (Nes *et al.*, 1988) and bramble tissue culture (Heintz *et al.*, 1972) and during the process one (pro-1*S*), and only one, of the four allylic methylene hydrogens (**88**) is lost and is replaced in the C_{30} product by the hydrogen from one molecule of labelled NADPH (Fig. 8-71) (Popjak *et al.*, 1961, 1975; Edmond *et al.*, 1971; Cornforth, 1973). This indicates an unusual biosynthetic process. Rilling (Rilling, 1966; Epstein and Rilling, 1970; Muscio *et al.*, 1974) observed that if NADPH was omitted an intermediate built up which could be isolated and converted onward to squalene when treated with the enzyme system including NADPH. The structure of the intermediate proposed by Rilling and Epstein (1969) was confirmed by

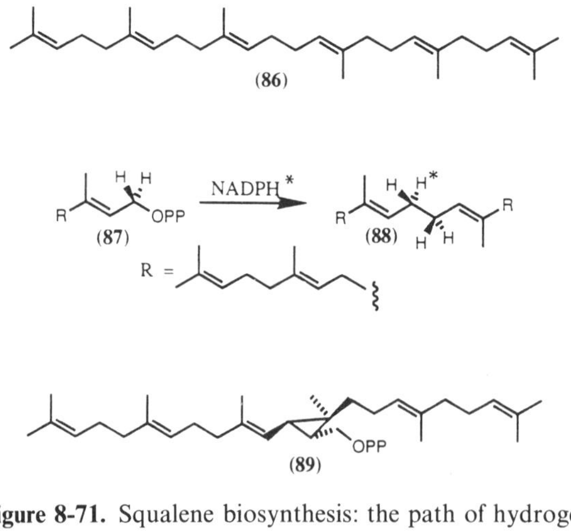

(86)

(87) (88) (89)

Figure 8-71. Squalene biosynthesis: the path of hydrogen.

synthesis (Campbell *et al.*, 1971, 1975; Altman *et al.*, 1971, 1978; Coates and Robinson, 1971) to be the pyrophosphate of a cyclopropane alcohol, presqualene alcohol (**89**). Apart from containing one extra chiral center and longer prenylated chains, its structure and stereochemistry (1*R*,2*R*,3*R*) (Popjak *et al.*, 1973; Altman *et al.*, 1978) is highly reminiscent of chrysanthemyl alcohol and it seems likely that the enzymic formation processes for the latter are similar. The carotenoid pathway also involves an analogous C_{40}-cyclopropane, prephytoene pyrophosphate (Poulter and Rilling, 1981; Crombie *et al.*, 1972b). Pattenden and Storer (1973b) have examined the incorporation of (3*R*,4*R*)-[4-^{3}H]- and (3*R*,4*S*)-[4-^{3}H]-mevalonic acid into chrysanthemic acid, but the results seem unclear at the present time.

Unfortunately, important though it is, the mechanism for the formation of presqualene alcohol has not been established with complete certainty, though it has been the subject of much discussion (for a useful summary see Poulter and Rilling, 1981). Fig. 8-72 outlines one possible mechanism for chrysanthemate formation analogous to that of presqualene, being initiated by a nucleophilic center X from the enzyme. The stages, all presumably enzyme mediated, leading

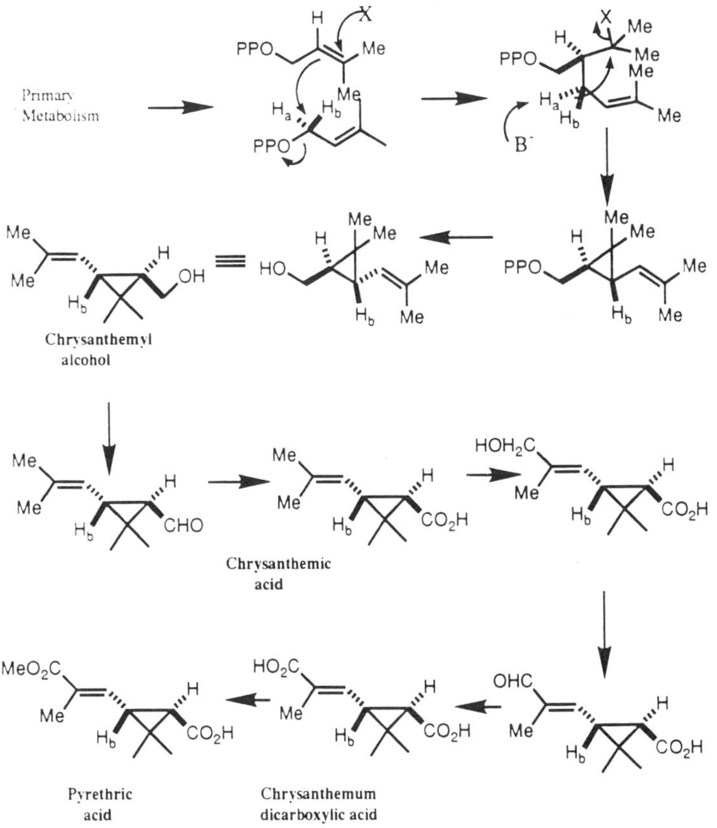

Figure 8-72. Proposed biosynthetic formation of chrysanthemic and pyrethric acids.

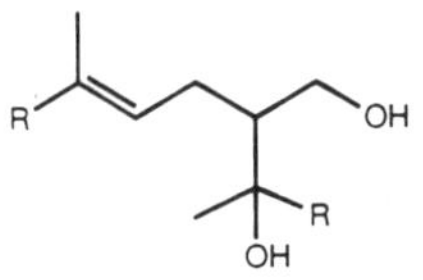

(90) 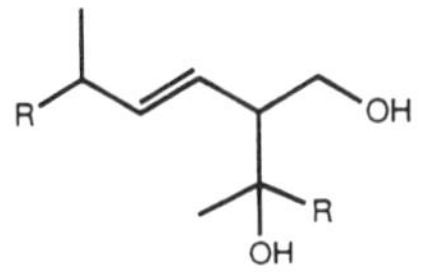 (91)

(a) R =

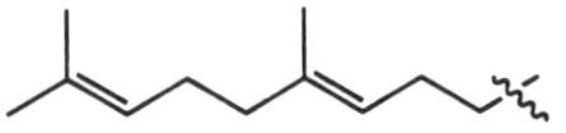

(b) R = Me

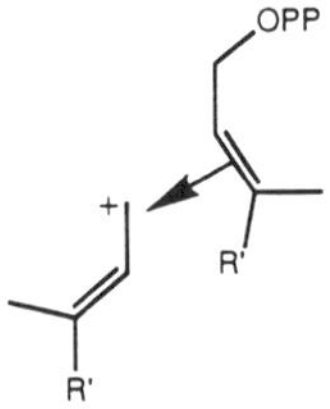

(93)

Figure 8-73. Presqualene compounds as possible intermediates in pyrethrin biosynthesis.

via chrysanthemic acid, to pyrethric ester are outlined. The possibility of there being an isolable prepresqualene (**90**) or (**91**) (Fig. 8-73) has been considered. Both compounds have been synthesized. The isomerized form (**91**), as either the (*E*)- or (*Z*)-compound, is not a squalene intermediate in *Rhizopus arrhius*, nor is its lower prenylogue a chrysanthemyl alcohol intermediate in *Artemisia annua* (Shirley *et al.*, 1982; Boulton *et al.*, 1986). However, bifarnesol (**90**) (van Tamelen and Schwartz, 1971; van Tamelen and Leopold, 1985) is reported to be incorporated into presqualene alcohol, squalene, and steroids by *Gibberella fujikuroi* (Nes *et al.*, 1988, 1990). In an interesting biomimetic model (Fig. 8-74), the Julia group (Babin *et al.*, 1981) were able to convert the alcohol (**92**) into chrysanthemyl alcohol (15% yield) by using the phenylthio as the anionic leaving group. Dewar and Ruiz (1987) in a theoretical paper have revived their earlier ideas on π-complexes and suggest two routes by which such a complex (**93**) (Fig. 8-73), which is deprotonated to give presqualene pyrophosphate, might form. Other biological cyclopropanation models have been proposed, such as

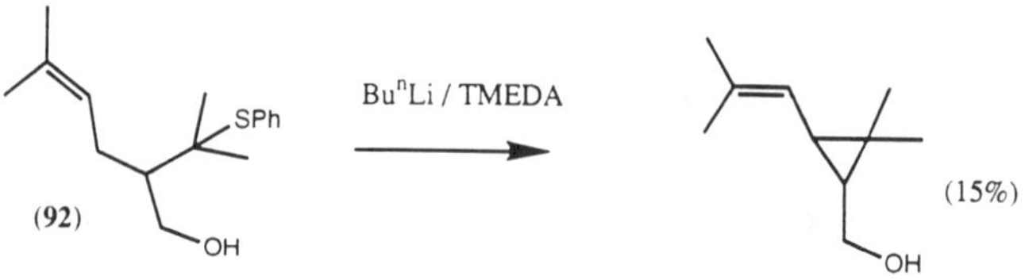

(92) Bu^nLi / TMEDA (15%)

Figure 8-74. Biomimetic model for chrysanthemyl alcohol and presqualene alcohol biosynthesis (Babin *et al.*, 1981).

the involvement of sulphur ylides in a copper catalysed carbene reaction (Cohen *et al.*, 1974), but do not seem of likely relevance. The decomposition of presqualene alcohol (as pyrophosphate) to squalene also presents mechanistic difficulties (Rilling *et al.*, 1971; Poulter and Rilling, 1981) and cannot be satisfactorily modelled *in vitro* using chrysanthemyl alcohol.

C. Biosynthesis of the Rethrolones

If experimental knowledge of the biosynthesis of the chrysanthemum acids is limited, there is still less certainty about the rethrolones. Circumstantial evidence suggests that they originate from fatty acid metabolism. Thus the pyrethrins are localized in the achenes of the flower, a particularly fat-rich environment containing substantial quantities of linoleic and linolenic acids (Head, 1968a). It is reported that the fatty acids represent 40% by weight of pyrethrum extract (containing 30% pyrethrins), with linoleic acid comprising 33% and linolenic 15% of the total fatty acids (Head, 1968a). Indeed pyrethrolone occurs combined with linoleic acid (**94**) (Fig. 8-75), the fatty acid acting as a replacement for chrysanthemic or pyrethric acid (Acree and LaForge, 1937).

Experiments using a crude enzyme preparation from *C. cinerariaefolium* achenes, designed to ascertain which segment of the isotopically labelled fatty acid chain might form the carbon chain of the rethrolones, have not been successful. The crude mixed enzyme preparation degraded the fatty acid mainly to acetate before incorporation, so labelling was scrambled and appeared in the chrysanthemate portion as well as being distributed over the rethrolone chain in a rather unmeaningful way (Crombie and Holloway, 1985).

However considerable progress has been made recently on a dehydratase mediated pathway (Fig. 8-76), widely distributed in plants, which leads from linolenic acid 13-hydroperoxide (**95**) to two ketols (**96**) and (**97**) and a cyclopentenone, 12-oxophytodienoic acid (12-oxoPDA) (**98**) (Zimmerman and Feng, 1978; Vick and Zimmerman, 1983, 1984, 1987; Crombie and Morgan, 1991; Crombie and Mistry, 1991). The formation of the latter, involving an unusual allene epoxide intermediate (**100**) produced via an epoxycarbonium ion (**99**) and firmly established experimentally (Crombie and Morgan, 1987, 1988; Hamberg, 1987; Brash *et al.*, 1988), is shown in Fig. 8-77. It is known experimentally (Fig. 8-78) that 12-oxoPDA undergoes 11,12-hydrogenation and three β-oxidations in plant systems to form epi-jasmonic acid (**101**), easily epimerized to jasmonic acid (**102**) (Vick and Zimmerman, 1983, 1984). Jasmonic acid, jasmone (**105**) and jasmine lactone (**106**) are well established components of the perfume secretions of the jasmine flower (Demole, 1982; Demole *et al.*, 1964) and just as epi-jasmonic acid originates from 12-oxoPDA, a pathway to jasmone can be readily envisaged. If, instead of reduction, the 11, 12-double

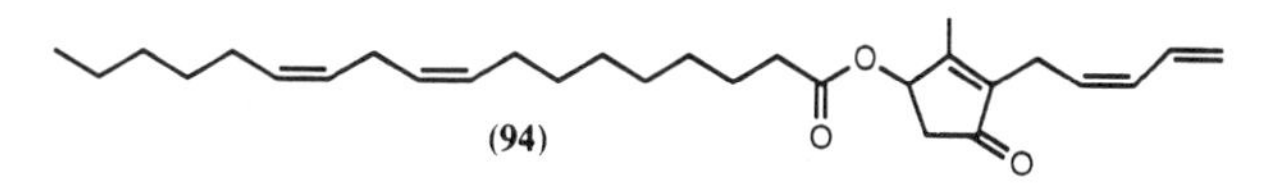

Figure 8-75. Pyrethrolone ester of linoleic acid.

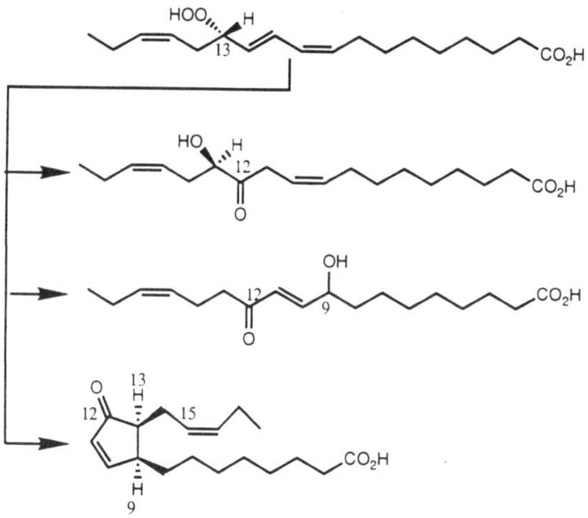

(95)

(96)

(97)

(98)

Figure 8-76. Products from the treatment of linolenic acid 13-hydroperoxide with flax enzyme preparation.

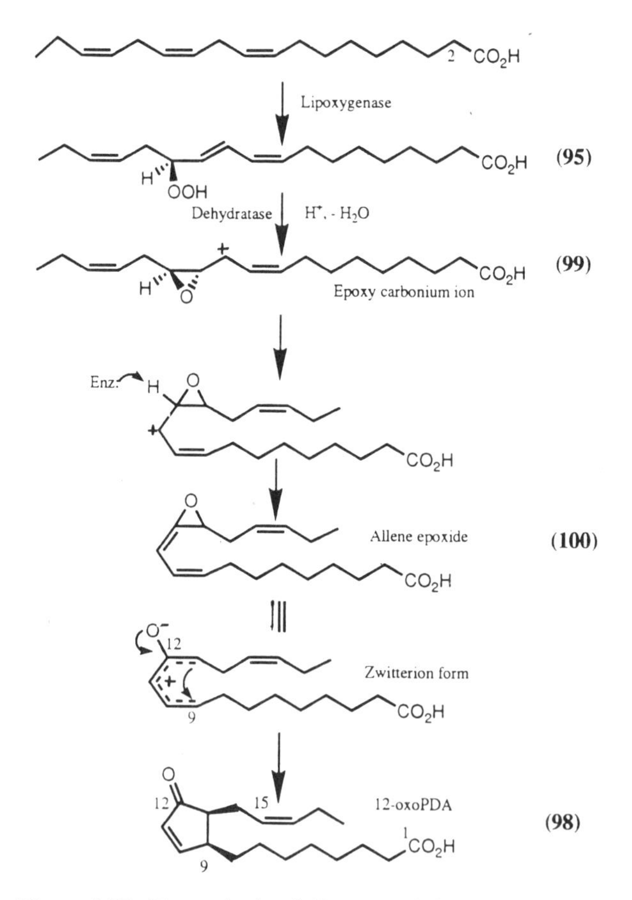

Figure 8-77. Biosynthesis of 12-oxoPDA from linolenic acid.

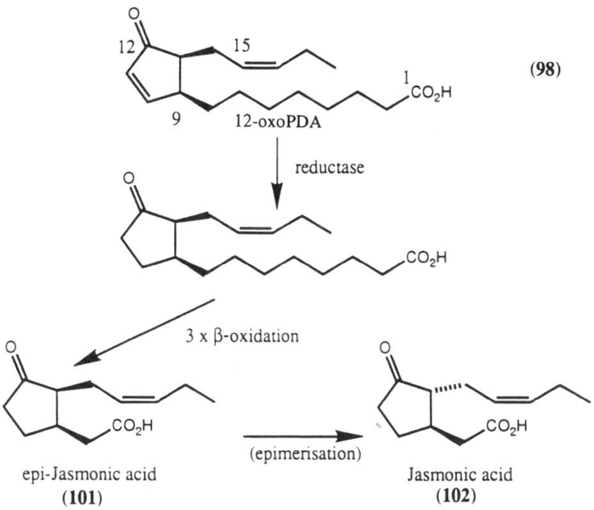

O
12
15
1
CO_2H
(98)
9
12-oxoPDA
reductase
O
CO_2H
3 x β-oxidation
O
CO_2H
epi-Jasmonic acid
(101)
(epimerisation)
O
CO_2H
Jasmonic acid
(102)

Figure 8-78. Reduction and β-oxidation of 12-oxoPDA leading to epi-jasmonic acid and jasmonic acid.

bond is isomerized into the thermodynamically stable position, and the three β-oxidations take place, the product is a vinylogous β-keto-acid which will decarboxylate readily to give jasmone (Fig. 8-79). The significance of jasmine lactone is that it demonstrates that enzymic oxidative entry into the terminal carbon of the (*Z*)-pentenyl side chain can occur successfully.

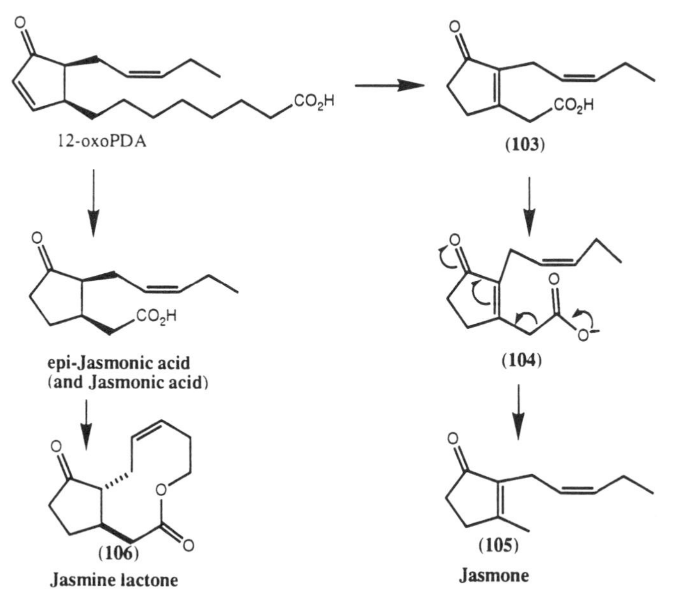

Figure 8-79. Proposal for the formation of epi-jasmonic acid, jasmonic acid, jasmine lactone, and jasmone, components of jasmine flower perfume.

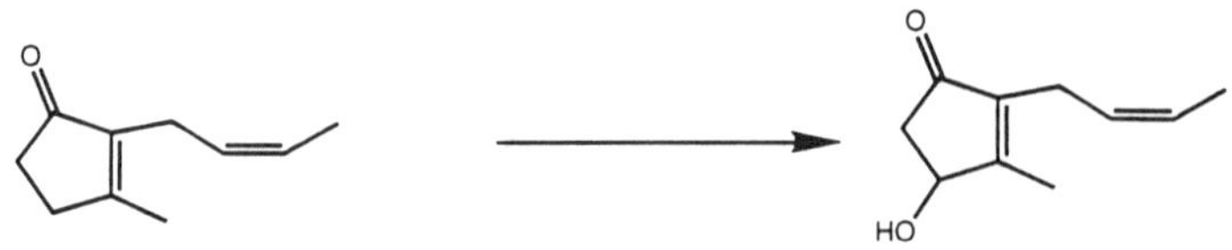

Cinerone (107) **Cinerolone (11)**

Figure 8-80. Microbiological introduction of the 4-hydroxyl group.

Enzymic oxidative entry at the rethrolone C-4 position is also well established (Fig. 8-80). Thus cinerone and allethrone are converted into cinerolone and allethrolone, respectively, by *Aspergillus niger* (LeMahieu *et al.*, 1968, 1970; Tabenkin *et al.*, 1969). It is also significant that the characteristic pyrethrolone side-chain is formed in connection with the epi-jasmonic acid pathway. During the formation of the latter compound by the fungus *Botryodiplodia theobromae* pat (syn. *Lasiodiplodia theobromae* Griff. and Mabl.) (Fig. 8-81) dehydro epi-jasmonic acid (**108**) was also isolated (Miersch *et al.*,1987, 1989). On this basis, an hypothesis for the biosynthetic pathway to jasmolone and pyrethrolone is suggested in Fig. 8-82. Since jasmone is apparently capable of ω-oxidative attack at the (*Z*)-pentenyl side chain to form jasmine lactone, the origins of cinerolone (which like cinerone is known elsewhere in Nature) might lie in ω-oxidation of the jasmolone side-chain followed by decarboxylation. Such proposals attractive though they seem, must await the tests of experimentation.

The final phase in rethrin biosynthesis is esterification of the appropriate acid and ketol, but whether this is carried out by a single acyl transferase or by specific acyl transferases leading to each of the six natural rethrins, is not known.

epi-Jasmonic acid

Dihydro epi-Jasmonic acid

Cucurbic acid

Dehydro epi-Jasmonic acid
(**108**)

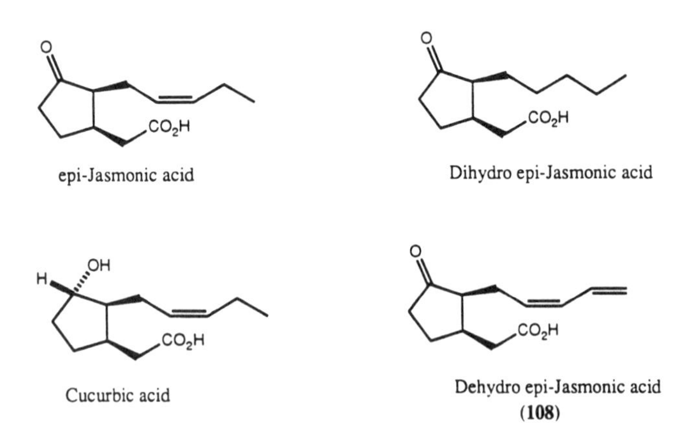

Figure 8-81. Products from the fungus *Botryodiplodia theobromae* Pat. (syn. *Lasiodiplodia theobromae* Griff and Maubl).

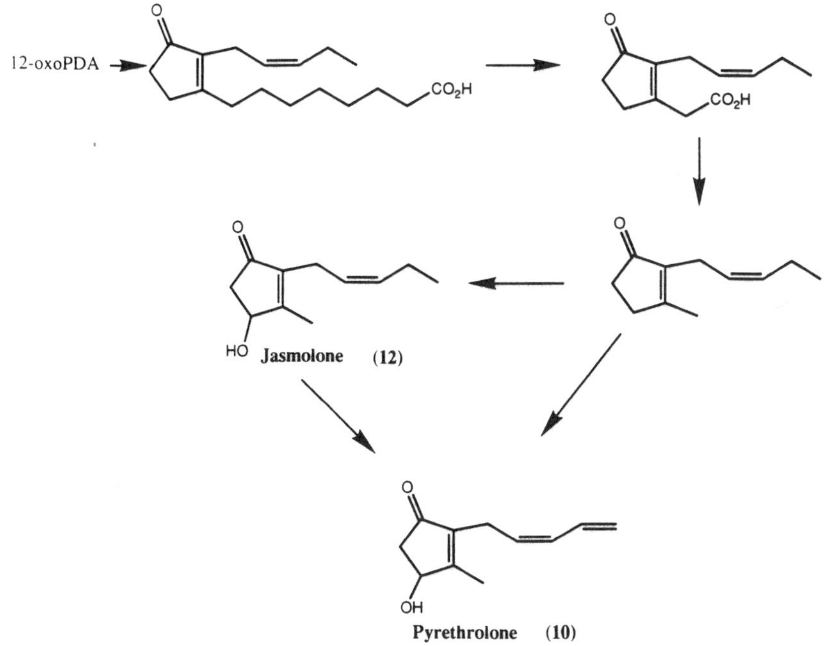

Figure 8-82. Hypotheses for biosynthetic pathways to pyrethrolone and jasmolone.

VI. SELECTED CHEMICAL NAMES AND REGISTRY NUMBERS

Chrysanthemum monocarboxylic (Chrysanthemic) acid (7). Registry number [10453-89-1].
Cyclopropanecarboxylic acid,-2,2-dimethyl-3-(2-methyl-1-propenyl)-
Other Registry numbers: *cis-* [15259-78-6]; *cis-*(±)-[2935-23-1]; *trans-* [827-90-7]; *trans-*(±)- [705-16-8]; (1*R-cis*)- [26771-11 9]; (1*R-trans*)- [4638-92-0]; (1*S-cis*)- [26771-06-2]; (1*S-trans*)- [2259-14-5]]

Chrysanthemum dicarboxylic acid (9). Registry number [497-95-0]
Cyclopropanecarboxylic acid, 3-(2-carboxy-1-propenyl)-2,2-dimethyl-

Chrysanthemyl alcohol (20). Registry number [5617-92-5].
Cyclopropanemethanol, 2,2-dimethyl-3-(2-methyl-1-propenyl)-

Pyrethrolone (10). Registry number [22054-38-2].
2-Cyclopenten-1-one, 4-hydroxy-3-methyl-2-(2,4-pentadienyl)-
Other Registry numbers: (*Z*)-(+)-[487-67-2].

Cinerolone (11). Registry number [487-64-9].
2-Cyclopenten-1-one, 2-(2-butenyl)-4-hydroxy-3-methyl-
Other Registry numbers: [*S*-(*Z*)]-[3894-82-4].

Jasmolone (12). Registry number [22054-39-3].
2-Cyclopenten-1-one, 4-hydroxy-3-methyl-2-(2-pentenyl)-
Other Registry numbers: [*S*-(*Z*)]-83541-04-2].

REFERENCES

Abou Donia, S.A., Doherty, C.F., and Pattenden, G. (1973). Biosynthesis of chrysanthemum-dicarboxylic acid, and the origin of the pyrethrin II's. *Tetrahedron Lett.*, 3477–3481.

Acree Jr., F. and LaForge, F.B. (1937). Constituents of pyrethrum flowers. X. Identification of the fatty acids combined with pyrethrolone. *J. Org. Chem.* **2**, 308–313.

Alexander, K., and Epstein, W.W. (1975). Biogenesis of non-head-to-tail monoterpenes. Isolation of (1*R*, 3*R*)-chrysanthemol from *Artemisia ludoviciana*. *J. Org. Chem.* **40**, 2576.

Allen, K.G., Banthorpe, D.V., Charlwood, B.V., and Voller, C.M. (1977). Biosynthesis of Artemisia ketone in higher plants. *Phytochemistry* **16**, 79–83.

Altman, L.J., Kowerski, R.C., and Rilling, H.C. (1971). Synthesis and conversion of presqualene alcohol to squalene. *J. Amer. Chem. Soc.* **93**, 1782–1783.

Altman, L.J., Kowerski, R.C., and Laungani, D.R. (1978). Studies in terpene biosynthesis. Synthesis and resolution of presqualene and prephytoene alcohols. *J. Amer. Chem. Soc.* **100**, 6174–6182.

Ando, T., and Casida, J.E. (1983). Conversion of (*S*)-allethrolone into pyrethrin I, jasmolin I, cinerin I and [propenyl-3-^{13}C]- and [propenyl-3-^{14}C]-(*S*)-bioallethrin. *J. Agric. Food. Chem.* **31**, 151–156.

Ando, T., Ruzo, L., Engel, J.L., and Casida, J.E. (1983). 3-(3,3-Dihalo-2-propenyl) analogues of allethrin and related pyrethroids: synthesis, biological activity and photostability. *J. Agric. Food Chem.* **31**, 250–253.

Ando, T., Jacobsen, N.E., Toia, R.F., and Casida, J.E. (1991a). Epoxychrysanthemates: two dimensional NMR analyses and stereochemical assignments. *J. Agric. Food Chem.* **39**, 600–605.

Ando, T., Toia, R.F., and Casida, J.E. (1991b). Epoxy and hydroxy derivatives of (*S*)-bioallethrin and pyrethrins I and II. Synthesis and metabolism. *J. Agric. Food Chem.* **39**, 606–611.

Ando, T., Kurotsu, Y., and Uchiyama, M. (1986). Hplc separation of the stereoisomers of natural pyrethrins and related compounds. *Agric. Biol. Chem.* **50**, 491–493.

d'Angelo, J., and Revial, G. (1983). Synthese de l'acide *cis*-chrysanthemique. *Tetrahedron Lett.*, 2103–2106.

d'Angelo, J., Revial, G., Azerad, R., and Buisson, D. (1986). A short, efficient, highly selective synthesis of (1*R*,3*S*)-*cis*-chrysanthemic acid through microbiological reduction of 2,2,5,5-tetramethyl-1,4-cyclohexanedione. *J. Org. Chem.* **51**, 40–45.

Aoki, S., Kaneto, K., Hasimoto, S., and Oogai, H. (1978). Production of pyrethrins by tissue culture. Japan Kokai 78 24,097 (Cl. C12D13100) 06Mar. 1987 Appl. JP 76-97868 (16 Aug. 1976); *Chem. Abstr.* **89**, 20589q (1978).

Aratani, T., Yoneyoshi, Y., and Nagasse, T., (1975). Asymmetric synthesis of chrysanthemic acid. An application of copper carbenoid reaction. *Tetrahedron Lett.*, 1707–1710.

Aratani, T., Yoneyoshi, Y., and Nagase, T., (1977). Asymmetric synthesis of chrysanthemic acid. An application of a copper carbenoid reaction. *Tetrahedron Lett.*, 2599–2602.

Arlt, D., Jautelat, M., and Lantzsch, R. (1981). Synthesis of pyrethroid acids. *Angew. Chem.* **93**, 719–738.

Babin, D., Fourneron, J.D., Harwood, L.M., and Julia, M. (1981). A biomimetic synthesis of chrysanthemol. *Tetrahedron* **37**, 325–332.

Babler, J.H., and Invergo, B.J. (1981). A facile route to a versatile synthon for the preparation of *cis*-pyrethroids. *Tetrahedron Lett.*, 2743–2746.

Babler, J.H., and Spina, K.P. (1985). A tandem Michael reaction—cycloalkylation utilising 2-nitropropane: a facile route to the acid components of insecticidal pyrethroids. *Tetrahedron Lett.*, 1923–1926.

Babler, J.H., and Tortorello, A.J. (1976). Regioselectivity in the cyclization of $\beta\gamma$-epoxy-carbanions. Application to the total synthesis of *trans*-chrysanthemic acid. *J. Org. Chem.* **41**, 885–887.

Baeckström, P. (1978). Photochemical formation of chrysanthemic acid and cyclopropylacrylic acid derivatives. *Tetrahedron* **34**, 3331–3335.

Baig, M.A., Banthorpe, D.V., and Branch, S.A. (1989). Hemi- and mono-terpenes from callus cultures of *Santolina chamaecyparissus*. *Fitoterapia* **60**, 184–186; *Chem. Abstr.* **111**, 230640y (1989).

Baig, M.A., Banthorpe, D.V., Carr, G., and Whittaker, D. (1990). The rearrangement of chrysanthemyl alcohol in fluorosulfuric acid. *J. Chem. Soc., Perkin Trans.* 2, 163–167.

Bakshi, D., Mahindroo, V.K., Soman, R., and Dev, Sukh. (1989). Monoterpenoids. VII. A simple synthesis of (−)-*cis*-caronaldehydic acid hemiacetal from (+)-car-3-ene. *Tetrahedron* **45**, 767–774.

Balbaa, S.I., Abdel-Kader, E.M., Abdel-Wahat, S.M., Zaki, A.Y., and El-Shamy, A.M. (1972). Comparative study between different methods used for estimation of pyrethrins, and a new suggested method. *Planta Medica.* **21**, 347-352.

Banthorpe, D.V., Doonan, S. and Gutowski, J.A. (1977). Biosynthesis of irregular terpenes in

extracts from higher plants *Artemisia annua* and *Santolina chamaecyparissus* containing *trans*-chrysanthemyl alcohol. *Phytochemistry* **16**, 85–92.
Barthel, W.F., and Haller, H.L. (1945). Purification of pyrethrum extract with nitromethane. U.S. Patent 2,372,183.
Barthel, W.F., Haller, H.L., and LaForge, F.B. (1944). The preparation of 98% pure pyrethrins for use in freon aerosol bombs. *Soap Sanit. Chem.* **20**, 121–135.
Bates, R.B., and Paknikar, S. (1965). Terpenes. IX. Biogenesis of some monoterpenoids not derived from a geranyl precursor. *Tetrahedron Lett.*, 1453–1457.
Beevor, P.S., Godin, P.J., and Snarey, M. (1965). Jasmolin I, cinerin I, and a new method for isolating research quantities of the pyrethroids. *Chem. Ind. (London)*, 1342–1343.
Begley, M.J., Crombie, L., Simmonds, D.J., and Whiting, D.A. (1972). Absolute configuration of the pyrethrins. Configuration and structure of (+)-allethronyl (+)-*trans*-chrysanthemate 6-bromo-2,4-dinitrophenylhydrazone by X-ray methods. *J. Chem. Soc., Chem. Commun.*, 1276–1277.
Begley, M.J., Crombie, L., Simmonds, D.J., and Whiting, D.A. (1974). X-Ray analysis of synthetic (4*S*)-2-(prop-2′-enyl)rethron-4-yl (1*R*),(3*R*)-chrysanthemate 6-bromo-2,4-dinitrophenylhydrazone and chiroptical correlation with the six natural pyrethrin esters. *J. Chem. Soc., Perkin Trans.* 1, 1230–1235.
Bevenue, A., Kawano, Y., and DeLano, F. (1970). Analytical studies of pyrethrin formulations by gas chromatography. *J. Chromatogr.* **50**, 49–58.
Bhat, N.G., Joshi, G.D., Gore, K.G., Kulkarni, G.H., and Mitra, R.B. (1981). New synthesis of methyl-1*R*-*cis*-2,2-dimethyl-3-(2-phenylprop-1-enyl)cyclopropanecarboxylate and methyl (+)-*cis*-chrysanthemate from (+)-car-3-ene. *Indian J. Chem. Sect. B* **20B**, 558–561.
Bhosale, S.S., Kulkarni, G.H., and Mitra, R.B. (1985). A new approach to the synthesis of optically active (+)-(1*R*)-*trans*-pyrethroids. *Indian J. Chem. Sect. B* **24B**, 543–546.
Bhosale, S.S., Joshi, G.S., and Kulkarni, G.H. (1988). Oxidation studies using pyridinium chlorochromate on (+)-3-carene derivatives. *Curr. Sci.* **57**, 478–479.
Bohlmann, F., and Knoll, K.H. (1978). Naturally occurring terpene derivatives: new farnesol derivatives from *Tanacetum odessanum*. *Phytochemistry* **17**, 319–320.
Boulton, K., Shirley, I.H., and Whiting, D.A. (1986). Mechanism of formation of natural cyclopropanes. Synthesis of postulated intermediates in presqualene and chrysanthemyl alcohol biosynthesis. *J. Chem. Soc., Perkin Trans.* 1, 1817–1824.
Brady, W.T., Norton, S.J., and Ko, J. (1983). A versatile new synthesis of pyrethroid acids. *Synthesis*, 1002–1005.
Bramwell, A.F., Crombie, L., Hemesley, P., Pattenden. G., Elliott, M., and Janes, N.F. (1969). Nuclear magnetic resonance spectra of the natural pyrethrins and related compounds. *Tetrahedron* **25**, 1727–1741.
Brash, A.R., Baertschi, S.W., Ingram, C.D., and Harris, T.M. (1988). Isolation and characterisation of natural allene oxides: unstable intermediates in the metabolism of lipid hydroperoxides. *Proc. Natl. Acad. Sci.* **85**, 3382–3386.
Brunke, E.-J., Hammerschmidt, F.-J., and Schmaus, G. (1992). The essential oil of *Santolina chamaecyparissus* L. *Dragoco Report*, 151-167.
Büchel, K.H., and Korte, F. (1962). Eine synthese der *DL*-chrysanthemumsäure. *Z. Naturforsch. B* **17**, 349–350.
Büchi, G., Minster, D., and Young, J.C.F. (1971). A new synthesis of rethrolones. *J. Amer. Chem. Soc.* **93**, 4319–4320.
Bullivant, M.J., and Pattenden, G. (1971). Photochemical decomposition of chrysanthemic acid and its alkyl esters. *Pyrethrum Post* **11**, 72–76.
Bullivant, M.J., and Pattenden, G. (1972). Triplet di-π-methane rearrangement of a "free-rotor" 2-prop-2-enylcyclo-pent-2-enone. *J. Chem. Soc., Chem. Commun.*, 864–865.
Bullivant, M.J., and Pattenden. G. (1976a). Photodecomposition of natural pyrethrins and related compounds. *Pestic. Sci.* **7**, 231–235.
Bullivant, M.J., and Pattenden, G. (1976b). Photochemistry of 2-(prop-2-enyl)-cyclopent-2-enones. *J. Chem. Soc., Perkin Trans.* 1, 249–256.
Bullivant, M.J., and Pattenden, G. (1976c). A photochemical di-π-methane rearrangement leading to methyl chrysanthemate. *J. Chem. Soc., Perkin Trans* 1, 256–258.
Bushway, R.J., Johnson, H., and Scott, D.W. (1985). Simultaneous liquid chromatographic determination of rotenone and pyrethrins in formulations. *J. Assoc. Off. Anal. Chem.* **68**, 580–582.
Cameron, A.G., and Knight, D.W. (1985). Stereospecific synthesis of *cis*-2-alkenylcyclopropane carboxylic acids. A total synthesis of (±)-*cis*-chrysanthemic acid . *Tetrahedron Lett.*, 3503–3506.
Cameron, A.F., Ferguson, G., and Hannaway, C. (1975). Absolute configuration of

(+)-*trans*-chrysanthemic acid. Crystal structure analysis of a *p*-bromoanilide derivative. *J. Chem. Soc., Perkin Trans.* 2, 1567–1568.
Cameron, A.G., Hewson, A.T., and Osammor, M.L. (1984). Synthesis of natural cyclopentanone products via vinylphosphonium salts. *Tetrahedron Lett.*, 2267–2270.
Campbell, I.G.M., and Harper, S.H. (1945). Experiments on the synthesis of the pyrethrins. I. Synthesis of chrysanthemum monocarboxylic acid. *J. Chem. Soc.* (*London*), 283–286.
Campbell, I.G.M., and Harper S.H. (1952). The chrysanthemum carboxylic acids. IV. Optical resolution of the chrysanthemic acids. *J. Sci. Food. Agric.* **3**, 189–192.
Campbell, R.V.M., Crombie, L., and Pattenden, G. (1971). Synthesis of presqualene alcohol. *J. Chem. Soc., Chem. Commun.* 218–219.
Campbell, R.V.M., Crombie, L., Findley, D.A.R., King, R.W., Pattenden, G., and Whiting, D.A. (1975). Synthesis of (±)-presqualene alcohol, (±)-prephytoene alcohol and structurally related compounds. *J. Chem. Soc., Perkin Trans.* 1, 897–913.
Cashyap, M.M., Kueh, J.S.H., Mackenzie, I.A., and Pattenden, G. (1978). *In vitro* synthesis of pyrethrins from the tissue culture of *Tanacetum cinerariaefolium. Phytochemistry* **17**, 544–545.
Casida, J.E. (ed.). (1973). "Pyrethrum, The Natural Insecticide." Academic Press, New York, and London.
Casida, J.E., and Ruzo, L.O. (1980). Metabolic chemistry of pyrethroid insecticides. *Pestic. Sci.* **11**, 257–269
Castellani, G., Scordamaglia, R., and Tosi, C. (1980). Theoretical calculations on the conformations of pyrethroids. 1. Chrysanthemic acid. *Gazz. Chim Ital.* **110**, 457–464.
Chen, Y.-L., and Casida, J.E. (1969). Photodecomposition of pyrethrin I, allethrin, phthalthrin and dimethrin. *J. Agric. Food Chem.* **17**, 208–215.
Class, T.J. (1991). Optimised g.c. analysis of natural pyrethrins and pyrethroids. *J. High Resol. Chromatogr.* **14**, 48–51.
Class, T.J., and Fresinius, J. (1992). G.c. and m.s studies on pyrethroid photo- and bio-transformation. *Anal. Chem.* **342**, 805–808.
Coates, R.M., and Robinson, W.H. (1971). Stereoselective total synthesis of (±)-presqualene alcohol. *J. Amer. Chem. Soc.* **93**, 1785–1786.
Cocker, W., Lauder, H.St.J., and Shannon, P.V.R. (1975). The chemistry of terpenes. XVIII. Synthesis of methyl (−)-*cis*-chrysanthemate and of methyl (+)-*cis*-homo-chrysanthemate from (+)-car-3- and -2-ene. *J. Chem. Soc. C.*, 332–335.
Cohen, T., Herman, G., Chapman, T.M., and Kuhn, D. (1974). A laboratory model for the biosynthesis of cyclopropane rings. Copper-catalysed cyclopropanation of olefins by sulfur ylides. *J. Amer. Chem. Soc.* **96**, 5627–5628.
Corey, E.J., and Jautelat, M. (1967). Construction of ring systems containing the gem-dimethylcyclopropane unit using diphenylsulfonium isopropylide. *J. Amer. Chem. Soc.* **89**, 3912–3913.
Cornforth, J.W. (1973). The logic of working with enzymes. *Chem. Soc. Rev.* **2**, 1.
Crombie, L. (1979), Unpublished work with Wellcome Research Laboratories.
Crombie, L. (1980). Chemistry and biosynthesis of the natural pyrethrins. *Pestic. Sci.* **11**, 102–118.
Crombie, L. (1988). The natural pyrethrins: a chemist's view. *In* "Neurotox 1988, Molecular basis of drugs and pesticide action" (G.G. Lund, ed.), pp. 3–25. *Excerpta Medica*, Amsterdam, New York and Oxford.
Crombie, L., and Crossley, J. (1963). Preparation of α-cyclopropane aldehydes and cyclopropyl ketones. *J. Chem. Soc.* (*London*), 4983–4984.
Crombie, L., and Elliott, M., (1961). Chemistry of the natural pyrethrins. *Fortsch. Chem. Org. Naturst.* **19**, 120–164.
Crombie, L., and Harper, S.H., (1949). Synthesis of cinerone, cinerolone, and cinerin I. *Nature* (*London*) **164**, 534.
Crombie, L., and Harper, S.H., (1950). Experiments on the synthesis of the pyrethrins. Part IV. Synthesis of cinerone, cinerolone, and cinerin I. *J. Chem. Soc.* (*London*), 1152–1160.
Crombie, L., and Harper, S.H., (1952). Experiments on the synthesis of the pyrethrins. Part VIII. Stereochemistry of jasmone and identity of dihydropyrethrone. *J. Chem. Soc.* (*London*), 869–875.
Crombie, L., and Harper, S.H. (1954). The chrysanthemum carboxylic acids. IV. The configuration of the chrysanthemic acids. *J. Chem. Soc.* (*London*), 470.
Crombie, L., and Harper, S.H. (1958). Spectroscopic assignment of geometrical configuration to rethrins I. *Chem. Ind.* (*London*), 1001–1002.
Crombie, L., and Holloway, S.J. (1985). Biosynthesis of the pyrethrins. Unsaturated fatty acids and the origins of the rethrolone fragments. *J. Chem. Soc., Perkin Trans.* 1, 1393–1400.
Crombie, L., and Mistry, K.M. (1991). Synthesis of 12-oxophytodienoic acid (12-oxoPDA) and

the compounds of its enzymic degradation cascade in plants, OPC-8:0, -6:0, -4:0 and -2:0 (epijasmonic acid), as their methyl esters. *J. Chem. Soc., Perkin Trans.* 1, 1981–1991.
Crombie, L., and Morgan, D.O. (1987). Experiments with [9,10,12,13-$^2H_4$]-linoleic acid on the formation of 9-[nona-(1′*E*),(3′*Z*)-dienyloxy]non-(8*E*)-enoic (colneleic) acid and (13*R*)-hydroxy-12-oxo-octadec-(9*Z*)-enoic acid by plant enzymes. *J. Chem. Soc., Chem. Commun.* 503–504.
Crombie, L., and Morgan, D.O. (1988). The conversion of linoleic acid (13*S*)-hydroperoxide into (13*R*)-hydroxy-12-oxo-octadec-(9*Z*)-enoic acid and 9-hydroxy-12-oxo-octadec-(10*E*)-enoic acid by flax enzyme. Isotopic evidence for allene epoxide intermediates. *J. Chem. Soc., Chem. Commun.* 556–557.
Crombie, L., and Morgan, D.O. (1991). Synthesis of [14,14-$^2H_2$]-linolenic acid and its use to confirm the pathway to 12-oxophytodienoic acid (12-oxoPDA) in plants: a conspectus of the epoxy-carbonium ion derived family of metabolites from linoleic and linolenic hydroperoxides. *J. Chem. Soc., Perkin Trans.* 1, 581–587.
Crombie, L., Elliott, M., Harper, S.H., and Reed, H.W.B. (1948). Total synthesis of some pyrethrins. *Nature (London)* **162**, 222.
Crombie, L., Edgar, A.J.B., Harper, S.H., Lowe, M.W.Q. and Thompson, D. (1950a). Experiments on the synthesis of the pyrethrins. Part V. Synthesis of side-chain isomers and analogues of cinerone, cinerolone, and cinerin I. *J. Chem. Soc. (London)*, 3552–3563.
Crombie, L., Elliott, M., and Harper, S.H. (1950b). Experiments on the synthesis of the pyrethrins. Part III. Synthesis of dihydrocinerin I and tetrahydropyrethrin I; a study of the action of *N*-bromosuccinimide on 3-methyl-2-n-alkyl (and alkenyl) cyclopent-2-en-1-ones. *J. Chem. Soc. (London)*, 971–978.
Crombie, L., Harper, S.H., and Thompson, R.A. (1951a). Lactonisation of the chrysanthemic acids. *J. Sci. Food Agric.* **9**, 421–428.
Crombie, L., Harper, S.H., Stedman, R.E., and Thompson, D. (1951b). Experiments on the synthesis of the pyrethrins. Part VI. New syntheses of the cinerolones. *J. Chem. Soc. (London)*, 2445–2449.
Crombie, L., Harper, S.H., and Thompson, D. (1951c). Synthesis of *trans*-pyrethrone, *trans*-pyrethrolone, and a pyrethrin I. *J. Chem. Soc. (London)*. 2906–2915.
Crombie, L., Harper, S.H., Newman, F.C., Thompson, D., and Smith, R.J.D.S. (1956a). Experiments on the synthesis of the pyrethrins. Part X. Intermediates for the synthesis of *cis*-pyrethrolone. *J. Chem. Soc. (London)*, 126–135.
Crombie, L., Harper, S.H., and Newman, F. C. (1956b). Experiments on the synthesis of the pyrethrins. XI. Synthesis of *cis*-pyrethrolone and pyrethrin I. Introduction of the *cis*-penta-2,4-dienyl system by selective hydrogenation. *J. Chem. Soc. (London)*, 3963–3971.
Crombie, L., Harper, S.H., and Sleep, K.C. (1957). Experiments on the synthesis of the pyrethrins. XIII. Total synthesis of (±)-*cis*- and *trans*-chrysanthemumdicarboxylic acid, (±)-*cis*- and *trans*-pyrethric acid and rethrins II. *J. Chem. Soc. (London)*, 2743–2754.
Crombie, L., Crossley, J., and Mitchard, D.A. (1963). Synthesis, absolute configuration, and ring fission of *cis*- and *trans*-homocaronic acid. Their configurative relation to natural terpenes. *J. Chem. Soc. (London)*, 4957–4969.
Crombie, L. Houghton R.P., and Woods, D.K. (1967). Requirements for the artemisyl, santolinyl, and lavandulyl fission of chrysanthemic acid relatives under carbonium ion type initiation. *Tetrahedron Lett.*, 4553–4557.
Crombie, L., Hemesley, P., and Pattenden, G. (1968). Synthesis of natural ketols of the pyrethrin series. *Tetrahedron Lett.*, 3021–3024.
Crombie, L., Hemesley, P., and Pattenden, G. (1969a). Synthesis of ketols of the natural pyrethrins. *J. Chem. Soc. C.*, 1016–1024.
Crombie, L., Hemesley, P., and Pattenden, G. (1969b). Synthesis of *cis*-jasmone and other *cis*-rethrones. *J. Chem. Soc. C.*, 1024–1027.
Crombie, L., Doherty, C.F., and Pattenden, G. (1970). Syntheses of ^{14}C-labelled (+)-*trans*-chrysanthemum mono- and dicarboxylic acids, and of related compounds. *J. Chem. Soc. C.*, 1076–1080.
Crombie, L., Ellis, J.A., Gould, R., Pattenden, G., Elliott, M., Janes, N.F., and Jeffs, K.A. (1971a). Oxidative dimerisations of natural rethrolones and related compounds with manganese dioxide. *J. Chem. Soc. C.*, 9–13.
Crombie, L., Doherty, C.F., Pattenden, G., and Woods, D.K. (1971b). The acid thermal decomposition products of natural chrysanthemum dicarboxylic acid. *J. Chem. Soc. C.*, 2739–2743.
Crombie, L., Findley, D.A.R., and Whiting, D.A. (1972a). Lanthanide induced chemical shifts in natural cyclopropanes: sterochemistry of chrysanthemyl and presqualene alcohols and esters. *Tetrahedron Lett.*, 4027–4028.

Crombie, L., Findley, A.R., and Whiting, D.A. (1972b). Synthesis of prephytoene alcohol. *J. Chem. Soc., Chem. Commun.* 1045–1046.

Crombie, L., Firth, P.A., Houghton, R.P., Whiting, D.A., and Woods, D.K. (1972c). Cyclopropane cleavage of chrysanthemic acid relatives to santolinyl, artemisyl, and lavandulyl structures: acid catalysed and biosynthetic experiments. *J. Chem. Soc., Perkin Trans.* 1, 642–652.

Crombie, L., Pattenden, G., and Simmonds, D.J. (1975a). Carbon-13 nuclear magnetic resonance spectra of the natural pyrethrins and related compounds. *J. Chem. Soc., Perkin Trans.* 1, 1500–1502.

Crombie, L., King, R.W., and Whiting, D.A. (1975b). Carbon-13 magnetic resonance spectra. Synthetic presqualene esters, related cyclopropanes, and isoprenoids. *J. Chem. Soc., Perkin Trans.* 1, 913–915.

Crombie, L., Pattenden, G., and Simmonds, D.J. (1976a). Structural and physical aspects of pyrethrum chemistry. *Pestic. Sci.* **7**, 225–230.

Crombie, L., Darnborough, G., and Pattenden, G. (1976b). Wittig rearrangements of allyl vinylcyclopropylmethyl ethers. *J. Chem. Soc., Chem. Commun.*, 684.

Crombie, L., Kneen, G., Pattenden, G., and Whybrow, D. (1980). Total synthesis of the macrocyclic diterpene (−)-casbene, the putative biogenetic precursor of lathyrane, tigliane, ingenane and related terpenoid structures. *J. Chem. Soc., Perkin Trans.* 1, 1711–1717.

Crowley, M.P., Inglis, H.S., Snarey, M., and Thain, E.M. (1961). Biosynthesis of the pyrethrins. *Nature (London)* **191**, 281–282.

Crowley, M.P., Godin, P.J., Inglis, H.S., Snarey, M., and Thain, E.M. (1962). The biosynthesis of the "pyrethrins" I. The incorporation of ^{14}C-labelled compounds into the flowers of *Chrysanthemum cinerariaefolium* and the biosynthesis of chrysanthemum monocarboxylic acid. *Biochim. Biophys. Acta* **60**, 312–319.

Curran, D.P. (1983). A short synthesis of γ-hydroxycyclopentenones. *Tetrahedron Lett.*, 3443–3446.

Danda, H., Maehara, A., and Umemura, T. (1991). Preparation of (4*S*)-4-hydroxy-3-methyl-2-(2′-propynyl)-2-cyclopentenone by combination of enzymic hydrolysis and chemical transformation. *Tetrahedron Lett.*, 5119–5122.

Dauben, Jr., H.J., and Wenkert, E. (1947). Synthesis and structure of tetrahydropyrethrolone. *J. Amer. Chem. Soc.* **69**, 2074–2075.

Davis, P.J., and Miski, M. (1988). Microbial enzymatic *trans*formations of chrysanthemol, lavandulol, and analogous alcohols to acids. Eur. Pat. Appl. E.P. 258,666. (Cl C12P7/40). 09 Mar. 1988. U.S. appl. 895,050. 11 Aug. 1986; *Chem. Abstr.* **109**, 72067u (1988).

Dembitskii, A.D., Yurina, R.A., Krotova, B.I., and Suleeva, R. (1984). Composition of the essential oil of *Tanacetum boreale* Fisch. *Maslo-Zhir Prom-st.* 19–21; *Chem. Abstr.* **101**, 177243m (1984).

Demole, E.P. (1982). The fragrance of jasmine. *In* "Fragrance Chemistry" (E.T. Theimer, ed.), pp. 349–395. Academic Press, New York and London.

Demole, E., Willhalm, B, and Stoll, M. (1964). Properties and structure of the ketolactone $C_{12}H_{16}O_3$ from jasmine (*Jasminium grandiflorum* L.) essence. *Helv. Chim. Acta* **47**, 1152–1159.

De Vos, M.J., and Krief, A. (1978). Pyrocine synthesis — a new approach to chrysanthemic ester. *Tetrahedron Lett.*, 1845–1846.

De Vos, M.J., and Krief, A. (1979a). On the mechanism of the chrysanthemic ester synthesis. *Tetrahedron Lett.*, 1511–1515.

De Vos, M.J., and Krief, A. (1979b). One pot synthesis of chrysanthemic esters analogs. *Tetrahedron Lett.*, 1515–1518.

De Vos, M.J., and Krief, A. (1979c). Simple synthetic route to chrysanthemic acid and hemicaronic aldehyde. *Tetrahedron Lett.*, 1891–1892.

De Vos, M.J., and Krief, A. (1983). Enantioselective synthesis of chiral caronic esters: application to the synthesis of (1*R*)-*trans*-chrysanthemic acid and its (1*R*)-*cis*-dibromovinyl analogue from dimenthyl fumarate. *Tetrahedron Lett.*, 103–106.

De Vos, M.J., Hevesi, L., Bayet, P., and Krief, A. (1976). A new design for the synthesis of chrysanthemic esters and analogues and for the "pear ester" synthesis. *Tetrahedron Lett.*, 3911–3914.

De Vos, M.J., Denis, J.N., and Krief, A. (1978). New stereospecific synthesis of *cis*- and *trans-d,l*-chrysanthemic esters and analogues via a common intermediate. *Tetrahedron Lett.*, 1847–1850.

Dewar, M.J.S., and Ruiz, J.M. (1987). Mechanism of the biosynthesis of squalene from farnesyl pyrophosphate. *Tetrahedron* 43, 2661–2674.

Dhillon, R.S., Gautam, V.K., Singh, S., and Singh, J. (1991). A convenient and efficient synthesis of (+)-(1*R*)-*cis*-chrysanthemic acid from (+)-Δ^3-3-carene. *Indian J. Chem. Sect. B* **30B**, 574–578.

Donegan, L., Godin, P.J., and Thain, E.M. (1962). The separation and estimation of the insecticidal constituents of pyrethrum extract by gas-liquid chromatographic analysis. *Chem. Ind.* (*London*), 1420.

Doyle, M.P., Tamblyn, W.H., Buhro, W.E., and Dorow., R.L. (1981). Exceptionally effective catalysis of cyclopropanation reactions by the hexarhodium carbonyl cluster. *Tetrahedron Lett.*, 1783–1786.

Edmond, J., Popjak, G., Wong, S.-M., and Williams, V.P. (1971). Presqualene alcohol. Further evidence on the structure of a C30 precursor of squalene. *J. Biol. Chem.* **246**, 6254–6271.

Elliott, M. (1961). The pyrethrins and related compounds. II. Infra-red spectra of the pyrethrins and of other constituents of pyrethrum extract. *J. Appl. Chem.* **11**, 19–23.

Elliott, M. (1964a). The pyrethrins and related compounds. III. Thermal isomerisation of *cis*-pyrethrolone and its derivatives. *J. Chem. Soc.* (*London*), 888–892.

Elliott, M. (1964b). The pyrethrins and related compounds. IV. The ultraviolet absorption of the conjugated *cis*-pentadiene in pyrethrolone. *J. Chem. Soc.* (*London*), 1854–1855.

Elliott, M. (1964c). The pyrethrins and related compounds. V. Purification of (+)-pyrethrolone as the monohydrate, and the nature of "pyrethrolone C," *J. Chem. Soc.* (*London*), 5225–5228.

Elliott, M. (1965). The pyrethrins and related compounds. VI. The structures of the "enols" of pyrethrolone. *J. Chem. Soc.* (*London*), 3097–3101.

Elliott, M., and Casida, J.E. (1972). Optically pure pyrethroids labelled with deuterium and tritium in the methylcyclopentenonyl ring. *J. Agric. Food Chem.* **20**, 295–299.

Elliott, M., and Janes, N.F. (1969). Pyrethrin II and related esters obtained by reconstitution. *Chem. Ind.* (*London*), 270–271.

Elliott, M., and Janes, N.F. (1973). Chemistry of the natural pyrethrins. *In* "Pyrethrum, The Natural Insecticide" (J.E. Casida, ed.), pp. 55–100. Academic Press, New York and London.

Elliott, M., and Janes, N.F. (1977). Preferred conformations of pyrethroids. *In* "Synthetic Pyrethroids Symposium" (M. Elliott and R.F. Gould, eds.), pp. 29–36. ACS Symposium Series **42**; *Chem. Abstr.* **87**, 79602a (1977).

Elliott, M., Harper, S.H., and Kazi, M.A. (1967). Experiments on the synthesis of the pyrethrins. XIV. Rethrins and the cyclopentadienone related to 3-methylcyclopent-2-enone. *J. Sci. Food Agric.* **18**, 167–171.

Elliott, M., Janes, N.F., Jeffs, K.A. , Crombie, L., Gould, R., and Pattenden, G. (1969). Oxidative dimerisation of 2-alkenylcyclopent-2-ene-1,4-diones by manganese dioxide. *Tetrahedron Lett.*, 373–374.

Elliott, M., Janes, N.F., Kimmel, E.C., and Casida, J.E. (1972). Metabolic fate of pyrethrin I, pyrethrin II, and allethrin administered orally to rats. *J. Agric. Food Chem.* **20**, 300–312.

Elmore, N.F., Roberts, J.E., and Whitham, G.H. (1985). Acid-catalysed rearrangement of two esters of *cis*-chrysanthemic acid. *J. Chem. Res.* (*S*), 98.

Epstein, W.W., and Poulter, C.D. (1973). A survey of some irregular monoterpenes and their biogenetic analogies to presqualene alcohol. *Phytochemistry* **12**, 737–747.

Epstein, W.W., and Rilling, H.C. (1970). Studies on the mechanism of squalene biosynthesis. The structure of presqualene pyrophosphate. *J. Biol. Chem.* **245**, 4597–4605.

Epstein, W.W., Klobus, M.A., and Edison, A.S. (1991). Irregular monoterpene constituents of *Artemisia tridentata cana*. The isolation, characterisation and synthesis of two new chrysanthemyl derivatives. *J. Org. Chem.* **56**, 4451–4456.

Ficini, J., and d'Angelo, J. (1976). Une nouvelle voie d'acces a l'acide (±)-chrysanthemique *trans*-. *Tetrahedron Lett.*, 2441–2444.

Ficini, J., and Genêt, J.P. (1975a). Synthesis of (±)-(*Z*)-cinerolone and (±)-(*Z*)-jasmolone. *Tetrahedron Lett.*, 2633–2636.

Ficini, J., and Genêt, J.P. (1975b). New route to α-hydroxy-γ-diketones and 4-hydroxycyclopentenones. Synthesis of (±)-(*Z*)-cinerolone and (±)-(*Z*)-jasmolone. *Bull. Soc. Chim. Fr.*, 1811–1813.

Ficini, J., Falou, S., and d'Angelo, J. (1983). Revision de structure d'un benzoate allylique, intermediaire de synthese d'acides chrysanthemique. *Tetrahedron Lett.*, 375–376.

Fitzsimmons, B.J., and Frazer-Reid, B. (1979). Annulated pyranosides as chiral synthons for carbocyclic systems. Enantiospecific routes to both (+)- and (−)-chrysanthemum dicarboxylic acid from a single precursor. *J. Amer. Chem. Soc.* **101**, 6123–6125.

Fitzsimmons, B.J., and Fraser-Reid, B. (1984). Annulated pyranosides. V. An enantiospecific route to (+)- and (−)-chrysanthemum dicarboxylic acids. *Tetrahedron* **40**, 1279–1287.

Franck-Neumann, M., Martina, D., and Heitz, M.P. (1982). Synthese stereo et enantioselective par l'intermediaire de complexes du fer de derives chrysanthemique dieniques et d'aldehydes hemicaronique *cis*- and *trans*-. *Tetrahedron Lett.*, 3493–3496.

Franck-Neumann, M., Sedrati, M., Vigneron, J.P., and Bloy, V. (1985). Pyrethroids. 5. Stereo- and enantioselective syntheses of both *cis*-methyl chrysanthemates in high optical purity. *Angew. Chem.* **97**, 995–6; *Chem. Abstr.* **103**, 215579b (1985).

Franck-Neumann, M., Miesch, M., and Kempf, H. (1987a). Stereospecific synthesis of pyrethroids using a carbanionic synthon. I. Gem-pyrazolenines. *Tetrahedron* **43**, 845–852.

Franck-Neumann, M., Miesch, M., and Kempf, H. (1987b). Stereospecific synthesis of *cis*-pyrethroids using a carbanionic synthon. II. Access to *cis*-chrysanthemic and *cis*-pyrethric derivatives. *Tetrahedron* **43**, 853–858.

Fujisawa, T., Ohta, H., Kobori, T., Hata, K., and Sakai, K. (1976). The regiospecific synthesis of vicinally dialkylated cyclopentadienes and its application to a synthesis of allethrolone. *Chem. Lett.*, 943–946.

Fujitani, J. (1909). Chemistry and pharmacology of insect powder. *Arch. Exp. Pathol. Pharmakol.* **61**, 47.

Funk, R.L., and Munger, J.D. (1985). The stereospecific total synthesis of (±)-*cis*-chrysanthemic acid via the alicyclic Claisen rearrangement. *J. Org. Chem.* **50**, 707–709.

Garbers, C.F., Beukes, M.S., Ehlers, C., and McKenzie, M.J. (1978). Electrophilic addition reactions in terpenoid synthesis. VI. (±)-Chrysanthemic acid and analogues thereof. *Tetrahedron Lett.*, 77–80.

Gaughan, R.G. , and Poulter, C.D. (1979). Non-head-to-tail monoterpenes. Synthesis of (*S*)-lyratol and (*S*)-lyratyl acetate from (1*R*,3*R*)-chrysanthemic acid. *J. Org. Chem.* **44**, 2441–2444.

Genêt, J.P., Piau, F., and Ficini, J., (1980). A novel synthesis of (±)-*trans*-chrysanthemic acid. *Tetrahedron Lett.*, 3183–3186.

Genêt, J.P., Denis, A., and Charbonnier, F. (1986). Synthesis of ethyl (±)-*cis*-3-(2,2-dihalogenovinyl)-2,2-dimethylcyclopropanecarboxylate. *Bull. Soc. Chim. Fr.*, 793–796.

Gnadinger, C.B. (1936–1945). "Pyrethrum Flowers," Supplement to 2nd Ed. McLaughlin Gormley King Co., Minneapolis, Minnesota.

Gnadinger, C.B. (1936). "Pyrethrum Flowers," 2nd Ed. McLaughlin Gormley King Co., Minneapolis, Minnesota.

Godin, P.J., Inglis, H.S., Snarey, M., and Thain, E.M. (1963). Biosynthesis of the pyrethrins. Part II. Pyrethric acid and the origin of the ester methyl group. *J. Chem. Soc.* (*London*), 5878–5880.

Godin, P.J., Sleeman, R.J., Snarey, M., and Thain, E.M. (1966). The jasmolins, new insecticidally active constituents of *Chrysanthemum cinerariaefolium* Vis. *J. Chem. Soc. C.*, 332–334.

Godin, P.J., King, T.A., Stahl, E., and Pfeifle, J. (1967). "Pyrethrum" in peonies. *Nature* **214**, 319.

Goffinet,R., and Locatelli, A. (1968). Separation of *d-trans*-chrysanthemic acid from its optical and geometric isomers. French Patent 1,536,458; *Chem. Abstr.* **71**, 90923w (1969).

Goldschmidt, Z., Crammer, B., and Ikan, R. (1984). Mechanistic study of the thermal acid-catalysed rearrangement of *trans*-methyl chrysanthemate to lavandulyl derivatives. *J. Chem Soc., Perkin Trans.* 1, 2697–2705.

Goldschmidt, Z., Crammer, B., and Gottlieb, H.E. (1987). The formation and X-ray crystal structure of a novel bridged dilactone derived from chrysanthemic acid by dimerisation of lavandulylic acids. *Heterocycles* **26**, 607–611; *Chem. Abstr.* **108**, 56342d (1988).

Gopichand, Y., Khanra, A.S., Mitra, R.B., and Chakravarti K.K. (1975). New efficient synthesis of (+)-*cis*- and (+)-*trans*-chrysanthemic acids from (+)-Δ^3-3-carene. *Indian J. Chem.* **13**, 433–436.

Green, G., Griffith, W.P., Hollinshead, D.M., Ley, S.V., and Schroder, M. (1984). Oxo complexes of ruthenium (VI) and (VII) as organic oxidants. *J. Chem. Soc., Perkin Trans.* 1, 681–686.

Greuter, H., Dingwall, J., Martin, P., and Bellus, D. (1981). Synthesis and reactivity of four-membered compounds. Part 17. Synthesis of (1*R*)-*cis*-3-(2′,2′-dihalovinyl)-2,2-dimethylcyclopropane carboxylic acids via Favorskii rearrangement of optically active cyclobutanes. *Helv. Chim. Acta* **64**, 2812–2820.

Griffith, W.P., Ley, S.V., Whitcombe, G.P., and White, A.D. (1987). Preparation and use of tetra-*n*-butylammonium per-ruthenate (TBAP reagent) and tetra-*n*-propylammonium perruthenate (TPAP reagent) as new catalytic oxidants for alcohols. *J. Chem. Soc., Chem. Commun.* 1625–1627

Groneman, A.F., Posthumus, M.A., Tuinstra, L.G.M.T., and Traag, W.A. (1984). Identification and determination of metabolites in plant cell biotechnology by g.c/m.s. Application to non-polar products of *Chrysanthemum cinerariaefolium* and *Tagetes* species. *Anal. Chim. Acta* **163**, 43–54.

Hamberg, M. (1987). Mechanism of corn hydroperoxide isomerase: detection of 12,13(*S*)-oxido-9(*Z*),11-octadecadienoic acid. *Biochem. Biophys. Acta* **920**, 76–84.

Hanafusa, T., Ohnishi, M., Mishima, M., and Yakawa, Y. (1970) Thermal *cis-trans*-isomerisation of methyl chrysanthemate. *Chem. Ind.* (*London*), 1050–1052.

Harper, S.H. (1954). The chrysanthemum carboxylic acids. VII. Catalytic hydrogenation of the chrysanthemic acids. *J. Sci. Food Agric.* **5**, 529–533.

Harper, S.H., and Thompson, R.A. (1952). The chrysanthemum carboxylic acids. V. Hydration of chrysanthemic acids. *J. Sci. Food Agric.* **3**, 230–234.
Harper, S.H., Reed, H.W.B., and Thompson, R.A. (1951). The chrysanthemum carboxylic acids. I. Preparation of the chrysanthemic acids. *J. Sci. Food Agric.* **2**, 94–100.
Harper, S.H., Sleep, K.C., and Crombie, L. (1954). Synthesis of the naturally derived geometrical isomer of chrysanthemum dicarboxylic acid. *Chem. Ind.* (*London*), 1538–1539.
Head, S.W. (1964). The identification of the active constituents of pyrethrum extract when separated by gas-liquid chromatography. *Pyrethrum Post* **7**(4), 12–14.
Head, S. W. (1966a). The quantitative determination of "pyrethrins" by gas-liquid chromatography. Part I. Detection by electron capture. *Pyrethrum Post* **8**(4), 3–7.
Head, S. W. (1966b). A study of the insecticidal constituents in *Chrysanthemum cinerariaefolium*. (1). Their development in the flower head. (2). Their distribution in the plant. *Pyrethrum Post* **8**(4), 32–37.
Head, S.W. (1967). The quantitative determination of "pyrethrins" by gas-liquid chromatography. Part II. Detection by hydrogen flame ionisation. *Pyrethrum Post* **9**(1), 12–17.
Head, S.W. (1968a). Fatty acid composition of extract from pyrethrum flowers (*Chrysanthemum cinerariaefolium*). *J. Agric. Food Chem.* **16**, 762–765.
Head, S. W. (1968b). The quantitative determination of the "pyrethrins" by gas-liquid chromatography. Part III. Through chromatography of the methyl esters of chrysanthemic acid and chrysanthemum dicarboxylic acid. *Pyrethrum Post* **9**, 31–36.
Heintz, R., Benveniste, P., Robinson, W.H., and Coates, R.M. (1972). Plant sterol metabolism. Demonstration and identification of a biosynthetic intermediate between farnesyl PP and squalene in a higher plant. *Biochem. Biophys. Res. Commun.* **49**, 1547–1553.
Ho, T.L. (1983). Routes to *trans*-chrysanthemic acid . IV. Synthesis from 4-acetyl-2-carene. *Synth. Commun.* **13**, 761–764.
Ho, T.L., and Din, Z.U. (1982). Routes to *trans*-chrysanthemic acid. II. Cyclomutative transformation of (−)-carvone. *Synth. Commun.* **12**, 257–259.
Hogstad, S., Loehre, J.G., and Anthonsen, T. (1984). Possible confusion of pyrethrins with thiophenes in *Tagetes* species. *Acta Chem. Scand. Ser. B* **B38**, 902–904.
Holm, K.H., and Skattebol, I. (1985). Further evidence for the vinylcyclopropylidene-cyclopentenylidene rearrangement. *Acta Chem. Scand. Ser. B* **B39**, 549–561.
Holmstead, R.L., and Soderlund, D.M. (1977). Separation and analysis of the pyrethrins by combined gas-liquid chromatography and chemical ionisation mass spectrometry. *J. Assoc. Off. Anal. Chem.* **60**, 685–689.
Horiuchi, F., and Matsui, M. (1973). Chrysanthemic acid. XXIII. New resolving agents. 2-Benzylamino alcohols synthesised from natural amino acids. *Agric. Biol. Chem.* **37**, 1713–1716.
Hunsdiecker, H. (1947). On the preparation of γ-diketones. The cyclopentenone ring closure of γ-diketones of the type $CH_3COCH_2CH_2COCH_3$. *Ber. Dtsch. Chem. Gess.* **75**, 455–460, 460–468.
Inokuchi, T., Tsujiyama, H., Kusumoto, M., and Torii, S. (1988). A facile synthesis of methyl (+)-*cis*-chrysanthemate from (+)-3-carene. *Chem. Express* **3**, 623–626; *Chem. Abstr.* **111**, 134494h (1989).
Inouye, Y., and Ohno, M. (1958). Absolute configuration of natural pyrethrins. *Kagaku* (*Tokyo*) **28**, 636.
Jacobson, M., and Crosby, D.G. (Eds) (1971). "Naturally Occurring Insecticides," Dekker, New York.
Jain, S.C. (1977). Chemical investigation of *Tagetes* tissue cultures. *Planta Medica* **31**, 68–70.
Joshi, G.S., and Kulkarni, G.H. (1988). Synthesis of 3-phenyl-3,4-dihydro-7,7-dimethylbicyclo-[4.1.0]heptane: a useful synthon for 1*R*-*cis*-pyrethroids from (+)-3-carene. *Chem. Ind.* (*London*), 370–371.
Julia, M., and Guy-Rouault, A. (1967). Synthèse de cyclopropanes à partir de sulfones. Application a l'acide chrysanthèmique. *Bull. Soc. Chim. Fr.*, 1411.
Julia, M., Julia, S., Bemont, B., and Tchernoff, C. (1959). Sur l'isomerisation des esters cyclopropaniques *cis*- en *trans*-. *C. R. Acad. Sci.* **248**, 242–244.
Julia, M., Julia, S., Jeanmart, C., and Langlois, M. (1962). Synthèses de la pyrocine et de l'acide *trans*-chrysanthèmique. *Bull. Soc. Chim. Fr.*, 2243–2246.
Julia, M., Julia, S., and Cochet, B. (1964a). Synthèse de l'acide (±)-*trans*-chrysanthèmique à partir du β-ethoxy-isovaleraldehyde et de l'acide (±)-nor-*trans*-chrysanthèmique à partir du β-ethoxy-butyraldehyde. *Bull. Soc. Chim. Fr.*, 1476–1486.
Julia, M., Julia, S., and Cochet, B. (1964b). Synthèses de la β,β-dimethyl-γ-isobuténytl-γ-butyrolactone et *trans*formation en acide chrysanthèmique. *Bull. Soc. Chim. Fr.*, 1487–1492.
Julia, M., Julia, S., and Linstrumelle, G. (1964c). Synthèse des acides (±)-*cis*-homocaronique et

(±)-*trans*-chrysanthemique par l'intermediaire de bicyclo[3,1,0]hexanones-2-substituées. *Bull. Soc. Chim. Fr.*, 2693–2694.

Julia, M., Julia, S., and Langlois, M. (1965a). Synthèses de l'acide *trans*-chrysanthèmique à partir de la dimethylacroleine et de l'acide *trans*-nor-chrysanthèmique à partir du crotonaldehyde. *Bull. Soc. Chim. Fr.*, 1007–1014.

Julia, M., Julia, S., and Langlois, M. (1965b). Synthèse de l'acide chrysanthèmique a partir de l'acide lévulique. *Bull. Soc. Chim. Fr.*, 1014–1019.

Kamal, R. (1977). Isolation of pyrethrins from *Tagetes erecta*. Indian 142,643(Cl. CO7C5/00). 06 Aug. 1977. Appl. 76/DE 10, 18 Oct, 1976; *Chem. Abstr.* **92**, P55353x (1980).

Katsuda, Y., Chikamoto, T., and Inouye, Y. (1958). The absolute configuration of naturally derived pyrethrolone and cinerolone. *Bull. Agric. Chem. Soc. Jap.* **22**, 427–428.

Katsuda, Y., Chikamoto, T., and Inouye, Y. (1959). Relationship between stereoisomerism and biological activity of pyrethroids. V. The absolute configuration of (+)-pyrethrolone and (+)-cinerolone. *Bull. Agric. Chem. Soc. Jap.* **23**, 174–178.

Kawamoto, I., Muramatsu, S., and Yura, Y. (1975). Cyclopentenones. II. A new synthesis of allethrolone. *Synth. Commun.* **5**, 185–191.

Kawano, Y., Yanagihara, K.H., and Bevenue, A. (1974). Analytical studies of pyrethrin formulations by gas chromatography. III. Analytical results on insecticidally active components from various world sources. *J. Chromatogr.* **90**, 119–128.

Kawano, Y., Yanagihara, K., Miyamoto, T., and Yamamoto, I. (1980). Examination of the conversion products of pyrethrins and allethrin formulations exposed to sunlight, by gas chromatography and mass spectrometry. *J. Chromatogr.* **198**, 317–328.

Khanna, P., and Khanna, R. (1976). Endogenous free ascorbic acid and effect of exogenous ascorbic acid on growth and production of pyrethrins from *in vitro* tissue culture of *Tagetes erecta* L. *Indian. J. Exp. Biol.* **14**, 630–631.

Khanna, P., Sharma, R., and Khanna, R. (1975). Pyrethrins from *in vivo* and *in vitro* tissue culture of *Tagetes erecta*. *Indian J. Exp. Biol.* **13**, 508–509.

Khanra, A.S., and Mitra, R.B. (1976). A new synthesis of (+)-*trans*-chrysanthemic acid from (+)-car-3-ene. *Indian J. Chem. Sect B* **14B**, 716–718.

King, T.A., and Paisley, H.M. (1969). Mass spectra and isolation of pyrethroids and related compounds. *J. Chem. Soc. C.*, 870–874.

Krief, A., Hevesi, L., Chaboteaux, G., Mathy, P., Sevrin, M., and De Vos, M.J. (1985). 2-Metallo-2-nitropropanes as isopropylidene transfer reagents for the cyclopropanation of electrophilic alkenes. Application to the synthesis of *trans*-chrysanthemic acid. *J. Chem. Soc., Chem. Commun.*, 1693–1695.

Krief, A., Dumont, W., and Pasau, P. (1988a). From tartaric acid to the most biologically active insecticides. Straightforward enantioselective synthesis of the pyrethrins. *Tetrahedron Lett.*, 1079–1082.

Krief, A., Surleraux. D., and Frauenrath, H. (1988b). Novel stereospecific route to *cis*-chrysanthemic acid. *Tetrahedron Lett.*, 6157–6160.

Krief, A., Dumont, W., Pasau, P., and Lecomte, P. (1989). Stereoselective synthesis of methyl (1*R*)-*trans*- and (1*R*)-*cis*-hemicaronaldehydes from natural tartaric acid: application to the synthesis of *S*-bioallethrin and deltamethrin insecticides. *Tetrahedron* **45**, 3039–3052.

Krief, A., Surleraux, D., Dumont, W., Pasau, P., and Lecomte, P. (1990). From grape to grape: novel stereoselective syntheses of chiral pyrethroids — synthesis of the most potent commercially available insecticides. *Pure Appl. Chem.* **62**, 1311–1318.

Kueh, J.S.H., MacKenzie, I.A., and Pattenden, G. (1985). Production of chrysanthemic acid and pyrethrins by tissue cultures of *Chrysanthemum cinerariaefolium*. *Plant Cell Rep.* **4**, 118–119.

LaForge, F.B., and Barthel, W.F. (1944). Heterogeneous nature of pyrethrolone. *J. Org. Chem.* **9**, 242–249. See also *idem, ibid.* (1945), **10**, 106–113; 114–120.

LaForge, F.B., and Green, N. (1952). Constituents of pyrethrum flowers. XXV. The synthesis of *d*-cinerolone, cinerin I, and its optical isomers. *J. Org. Chem.* **17**, 1635–1640.

LaForge, F.B., Green, N., and Schechter, M.S. (1952). Dimerized cyclopentadienones from esters of allethrolone. *J. Amer. Chem. Soc.* **74**, 5392–5394.

LaForge, F.B., Green, N., and Schechter, M.S. (1954). Allethrin. Resolution of *dl*-allethrolone and synthesis of four optical isomers of *trans*-allethrin. *J. Org. Chem.* **19**, 457–462.

Lehmkuhl, H., and Mehler, K. (1978). Anlagerungen von Alkylmagnesiumhalogeniden an 3,3-Dimethylcyclopropen: ein neuer weg zu *cis*- oder *trans*-chrysanthemumsäure. *Liebig's Ann. Chem.*, 1841–1853.

Lehmkuhl, H., and Mehler, K. (1982). Notiz über eine Chrysanthemumsure synthese. *Liebig's Ann. Chem.*, 2244.

LeMahieu, R.A., Carson, M., and Kierstead, R.W. (1968). The conversion of cinerone into cinerolone. *J. Org. Chem.* **33**, 3660–3662.

LeMahieu, R.A., Tabenkin, B., Berger, J., and Kierstead, R.W. (1970). Microbiological hydroxylation of allethrone. *J. Org. Chem.* **35**, 1687–1688.

Lochynski, S., Jarosz, B., Walkowicz, M., and Piatkowski, K. (1988). Modification of the synthesis of dihydrochrysanthemolactone from (+)-3-carene. *J. Prakt. Chem.* **330**, 284–288.

Lowenthal, R.E., and Masamune, S. (1991). Asymmetric copper catalysed cyclopropanation of trisubstituted and unsymmetrical *cis*-1,2-disubstituted olefins: modified bis-oxazoline ligands. *Tetrahedron Lett.*, 7373–7376.

Madeleyn, E., and Vandewalle, M. (1973). Cyclopentanones. VII. Synthesis of the rethrolones. *Bull. Soc. Chim. Belg.* **82**, 293–297.

Majewski, M., and Snieckus, V. (1984). Synthesis of pyrethroid amides via epoxy amide cyclisation. *J. Org. Chem.* **49**, 2682–2687.

Mancuso, A.J., Brownfain, D.S., and Swern, D. (1979). Structure of the dimethyl sulfoxide–oxalyl chloride reaction product. Oxidation of heteroaromatic and diverse alcohols to carbonyl compounds. *J. Org. Chem.* **44**, 4148–4150.

Mann, J., and Thomas, A. (1986). A novel approach to *cis*-chrysanthemic acids. *Tetrahedron Lett.*, 3533–3534.

Mann, J., and Weymouth-Wilson, A. (1991). A new approach to *cis*-chrysanthemic acid. *Carbohydrate Res.* **216**, 511–515.

Martel, J. (1969). *d-trans*-Pyrethric acid. German Patent 1,807,091 (and 1,935,986); *Chem. Abstr.* **72**, 89876x (1970).

Martel, J., and Buendia, J. (1970a). Racemic and optically active *cis*-chrysanthemumic acids. German Patent 2,010,182.; *Chem. Abstr.* **73**, 109362c (1970).

Martel, J., and Buendia, J. (1970b). New syntheses of the natural pyrethric acid. *Proc. IUPAC Riga Symp. Natur. Prod.*, Abstr. E112, 572.

Martel, J., and Goffinet, B. (1968). (±)-*trans*-Chrysanthemic acid and its esters. *French Addition* 90,564.; *Chem. Abstr.* **70**, 37280n (1969).

Martel, J., and Huynh, C. (1967). Synthèse de l'acid chrysanthèmique. II. Accès stéréosélectif au (±)-*trans*-chrysanthemate d'éthyl. *Bull. Soc. Chim. Fr.*, 985–986.

Martel, J., Huynh, C., and Nomine, G. (1966). Unsaturated cyclopentenones. French Patent 1,434,224; *Chem. Abstr.* **65**, 16878f (1966).

Martel, J., Huynh, C., Toromanoff, E., and Nomine, G. (1967). Accés aux β-thianones et β-thianonols dioxyde-1,1 par l'emploi des dialcoylsulfones. Application à la préparation de la β,β-diméthyl-γ-isobutenyl-γ-butyrolactone. Bull. Soc. Chim. Fr. 982–985.

Martin, P., Greuter, H., Rihs, G., Winkler, T., and Bellus, D. (1981). Synthesis and reactivity of four-membered ring compounds. Part 16. Rearrangement of α-halo- to α′-halocyclobutanones, key step of a highly versatile synthesis of pyrethroids. *Helv. Chim. Acta* **64**, 2571–2586.

Matsui, M. (1950). Pyrocin (a new insecticide). *Botyu-Kagaku* **15**, 1–20.

Matsui, M., and Horiuchi, F. (1971). Resolution of the chrysanthemic acids with L-lysine. *Agric. Biol. Chem.* **35**, 1984–1985.

Matsui, M., and Meguro, H. (1963). Studies on chrysanthemic acid. IX. Alternate preparation of (+)-*trans*-pyrethric acid and methyl (+)-*trans*-2,2-dimethyl-3-(2-carboxy-1-propyl)cyclopropane carboxylate. *Agric. Biol. Chem.* **27**, 379–380.

Matsui, M., and Uchiyama, M. (1962). Studies on chrysanthemic acid. VII. Selective synthesis of (±)-*trans*-chrysanthemic acid from (±)-pyrocin. *Agric. Biol. Chem.* **26**, 532–534.

Matsui, M., and Yamada, Y. (1963). Studies on chrysanthemic acid. VIII. Synthesis of pyrethric acids. *Agric. Biol. Chem.* **27**, 373–378.

Matsui, M., and Yamada, Y. (1965). Studies on chrysanthemic acid. Part XV. Selenium dioxide oxidation of tert-butyl (±)-chrysanthemate. The isolation of tert-butyl (±)-*trans*-2,2-dimethyl-3-(2′-formyl-3′-hydroxyl-1′-propenyl)cyclopropanecarboxylate. *Agric. Biol. Chem.* **29**, 956–958.

Matsui, M., and Yoshioka, H. (1965). *trans*-Chrysanthemummonocarboxylic acid. Japanese Patent 6457 (1965); *Chem. Abstr.* **63**, 1822c (1965).

Matsui, M., Miyano, M., Yamashita, K., Kubo, H., and Tomita, K. (1957). Synthesis of pyrethridic (chrysanthemumdicarboxylic) acid. *Bull. Agric. Chem. Soc. Jap.* **21**, 22–29.

Matsui, M., Uchiyama, M., and Yoshioka, H. (1963a). Studies on chrysanthemic acid. Part X. Oxidation and rearrangement products from chrysanthemic acid, 2,6-dimethylhepta-2,4,6-triene-5-carboxylic acid. *Agric. Biol. Chem.* **27**, 549–553.

Matsui, M., Uchiyama, M., and Yoshioka, H. (1963b). Studies on chrysanthemic acid. Part XI. Oxidation products from chrysanthemic acid: 2,2-dimethyl-3-(1′-oxo-2′-hydroxy-2′-methyl)propyl-1,3-*trans*-cyclopropane-1-carboxylic acid. *Agric. Biol. Chem.* **27**, 554–557.

Matsui, M., Uchiyama, M., and Yoshioka, H. (1963c). Studies on chrysanthemic acid. Part XII. Racemisation of optically active 2,2-dimethyl-3-(1′-oxo-2′-hydroxy-2′-methyl)propyl-1,3-*trans*-cyclopropane-1-carboxylic acid and reduction to give (±)-*trans*-chrysanthemic acid. *Agric. Biol. Chem.* **27**, 558–561.
Matsui, M., Yoshioka, H., Sakamoto, H., Yamada, Y., and Kitahara, T. (1965). Selective synthesis of optically pure (+)- and (−)-*trans*-chrysanthemic acids from (+)-Δ^3-3-carene. *Agric. Biol. Chem.* **29**, 784–786.
Matsui, M., Yoshioka, H., Sakamoto, H., Yamada, Y., and Kitahara, T. (1965). Studies on chrysanthemic acid. Part XVII. Stereospecific synthesis of optically pure (+)-*trans*-chrysanthemic acid from (+)-Δ^3-carene. *Agric. Biol. Chem.* **31**, 33–37.
Matsumoto, T., Nagai, A., and Takahashi, Y. (1963). The stereoselective synthesis of *trans*-chrysanthemum monocarboxylic acid. *Bull. Chem. Soc. Jap.* **36**, 481–482.
Matsuo, N., Takagaki, T., Watanabe, K., and Ohno, N. (1993). The first practical synthesis of (*S*)-pyrethrolone, an alcohol moiety of natural pyrethrins I and II. *Biosci. Biotech. Biochem.* **57**, 693–694.
Matsuo, T., Mori, K., and Matsui, M. (1976). Synthesis of optically active *trans*-chrysanthemic acid from optically active pantolactone. *Tetrahedron Lett.*, 1979–1982.
Meinen, V.J., and Kassera, D.C. (1982). Gas-liquid chromatographic determination of pyrethrins and piperonyl butoxide: collaborative study. *J. Assoc. Off. Anal. Chem.* **65**, 249–255.
Miersch, O., Priess, A., Sembdner, G., and Schreiber, K. (1987). (+)-7-Isojasmonic acid and related compounds from *Botryodiplodia theobromae*. *Phytochemistry* **26**, 1037–1039.
Miersch, O., Schmidt, J., Sembdner, G., and Schreiber, K. (1989). Jasmonic acid-like substances from *Botryodiplodia theobromae*. *Phytochemistry* **28**, 1303–1305.
Mills, R.W., Murray, R.D.H., and Raphael, R.A. (1973). A new stereoselective synthesis of *trans*-chrysanthemic acid [2,2-dimethyl-3-(2-methylprop-1-enyl)]cyclopropanecarboxylic acid. *J. Chem. Soc., Perkin Trans.* 1, 133–137.
Miski, M., and Davis, P.J. (1988a). Microbiologically catalysed enantio- and diastereo selective oxidation of chrysanthemol stereoisomers to chrysanthemic acids. *Appl. Environ. Microbiol.* **54**, 2268–2272.
Miski, M., and Davis, P. J. (1988b). Gc determination of racemic *cis*- and *trans*-chrysanthemols and their potential aldehyde and carboxylic acid microbial metabolites. *J. Chromatogr.* **437**, 436–441.
Mitra, R.B., and Khanra, A.S. (1977). A stereospecific synthesis of methyl (+)-*trans*-chrysanthemate from (+)-pinene. *Synth. Commun.* **7**, 245–250.
Mitra, R.B., Kulkarni, G.H., and Khanna, P.N. (1987). A novel approach to synthesis of (1*R*)-*cis*-caronaldehyde, a key intermediate for photostable pyrethroid insecticides. *Synth. Commun.* **17**, 1089–1084.
Mitsuta, M., Komaki, R., Ando, Y., and Hirohara, H. (1985). Biochemical preparation of optically active chrysanthemic acid. Jpn. Kokai Tokyo Koho JP 60,199,393. [85, 199, 393] (Cl C12p7/62) 08 Oct 1985.
Miyano, M., and Dorn, C.R. (1973). Prostaglandins. VI. Correlation of the absolute configuration of pyrethrolone with that of the prostaglandins. *J. Amer. Chem. Soc.* **95**, 2664–2669.
Monpert, A., Martelli, J. Gree, R., and Carrie, R. (1983). Synthesis of optically active hemicaronaldehydes using chiral butadiene-iron tricarbonyl complexes. *Nouveau J. Chim.* **7**, 345–346; *Chem Abstr.* **100**, 22808e (1984).
Mourot, D., Boisseau, J., and Gayot, G. (1978). Separation of the pyrethrins by hplc. *Anal. Chim. Acta* **97**, 191–195.
Mukaiyama, T., Yamashita, H., and Asami, M. (1983). An asymmetric synthesis of bicyclic lactones and its application to the asymmetric synthesis of (1*R*, 3*S*)-*cis*-chrysanthemic acid. *Chem. Lett.*, 385–388.
Murano, A. (1972). Determination of the optical isomers of insecticidal pyrethroids by gas-liquid chromatography. *Agric. Biol. Chem.* **36**, 917, 2203–2211.
Muscio, F., Carlson, J.P., Kuehl, L., and Rilling, H.C. (1974). Presqualene pyrophosphate. Normal intermediate in squalene biosynthesis. *J. Biol. Chem.* **249**, 3746–3749.
Naik, R.H., and Kulkarni, G.H. (1983). Synthesis of methyl (+)-*cis*-chrysanthemate and (+)-*cis*- homochrysanthemate from (+)-3-carene. *Indian J. Chem. Sect. B.* **22B**, 859–863.
Nakada, Y., Yura, Y., and Murayama, K. (1972). Chrysanthemate derivatives. II. Pyrolysis of pyrethrin I. *Bull. Chem. Soc. Jap.* **45**, 2243.
Nelson, R.H., (ed.) (1945–1972). "Pyrethrum Flowers," 3rd Ed. McLaughlin Gormley King Co., Minneapolis, Minnesota.
Nes, W.D., van Tamelen, E.E., and Leopold, E.J. (1988). Presqualene alcohol from *Gibberella fujikuroi*. *Phytochemistry* **27**, 628–629.

Nes, W.D., Phu, Le, van Tamelen, E.E., and Leopold, E.J. (1990). Presqualene alcohol, squalene, and sterol biosynthesis from bifarnesol. *Exp. Mycol.* **14**, 74–77.
Nesmeyanova, O.A., Rudashevskaya, T.A., Dyachenko, A.I., Savilova, S.F., and Nefedov, O.M. (1982). A new synthesis of *cis*-chrysanthemic acid. *Synthesis*, 296.
Nieass, C.S., Wainwright, M.S., and Chaplin, R.P. (1984). Applications of a technique for the hplc analysis of liquid carbon dioxide solutions. *J. Liquid Chromatogr.* **7**, 493–508.
Nikiforov, A., and Kohlmann, H. (1983). Characterisation of natural and synthetic pyrethrins by EI, CI, and FI/FD mass spectra and computer averaged integrated FD mass spectrometry. *Spectrom. Ion Phys.* **48**, 141–144.
Nishizawa, Y., and Casida, J.E. (1965). Synthesis of *d-trans*-chrysanthemic acid -1-^{14}C and its antipode on a micro scale. *J. Agric. Food Chem.* **13**, 525–527.
Nohira, H., Nohira, M., Yoshida, S., Osada, A., and Terunuma, D. (1988). Optical resolution of α-ethyl benzylamine and its application as a resolving agent. *Bull. Chem. Soc. Jap.* **61**, 1395–1396.
Ohkata, K., Isako, T., and Hanafusa, T. (1978). Syntheses of chrysanthemum monocarboxylic acid and its related compounds. *Chem. Ind. (London)*, 274–275.
Ohloff, G. (1965). Thermische Umlagerung von Derivaten des (+)-Δ^4-Carene und der (±)-*cis-trans*- Chrysanthemumsäure. *Tetrahedron Lett.*, 3795–3800.
Ohno, M., Matsuoka, S., and Eguchi, S. (1986). Catalytic electrophilic reactions of chrysanthemic acid derivatives with unsaturated organosilanes. An application to synthesis of modified types of C_{15} and C_{20} isoprenoids with non-head-to-tail linkages. *J. Org. Chem.* **51**, 4553–4558.
Okada, K., Nozaki, M., Takashima, Y., Nakatani, N., Nakatani, Y., and Matsui, M. (1977). Synthesis of pyrethrolone derivatives. *Agric. Biol. Chem.* **41**, 2205–2208.
Olive, B.M. (1973). Colour specific reagent for the identification and semiquantitation of pyrethrins and piperonyl butoxide by tlc. *J. Assoc. Off. Anal Chem.* **56**, 915–918.
Otieno, D.A., Pattenden, G., and Popplestone, C.R. (1977). Thermal acid catalysed rearrangements of natural chrysanthemic acid. *J. Chem. Soc., Perkin Trans.* 1, 196–201.
Otieno, D.A., Jondiko, I.J., McDowell, P.G., and Kezdy, F.J. (1982). Quantitative analysis of the pyrethrins by hplc. *J. Chromatogr. Sci.* **20**, 566–570.
Pattenden, G. (1970). Biosynthesis of the pyrethrins. *Pyrethrum Post* **10**(4), 2–5.
Pattenden, G., and Storer, R. (1973a). Isomerization of 4-hydroxy-2-(prop-2-enyl)cyclopent-2-enones to 2-*n*-propylcyclopent-2-ene-1,4-diones. *J. Chem. Soc., Chem. Commun.*, 875–876.
Pattenden, G., and Storer, R. (1973b). Studies on the biosynthesis of chrysanthemum monocarboxylic acid. *Tetrahedron Lett.*, 3473–3476.
Pattenden, G., and Storer, R. (1974a). Synthesis of *Z*-rethrolones and *Z*-rethrones. *J. Chem. Soc., Perkin Trans* 1, 1603–1606.
Pattenden, G., and Storer, R. (1974b). Acid catalysed *trans*formations of substituted 4-hydroxy-2-(prop-2-enyl)cyclopent-2-enones. *J. Chem. Soc., Perkin Trans.* 1, 1606–1611.
Pattenden G., and Storer, R. (1976). Asymmetric synthesis of [C-1-^{3}H]-labelled (±)-*trans*-[1*R*,3*R*]-chrysanthemic acid. *J. Labelled Compd. Radiopharm.*, 551–556.
Pattenden, G., Crombie, L., and Hemesley, P. (1973). Mass spectra of the pyrethrins and related compounds. *Org. Mass. Spectrom.* **7**, 719–735.
Pattenden, G., Popplestone, C.R., and Storer, R. (1975). Investigation of the role of chrysanthemyl, lavandulyl and artemisyl alcohols in the biosynthesis of chrysanthemic acid. *J. Chem. Soc., Chem. Commun.*, 290–291.
Pellegrini Jr., J.P., Miller A.C., and Sharpless, R.V. (1952). Biosynthesis of radioactive pyrethrins using $^{14}CO_2$. *J. Econ. Entomol.* **45**, 532–536.
Pierre, J.-L., Perraud, R., and Arnaud, P. (1970). Résonance magnétique nucléaire des petits cycles. XII. Étude conformationelle des esters et alcools (primaires) *cis*- and *trans*-chrysanthémiques. *Bull. Soc. Chim. Fr.*, 1539–1550.
Popjak, G., Goodman, D.S., Cornforth, J.W., Cornforth, R.H, and Ryhage, R. (1961). Studies on the biosynthesis of cholesterol. XV. Mechanism of squalene biosynthesis from farnesyl pyrophosphate and from mevalonate. *J. Biol. Chem.* **236**, 1934–1947.
Popjak, G., Edmond, J., and Wong, S. (1973). Absolute configuration of presqualene alcohol. *J. Amer. Chem. Soc.* **95**, 2713–2714.
Popjak, G., Ngan, H.-L., and Agnew, W. (1975). Stereochemistry of the biosynthesis of presqualene alcohol. *Bioorg. Chem.* **4**, 279–289.
Poulter, C.D. (1972). Model studies of terpene biosynthesis. Stereoselective ionisation of *N*-methyl-4-[(α-*S*,1*R*,3*R*)-chrysanthemyloxy]pyridinium-diiodide. *J. Amer. Chem. Soc.* **94**, 5515–5516.
Poulter, C.D., and Hughes, J.M. (1977a). Model studies of non-head-to-tail terpenes. Stereochemistry of ionisation for *N*-methyl-4-[(1*S*,1′*R*,3′*R*)-[1-^{2}H]-chrysanthemyloxy]- pyridinium iodide. *J. Amer. Chem. Soc.* **99**, 3824–3829.

Poulter, C.D., and Hughes, J.M. (1977b). Model studies of the biosynthesis of non-head-to-tail terpenes. Stereochemistry of the head-to-head rearrangement. *J. Amer. Chem. Soc.* **99**, 3830–3837.
Poulter, C.D., and Rilling, H.C. (1981). Prenyl transferases and isomerase. *In* "Biosynthesis of Isoprenoid Compounds" (J.W. Porter and S.L. Spurgeon, eds.), Vol. 1, p. 413. John Wiley, New York and Chichester.
Poulter, C.D., Rilling, H.C., Epstein, W.W., and Larsen, B. (1971). Mechanism of squalene biosynthesis. Presqualene pyrophosphate, stereochemistry, and a mechanism for its conversion to squalene. *J. Amer. Chem. Soc.* **93**, 1783–1785.
Poulter, C.D., Marsh, L.L., Hughes, J.M., Argyle, J.C., Satterwhite, D.M., Goodfellow, R.J., and Moesinger, S.G. (1977). Model studies of the biosynthesis of non-head-to-tail terpenes. Rearrangements of chrysanthemyl systems. *J. Amer. Chem Soc.* **99**, 3816–3823.
Ravishankar, G.A., Rajasekaran, T., Sarma, K.S., and Venkataraman, L.V. (1980). Production of pyrethrins in cultured tissues of pyrethrum (*Chrysanthemum cinerariaefolium* Vis.). *Pyrethrum Post* **17**, 66–69.
Reichelt, I., and Reissig, H.-U. (1985). Neue Chrysanthemumsäure Derivate durch Deprotoneirung-Alkylierung. *Liebig's Ann. Chem.*, 650–652.
Rickett, F.E. (1972). Preparative scale separation of pyrethrins by liquid-liquid partition chromatography. *J. Chromatogr.* **66**, 356–360.
Rickett, F.E. (1973). Determination of the enantiomeric purity of synthetic pyrethroids. I. Chrysanthemic acid moiety. *Analyst* (*London*) **98**, 687–691.
Rickett, F.E., and Henry, P.B. (1974). Quantitative determination of the enantiomeric purity of synthetic pyrethroids. II. *S*-Bioallethrin. *Analyst* (*London*) **99**, 330–337.
Rilling, H.C. (1966). A new intermediate in the biosynthesis of squalene. *J. Biol. Chem.* **241**, 3233–3236.
Rilling, H.C., and Epstein, W.W. (1969). Studies on the mechanism of squalene biosynthesis. Presqualene, a pyrophosphorylated precursor to squalene. *J. Amer. Chem. Soc.* **91**, 1041–1042.
Rilling, H.C., Poulter, C.D., Epstein, W.W., and Larsen, B. (1971). Studies on the mechanism of squalene biosynthesis. Presqualene pyrophosphate, stereochemistry and a mechanism for its conversion to squalene. *J. Amer. Chem. Soc.* **93**, 1783–1785.
Romanet, R.F., and Schlessinger, R.H. (1974). New and highly efficient synthesis of rethrolones. *J. Amer. Chem. Soc.* **96**, 3701–3702.
Rukavishnikov, A.V., Tkachev, A.V., Volodarskii, A.V., and Pentegova, V.A. (1989). New approach to the synthesis of (1*R*)-*cis*- and (1*R*)-*trans*-chrysanthemic acid from (+)-3-carene. *Zh. Org. Khim.* **25**, 1665–1671.
Saljoughian, I. (1985). A convenient regiospecific synthesis of ^{2}H-labelled methyl *trans*-chrysanthemate. *J. Labelled. Compd. Radiopharm.* **22**, 1093–1095.
Sasaki, T., Eguchi, S., and Ohno, M. (1968a). Studies on chrysanthemic acid. I. Some reactions of the isobutenyl group in chrysanthemic acid. *J. Org. Chem.* **33**, 676–679.
Sasaki, T., Eguchi, S., and Ohno, M. (1968b). Studies on chrysanthemic acid. III. Synthesis and reactions of isocyanates from chrysanthemic acid. *Tetrahedron* **25**, 2145–2153.
Sasaki, T., Eguchi, S., and Ohno, M. (1970). Studies on chrysanthemic acid. IV. Photochemical behaviour of chrysanthemic acid and its derivatives. *J. Org. Chem.* **35**, 790–793.
Sasaki, T., Eguchi, S., Ohno, M., and Umemura, T. (1973). Chrysanthemyl carbenes. Isobutenyl substituent effect and conformational control in cyclopropylcarbene rearrangements. *J. Org. Chem.* **38**, 4095–4100.
Sasaki, M., Okada, K., and Matsui, M. (1979). A convenient conversion of allethrolone into pyrethrolone. *Agric. Biol. Chem.* **43**, 379–381.
Sato, T., Kawara, T., Sakata, K., and Fujisawa, T. (1981). Jasmonoid synthesis from *cis*-4-heptenoic acid. *Bull. Chem. Soc. Jap.* **54**, 505–508.
Scharf, K.H. (1979). Isolation and characterisation of pyrethrins from the petals of *Chrysanthemum cinerariaefolium* and insect sprays. *Prax. Naturwiss. Biol.* **28**, 309–315.
Scharf, H.-D., and Mattay, J. (1978). Revision der Chrysanthemumdicarbonsäure Synthese nach Inouye *et al. Chem. Ber.* **111**, 2206–2222.
Schechter, M.S., Green, N., and LaForge, F.B. (1949). Constituents of pyrethrum flowers. XXIII. Cinerolone and the synthesis of related cyclopentenolones. *J. Amer. Chem. Soc.* **71**, 3165–3173.
Schechter, H.G., Green, N., and LaForge, F.B. (1952). Constituents of pyrethrum flowers. XXIV. Synthetic *dl-cis*-cinerolone and other cyclopentenolones. *J. Amer. Chem. Soc.* **74**, 4902.
Schmidt, H.G. (1985). New intermediates for the synthesis of pyrethroids. *In* "Recent Advances in the Chemistry of Insect Control" (N.F. Janes, ed.), pp. 178–1791. The Royal Society of Chemistry, London.
Scott, F., and Nkwelo, M.M. (1985). Convenient one pot synthesis of *dl*-pyrocin. *Synth. Commun.* **15**, 1051–1056.

Seda, L., Toninelli, G., and Sartorel, B. (1983). G.c. analysis of pyrethrum extracts using glass capillary columns. *Riv. Ital. Sostanze Grasse* **60**, 133–137; *Chem. Abstr.* **100**, 116315u (1984).

Sevrin, M., Hevesi, L., and Krief, A. (1976). A total stereospecific synthesis of *d,l-cis-* and *d,l-trans*-chrysanthemic esters. *Tetrahedron Lett.*, 3915–3918.

Shaha, S.C., Joshi, G.D., Pai, P.P., Deshmukh, A.R.A.S., and Kulkarni, G.H. (1989). An improved procedure for 3,7,7-trimethyl-3-α-hydroxy-*bicyclo*[4.1.0]hept-4-ene: an important intermediate for pyrethroids. *Chem. Ind. (London)*, 568–569.

Shirley, I., Smith, I.H., and Whiting, D.A., (1982). Synthesis of pre-presqualene, a predicated intermediate in presqualene biosynthesis, and of prenylogues. *Tetrahedron Lett.*, 1501–1504.

Shono, T., Matsumura, Y., Hamaguchi, H., and Nakamura, K. (1976). A facile and general synthesis of 4-hydroxycyclopentenones. *Chem. Lett.*, 1249–1252.

Sims, M. (1981). Pyrethrin extraction with liquid carbon dioxide. Jpn. Kokai Tokkyo Koho 81 20,548 (Cl C07C69/747) 26 Feb 1981; *Chem. Abstr.* **95**, 19728b (1981).

Singh, R.P., Subbarao, H.N. and Dev, S. (1979). Organic reactions in a solid matrix. V. Silica gel supported chromic acid reagents. *Tetrahedron* **35**, 1789–1793.

Smith, I.H., and Casida, J.E. (1981). Epoxychrysanthemic acid as an intermediate in metabolic decarboxylation of chrysanthemate insecticides. *Tetrahedron Lett.*, 203–206.

Sobti, R., and Sukh Dev. (1974). (+)-*trans*-Chrysanthemic acid from (+)-Δ^3-carene. *Tetrahedron* **30**, 2927–2929.

Staba, E.J., Nygaard, B.G., and Zito, S.W. (1984). Light effects on pyrethrum shoot cultures. *Plant Cell, Tissue Organ Cult.* **3**, 211–214.

Stahl, E., and Pfeifle, J. (1966). Thin layer and gas chromatography of pyrethrum based insecticides. *Pyrethrum Post.* **8**(4), 8.

Stahl, E., and Schuetz, E. (1980). Extraction of natural compounds with supercritical gases. 3. Pyrethrum extracts with liquefied and supercritical carbon dioxide. *Planta Medica* **40**, 12–21.

Staudinger, H., and Ruzicka, L. (1924). Insektentötende Stoffe. I–VI and VIII–X. *Helv. Chim. Acta* **7**, 177–201, 201–211, 212–235, 236–244, 245–259, 377–406, 406–441, 442–448, and 448–458.

Staudinger, H., Muntwyler, O., Ruzicka, L., and Seibt, S. (1924). Insektentötende Stoffe. VII. Synthesen der Chrysanthemumsäure und Trimethylencarbonsäure mit Ungesättigen Seitenkette. *Helv. Chim. Acta* **7**, 390–406.

Suzukamo, G., Fukao, M., and Tamura, M. (1984). C_3-Epimerisation and selective C_2–C_3 bond fission of alkyl chrysanthemates. *Tetrahedron Lett.*, 1595–1598.

Suzuki, Y., Hirai, H., Toyoura, A., and Magara, O. (1970). *trans*-Chrysanthemic acid from *cis*-chrysanthemoyl chloride. German Patent 2,003, 065; *Chem. Abstr.* **73**, 76735w (1970).

Szekléy, I., Lovász-Gáspár, M,. and Kovács, G. (1980). A new synthesis of natural pyrethrins. *Pestic. Sci.* **11**, 129–133; *Chem. Abstr.* **94**, 102883y (1981).

Tabenkin, B., LeMahieu, R., Berger, J., and Kiersted, R.W. (1969). Microbiological hydroxylation of cinerone to cinerolone. *Appl. Microbiol.* **17**, 714–717.

Takagaki, T., and Matsuo, N. (1991). Preparation of 4-oxo-2-methyl-3-(4-penten-2-ynyl)-2-cyclopentenyl esters as intermediates for pyrethrins. Jpn. Kokai Tokkyo Koho. JP03 14,541 [91 14,541] (Cl. CO7C69/145) 23 Jan 1991; *Chem. Abstr.* **114**, 246862v (1991).

Takahashi, T., Hori, K., and Tsuji, J. (1981). 2-Methylene-3-alkoxycyclopentanones as reactive intermediates, and their application to the synthesis of dihydrojasmone, *cis*-jasmone and (±)-jasmolone. *Chem. Lett.*, 1189–1192.

Takano, S., Tanaka, M., Seo, K., Hirama, M., and Ogasawara, K. (1985). General chiral route to irregular monoterpenes via a common intermediate: synthesis of (*S*)-lavandulol, *cis*-(1*S*,3*R*)-chrysanthemol, (1*S*,2*R*)-rothrockene and (*R*)-santolinatriene. *J. Org. Chem.* **50**, 931–936.

Takeda, A., Sakai, T., Shinohara, S., and Tsuboi, S. (1977). A new synthesis of pyrocin and related compounds. *Bull. Chem. Soc. Jap.* **50**, 1133–1136.

Takei, S., Sugita, T., and Inouye, Y. (1958). Eine neue Synthese der Chrysanthemumdicarbonsäure. *Liebigs Ann. Chem.* **618**, 105–109.

Thomas, A. (1973). The synthesis of monoterpenes. *In* "Total Synthesis of Natural Products" (J. ApSimon, ed.), Vol. 2, pp. 1–195. Wiley-Interscience, New York and London.

Torii, S., Tanaka, H., and Nagai, Y. (1977). Cyclopropanes from allylic halides. 1. Synthesis of dimethyl 3-(2-methyl-1-propenyl)-2,2-dimethylcyclopropane-1,1-dicarboxylate and pyrocin as precursors of chrysanthemic acid. *Bull. Chem. Soc. Jap.* **50**, 2825–2826.

Torii, S., Inokuchi, T., and Oi, R. (1983). Electrooxidative cleavage of carbon-carbon bonds. 2. Double cleavage of α,β-epoxyalkanones and enantiospecific syntheses of chiral methyl *cis-* and *trans*-chrysanthemates from (+)- and (−)-carvone. *J. Org. Chem.* **48**, 1944–1951.

Tosi, C., Barino, L., Castellani, G., and Scordamaglia, R. (1982). Theoretical calculations on the conformations of pyrethroids. Part II. The energetics of 2-arylalkanoates and their topographical similarity to cyclopropane carboxylates. *Theochem.* **4**, 315–325.

Tsuji, J., Yamakawa, T., and Mandai, T. (1979). Application of palladium catalysis to a simple synthesis of (±)-pyrethrolone. *Tetrahedron Lett.*, 3741–3744.

Ueda, K., and Matsui, M. (1970a). Studies on chrysanthemic acid. Part XIX. Conversion of optically active *trans*-chrysanthemic acid to the racemic one via pyrocine. *Agric. Biol. Chem.* **34**, 1115–1118.

Ueda, K., and Matsui, M. (1970b). Studies on chrysanthemic acid. Part XX. Synthesis of four geometrical isomers of (±)-pyrethric acid. *Agric. Biol. Chem.* **34**, 1119–1125.

Ueda, K., and Matsui, M., (1971). Studies on chrysanthemic acid. Part XXI. Photochemical isomerisation of chrysanthemic acid and its derivatives. *Tetrahedron* **27**, 2771–2774.

Ueda, K., and Suzuki, Y. (1971). (+)-*trans*-Chrysanthemumic acid by resolution of (±)-*trans*-chrysanthemumic acid. German Patent 2,032,097; *Chem. Abstr.* **74**, 87484k (1971).

Vandewalle, M., and Madeleyn, E. (1970). Cyclopentanones. III. A new synthesis of (±)-allethrolone. *Tetrahedron* **26**, 3551–3554.

van Tamelen E.E., and Leopold, E.J. (1985). Mechanism of presqualene pyrophosphate-squalene biosynthesis. II. Synthesis of bifarnesol. *Tetrahedron Lett.*, 3303–3306.

van Tamelen, E.E., and Schwartz, M.A. (1971). Mechanism of presqualene pyrophosphate-squalene biosynthesis. *J. Amer. Chem. Soc.* **93**, 1780–1782.

Vick, B.A., and Zimmerman, D.C. (1983). The biosynthesis of jasmonic acid: a physiological role for plant lipoxygenase. *Biochem. Biophys. Res. Commun.* **111**, 470–477.

Vick, B.A., and Zimmerman, D.C. (1984). Biosynthesis of jasmonic acid by several plant species. *Plant. Physiol.* **75**, 458–461.

Vick, B.A., and Zimmerman, D.C., (1987). Oxidative systems for modification of fatty acids: the lipoxygenase pathway. *In* "The Biochemistry of Plants" (P.K. Stumpf and E.E. Conn, eds.), pp. 53–90. Academic Press, New York.

Wambugu, F.M., and Rangan, T.S. (1981). *In vitro* clonal multiplication of pyrethrum *Chrysanthemum cinerariaefolium* Vis) by micropropagation. *Plant Sci. Lett.* **22**, 219–226.

Warszawki, T., Cieniecka-Roslonkiewicz, A., and Gwiazda, M. (1988). The possibilities of pyrethrins biosynthesis in tissue cultures of *Chrysanthemum cinerariaefolium* Vis. *Pestycydy* (*Warsaw*), 27–31; *Chem. Abstr.* **109**, 188656q (1988).

Welch, S.C., and Valdes, T.A. (1977). A synthesis of (±)-*trans*-chrysanthemic acid. *J. Org. Chem.* **42**, 2108–2111.

West, T.F. (1944). The structure of pyrethrolone and related compounds. Part IV. *J. Chem. Soc.* (*London*), 239–242.

Wieboldt, R.C., and Smith, J.A. (1988). Supercritical fluid chromatography with Fourier transform infra-red detection. *In* "Supercritical Fluid Extraction Chromatography" pp. 229–242, *ACS Symp. Ser.* **336**; *Chem. Abstr.* **109**, 34658e (1988).

Wieboldt, R.C., Kempfort, K.D., Later, D.W., and Campbell, E.R. (1989). Analysis for pyrethrins using capillary supercritical fluid chromatography and capillary g.c. with Fourier transform I.R. detection. *J. High Resol. Chromatogr.* **12**, 106–111.

Williams, J.L., and Rettig, M.F. (1981). Dichlorobis(organonitrile)palladium (II) catalysis of *cis*- to *trans*-ethyl chrysanthemate and chrysanthemic acid. *Tetrahedron Lett.*, 385–388.

Winteringham, F.P.W., Harrison, A., and Bridges, P.M. (1955). Absorption and metabolism of ^{14}C-pyrethroids by the adult fly *Musca domestica in vivo*. *Biochem. J.* **61**, 359–367.

Woessner, W.D., and Ellison, R.A. (1972). Synthesis of 4-hydroxy-2-cyclopenten-1-ones via 1,3-dithianes. *Tetrahedron Lett.*, 3735–3738.

Yadav, J.S., Mysorekar, S.V., and Rama Rao, A.V. (1989). Synthesis of (1*R*)-(+)-*cis*-chrysanthemic acid. *Tetrahedron* **45**, 7353–60.

Yamamoto, R. (1919). The insecticidal principle in *Chrysanthemum cinerariaefolium*. Part I. *J. Chem. Soc. Jap.* (*Tokyo*) **40**, 126.

Yamamoto, R. (1923). The insecticidal principle in *Chrysanthemum cinerariaefolium*. Parts II and III. On the constitution of pyrethronic acid. *J. Chem. Soc. Jap.* (*Tokyo*) **44**, 311.

Yamamoto, I., and Casida, J. E. (1968). Synthesis of ^{14}C-labelled pyrethrin I, allethrin, phthalthrin, and dimethrin on a submillimole scale. *Agric. Biol. Chem.* **32**, 1382–1391.

Zang, E., and Chow, C. (1991). Synthesis of chiral mixed-ligand complexes and their use in asymmetric synthesis of chrysanthemic acid. *Chin. Chem. Lett.* **2**, 169–170.

Zieg, R.G., Zito, S.W., and Staba, E.J. (1983). Selection of high pyrethrin producing tissue cultures. *Planta Medica* **48**, 88–91.

Zimmerman, D.C., and Feng, P. (1978). Characterisation of a prostaglandin-like metabolite of linolenic acid produced by flaxseed extract. *Lipids* **13**, 313–316.

Zito, S.W., and Staba, E.J. (1984). A cell-free homogenate of *Chrysanthemum cinerariaefolium*.

(McLaughlin Gormley King Co.). Eur. Pat. Appl. 124,049 (Cl. C12P7/62), 07 Nov. 1984. U.S. Appl. 489,957, 29 Apr. 1983. 31 pp; *Chem. Abstr.* **102**, P77291c (1985).
Zito, S.W., and Tio, C.D. (1990). Constituents of *Chrysanthemum cinerariaefolium* in leaves, regenerated plantlets and callus. *Phytochemistry* **29**, 2533–2534.
Zito, S.W., Srivastava, V., and Adebayo-Olojo (1991). Incorporation of [1-^{14}C]-isopentenyl pyrophosphate into monoterpenes by a cell-free homogenate prepared from callus cultures of *Chrysanthemum cinerariaefolium*. *Planta Medica* **57**, 425–427.

9

Environmental Fate of Pyrethrins

DONALD G. CROSBY

I. INTRODUCTION

Each year about 200,000 kg of pyrethrins are used as an insecticide. Clearly they enter the environment, but their subsequent fate is largely unknown. For most other pesticides, including synthetic pyrethroids, the US Environmental Protection Agency requires fate information for registration, but the pyrethrins historically have been relieved of this requirement. In addition, the difficulty of isolating and keeping the pure individual congeners probably inhibited the academic pursuit of such data.

The transport (dissolution, volatility, bioconcentration, and adsorption) and environmental transformations of natural pyrethrins may offer useful insights into toxic action, spray drift, and persistence as well as practical applications in their extraction, storage, and utility. Recent advances in chemical property estimation methods (Lyman *et al.*, 1990), formulas for which are shown in Table 9-1, now allow realistic calculation of key values, and precedent as well as data from synthetic analogs allow the accuracy of these calculations to be gauged. This review presents the reasoning and results behind the prediction of pyrethrins fate and indicates some possible extensions and applications.

II. TRANSPORT

A. Solvent Partitioning

The distribution of a substance between two immiscible phases forms the basis of its environmental transport. Its distribution between any two immiscible liquid solvents is constant and provides a partition coefficient:

$$K_p = \frac{C_{\text{organic solvent}}}{C_{\text{water}}}$$

The common organic solvent for environmental considerations today is 1-octanol, and the octanol-water partition coefficient is termed K_{ow}. Hansch and Leo (1979) found that K_{ow} was intimately related to chemical structure and that the contribution of each structural element or atom was additive.

Table 9-1. Equations for Chemical Property Estimation[a]

Partition Coefficients

$$\log\frac{1}{S}=1.214\log K_{ow}-0.850 \text{ moles/l} \quad (1)$$

$$\log \text{BCF}=0.76\log K_{ow}-0.23 \quad (2)$$

Volatilization

$$\ln P=\frac{\Delta H_v}{\Delta Z_b R T_b}\left(f\frac{T}{T_b}\right) \quad (3)$$

$$\log V=1.02\log P-4.42\ \mu\text{g/cm}^2\text{/h} \quad (4)$$

$$H'=\frac{C_{air}}{C_{water}}=\frac{16.04PM}{TS} \quad (5)$$

Soil Adsorption

$$\log K_{oc}=0.544\log K_{ow}-1.377 \quad (6)$$

$$t_{1/2v}=1.58\times10^{-8}\left[\frac{K_{oc}S}{P}\right] \text{days}^{b} \quad (7)$$

[a] S = aqueous solubility, BCF = bioconcentration factor, P = vapor pressure (atm), V = volatilization rate, H' = Henry's constant, M = molecular weight, T = absolute temperature (K), K_{oc} = organic carbon soil absorption coefficient, ΔH_v = heat of vaporization, ΔZ_b = compressibility factor, R = gas constant, T_b = normal boiling point. Adapted from Lyman *et al.* (1990).
[b] P in torr.

Therefore, knowing the formula of pyrethrin I (Fig. 9-1) (**1**, R= CH=CH_2) to be $C_{21}H_{28}O_3$ and the structure to contain an ester, ketone, rings, and unsaturation, the appropriate fragmentation constants can be added to provide an estimate of log K_{ow} (Table 9-2). Fortunately, the measured K_{ow} of bioallethrin (Fig. 9-1) (**1**, R=H) has been reported (Worthing and Hance, 1991) and is identical to its calculated log K_{ow}, providing confidence that fragment additivity works in this case. It also is comforting that the calculated K_{ow} values for these four closely-related compounds lie well within an order of magnitude, that is, around 100,000; they are very lipophilic.

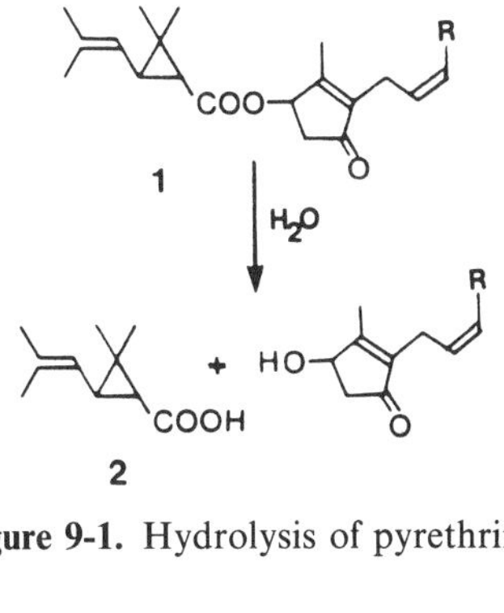

Figure 9-1. Hydrolysis of pyrethrins.

Table 9-2. Predicted Aqueous Partition Coefficients of Pyrethrins

Chemical	log K_{ow} (Calc'd)	log K_{ow} (Lit)	S (mg/l) (Calc'd)	S (mg/l) (Lit)	BCF
Bioallethrin	4.68	**4.68**[a]	4.45	**4.6**[a]	2,100
Pyrethrin I	5.62	5.9[b]	0.35	0.2[b]	11,000
Cinerin I	4.77	5.6[b]	3.62	30.6[b]	2,500
Jasmolin I	5.43	—	0.60	—	4,700
Pyrethrin II	3.56	4.3[b]	125.6	9.0[b]	300
Cinerin II	2.71	—	1038.0	—	70
Jasmolin II	3.37	—	214.8	—	210

[a]Measured (Worthing and Hance, 1991).
[b]Calculated (Briggs *et al.*, 1983).

In contrast to the pyrethrins I series, the K_{ow} values for the pyrethrins II series show them to be much more hydrophilic, with K_{ow} averaging around 1,600 (that calculated for cinerin II is only 513). This apparent difference between the pyrethrins I series and the pyrethrins II series suggests that inefficient hexane extraction of pyrethrum flowers would tend to leave the former behind, and that manipulation of solvent composition might afford a practical means to separate or preferentially extract the two series. The remaining literature values of log K_{ow} listed in Table 9-1 are those presented by Briggs *et al.* (1983), based primarily on the calculations of Briggs (1981). Although derived from the same additivity factors (Hansch and Leo, 1979), calculation of our values started with the known K_{ow} of bioallethrin (the same measured and calculated), a big advantage not available to Briggs' pioneering work.

Given log K_{ow}, the aqueous solubility can be calculated (Table 9-2). Lyman *et al.* (1990) provide several regression equations which might apply and Briggs (1981) suggests others. The one chosen [Equation (1)] represents the regression from values for 140 organic liquids (Hansch *et al.*, 1968), covering a variety of structural types, and predicts the measured solubility of bioallethrin very well (Table 9-2). The equation representing only aliphatic esters is much less satisfactory (calculated $S = 18.16$ mg/l for bioallethrin) (Hansch *et al.*, 1968). While the pyrethrins I series has low solubilities (pyrethrin I is predicted to be the least soluble of all), the pyrethrins II series is relatively soluble (cinerin II is predicted to have an aqueous solubility of more than a gram per liter), suggesting that the two series might be separable by aqueous extraction alone.

The bioconcentration factor (BCF), the ratio of the concentration of a chemical in an animal to that in the animal's environment, also represents partitioning and is related to K_{ow}. BCF in aquatic animals such as fish has received particular attention, with the result that a number of regression equations are available for its calculation (Lyman *et al.*, 1990). The one presented here [(Equation (2)] is derived from a broad range and large number of chemical types (Veith *et al.*, 1980) and is generally conceded to be the most serviceable. As expected from K_{ow}, members of the pyrethrins I series are seen to be the more highly bioconcentrated (Table 9-2) by at least an order of magnitude compared to the pyrethrins II series, although even the value of 11,000 for pyrethrin I seems small compared to that for DDT (100,000). No measured BCF values for the natural pyrethrins appear in the literature, probably because of their relative ease of biodegradation and the comparative difficulty of tissue analysis. It should be emphasized that the BCF values in Table 9-2 are calculated from partitioning only; as the pyrethrins are readily biodegradable, the measured values probably will be much lower.

B. Volatilization

Volatilization, the movement of a chemical from a solid or liquid surface into the gaseous state, often represents the most important transport route. It is a function of vapor pressure (P), the tendency of molecules to escape the condensed phase. In order to calculate vapor pressure, one must know the heat of vaporization (ΔH_v) and boiling point at nominal atmospheric pressure (T_b), in addition to the more accessible compressibility factor (ΔZ_b), universal gas constant (R), and ambient temperature (T) [(Equation (3)]. Normal boiling points for high-boiling or unstable organic compounds (such as pyrethrins) usually are not known but can themselves be calculated (Lyman *et al.*, 1990), given the molar refraction and parachor (both of which are additive properties) (Table 9-3). The ratio of H_v to T_b is estimated simply:

$$\frac{\Delta H_b}{T_b} = 8.75 + \mathrm{R} \ln T_b$$

while P is a much more complex function of T/T_b for which reference to Lyman *et al.* (1990) is recommended. The measured vapor pressure of bioallethrin (Table 9-3) is estimated quite accurately by this calculation, lending a certain amount of credibility to the other pyrethrins values. The result is that pyrethrin I and pyrethrin II are seen to have the lowest vapor pressures of the group, although none is far from 10^{-4} torr (13 mPa). However, vapor pressure is only a measure of escaping *tendency*. What one generally is interested in is volatilization rate (V). Volatilization rates from a solid surface seem especially important for pyrethrins, as the insecticides often are applied to floors, baseboards, and walls. Gückel *et al.* (1973) experimentally developed relationships between vapor pressure (P) and volatilization rates from a glass surface. Interpolation on their regression line provided estimates (Table 9-3) of the rate at which a thin film of pure pyrethrins would evaporate from a solid surface [Equation (4)].

In a typical indoor application, a film on the order of 1 $\mu g/cm^2$ of mixed pyrethrins was laid down from a sprayer or "bug bomb" (Class and Kintrup, 1991). As seen from Table 9-3, such a film could be expected to last no more than a few hours at best, even if volatilization were the only route of dissipation (see Section III.B). Volatilization also is predicted to be congener-specific; cinerin I and jasmolin II would evaporate the most rapidly, while pyrethrin I and pyrethrin II would be the most persistent. However, the values in the Tables refer to the individual, pure congeners only. In the liquid pyrethrins mixture, each congener can be expected to volatilize approximately according to its mole

Table 9-3. Predicted Volatilization of Pyrethrins

Chemical	P_{vp} (torr) (Calc'd)	P_{vp} (torr) (Lit)	V ($\mu g/cm^2/h$)	H'
Bioallethrin	3.5×10^{-4}	3.3×10^{-4} [a]	3.40	1.2×10^{-3}
Pyrethrin I	8.6×10^{-5}	2×10^{-5} [b]	0.89	4.3×10^{-3}
Cinerin I	2.0×10^{-4}	4×10^{-5} [b]	1.98	9.4×10^{-4}
Jasmolin I	1.1×10^{-4}	—	1.18	3.3×10^{-3}
Pyrethrin II	5.6×10^{-5}	4×10^{-7} [b]	0.65	8.9×10^{-6}
Cinerin II	1.2×10^{-4}	—	1.38	2.2×10^{-6}
Jasmolin II	1.5×10^{-4}	—	1.80	1.4×10^{-5}

[a]Measured (Worthing and Hance, 1991).
[b]Calculated (Briggs *et al.*, 1983).

fraction (Raoult's law), so the volatilization rate will be reduced correspondingly. In addition, adsorption to solid surfaces (such as glass or painted wood) (Section II.C) may retard volatilization to some degree (Class and Kintrup, 1991).

Volatilization from water is governed by Henry's law, which relates vapor pressure to aqueous solubility [Equation (5)]. The nondimensional Henry's law constant (H', Table 9-3) relates the equilibrium concentration of chemical in air to that in water; an H' of greater than 10^{-2} indicates rapid volatilization, while one below about 10^{-5} indicates slow loss, if any. As seen from Table 9-3, insecticides of the pyrethrins I series are predicted to volatilize rather rapidly, while those of the pyrethrins II series are expected to be almost nonvolatile. This finding may be significant in at least two regards: as damp surfaces actually are covered with a very thin film of water, volatility of certain pyrethrins (such as cinerin I) from a solid surface may be much more rapid than expected, while others (cinerin II) may be slower to evaporate than would be expected from V alone; and, second, the data suggest steam distillation as a possible method to enrich the volatile fraction in pyrethrins I congeners in comparison to those of the pyrethrins II series.

C. Adsorption

Reversible binding to soil surfaces can be expected to inhibit pyrethrins movement via both volatility and dissolution but, conversely, to enhance their bulk movement in water and on airborne particles. Contrary to earlier thinking, the organic fraction of soil is primarily responsible for adsorption (rather than clay), the distribution coefficient based on organic carbon content giving rise to K_{oc} as the partition coefficient which relates chemical concentration in water to that in sediment (Table 9-4). K_{oc}, then, is related to K_{ow} and may be estimated from it [Equation (6)]; the smaller the K_{oc}, the weaker the binding to soil. The calculated adsorption in the pyrethrins II series is comparatively weak ($K_{oc} < 2{,}000$), while that of the pyrethrins I congeners is strong; pyrethrin I, at 27,000, would approach the chlorinated hydrocarbons in adsorptive binding to soil.

Volatilization from moist soil is a function of not only K_{oc} but also of the

Table 9-4. Predicted Soil Adsorption of Pyrethrins

Chemical	log K_{oc}	K_{oc}	Volat. $t_{1/2}$ (d)
Bioallethrin	3.92	8,400	1.8
Pyrethrin I	4.43	27,200	1.8
Cinerin I	3.97	9,400	2.7
Jasmolin I	4.33	21,400	1.9
Pyrethrin II	3.31	2,100	73.2
Cinerin II	2.85	700	97.0
Jasmolin II	3.21	1,600	36.8

Henry's law constant (P/S) [Equation (7) and Table 9-4]. Therefore, despite their weaker adsorption, their low H' and S will tend to hold pyrethrins II congeners in the soil (Table 9-4), resulting in a relatively slow physical loss (loss of half of the adsorbed pyrethrin, or $t_{1/2}$, requiring 1–3 months) [Equation (7)]. As the congeners ideally are adsorbed as individuals, Raoult's law should not apply here, but losses due to volatilization may be accelerated by rainfall or irrigation. Adsorption to other surfaces, such as wood, plastic, and glass, also will retard both volatilization and dissolution to some extent.

III. TRANSFORMATIONS

A. Dark Reactions

Xenobiotics are subject to environmental oxidation, reduction, hydrolysis, and other chemical transformations. Certain reactions with environmental reagents can occur under ambient conditions, even in the dark. As carboxylic esters, the natural pyrethrins may be expected to undergo hydrolysis, especially in alkaline waters (Fig. 9-1). However, as the ester linkage is between aliphatic acids and alcohols, the rate will be low; at pH 7 and 25°C, the hydrolysis $t_{1/2}$ of the analogous isopropyl acetate is 8.4 years and that of allyl acetate 9.6 years (Maybe and Mill, 1978); the pK_a of cyclopropanecarboxylic acid is 4.6, close to that of acetic acid. Neither the hydrolysis rates of chrysanthemate esters nor the pK_a of chrysanthemic acid (**2**) have been reported, although the carbomethoxy group of the pyrethrins II series is much more stable to alkaline hydrolysis than is the cyclopropane ester, allowing the isolation of pyrethric acid (Staudinger and Ruzicka, 1924a; Crombie *et al.*, 1957). A three-dimensional model of pyrethrin I (Crombie *et al.*, 1976) does not indicate anything structurally unusual about its ester group.

Treatment of ethyl chrysanthemate with 50% aqueous sulfuric acid at room temperature results in rapid and extensive reactions of the cyclopropane ring (Goldschmidt *et al.*, 1984); the same reactions take place with chrysanthemic acid at 210°C in the presence of an acid catalyst (Otieno *et al.*, 1977), and the purely thermal degradation of pyrethrin II probably follows a similar course at 400°C (Nakada *et al.*, 1972). These are hardly "environmental"

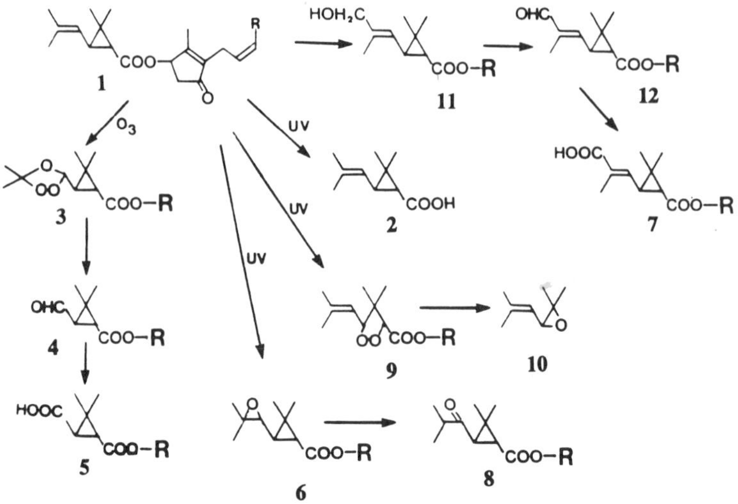

Figure 9-2. Predicted photodegradation and ozonolysis pathways of pyrethrins. R = rethronyl.

transformations. However, the thermal elimination of chrysanthemic acid and its relatives from the rethrins, driven by conjugation of the generated rethrolone double bond with the ketone carbonyl (Elliott, 1964), would be expected at lower temperatures and may be responsible for thermal instability of pyrethrins upon gas chromatography. In fact, this elimination does take place at room temperature when allethrin is treated with sodium hydroxide in aqueous ethanol (LaForge *et al.*, 1952).

Simple chrysanthemic acid derivatives are known to react with gaseous ozone at the olefinic double bond to produce the corresponding ozonide (**3**), aldehyde (**4**), acid (**5**), and epoxide (**6**) (Staudinger and Ruzicka, 1924a; Ruzo *et al.*, 1986) (Fig. 9-2), although intact natural pyrethrins appear not to have been examined except for jasmolin I (Godin *et al.*, 1966). With rethrolones, only the aldehydes associated with side-chain cleavage have been reported (Staudinger and Ruzicka, 1924b). As our atmosphere contains traces of ozone (often detectable in smoggy air by its odor), the environmental ozonolysis of natural pyrethrins must take place in ambient air, at a low rate at least (Class, 1991). Nambu *et al.* (1980) also reported that the ozonide, aldehyde, and acid of the synthetic chrysanthemate ester, phenothrin (Fig. 9-3) (**13**), were detected on the leaves of treated plants exposed to air and sunlight.

B. Photodegradation

The breakdown (deactivation) of natural pyrethrins when exposed to daylight is perhaps their most prominent and best-recognized chemical characteristic (Chen and Casida, 1969). Many of their classical photochemical reactions already have been reviewed (Bullivant and Pattenden, 1976a; Otieno and Pattenden, 1980; Elliott, 1973), but the experimental conditions for

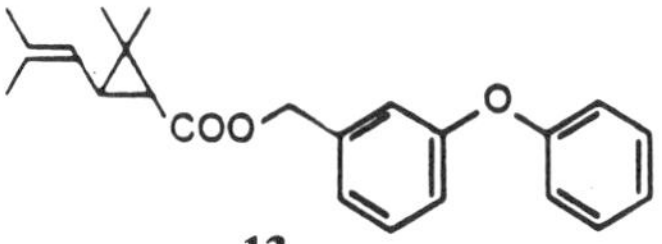

Figure 9-3. Structures of phenothrin (**13**), diradical (**14**), and butoxyethyl 2,4-dichlorophenoxyacetate (**15**).

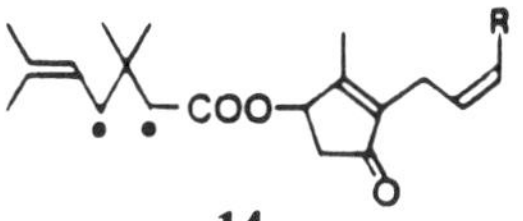

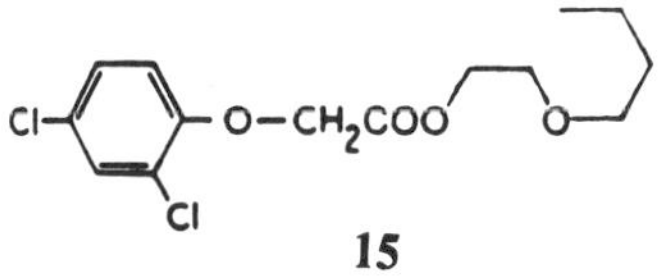

most of these studies had little relation to the natural environment. This presentation includes only the photochemistry applicable to environmental reagents and conditions [e.g., ultraviolet radiation (UV) in the sunlight spectrum (290–400 nm), aerobic conditions, ambient temperatures].

Pyrethrins photodegradation is rapid (Fig. 9-4). In dilute solution in an organic solvent, but in the virtual absence of atmospheric oxygen, the principal reaction is isomerization of the pyrethrolone side-chain, from a *cis*-(*Z*-) to a *trans*-(*E*-) configuration (Kawano *et al.*, 1980; Dickinson, 1982) (Fig. 9-5). The formation of a cyclopropane ring from the three carbons of the allethrin and bioallethrin side-chain, observed under both the more energetic 254 nm Hg arc

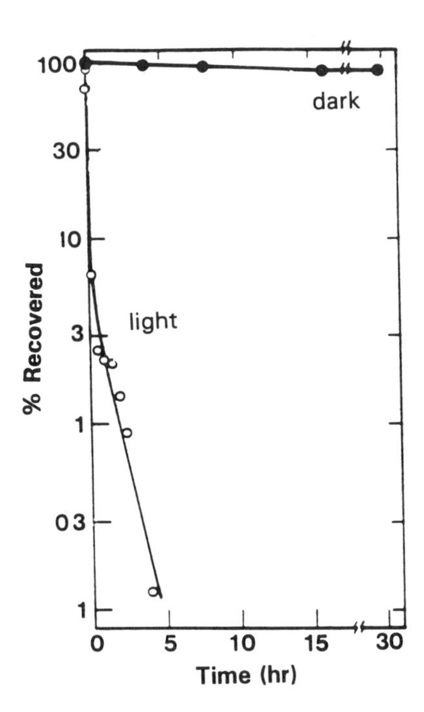

Figure 9-4. Photodegradation rate of pyrethrin I. Adapted from Chen and Casida (1969).

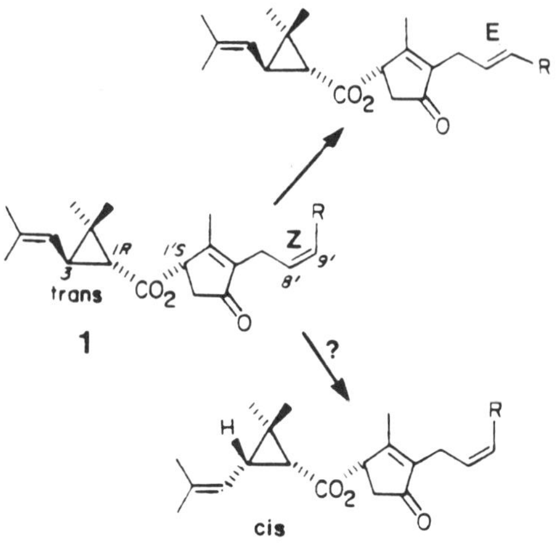

Figure 9-5. Photoisomerization of pyrethrins.

radiation and milder sunlight conditions (Bullivant and Pattenden, 1976b; Ruzo *et al.*, 1980), does not seem to occur with natural pyrethrins. Isomerization from *trans* to *cis* in the cyclopropanecarboxylic acid moieties (Fig. 9-5) also is observed with 254 nm radiation (Bullivant and Pattenden, 1971), but the literature is unclear as to whether this reaction actually takes place under solar UV. However, the closely-related bioallethrin clearly undergoes isomerization on the cyclopropane ring in sunlight (Ruzo *et al.*, 1980).

Under aerobic conditions, photooxidation predominates. Although Stahl (1960) indicated that pyrethrins adsorbed on Kieselguhr formed "peroxides" by the action of sunlight or long wavelength UV, it remained for Chen and Casida (1969) to demonstrate in detail that exposure of a thin film of pyrethrin I to a 275 W RS-sunlamp (297 nm) produced an assortment of oxidized products, primarily carboxylic acids. Although the work of Freeman (1956) suggested that most photodegradation took place in the rethrolone part of the molecules, and Head *et al.* (1968) indicated that the degradation occurred only in the acid part, Chen and Casida (1969) isolated products showing degradation in both parts. Thin-layer chromatography revealed 14 photoproducts derived from pyrethrin I in addition to an undetermined number remaining at the origin. Quantitating by ^{14}C, no separated product represented more than 4% of the original radioactivity, the origin contained 31%, and 37% of the ^{14}C remained unaccounted for (and lost at least partly by volatilization). Hydrolysis did not appear to be involved, although Ruzo *et al.* (1982) showed that chrysanthemic acid was a major product in the photolysis of the synthetic pyrethroid, phenothrin (**13**). Although none of the ester photoproducts was identified, hydrolysis with methanolic sodium hydroxide resulted in an acid fraction from which chrysanthemic acid (**2**) and *trans*-caronic acid (Fig. 9-2) (**5**, R=H) were identified as major products (Chen and Casida, 1969); other acids tentatively identified included those in which a butenyl methyl of **2** had been oxidized to the corresponding alcohol (**11**), aldehyde (**12**), and acid (**7**, chrysanthemum dicarboxylic acid), and in which the double bond was replaced by a ketone carbonyl (**8**) (Fig. 9-2, R=H in each instance).

Isobe *et al.* (1984) irradiated allethrin with a 15 W fluorescent lamp and detected many polar products including allethrolone, allethronyl glyoxylate, and allethronyl senecioate. By analogy, it seems probable that with pyrethrolone esters the corresponding peroxy acid (**9**), glyoxylic acid, and the epoxide (**10**), also would be products of pyrethrin I photodegradation (Fig. 9-2).

All of the other pyrethrins also undergo photodegradation; pyrethrins were degraded more rapidly than cinerins (Freeman, 1956; Brown *et al.*, 1957), and pyrethrin I and cinerin I were degraded more rapidly than pyrethrin II and cinerin II (Brown *et al.*, 1957). Exposure of a highly purified pyrethrum concentrate in petroleum ether to direct sunlight in a closed container for 3 months produced only a resin which, upon hydrolysis, provided chrysanthemic acid and chrysanthemum dicarboxylic acid (Campbell and Mitchell, 1950), indicating that polymerization of the rethrolones plays an important role in eventual pyrethrins photodegradation.

A major mystery remains. Ultraviolet absorption spectra of the natural pyrethrins show a strong maximum between 220 and 230 nm (Fig. 9-6) (Crombie and Elliott, 1961); no appreciable UV absorption is observed above 290 nm,

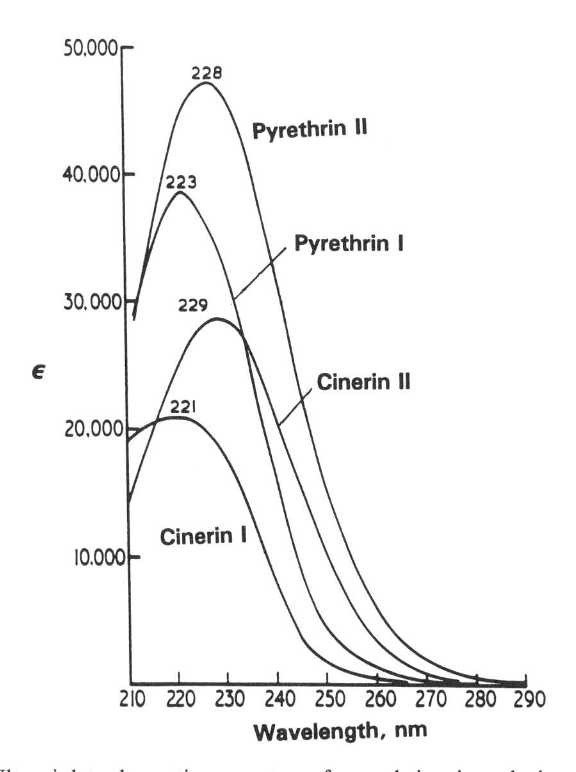

Figure 9-6. Ultraviolet absorption spectra of pyrethrins in solution. Adapted from Crombie and Elliott (1961).

the low-wavelength cutoff for solar UV. As the first principle of photochemistry states that energy must be absorbed in order for a chemical reaction to take place, how is one to explain the rapid photodegradation of the pyrethrins in sunlight? Several investigators have shown that natural pigments in pyrethrum extract catalyze the photodegradation (Brown and Phipers, 1955; Brown *et al.*, 1957; Head *et al.*, 1968), but synthetic pyrethrin I as well as the synthetic pyrethroids allethrin and bioallethrin also undergo rapid photodegradation (Chen and Casida, 1969; Ruzo *et al.*, 1980), so this catalysis clearly does not provide the explanation. Although small but measurable UV absorption actually does occur at 300 nm (Crosby, 1992), it seems unlikely that this could result in such a rapid degradation, except possibly for the low-energy isomerization of the olefinic rethrolone side-chain. However, the molecular-orbital calculations of Katagi *et al.* (1988) suggest that *cis-trans* isomerization may also be possible.

The rapid and extensive decomposition of the pyrethrins very likely is due primarily to UV-energized autoxidation (direct reaction with atmospheric triplet oxygen). In a typical autoxidation (Finlayson-Pitts and Pitts, 1986), a carbon free radical, generated through hydrogen abstraction by an "initiator," reacts with diradical oxygen to produce a peroxy radical; this subsequently reacts with olefins to form epoxides or abstracts a hydrogen to give a hydroperoxide, leading sequentially to the observed alcohols, aldehydes, ketones, and carboxylic acids. As the peroxy radicals then serve as initiators, the resulting chain reactions quickly propagate, resulting in a rapid increase in degradation rate. But what serves as the first radical initiator for the pyrethrins? Obviously UV radiation is necessary in this case; pyrethrins are stable in the dark (Chen and Casida, 1969; Dickinson, 1982). One plausible candidate is the diradical (**14**) (Fig. 9-3) generated photochemically from cyclopropanecarboxylates. This intermediate frequently has been invoked to explain the photodegradation of cyclopropane derivatives (e.g., Griffin *et al.*, 1965; Jorgenson, 1969), including natural and synthetic pyrethrins and their acids (e.g., Sasaki *et al.*, 1970; Ueda and Matsui, 1971; Bullivant and Pattenden, 1976a) and extending to irradiation at sunlight wavelengths (Ruzo *et al.*, 1982; Isobe *et al.*, 1984). The key is the highly strained cyclopropane ring, which is further weakened by conjugation with the adjacent carboxylate ester and olefinic side-chain. The calculated C—C bond dissociation energy for unsubstituted cyclopropane, 70 kcal/mole (Bernett, 1967), would correspond to a wavelength of over 400 nm, well into the sunlight range if a chromophore were available to absorb energy.

Another initiation mechanism may involve the natural reagents NO_2 and O_3 found in all atmospheric samples. NO_2 is photolyzed by solar UV to NO and ^3P oxygen atoms, the latter known to react with olefins to form epoxides, carbonyl compounds, and free radicals (Finlayson-Pitts and Pitts, 1986). Ozone is photolyzed to even more reactive oxygen atoms (^1D) which can act as initiators directly or, with water, generate highly reactive hydroxyl radicals. Although atmospheric concentrations of these two reagents are relatively low, their photolysis would open the way to the extensive chain reactions involving molecular oxygen.

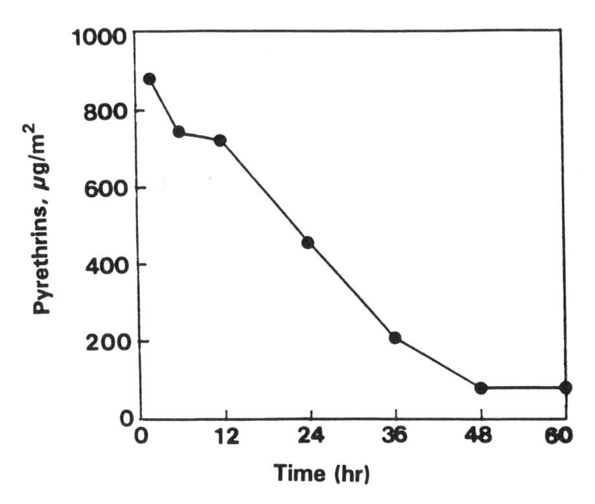

Figure 9-7. Dissipation rate of pyrethrins from thin films on glass plates indoors. Adapted from Class and Kintrup (1991).

It appears, then, that two photodegradation mechanisms operate simultaneously under longwave UV (solar) irradiation. The first, requiring only relatively low energy absorption, results in slow isomerization and polymerization in the rethrolone side-chain; the second involves the rapid and extensive free-radical oxidation of the carboxylic acid moiety (and probably of the rethrolone also). A practical example of the significance of this photooxidation was reported by Class and Kintrup (1991) who sprayed a commercial pyrethrum formulation into the air of a moderately large sunlit room (20 m^2 surface area, 50 m^3 volume). The initial room air sample (15 minutes) contained 125 $\mu g/m^3$ of pyrethrins which declined to 48 $\mu g/m^3$ during the next 15 minutes as the insecticide particles settled out. Horizontal glass plates registered about 900 $\mu g/m^2$, the level declining about 90% during 48 h, primarily in daylight (Fig. 9-7). The detected degradation products were shown to be the ozonide at the chrysanthemic acid double bond (**3**) and the corresponding caronaldehyde (**4**, R=rethronyl) (Fig. 9-2). Obviously, photooxidation is a major degradation route even in indirect sunlight filtered through window glass (>320 nm).

C. Microbial Degradation

Soil is the final repository of most pesticides and their degradation products. It is composed of an inorganic fraction (sand, clay, minerals, air, water), an organic fraction (simple organic compounds, macromolecules such as humic acid, and partially decomposed plant and animal fragments), and living organisms (especially bacteria, fungi, and yeasts). This microbial component is capable of carrying out a wide variety of metabolic reactions with xenobiotics (Alexander, 1980; Bollag and Liu, 1990), of which oxidation and hydrolysis are generally the most pronounced. However, numerous organic and inorganic reagents, including extracellular oxidases and hydrolases and free-radicals (Crosby, 1976), also allow pesticide degradation even in sterile soil.

Although the degradation of natural pyrethrins in soil has not been reported, the available information on phenothrin (**13**) (Nambu *et al.*, 1980) and other pyrethroids (Roberts, 1981) reveals a lot about the microbial degradation to be expected of other chrysanthemate esters. Permethrin was stable in sterile Hagerstown silty clay loam (Kaufman *et al.*, 1977), indicating that any degradation probably was microbial; as expected, ester hydrolysis predominated in nonsterile soil. In two Japanese soils, both the 1*R*,*trans*- and 1*R*,*cis*-isomers were rapidly degraded under upland (dry) conditions, with half-lives of less than 2 days; although (unlabeled) chrysanthemic acid and its degradation products were not reported, the alcohol hydrolysis product, [^{14}C]3-phenoxybenzyl alcohol, was identified and measured together with numerous of its oxidation products (Nambu *et al.*, 1980). Under flooded (anaerobic) field conditions, the degradation half-life was much longer—2 to 4 weeks for the *trans*-isomer and 1 to 2 months for the *cis*-isomer (Nambu *et al.*, 1980). Under both flooded and upland conditions, $^{14}CO_2$ was formed rapidly from the benzyl-labeled ester ($>50\%$ of applied ^{14}C within 30 days), although more was formed under the drier conditions and from the 1*R*,*trans*-isomer than otherwise. Radiolabeled residues also were incorporated into the soil-bound and unextractable fractions associated with soil humic and fulvic acids; only 1 day after incorporation at 1 mg/kg, 18% of the *trans*-isomer ^{14}C was unextractable from upland soil, and this increased to a maximum of 35% within 7 days. Residues of *trans*-phenothrin fell to <10 ppb within 45 days in aerobic soil but still remained close to 300 ppb after 60 days in anaerobic soil.

Microbial degradation of natural pyrethrins in field water has not been reported, either. However, that of another ester, butoxyethyl 2,4-dichlorophenoxyacetate (**15**) (Fig. 9-3) has been investigated (Paris *et al.*, 1975); hydrolysis was rapid (75–91% in 15 minutes by fungi, 50–91% by bacteria) and could be carried out by bacteria, fungi, and yeasts. The hydrolysis products subsequently were subjected to microbial oxidation. The closely-related cinerone, 2-(2′-*cis*-butenyl)-3-methylcyclopenten-1-one, was both reduced and oxidized on the side-chain (Tabenkin *et al.*, 1969).

Although rats and mice readily hydrolyze the pyrethrate methyl ester (Class *et al.*, 1990), neither mammals nor arthropods appreciably hydrolyze pyrethrin C-4 esters (Yamamoto *et al.*, 1969; Casida, 1973). However, they do carry out rapid and extensive oxidative metabolism (Class *et al.*, 1990). Based on this, and on known microbial reactions (Bollag and Liu, 1990), one can easily predict that microbial degradation products such as those shown in Fig. 9-8 should be formed from pyrethrins. With the unsaturated side-chains, reactive methylene group, and secondary alcohol, extensive rethrolone degradation is to be expected; degradation of the cyclopropanecarboxylic portion may be slower, due to its highly branched structure (Dias and Alexander, 1971; Hammond and Alexander, 1972). Pyrethrins degradation in normal soils and natural waters should be rapid and mineralization largely complete.

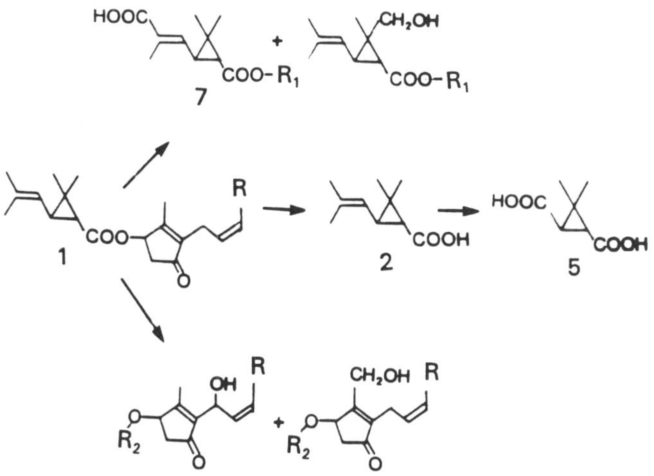

Figure 9-8. Predicted microbial degradation routes of pyrethrins. R = alkyl or alkenyl, R_1 = rethronyl, R_2 = chrysanthemyl.

IV. CONTROL OF TRANSPORT AND TRANSFORMATION

There have been many attempts to stabilize pyrethrum extract and pyrethrins formulations—that is, to control the transport and transformations which reduce insecticidal potency. Several brief reviews (Metcalf, 1955; Miskus and Andrews, 1972; Naumann, 1990) and a bibliography (Boleszny, 1970) have considered the subject.

Control over transport has received very little attention. There are several patents on the use of high-boiling organic solvents such as glyceryl esters, polysiloxanes, and squalene to reduce volatility (Naumann, 1990), and microencapsulation accomplishes the same purpose. However, with the expanding use of water as a pyrethrins carrier and the relatively high Henry's law constants predicted for the pyrethrins (Section II.B), means to reduce the volatilization from water may become necessary.

The most prominent route of transformation is photodegradation (Section III.B). Although maximum UV absorption by pyrethrins occurs near 230 nm, Miskus and Andrews (1972) found that the spectral range responsible for photodegradation in practice was 290–320 nm (from the short wavelength limit of sunlight to the UV absorption of soft glass); however, both Kawano *et al.* (1980) and Class and Kintrup (1991) observed that photochemical changes occur in pyrethrins exposed to sunlight through glass windows. A wide variety of chemicals have been tested as "sunscreens" which would absorb UV energy in this spectral region preferentially and so prevent or reduce absorption by the pyrethrins (Naumann, 1990), but few have been put into practical use. Many commercial sunscreens protect cinerin I and II against sunlight (Miskus and Andrews, 1972). While a 4 h irradiation in hexane allowed recovery of only 6% of the cinerin I and 2% of the pyrethrin I, mineral oil allowed recovery of

Table 9-5. Sunscreens and Antioxidants for Protection of Pyrethrins[a]

Chemical	Solvent	% Recovery (4 hrs)	
		Cinerin I	Pyrethrin I
None	Hexane	6	2
None	Mineral oil	67	59
None	Solvent A[b]	66	49
Amyl *p*-dimethylaminobenzoate (**16**)	Mineral oil	77	65
Benzyl cinnamate (**17**)	Mineral oil	80	71
Ethyl cinnamate (**18**)	Mineral oil	93	72
Sulisobenzone[c] (**19**)	Solvent A	96	34
Cyasorb[d] (**20**)	Solvent A	93	77
2,6-Dioctadecyl-*p*-cresol (**21**)	Mineral oil	94	78
16 + **21**	Mineral oil	102	83
17 + **21**	Mineral oil	103	92
18 + **21**	Mineral oil	97	80
19 + **21**	Solvent A	90	85

[a]Adapted from Miskus and Andrews (1972).
[b]Solvent A is 10% butoxyethanol in mineral oil.
[c]2-Hydroxy-4-methoxybenzophenone-5-sulfonic acid.
[d]4-(2,3-Dihydroxypropoxy)-2-hydroxybenzophenone methacrylate.

67% and 59%, respectively, and a solution of 10% butoxyethanol in mineral oil afforded about the same protection (Table 9-5). Addition of sunscreens commonly used for protection against human sunburn greatly increased the effectiveness of the mineral oil, and the highly UV-absorbing sesame or safflower oils probably would have been even better. However, in the years since 1972, many additional sunscreen chemicals have been developed for both human and industrial use (Grayson, 1983), providing new options for testing. It is important to note that many effective UV absorbers suffer other serious difficulties as practical sunscreens: 4-aminoazobenzene (carcinogenic), 2,4-dihydroxyazobenzene (Food Yellow 10, a yellow dye), and *p*-nitrophenol (acidic) represent early examples. More recently, Dureja *et al.* (1984) reported successful tests on a series of 2,6-dinitroanilines, but these compounds are brightly colored and some of them are highly phytotoxic.

As photooxidation actually is the principal abiotic transformation process for pyrethrins, chemical antioxidants also have been explored as stabilizers (Naumann, 1990). Bell and Kido (1956) showed that a low concentration (0.02%) of butylated hydroxyanisole (BHA) protected the effectiveness of pyrethrins in dusts for up to 8 weeks, as did 4-methoxy-2-propenylphenol. Other common antioxidants, such as 2,6-dioctadecyl-*p*-cresol (Table 9-5), were very effective protectants (Miskus and Andrews, 1972), and a combination of UV screen, antioxidant, and mineral oil should provide complete protection for at least short periods of time. Field tests in the United States and Canada showed that such a combination with pyrethrum provided good control of hemlock looper and spruce budworm (cited by Miskus and Andrews, 1972). Results with other food-grade antioxidants [such as butylated hydroxytoluene (BHT) and ascorbic acid] and industrial antioxidants have not been reported.

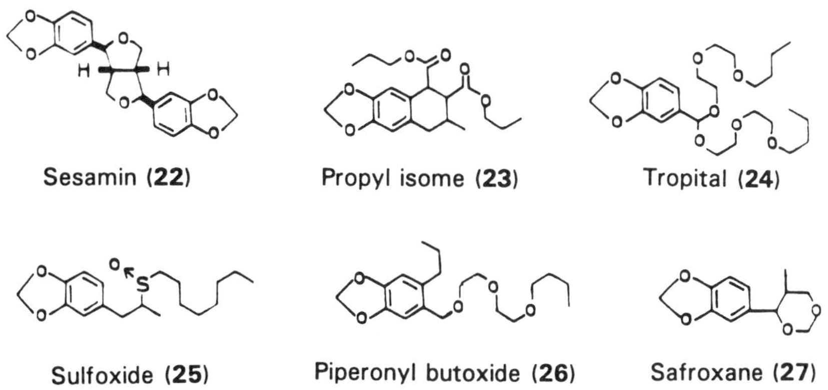

Sesamin (**22**) Propyl isome (**23**) Tropital (**24**)

Sulfoxide (**25**) Piperonyl butoxide (**26**) Safroxane (**27**)

Figure 9-9. Chemical structures of common pyrethrins synergists.

One form of antioxidant is an essential ingredient in pyrethrins applications. The synergistic effect of sesame oil on the insecticidal activity of pyrethrins was noted in 1940, and the responsible ingredient was shown to be sesamin (**22**) (Haller *et al.*, 1942). Other pyrethrins synergists are propyl isome (**23**), tropital (**24**), sulfoxide (**25**), piperonyl butoxide (**26**), and safroxane (**27**)—all of which possess a methylenedioxybenzene (1,3-benzodioxole) moiety (Fig. 9-9). The most important of these is piperonyl butoxide. Although it absorbs UV energy at the lower end of the sunlight region (λ_{max} 290 nm, ε 4.3×10^6) (Gore *et al.*, 1971) piperonyl butoxide remains stable to photolysis (Fishbein and Gaibel, 1970) and does not protect pyrethrins against light (Donaldson and Stevenson, 1960). The presence of any of these stabilizers sharply increases insecticidal potency by inhibiting insect mixed-function oxidases responsible for pyrethrins biodegradation (Esaac and Casida, 1969; Casida, 1970; Yamamoto, 1973). The methylenedioxy group is readily oxidized by the microsomal oxidase system (Casida *et al.*, 1966) as well as by hydride ion transfer (Hennessy, 1965) and reaction with hydroxyl radicals (Kumagai *et al.*, 1991).

Interestingly, piperonyl butoxide was effective as a foliar spray for protecting plants against damage due to atmospheric ozone (Koiwai *et al.*, 1974; Rubin *et al.*, 1980); ozone and other environmental oxidants are known to react readily with acetals (Deslongchamps *et al.*, 1974). Several other pyrethrins synergists (**23**, **25**, **27**) similarly were shown to prevent ozone damage, providing another area of interest for these antioxidants.

V. CONCLUSION

Throughout their long history, pyrethrins have been observed to be strikingly nonpersistent. However, little has been reported on their transport and transformation—the "environmental fate"—under actual environmental conditions. It appears that the physical properties associated with pyrethrins transport can be estimated satisfactorily, judging from the accurate prediction of those of the closely-related synthetic pyrethroid, bioallethrin; likewise, their chemical, photochemical, and microbial transformations can be predicted based on those of close relatives such as bioallethrin and phenothrin.

The pyrethrins I series (chrysanthemate esters) appears to have a higher octanol-water partition coefficient (K_{ow}), bioconcentration factor (BCF), volatility, and soil adsorption (K^{oc}), and lower aqueous solubility than does the pyrethrins II series (pyrethrate esters). On balance, this suggests that pyrethrin I, cinerin I, and jasmolin I may be more readily transported (more "mobile") in the environment than their pyrethrate counterparts, and that pyrethrin I, especially, may be bioconcentrated from even very dilute aqueous solutions and may evaporate readily from treated surfaces.

Aqueous hydrolysis is predicted to be very slow in many natural waters, but rapid in microbially-rich water. Microbial degradation in aerobic soil and water is expected to be faster than under anaerobic (waterlogged) conditions. Abiotic degradation will take place on the soil surface or, in fact, on any surface exposed to natural daylight.

In sunlight, the pyrethrins will be photodegraded rapidly, even though the light comes through soft-glass windows. Isomerization of the rethrolone side-chain, photooxidation to a wide variety of carboxylic acids, and probably *cis-trans* isomerization of the cyclopropane acids take place, and the action of atmospheric ozone seems to be important. This photodegradation can be effectively inhibited by incorporation of UV-absorbing agents (sunscreens) into the formulations, especially well if antioxidants are added. Common insecticide synergists such as piperonyl butoxide serve primarily to protect against biodegradation rather than photooxidation.

After enjoying decades of minimum government regulation, the pyrethrins must now conform to normal registration requirements as far as data on their movement and breakdown is concerned. These data are presently being collected, and it will be interesting and instructive to see how closely they coincide with the predictions made here.

REFERENCES

Alexander, M. (1980). Biodegradation of chemicals of environmental concern. *Science* **211**, 132–138.

Bell, A., and Kido, G.S. (1956). Hydroquinone and its derivatives as stabilizers for pyrethrum and allethrin. *J. Agric. Food Chem.* **4**, 340–343.

Bernett, W.A. (1967). A unified theory of bonding for cyclopropanes. *J. Chem. Ed.* **44**, 17–24.

Boleszny, I. (1970). "Pyrethrins and Pyrethrum Insecticides: The Application and Chemistry of Pyrethrin Insecticides," 2nd Ed. State Library of South Australia, Adelaide, SA, Research Bibliographies, Series 4, No. 95.

Bollag, J.-M., and Liu, S.-Y. (1990). Biological transformation processes of pesticides. *In* "Pesticides in the Soil Environment: Processes, Impacts, and Modeling" (H.H. Cheng, ed.), pp. 169–211. Soil Science Society of America, Madison, WI.

Briggs, G.G. (1981). Theoretical and experimental relationships between soil adsorption, octanol-water partition coefficients, water solubilities, bioconcentration factors, and the parachor. *J. Agric. Food Chem.* **29**, 1050–1059.

Briggs, G.G., Elliott, M., and Janes, N.F. (1983). Present status and future prospects for synthetic pyrethroids. *In* "Pesticide Chemistry: Human Welfare and the Environment" (J. Miyamoto and P.C. Kearney, eds.), Vol. 2, pp. 157–164. Pergamon Press, New York, NY.

Brown, N.C., and Phipers, R.F. (1955). The analysis of pyrethrins. Errors arising during the examination of partially degraded materials. *Pyrethrum Post* **3**, 23–26.

Brown, N.C., Hollinshead, D.T., Phipers, R.F., and Wood, M.C. (1957). Application of chromatography to analysis of pyrethrins. *Soap Chem. Specialties* **33**, 87, 91.
Bullivant, M.J., and Pattenden, G. (1971). Photochemical decomposition of chrysanthemic acid and its alkyl esters. *Pyrethrum Post* **11**, 72–76.
Bullivant, M.J., and Pattenden, G. (1976a). Photodecomposition of natural pyrethrins and related compounds. *Pestic. Sci.* **7**, 231–235.
Bullivant, M.J., and Pattenden, G. (1976b). Photochemistry of 2-(prop-2-enyl)cyclopent-2-enones. *J. Chem. Soc., Perkin Trans.* 1, 249–256.
Campbell, A., and Mitchell, W. (1950). Polymerized pyrethrins. *J. Sci. Food Agric.* **1**, 137–139.
Casida, J.E. (1970). Mixed-function oxidase involvement in the biochemistry of insecticide synergists. *J. Agric. Food Chem.* **18**, 753–772.
Casida, J.E. (1973). Biochemistry of the pyrethrins. *In* "Pyrethrum, The Natural Insecticide" (J.E. Casida, ed.), pp. 101–120. Academic Press, New York, NY.
Casida, J.E., Engel, J.L., Esaac, E.G., Kamienski, F.X., and Kuwatsuka, S. (1966). Methylene-C^{14}-dioxyphenyl compounds: Metabolism in relation to their synergistic action. *Science* **153**, 1130–1133.
Chen, Y.-L., and Casida, J.E. (1969). Photodecomposition of pyrethrin I, allethrin, phthalthrin, and dimethrin. *J. Agric. Food Chem.* **17**, 208–215.
Class, T.J. (1991). Determination of pyrethroids and their degradation products in indoor air and on surfaces by HRGC-ECD and HRGC-MS (NCI). *J. High Resol. Chromatogr.* **14**, 446–450.
Class, T.J., and Kintrup, J. (1991). Pyrethroids as household insecticides: analysis, indoor exposure, and persistence. *Fresenius J. Anal. Chem.* **340**, 446–453.
Class, T.J., Ando, T., and Casida, J.E. (1990). Pyrethroid metabolism: Microsomal oxidase metabolites of (*S*)-bioallethrin and the six natural pyrethrins. *J. Agric. Food Chem.* **38**, 529–537.
Crombie, L., and Elliott, M. (1961). Chemistry of the natural pyrethrins. *Progr. Chem. Org. Natural Prod.* **19**, 120–164.
Crombie, L., Harper, S.H., and Sleep, K.C. (1957). Experiments on the synthesis of pyrethrins. XIII. Total synthesis of (±)-*cis*- and *trans*-chrysanthemumdicarboxylic acid, (±)-*cis*- and *trans*-pyrethric acid, and rethrins II. *J. Chem. Soc.*, 2743–2754.
Crombie, L., Pattenden, G., and Simmonds, D. J. (1976). Structural and physical aspects of pyrethrum chemistry. *Pestic. Sci.* **7**, 225–230.
Crosby, D.G. (1976). Nonbiological degradation of herbicides in the soil. *In* "Herbicides: Physiology, Biochemistry, Ecology" (L.J. Audus, ed.), 2nd Ed., Vol. 2, pp. 65–97. Academic Press, New York, NY.
Crosby, D. G. (1992). Unpublished observations.
Deslongchamps, P., Atlani, P., Frehel, D., Malaval, A., and Moreau, C. (1974). Oxidation of acetals by ozone. *Can. J. Chem.* **52**, 1351–1364.
Dias, F.F., and Alexander, M. (1971). Effect of chemical structure on the biodegradability of aliphatic acids and alcohols. *Appl. Microbiol.* **22**, 1114–1118.
Dickinson, C.M. (1982). Stability of individual natural pyrethrins in solution after separation by preparative high-performance liquid chromatography. *J. Assoc. Off. Anal. Chem.* **65**, 921–926.
Donaldson, J.M., and Stevenson, J.H. (1960). Stabilizing effect of piperonyl butoxide on pyrethrins exposed to ultraviolet light. *J. Sci. Food Agric.* **11**, 370–373.
Dureja, P., Casida, J.E., and Ruzo, L.O. (1984). Dinitroanilines as photostabilizers for pyrethroids. *J. Agric. Food Chem.* **32**, 246–250.
Elliott, M. (1964). The pyrethrins and related compounds. III. Thermal isomerization of *cis*-pyrethrolone and its derivatives. *J. Chem. Soc.* 888–892.
Elliott, M. (1973). Chemistry of the natural pyrethrins. *In* "Pyrethrum, The Natural Insecticide" (J.E. Casida, ed.), pp. 55–100. Academic Press, New York, NY.
Esaac, E.G., and Casida, J.E. (1969). Metabolism in relation to mode of action of methylenedioxyphenyl synergists in houseflies. *J. Agric. Food Chem.* **17**, 539–550.
Finlayson-Pitts, B.J., and Pitts, J.N. (1986). "Atmospheric Chemistry: Fundamentals and Experimental Techniques." John Wiley and Sons, New York, NY.
Fishbein, L., and Gaibel, Z.L.F. (1970). Photolysis of pesticidal synergists. I. Piperonyl butoxide. *Bull. Environ. Contam. Toxicol.* **5**, 546–552.
Freeman, S.K. (1956). Stability of allethrin vs. pyrethrins. *Soap Chem. Specialties* **32**, 131, 150–151.
Godin, P.J., Sleeman, R.J., Snarey, M., and Thain, E.M. (1966). The jasmolins, new insecticidally active constituents of *Chrysanthemum cinerariaefolium* VIS. *J. Chem. Soc. C*, 332–334.
Goldschmidt, A., Crammer, B., and Ikan, R. (1984). Mechanistic study of the thermal acid-catalyzed rearrangement of *trans* methyl chrysanthemate to lavandulyl derivatives. *J. Chem. Soc., Perkin Trans.* 1, 2697–2705.

Gore, R.C., Hannah, R.W., Pattachini, S.C., and Porro, T.J. (1971). Infrared and ultraviolet spectra of seventy-six pesticides. *J. Assoc. Off. Anal. Chem.* **54**, 1040–1082.

Grayson, M. (1983). "Kirk-Othmer Encyclopedia of Chemical Technology," 3rd Ed., Vols. 7 and 23. John Wiley and Sons, New York, NY.

Griffin, G.W., Covell, J., Petterson, R.C., Dodson, R.M., and Klose, G. (1965). Photochemical interconversion of propenes and cyclopropanes. *J. Amer. Chem. Soc.* **87**, 1410–1414.

Gückel, W., Synnatschke, G., and Rittig, R. (1973). A method for determining the volatility of active ingredients used in plant protection. *Pestic. Sci.* **4**, 137–147.

Haller, H.L., LaForge, F.B., and Sullivan, W.N. (1942). Effect of sesamin and related compounds on the insecticidal action of pyrethrum on houseflies. *J. Econ. Entomol.* **35**, 247–248.

Hammond, M.W., and Alexander, M. (1972). Effect of chemical structure on microbial degradation of methyl-substituted aliphatic acids. *Environ. Sci. Technol.* **6**, 732–735.

Hansch, C., and Leo, A.J. (1979). "Substituent Constants for Correlation Analysis in Chemistry and Biology." John Wiley and Sons, New York, NY.

Hansch, C., Quinlan, J.E., and Lawrence, G.L. (1968). The linear free-energy relationships between partition coefficients and aqueous solubility of organic liquids. *J. Org. Chem.* **33**, 347–350.

Head, S.W., Sylvester, N.K., and Challinor, S.K. (1968). Effect of piperonyl butoxide on the stability of films of crude and refined pyrethrum extracts. *Pyrethrum Post* **9**, 14–22.

Hennessy, D.J. (1965). Hydride-transferring ability of methylenedioxybenzenes as a basis of synergistic activity. *J. Agric. Food Chem.* **13**, 218–220.

Isobe, N., Matsuo, M., and Miyamoto, J. (1984). Novel photoproducts of allethrin. *Tetrahedron Lett.*, 861–864.

Jorgensen, M.J. (1969). Photochemistry of α,β-unsaturated esters. VII. The photolytic behavior of vinylcyclopropanecarboxylates. *J. Amer. Chem. Soc.* **91**, 6432–6443.

Katagi, T., Kikuzono, Y., Mikami, N., Matsuda, T., and Miyamoto, J. (1988). A theoretical approach to photochemistry of pyrethroids possessing the cyclopropane ring. *J. Pestic. Sci. (Nippon Noyaku Gakkaishi)* **13**, 129–132.

Kaufman, D.D., Haynes, S.C., Jordan, E.G., and Kayser, A.J. (1977). Permethrin degradation in soil and microbial cultures. *ACS Sympos. Ser.* **42**, 147–161.

Kawano, Y., Yanagahara, K., Miyamoto, T., and Yamamoto, I. (1980). Examination of the conversion products of pyrethrins and allethrin formulations exposed to sunlight by gas chromatography and mass spectrometry. *J. Chromatogr.* **198**, 317–328.

Koiwai, A., Kitano, H., Fukuda, M., and Kisaki, T. (1974). Methylenedioxyphenyl and its related compounds as protectants against ozone injury to plants. *Agric. Biol. Chem.* **38**, 301–307.

Kumagai, Y., Lin, L.Y., Schmitz, D.A., and Cho, A.K. (1991). Hydroxyl radical mediated demethylenation of (methylenedioxy)phenyl compounds. *Chem. Res. Toxicol.* **4**, 330–334.

LaForge, F.B., Green, N., and Schechter, M.S. (1952). Dimerized cyclopentadienes from esters of allethrolone. *J. Amer. Chem. Soc.* **74**, 5392–5394.

Lyman, W.J., Reehl, W.F., and Rosenblatt, D.H. (1990). "Handbook of Chemical Property Estimation Methods." American Chemical Society, Washington, DC.

Maybe, W., and Mill, T. (1978). Critical reviews of hydrolysis of organic compounds in water. *J. Phys. Chem. Ref. Data* **7**, 383–415.

Metcalf, R.L. (1955). "Organic Insecticides: Their Chemistry and Mode of Action," pp. 76–101. *Interscience Publishers, New York, NY.*

Miskus, R.P., and Andrews, T.L. (1972). Stabilization of thin films of pyrethrins and allethrin. *J. Agric. Food Chem.* **20**, 313–315.

Nakada, Y., Yura, Y., and Murayama, K. (1972). Chrysanthemate derivatives. II. Pyrolysis of pyrethrin I. *Bull. Chem. Soc. Jap.* **45**, 2243.

Nambu, K., Ohkawa, H., and Miyamoto, J. (1980). Metabolic fate of phenothrin in plants and soils. *J. Pestic. Sci. (Nippon Noyaku Gakkaishi)* **5**, 177–197.

Naumann, K. (1990). "Synthetic Pyrethroid Insecticides: Structures and Properties." Springer Verlag, Berlin.

Otieno, D.A., and Pattenden, G. (1980). Degradation of the natural pyrethroids. *Pestic. Sci* **11**, 270–278.

Otieno, D.A., Pattenden, G., and Popplestone, C.R. (1977). Thermal acid-catalyzed rearrangements of natural chrysanthemic acids. *J. Chem. Soc., Perkin Trans.* 1, 196–201.

Paris, D.F., Lewis, D.L., Barnett, J.T., and Baughman, G.L. (1975). Microbial degradation and accumulation of pesticides in aquatic systems. EPA-660/3-75-007, pp. 27–29, U.S. Environmental Protection Agency, Corvallis, OR.

Roberts, T.R. (1981). The metabolism of the synthetic pyrethroids in plants and soils. *Prog. Pestic. Biochem.* **1**, 115–146.

Rubin, B., Leavitt, J.R.C., Penner, D., and Saettler, A.W. (1980). Interaction of antioxidants with ozone and herbicide stress. *Bull. Environ. Contam. Toxicol.* **25**, 623–629.
Ruzo, L.O., Gaughan, L.C., and Casida, J.E. (1980). Pyrethroid photochemistry: *S*-bioallethrin. *J. Agric. Food Chem.* **28**, 246–249.
Ruzo, L.O., Smith, I.H., and Casida, J.E. (1982). Pyrethroid photochemistry: Photooxidation reactions of the chysanthemates, phenothrin and tetramethrin. *J. Agric. Food Chem.* **30**, 110–115.
Ruzo, L.O., Kimmel, E.C., and Casida, J.E. (1986). Ozonides and epoxides from ozonization of pyrethroids. *J. Agric. Food Chem.* **34**, 937–940.
Sasaki, T., Eguchi, S., and Ohno, M. (1970). Studies on chrysanthemic acid. IV. Photochemical behavior of chrysanthemic acid and its derivatives. *J. Org. Chem.* **35**, 790–793.
Stahl, E. (1960). Zur Inaktivierung der Pyrethrine am Wirkungsort. *Arch. Pharm.* **293**, 531–537.
Staudinger, J., and Ruzicka, L. (1924a). Insektentötende Stoffe. II. Zur Konstitution der Chrysanthemum-monocarbonsäure und -dicarbonsäure. *Helv. Chim. Acta* **7**, 201–211.
Staudinger, J., and Ruzicka, L. (1924b). Insektentötende Stoffe. III. Zur Konstitution des Pyrethrolones. *Helv. Chim. Acta* **7**, 201–211.
Tabenkin, B., LeMahieu, R.A., Berger, J., and Kierstead, R.W. (1969). Microbiological hydroxylation of cinerone to cinerolone. *Appl. Microbiol.* **17**, 714–717.
Ueda, K., and Matsui, M. (1971). Studies on chrysanthemic acid. XXI. Photochemical isomerism of chrysanthemic acid and its derivatives. *Tetrahedron* **27**, 2771–2774.
Veith, G.D., Macek, K.J., Petrocelli, S.R., and Carroll, J. (1980). An evaluation of using partition coefficients and water solubility to estimate bioconcentration factors for organic chemicals in fish. *In* "Aquatic Toxicology" (J.G. Eaton, P.R. Parrish, and A.C. Hendricks, eds.), pp. 116–129, ASTM STP707. American Society for Testing and Materials, Philadelphia, PA.
Worthing, C.R., and Hance, R.J. (1991). "The Pesticide Manual," 9th Ed., pp. 76–77. British Crop Protection Council, Farnham, UK.
Yamamoto, I. (1973). Mode of action of synergists in enhancing the insecticidal activity of pyrethrum and pyrethroids. *In* "Pyrethrum, The Natural Insecticide" (J.E. Casida, ed.), pp. 195–210. Academic Press, New York, NY.
Yamamoto, I., Kimmel, E.C., and Casida, J.E. (1969). Oxidative metabolism of pyrethroids in houseflies. *J. Agric. Food Chem.* **17**, 1227–1236.

IV

Toxicology of Pyrethrins and Pyrethrum Extract

10

Mode of Action of Pyrethrins and Pyrethroids

DAVID M. SODERLUND

I. INTRODUCTION

The rapid knockdown action and lethal effects of pyrethrins and pyrethroids on insects depend on the ability of these compounds to disrupt the normal functioning of the insect nervous system. Despite the extensive use of the pyrethrins for insect control, there is very little information on the mode of action of the natural esters as neurotoxicants. However, a more detailed understanding of the mode of action of pyrethrins can be obtained by inference from the large body of literature describing the mode of action of the variety of pyrethroid insecticides that have been synthesized as analogs of the natural esters.

This chapter reviews studies of the fundamental action on nerves of pyrethrins and allethrin, a synthetic analog of pyrethrin I (Fig. 10-1). It also considers biophysical and biochemical studies with a wider structural variety of synthetic analogs that further illuminate the mode of action of pyrethrins and pyrethroids and provide information on the pyrethrins/pyrethroid binding site in the nervous system. Finally, this chapter evaluates the impact of recent studies of resistant and neurological mutant insect strains on the understanding of the mode of action of pyrethrins and pyrethroids. The perspective of this chapter is necessarily selective because of its focus on aspects of pyrethroid neurotoxicology that are particularly relevant to an understanding of the mode of action of pyrethrins. More comprehensive coverage of the literature on pyrethroid mode of action is available in four recent reviews (Sattelle and Yamamoto, 1988; Soderlund and Bloomquist, 1989; Narahashi, 1992; Bloomquist, 1993).

II. PYRETHRINS AND ALLETHRIN

A. Pyrethrins

Initial studies of the action of pyrethrins on insect nerves employed extracellular recordings of action potentials in the cercal nerve-giant fiber pathway of the

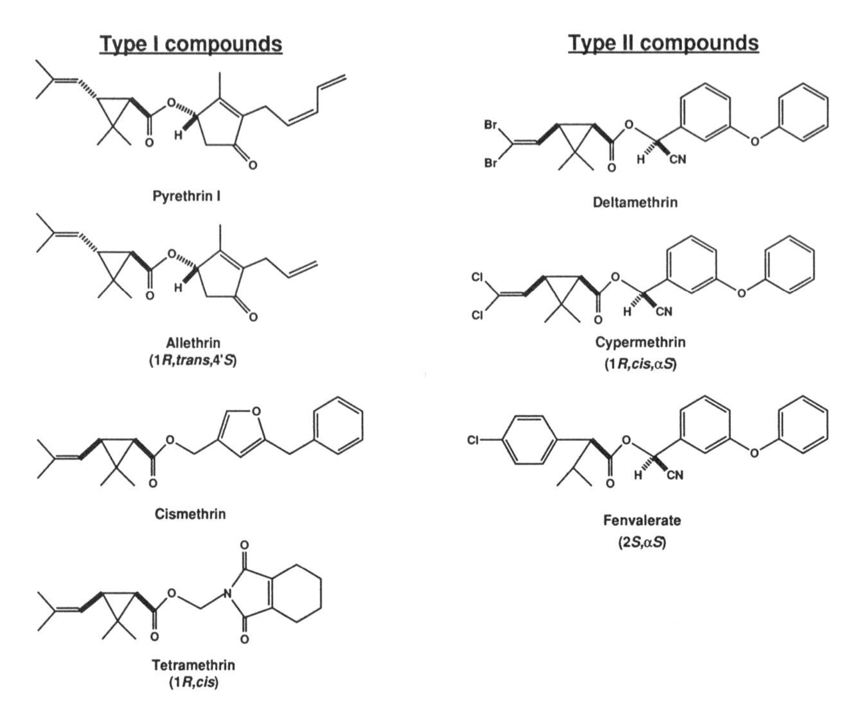

Figure 10-1. Structures of pyrethrin I, allethrin, and other pyrethroids studied. For compounds used as isomer mixtures, the structure of the most insecticidal isomer is shown.

cockroach *Blatta orientalis* (Lowenstein, 1942). Irrigation of these preparations with solutions of pyrethrum extract produced an initial phase of excitation, characterized by bursts of spontaneous and evoked action potentials in the giant fiber pathway and in other nerves, followed by blockade of the giant fiber response to cercal stimulation. The topical treatment of isolated American cockroach (*Periplaneta americana*) legs with pyrethrins also resulted in the induction of bursts of action potentials recorded from the crural nerves (LaLonde and Brown, 1954).

Other studies from the same period described the effects of pyrethrins on crayfish nerve and muscle preparations. In neuromuscular preparations, pyrethrins produced bursts of action potentials in the motor nerve after a single stimulus, which were followed under some conditions by nerve block (Ellis *et al.*, 1942; Welsh and Gordon, 1947). Both of these studies noted that the responses to pyrethrins in these preparations were qualitatively identical to the responses obtained following exposure to DDT. Pyrethrins were also found to block synaptic transmission in the crayfish giant fiber-motor nerve pathway (Schallek and Wiersma, 1948). A subsequent comparative study (Camougis and Davis, 1971) confirmed the stimulatory and blocking actions of pyrethrins in extracellular recordings from crayfish ganglia and peripheral nerve but found that frog and rat peripheral nerve preparations were less sensitive to pyrethrins.

Information on the physiological actions of the individual ester components of the pyrethrins is limited to a single report on the effects of pyrethrin I on axonal conduction and spontaneous ganglionic activity in the American cockroach (Burt and Goodchild, 1971). Exposure of isolated nerve preparations to solutions of pyrethrin I resulted in a concentration- and time-dependent block of action potentials in giant fiber axons and increased the endogenous electrical activity of the sixth abdominal ganglion. The actions of pyrethrin I *in vitro* were also compared to the effects of pyrethrin I applied topically to cockroaches at the LD_{95} on neurophysiological responses of poisoned insects. Results of these studies suggested that certain neuronal elements within the sixth abdominal ganglion were highly sensitive to pyrethrin I and that changes in the spontaneous activity of these elements were closely correlated with the onset of intoxication *in vivo*.

B. Allethrin

1. Extracellular and Intracellular Recordings

The mode of action of allethrin, a close structural analog of pyrethrin I, has been studied in much greater detail than that of the pyrethrins. Extracellular recordings of evoked compound action potentials in *P. americana* abdominal nerve cord preparations (Narahashi, 1962a) demonstrated that allethrin produced pronounced repetitive discharges followed by block, an effect qualitatively identical to that of the pyrethrins (Lowenstein, 1942). Further analyses (Narahashi, 1962a) revealed that the induction of repetitive discharges was associated with the development of a negative after-potential, a sustained period of depolarization evident at the end of the falling phase of the action potential. Intracellular recordings of the action of allethrin in these preparations demonstrated that both the height of the negative after-potential and the induction of repetitive discharges were temperature-dependent, thereby implicating the induction of the negative after-potential as the mechanism underlying allethrin-dependent burst discharges (Narahashi, 1962a), but did not identify the ionic basis of the negative after-potential (Narahashi, 1962b). Subsequent studies using intracellular recording techniques confirmed that allethrin produced similar hyperexcitatory and, in some cases, blocking effects on leech neurons (Leake, 1977), presynaptic nerve terminals of the frog motor endplate (Wouters *et al.*, 1977), and squid giant axons (Starkus and Narahashi, 1978).

Gammon (1978) used extracellular electrodes implanted in freely walking *P. americana* to assess the relationship between the neurophysiological actions of allethrin on central and peripheral nerves and the signs of allethrin intoxication *in vivo*. These studies showed that the appearance of excitatory effects in peripheral nerves was temporally correlated with the onset of incoordinated movements early in intoxication at both 15 and 32°C, whereas excitatory effects in the central nervous system were consistently noted only at 32°C and were correlated with the hyperactivity, tremors, and prostration that occurred later in the intoxication time course. These findings suggest that hyperexcitation of the peripheral nervous system is a critical element of allethrin intoxication. In contrast, allethrin-dependent block of peripheral and central nerve conduction did not occur until several hours after paralysis and was therefore not considered to be an essential cause of allethrin intoxication.

2. *Voltage Clamp Experiments*

The sodium and potassium currents that determine the shape and duration of action potentials in nerve membranes can be studied directly in nerve axons using the voltage clamp technique. A description of the principles and methods of voltage clamp analyses, together with a summary of the use of this method to characterize the actions of pyrethroids, is given by Narahashi (1984). Voltage clamping permits control over the nerve membrane potential, so that the nerve can be held at a defined hyperpolarized (inside negative) potential, rapidly depolarized, held at a defined depolarized potential, and then repolarized to the original holding potential. During this imposed depolarization step the transient inward (sodium) and steady-state outward (potassium) currents flowing through voltage-activated sodium and potassium channels can be observed and characterized. Further experimental control is obtained by replacing intracellular potassium or extracellular sodium with impermeant ions and by using selective sodium channel and potassium channel blockers to isolate individual ionic components of the membrane current.

The ionic mechanisms underlying the effects of allethrin on nerve excitability were elucidated using the voltage clamp technique with both invertebrate and vertebrate axon preparations (Narahashi and Anderson, 1967; Murayama *et al.*, 1972; Wang *et al.*, 1972; Vijverberg *et al.*, 1982). Although effects of allethrin on both sodium and potassium currents were observed, these studies identified the sodium current as the component of the nerve action potential that was most sensitive to allethrin. Fig. 10-2 summarizes the action of allethrin on experimentally-isolated sodium currents in these preparations. In an untreated nerve, depolarization elicits a transient inward sodium current that inactivates within a few milliseconds. In nerves treated with allethrin, the voltage-dependent activation, or opening, of sodium channels is virtually unaltered but the time

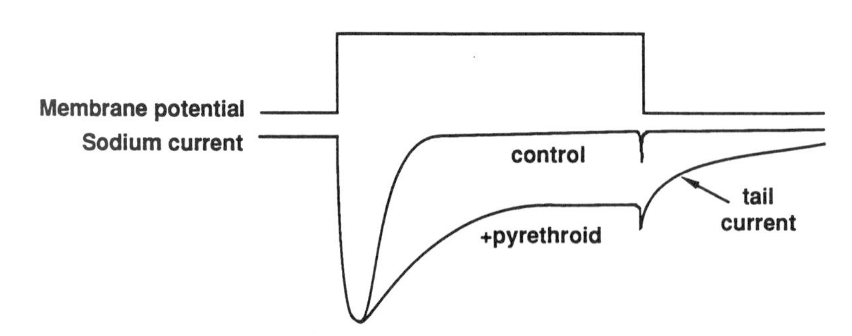

Figure 10-2. Diagram of typical sodium current during and after a pulsed depolarization under voltage clamp conditions recorded from untreated nerves and nerves exposed to a Type I pyrethroid (modified from Narahashi, 1992).

course of the sodium current during depolarization is prolonged and a residual, slowly-decaying current (the "tail current") is observed following repolarization. Thus, allethrin prolongs the time course of sodium permeability associated with the action potential by selectively altering the kinetics of sodium channel inactivation (closing). This effect directly accounts for the negative after-potential observed in intracellular microelectrode recordings. Moreover, the decay of allethrin-induced tail currents is prolonged at low temperatures, a finding that correlates with the negative temperature coefficient of allethrin intoxication. These findings establish the voltage-sensitive sodium channel as the principal target site for the action of allethrin and, by inference, the pyrethrins. The effects of allethrin in these experiments are also very similar to the effects of DDT (Narahashi, 1969), a finding that further underscores the close pharmacological relationship between pyrethrins, pyrethroids, and DDT analogs.

III. PYRETHROIDS

A. Two Types of Pyrethroid Action

The toxicological and pharmacological characterization of a variety of structurally diverse synthetic pyrethroids in both insects and mammals identified two distinct types of pyrethroid action, which are conventionally designated Type I and Type II. This distinction was originally based on the different signs of acute intoxication produced by these groups of compounds in mammals. Pyrethrins, allethrin, and a variety of other synthetic pyrethroids that are designated as Type I compounds produce a coarse whole body tremor (T syndrome) similar to that caused by DDT (Verschoyle and Barnes, 1972; Verschoyle and Aldridge, 1980; Lawrence and Casida, 1982). Deltamethrin (see Fig. 1) and other esters of α-cyano-3-phenoxybenzyl alcohol that are designated as Type II pyrethroids produce writhing convulsions (choreoathetosis) with profuse salivation (CS syndrome) (Barnes and Verschoyle, 1974; Verschoyle and Aldridge, 1980; Lawrence and Casida, 1982). Type I and Type II compounds also produce different signs of intoxication in insects (Gammon *et al.*, 1981; Scott and Matsumura, 1983; Bloomquist and Miller, 1985) and differential effects on the cercal nerve-giant fiber pathway of *P. americana* (Gammon *et al.*, 1981).

The characterization of the mode of action of pyrethroids exemplifying both Type I and Type II actions has been aided by the development of new biophysical and biochemical tools for the study of sodium channel pharmacology. Two important emphases of these studies have been to determine whether the pharmacological properties of allethrin are typical of the properties of a greater structural diversity of pyrethroids and whether the effects of pyrethroids on sodium channels are sufficient to account for the actions of both Type I and Type II compounds in insects and mammals. The following sections summarize important aspects of these studies; a more thorough coverage of this literature is available in recent reviews (Sattelle and Yamamoto, 1988; Soderlund and Bloomquist, 1989; Bloomquist, 1993).

B. Biophysical Characterization

1. Intracellular Recordings

The diversity of effects of DDT analogs and synthetic pyrethroids on the propagation of action potentials in invertebrate axons is clearly exemplified by the results of Lund and Narahashi (1983), which described the actions of a wide range of structures on the resting membrane potential and excitability of crayfish giant axons. DDT analogs and many Type I pyrethroids, including allethrin, produced effects anticipated from previous studies: the induction of a negative after-potential and repetitive bursts of action potentials elicited with a single stimulus, but little or no effect on the resting membrane potential. In contrast, many α-cyano-3-phenoxybenzyl esters with Type II effects did not cause repetitive discharges but instead produced a use-dependent block of nerve conduction coupled with a depolarization of the resting membrane potential. A few compounds were intermediate in their actions, producing bursts of action potentials of declining amplitude in response to a single stimulus in combination with use-dependent depolarization.

2. Voltage Clamp Experiments

The correlation of these effects of pyrethroids and DDT analogs on cellular excitability with effects on voltage-dependent sodium currents was explored using voltage clamped crayfish axons (Lund and Narahashi, 1983) and frog nodes of Ranvier (Vijverberg *et al.*, 1983). Modifications of sodium current induced by exposure of these preparations to DDT analogs and pyrethroids were consistent with the results of previous findings with DDT and allethrin (Narahashi, 1969), which are illustrated in Fig. 10-2: sodium inactivation during the depolarizing pulse was incomplete, and a sodium tail current was evident upon repolarization of the membrane. However, significant structure-dependent differences were found in rate of decay of the sodium tail current. Tail currents produced by DDT analogs and pyrethroids with Type I actions decayed rapidly, whereas tail currents produced by Type II compounds were much more prolonged.

These results suggested that the principal difference between Type I and Type II actions was in the lifetime of the pyrethroid-modified open sodium channel and that these differences were sufficient to explain the diversity of effects of pyrethroids and DDT analogs on nerve excitability. In this analysis, the transient prolongation of sodium permeability caused by DDT analogs and Type I pyrethroids produces the negative after-potential that is responsible for repetitive activity, but does not produce sufficient sodium influx to compromise the resting potential. In contrast, repetitive stimulation of nerves exposed to Type II compounds recruits increasing numbers of sodium channels into persistent pyrethroid-modified open states, which results in use-dependent depolarization, inactivation of unmodified channels, and block of conduction.

Recently, Bloomquist (1993) compared the mouse intracerebral toxicity (Lawrence and Casida, 1982) and the time constant for tail current decay in frog node of Ranvier (Vijverberg *et al.*, 1983) for eleven pyrethroids. This analysis identified a robust correlation of acute toxicity and persistence of sodium channel modification that spanned at least two orders of magnitude for each attribute and encompassed both Type I and Type II compounds. This result underscores the significance of sodium channel modification as the principal mechanism of toxic action of pyrethroids in mammals.

3. Patch Clamp Experiments

The development of patch clamp techniques, which allow the observation of the opening and closing of individual sodium channels in electrically-isolated membrane patches, permitted direct assessments of the lifetimes of pyrethroid-modified sodium channels. Most of these experiments have been performed using membrane patches isolated from mouse N1E-115 neuroblastoma cells. The results of these studies were consistent with the predictions from voltage clamp experiments. Tetramethrin (Fig. 10-1), a Type I pyrethroid, increased the mean sodium channel open time approximately 10-fold without altering the unitary conductance of these channels (Yamamoto *et al.*, 1983). In contrast, the Type II pyrethroids deltamethrin and fenvalerate (Fig. 10-1) increased the mean sodium channel open times up to 200-fold and produced channels that remained in the open state at the end of a voltage pulse (Chinn and Narahashi, 1986; Holloway *et al.*, 1989). In the case of deltamethrin, effects on channel open time were accompanied by a reduction in the incidence of channel opening, which was interpreted as evidence for deltamethrin-dependent stabilization of both closed and open channel states (Chinn and Narahashi, 1986). Thus, at the level of single sodium channels Type I and Type II pyrethroids share a common mode of action but are distinguished by the rate of inactivation of the pyrethroid-modified open channel.

C. Biochemical Characterization

1. Sodium Channel Pharmacology

A variety of naturally-occurring neurotoxins act at the sodium channel and play a role in the chemical ecology of many species. These toxins, which block sodium transport, enhance sodium channel activation, or prolong the time course of sodium channel inactivation, have been used extensively to characterize the pharmacological properties of sodium channels. Five principal neurotoxin recognition sites associated with the sodium channel have been identified in both functional assays and in radioligand binding experiments. These are designated as Sites 1–5, based on the classification of Catterall (1988) (Table 10-1). Each of these sites represents a physically-distinct domain of the sodium channel protein and can be labeled directly by a toxin-derived radioligand. These neurotoxins have been employed in combination with pyrethroids and DDT analogs to obtain insight into the identity of the insecticide binding site on the sodium channel and its relationship to established neurotoxin binding sites.

Table 10-1. Identified and Inferred Neurotoxin Binding Domains on the Voltage-Sensitive Sodium Channel[a]

Site	Active neurotoxins	Physiological effect	Allosteric coupling
1	Tetrodotoxin Saxitoxin	Inhibit ion transport	None
2	Veratridine Batrachotoxin Aconitine Grayanotoxins	Persistent activation	Sites 3, 5, and 6
3	α-Scorpion toxins Sea anemone toxins	Slow inactivation	Sites 2, 5, and 6
4	β-Scorpion toxins	Enhance activation	None
5	Brevetoxins Ciguatoxin	Persistent activation	Sites 2, 3, and 6
6	Pyrethrins Pyrethroids DDT and analogs	Slow inactivation	Sites 2, 3, and 5

[a]Sites 1–5 after Catterall (1988); Site 6 after Lombet *et al.* (1988).

2. *Sodium Uptake Experiments*

Assays of neurotoxin-dependent radiosodium uptake into cultured cells or mammalian brain vesicles provide a biochemical approach to the assessment of the action of these toxins on sodium channel function. The first studies of the action of pyrethroids using this technique employed cultured mouse neuroblastoma cells (Jacques *et al.*, 1980). In these assays pyrethroids did not affect sodium uptake in the absence of other neurotoxins, but some compounds (e.g., deltamethrin) enhanced the stimulation of sodium uptake produced by veratridine (VTD), batrachotoxin (BTX) and dihydrograyanotoxin II, all of which bind at Site 2 and cause persistent activation of the sodium channel. However, other neurotoxic compounds (e.g., cismethrin; Fig. 10-1) failed to enhance VTD-dependent sodium uptake. Subsequent studies with neuroblastoma cells described the enhancement of VTD-dependent sodium uptake by DDT and a greater variety of synthetic pyrethroids (Roche *et al.*, 1985; Lombet *et al.*, 1988) but did not find similar effects with a series of insecticidal DDT-pyrethroid hybrid structures (Holan *et al.*, 1985). Studies of the effects of pyrethroids and DDT analogs on sodium uptake into mouse brain synaptosomes (Ghiasuddin and Soderlund, 1985; Soderlund *et al.*, 1987; Bloomquist and Soderlund, 1988) also showed that these compounds did not affect sodium uptake in the absence of other neurotoxins but enhanced VTD- and BTX-stimulated sodium uptake.

of DDT and pyrethroids was found to consist of two separate effects: a small (up to three-fold) increase in the apparent affinity of the sodium channel for Site 2 activators; and an enhancement of the maximal sodium uptake obtained with saturating concentrations of VTD (Jacques *et al.*, 1980; Bloomquist and Soderlund, 1988). Because polypeptide toxins that bind to Site 3 typically produce these same effects on VTD-dependent sodium uptake (Tamkun and Catterall, 1981), the possible interactions of pyrethroids and DDT analogs with toxins acting at Site 3 are also of interest. In brain synaptosomes, DDT (but not cismethrin) acts synergistically with sea anemone (*Anemonia sulcata*) toxin to enhance VTD-dependent sodium uptake (Bloomquist and Soderlund, 1988), whereas a variety of pyrethroids enhance the stimulation of sodium uptake caused by sea anemone toxin in neuroblastoma cells (Jacques *et al.*, 1980). These results imply the existence of an insecticide-binding domain on the sodium channel that is distinct from Site 2 and Site 3 but allosterically coupled to both of these sites.

3. Radioligand Binding Experiments

Attempts to identify and characterize the pyrethroid binding site on the sodium channel directly by use of high specific activity pyrethroid radioligands have been frustrated by the extreme lipophilicity of pyrethroids, which results in the measurement of high levels of nonspecific and unsaturable binding to brain membrane preparations (Soderlund *et al.*, 1983). As a consequence, radioligand binding studies have been limited to the evaluation of the effects of pyrethroids and DDT analogs on the binding of radioligands that label the five well-characterized neurotoxin-binding domains of the sodium channel (Table 10-1).

The allosteric enhancement by pyrethroids and DDT of the affinity of sodium channels for Site 2 activators provided the impetus for examining the effect of these compounds on the binding of [^{3}H]batrachotoxinin A-20-α-benzoate (BTX-B), an analog of BTX that labels Site 2. The multiple allosteric coupling of Site 2 to other sites on the sodium channel (Brown, 1988) (Table 10-1) also permits the use of BTX-B to assess the interactions of multiple neurotoxins that bind to the different sites that are coupled to Site 2. Initial assessments of the effects of pyrethroids (Brown *et al.*, 1988; Lombet *et al.*, 1988) and DDT analogs (Payne and Soderlund, 1989) showed that active isomers and analogs increased the affinity of sodium channels for BTX-B two- to three-fold without affecting the binding capacity for this ligand. Kinetic experiments with DDT (Payne and Soderlund, 1989) and deltamethrin (Rubin *et al.*, 1993) showed that the enhancement of BTX-B binding resulted from the stabilization by these insecticides of the BTX-B - sodium channel complex. Other experiments (Lombet *et al.*, 1988) showed that the enhancement of BTX-B binding by pyrethroids was additive with the enhancing effects of neurotoxins acting at Site 3 and Site 5 either singly or in combination.

Studies of the action of pyrethroids on the binding of other radioligands, though less extensive, complement the results obtained with BTX-B. Pyrethroids do not affect the binding of radioligands that label Site 1 (Lombet *et al.*, 1988) or Site 4 (Barhanin *et al.*, 1982). It is somewhat surprising in view of the apparent allosteric interactions between insecticides and *Anemonia sulcata* toxin II in sodium uptake experiments (Jacques *et al.*, 1980; Bloomquist and Soderlund, 1988) that pyrethroids are reported to have no effect on the binding of a polypeptide toxin radioligand that labels Site 3 (Vincent *et al.*, 1980). However, pyrethroids enhance the binding of a tritiated brevetoxin derivative to Site 5 (Lombet *et al.*, 1987).

Results of radioligand binding assays demonstrate that the actions of DDT and pyrethroids on the sodium channel involve a unique binding domain (Lombet *et al.*, 1988) (Table 10-1). This insecticide recognition site is separable from Sites 1 and 4 on the basis of the lack of displacement of ligands that label these sites. The insecticide site is also functionally distinct from but allosterically coupled to Sites 2, 3, and 5. All of the biochemical evidence available to date is consistent with the existence of a single toxicologically-relevant binding site that recognizes pyrethrins, pyrethroids, and DDT analogs. However, these data also do not rule out the existence of multiple sites for neuroactive pyrethroid isomers, as has been suggested on the basis of physiological assays of the interactions of the 1*R*,*trans* and 1*R*,*cis* isomers of tetramethrin (Lund and Narahashi, 1982). A more rigorous characterization of the DDT/pyrethroid binding domain would require a specific radioligand capable of labelling this site specifically.

4. Structure-Activity Correlations

The development of biochemical assays for pyrethroid-sodium channel interactions was motivated in part by the need for methods capable of giving quantitative potency comparisons. For limited series of compounds, both the radiosodium uptake and radioligand binding approaches have yielded useful qualitative and quantitative correlations. For example, both deltamethrin and its noncyano analog were effective enhancers of VTD-dependent sodium uptake into mouse brain preparations, but the enantiomers of these compounds, which lack intrinsic toxicity to mice, were inactive in this assay (Ghiasuddin and Soderlund, 1985). Similarly, of the eight possible isomers of cypermethrin (Fig. 10-1) only the two neurotoxic isomers were effective as enhancers of BTX-B binding to rat brain sodium channels (Brown *et al.*, 1988). Despite the existence of strong quantitative correlations between the enhancement of sodium uptake and the enhancement of BTX-B binding for limited series of compounds (Lombet *et al.*, 1988), examination of a wider variety of structures revealed that a number of neurotoxic Type I compounds failed to enhance VTD-dependent sodium uptake into neuroblastoma cells (Jacques *et al.*, 1980) and also did not enhance the binding of BTX-B to mouse brain sodium channels (Rubin *et al.*, 1993). An intrinsic limitation of these approaches for developing structure-activity correlations lies in their absolute requirement for at least one other class of neurotoxin in order to observe a pyrethroid-dependent effect. Thus, there is no assurance that the structure-activity relationships for pyrethroids determined in these assays are unbiased reflections of the structure-activity relationships that govern the binding of pyrethroids to sodium channels in the absence of other neurotoxins.

IV. TARGET SITE-MEDIATED RESISTANCE

A. Target Site-Mediated Resistance as a Probe of Mechanism

Mutations that confer resistance to pyrethrins and pyrethroids may affect the rate of insecticide penetration, the rates of metabolic detoxication, or the sensitivity of the nervous system to intoxication (Soderlund and Bloomquist, 1990). Of these, mutations that affect neuronal sensitivity are of potential value for understanding mode of action because they may produce a structural modification of the target macromolecule that alters the pyrethrins/pyrethroid binding site. Two types of mutations have proven useful in elucidating the mode of action of pyrethrins and pyrethroids. The first type includes insect strains with defined resistance mechanisms that have been derived from populations exposed to DDT, pyrethrins, or pyrethroids in the field and further selected and characterized in the laboratory. The second type includes conditional neurological mutants of *Drosophila melanogaster* that are known to affect neuronal excitability and aspects of sodium channel function but have not been isolated using insecticide resistance as a criterion.

B. The *kdr* Resistance Mechanism of the House Fly

Resistance to the rapid paralytic effects of DDT was first documented in house flies (*Musca domestica*) (Busvine, 1951). Subsequent studies (reviewed in Oppenoorth, 1985) showed that this mechanism (termed "knockdown resistance" or *kdr*) confers resistance to the rapid paralytic actions of pyrethrins and pyrethroids as well as DDT. The *kdr* gene has been mapped to autosome 3 and includes alleles (designated *super-kdr*) that confer enhanced levels of resistance (Farnham, 1977; Sawicki, 1978; Farnham *et al.*, 1987). Neurophysiological studies with larval and adult house flies have demonstrated that the *kdr* trait confers reduced sensitivity to pyrethroids at the level of the nerve (Miller *et al.*, 1979; Salgado *et al.*, 1983).

Three hypotheses have been advanced to explain the reduced neuronal sensitivity conferred by the *kdr* mutation: reduced sodium channel density in nerve membranes (Chang and Plapp, 1983; Rossignol, 1988); altered lipid composition of nerve membranes (Chiang and Devonshire, 1982); and a structural alteration of sodium channels in resistant insects (Salgado *et al.*, 1983; Pauron *et al.*, 1989). Subsequent studies have failed to document reduced sodium channel density as an obligatory component of resistance in *kdr* and *super-kdr* house fly strains (Grubs *et al.*, 1988; Pauron *et al.*, 1989). However, physiological and biochemical assays using insect nerve preparations or subcellular fractions are unable to distinguish between mechanisms involving alterations in membrane environment or channel structure.

An alternative strategy to define the mechanism underlying *kdr* involves the identification of the gene product of the *kdr* locus. Recent progress in the molecular biology of vertebrate and insect sodium channels has permitted the development of conspecific molecular probes for a gene fragment in the house fly that exhibits a high degree of predicted amino acid sequence conservation with vertebrate sodium channel genes and with the product of *para* locus, a physiologically-important sodium channel of *D. melanogaster* (Knipple *et al.*, 1991). This probe was used to identify several restriction fragment length polymorphisms (RFLPs) for the *para*-homologous locus between susceptible and *kdr* house flies that were mapped to autosome 3, the linkage group that contains the *kdr* locus (K.E. Doyle, D.M. Soderlund, and D.C. Knipple, unpublished). The use of one of these RFLPs in high resolution mapping experiments demonstrated that the *kdr* trait and this sodium channel gene were genetically inseparable (Knipple *et al.*, 1994). This finding is in agreement with the results of a similar analysis (Williamson *et al.*, 1993) which used other RFLP markers within the same house fly sodium channel gene and evaluated both the *kdr* and *super-kdr* alleles of the *kdr* locus. These results provide convincing genetic evidence that the product of the *para*-homologous sodium channel gene is the site of action of pyrethrins and pyrethroids in the house fly.

C. Neurological Mutants of Drosophila Melanogaster

1. nap^{ts}

The *no-action-potential, temperature-sensitive* (*nap^{ts}*) mutation was isolated as a conditional, temperature-sensitive trait that produces nerve conduction block and rapid paralysis at elevated temperatures (Wu *et al.*, 1978). Membranes isolated from *nap^{ts}* flies contained a reduced density of sodium channels (measured as [^{3}H]saxitoxin binding sites), and the temperature-sensitive paralytic phenotype of *nap^{ts}* flies was mimicked by feeding tetrodotoxin, a sodium channel blocker, to wild-type flies (Jackson *et al.*, 1984). Subsequent studies showed that *nap^{ts}* flies were hypersensitive to tetrodotoxin intoxication but resistant to VTD (Jackson *et al.*, 1986). These altered responses to sodium channel-directed neurotoxins are consistent with the reduction in sodium channel density found in binding studies with *nap^{ts}* membrane preparations.

The biochemical phenotype of the *nap^{ts}* strain provided the opportunity to test the hypothesis that a reduction in sodium channel density could confer resistance to pyrethroids. Kasbekar and Hall (1988) found significant resistance to fenvalerate in *nap^{ts}* adults and demonstrated that feeding tetrodotoxin to wild-type flies mimicked the knockdown resistance to fenvalerate found in the *nap^{ts}* strain. Moreover, linkage analysis showed that fenvalerate resistance mapped to the *nap^{ts}* locus. Other studies (Bloomquist *et al.*, 1989) showed that resistance extended to DDT and a wide structural variety of pyrethroids and was correlated with an increased latency for pyrethroid-induced burst discharges in neurophysiological assays.

2. para^{ts}

The *paralytic, temperature-sensitive* (*para^{ts}*) mutant was also isolated as a conditional, temperature-sensitive trait (Suzuki *et al.*, 1971). Recently, detailed molecular and genetic studies of the *para* locus have shown that the *para$^+$* gene product is the principal sodium channel gene of *D. melanogaster* (Loughney *et al.*, 1989; Stern *et al.*, 1990). Using the same approach employed for studies of pyrethroid resistance in the *nap^{ts}* strain, Hall and Kasbekar (1989) examined the sensitivity of three allelic *para^{ts}* mutants to fenvalerate. Results of these studies showed that the *para^{ts1}* and *para^{ts4}* strains were resistant to fenvalerate whereas the *para^{ts2}* strain was hypersensitive. Because the *para* locus is known to encode a physiologically-important sodium channel gene, these findings demonstrate that mutation-dependent alterations in sodium channel structure are capable of modifying pyrethroid sensitivity in *D. melanogaster*.

V. CONCLUSION

A large body of evidence, encompassing results of physiological, biochemical, and genetic studies, implicates the neuronal voltage-sensitive sodium channel as the principal site of action of pyrethrins, pyrethroids, and DDT analogs. Information specifically on the mode of action of pyrethrins is very limited, but comparisons of the actions of pyrethrins, allethrin, other synthetic pyrethroids, and DDT analogs provide strong evidence for a continuum of actions on the sodium channel that encompasses all of these insecticides. Within this continuum of effects, substantial qualitative differences between the actions of Type I and Type II compounds on sodium channels appear to be sufficient to explain the different syndromes of intoxication observed in both insects and mammals. Although some pyrethroids, particularly those with Type II actions, have been examined for effects at other targets (e.g., ligand-gated ion channels, energy-dependent ion pumps), the actions documented *in vitro* in these other systems (reviewed in: Sattelle and Yamamoto, 1988; Soderlund and Bloomquist, 1989; Bloomquist, 1993) are not well correlated with intoxication at the level of the intact nerve and do not appear to contribute to the primary insecticidal activity of these compounds.

Pyrethrins, pyrethroids, and DDT analogs appear to affect sodium channel function by binding to a unique site on the channel that is distinct from the five well-characterized neurotoxin recognition sites but is allosterically coupled to three of these sites (Table 10-1). Whether there is a single binding domain on the sodium channel that recognizes this diverse group of insecticidal structures remains unclear. Perhaps the most compelling evidence in favor of a single site comes from studies of target site-mediated resistance, in which a single mutation at a sodium channel structural gene locus is capable of conferring global cross-resistance to pyrethrins, pyrethroids, and DDT analogs. However, one physiological study suggests the existence of two separate domains that bind neuroactive pyrethroid isomers (Lund and Narahashi, 1982), a finding that is not explicitly contradicted by the results of other pharmacological and genetic studies.

The fact that pyrethrins, pyrethroids, and DDT analogs share a common mode of action has an important implication for the continued effective use of pyrethrins for insect control. Because target site-mediated resistance appears to affect all of these insecticides, selection for resistance by one group of compounds is likely to select for cross-resistance to all other compounds. Moreover, prior selection for target site-mediated resistance by DDT may lead to the rapid reselection of the same trait by intensive treatment with either pyrethrins or synthetic pyrethroids. Thus, the vigilant detection and management of resistance to pyrethrins in target pest populations must be undertaken in the context of both the longstanding treatment history of these populations with DDT and the potential for contemporary or concurrent selection with synthetic pyrethroids. In the absence of such efforts, there is the danger that the long-term effective use of pyrethrins may be compromised.

REFERENCES

Barhanin, J., Giglio, J.R., Leopold, P., Schmid, A., Sampaio, V., and Lazdunski, M. (1982). *Tityus serrulatus* venom contains two classes of toxins: *Tityus* g toxin is a new tool with a very high affinity for studying the Na^+ channel. *J. Biol. Chem.* **257**, 12553-12558.

Barnes, J.M., and Verschoyle, R.D. (1974). Toxicity of new pyrethroid insecticide. *Nature* **248**, 711.

Bloomquist, J.R. (1993). Neuroreceptor mechanisms in pyrethroid mode of action and resistance. *In* "Reviews in Pesticide Toxicology" (M. Roe and R. J. Kuhr, eds.), Vol. 2, pp. 181–226. Toxicology Communications, Raleigh, NC.

Bloomquist, J.R., and Miller, T.A. (1985). Carbofuran triggers flight motor output in pyrethroid-blocked reflex pathways of the house fly. *Pestic. Biochem. Physiol.* **23**, 247-255.

Bloomquist, J.R., and Soderlund, D.M. (1988). Pyrethroid insecticides and DDT modify alkaloid-dependent sodium channel activation and its enhancement by sea anemone toxin. *Mol. Pharmacol.* **33**, 543-550.

Bloomquist, J.R., Soderlund, D.M., and Knipple, D.C. (1989). Knockdown resistance to dichlorodiphenyltrichloroethane and pyrethroid insecticides in the *nap^ts^* mutant of *Drosophila melanogaster* is correlated with reduced neuronal sensitivity. *Arch. Insect Biochem. Physiol.* **10**, 293-302.

Brown, G.B. (1988). Batrachotoxin: a window on the allosteric nature of the voltage-sensitive sodium channel. *Int. Rev. Neurobiol.* **29**, 77-116.

Brown, G.B., Gaupp, J.E., and Olsen, R.W. (1988). Pyrethroid insecticides: stereospecific allosteric interaction with the batrachotoxinin-A benzoate binding site of mammalian voltage-sensitive sodium channels. *Mol. Pharmacol.* **34**, 54-59.

Burt, P.E., and Goodchild, R.E. (1971). The site of action of pyrethrin I in the nervous system of the cockroach *Periplaneta americana*. *Entom. Exp. Appl.* **14**, 179-189.

Busvine, J.R. (1951). Mechanism of resistance to insecticide in house flies. *Nature* **168**, 193-195.

Camougis, G., and Davis, W.M. (1971). A comparative study of the neuropharmacological basis of action of pyrethrins. *Pyrethrum Post* **11**, 7-14.

Catterall, W.A. (1988). Structure and function of voltage-sensitive ion channels. *Science* **242**, 50-61.

Chang, C.P., and Plapp, F.W., Jr. (1983). DDT and pyrethroids: receptor binding in relation to knockdown resistance (*kdr*) in the house fly. *Pestic. Biochem. Physiol.* **20**, 86-91.

Chiang, C., and Devonshire, A.L. (1982). Changes in membrane phospholipids, identified by Arrhenius plots of acetylcholinesterase and associated with pyrethroid resistance (*kdr*) in houseflies (*Musca domestica*). *Pestic. Sci.* **13**, 156-160.

Chinn, K., and Narahashi, T. (1986). Stabilization of sodium channel states by deltamethrin in mouse neuroblastoma cells. *J. Physiol.* **380**, 191-207.

Ellis, C.H., Thienes, C.H., and Wiersma, C.A.G. (1942). The influence of certain drugs on the crustacean nerve-muscle system. *Biol. Bull.* **83**, 334-352.

Farnham, A.W. (1977). Genetics of resistance of house flies (*Musca domestica* L.) to pyrethroids. I. Knockdown resistance. *Pestic. Sci.* **8**, 631-636.

Farnham, A.W., Murray, A.W.A., Sawicki, R.M., Denholm, I., and White, J.C. (1987). Characterization of the structure-activity relationship of *kdr* and two variants of *super-kdr* to pyrethroids in the house fly (*Musca domestica* L.). *Pestic. Sci.* **19**, 209-220.

Gammon, D.W. (1978). Neural effects of allethrin on the free walking cockroach *Periplaneta americana*: an investigation using defined doses at 15 and 32°C. *Pestic. Sci.* **9**, 79-91.

Gammon, D.W., Brown, M.A., and Casida, J.E. (1981). Two classes of pyrethroid action in the cockroach. *Pestic. Biochem. Physiol.* **15**, 181-191.

Ghiasuddin, S.M., and Soderlund, D.M. (1985). Pyrethroid insecticides: potent, stereospecific enhancers of mouse brain sodium channel activation. *Pestic. Biochem. Physiol.* **24**, 200-206.

Grubs, R.E., Adams, P.M., and Soderlund, D.M. (1988). Binding of [^{3}H]saxitoxin to head membrane preparations from susceptible and knockdown-resistant house flies. *Pestic. Biochem. Physiol.* **32**, 217-223.

Hall, L.M., and Kasbekar, D.P. (1989). *Drosophila* sodium channel mutations affect pyrethroid sensitivity. *In* "Insecticide Action: From Molecule to Organism" (T. Narahashi and J.E. Chambers, eds.), pp. 99-114. Plenum Press, New York.

Holan, G., Frelin, C., and Lazdunski, M. (1985). Selectivity of action between pyrethroids and combined DDT-pyrethroid insecticides on Na^+ influx into mammalian neuroblastoma. *Experientia* **41**, 520-522.

Holloway, S.F., Salgado, V.L., Wu, C.H., and Narahashi, T. (1989). Kinetic properties of single sodium channels modified by fenvalerate in mouse neuroblastoma cells. *Pflugers Arch.* **414**, 613-621.

Jackson, F.R., Wilson, S.D., Strichartz, G.R., and Hall, L.M. (1984). Two types of mutants affecting voltage-sensitive sodium channels in *Drosophila melanogaster*. *Nature* **308**, 189-191.

Jackson, F.R., Wilson, S.D., and Hall, L.M. (1986). The *tip-E* mutation of *Drosophila* decreases saxitoxin binding and interacts with other mutations affecting nerve membrane excitability. *J. Neurogenet.* **3**, 1-17.

Jacques, Y., Romey, G., Cavey, M.T., Kartalovski, B., and Lazdunski, M. (1980). Interaction of pyrethroids with the Na^+ channel in mammalian neuronal cells in culture. *Biochim. Biophys. Acta* **600**, 882-897.

Kasbekar, D.P., and Hall, L.M. (1988). A *Drosophila* mutation that reduces sodium channel number confers resistance to pyrethroid insecticides. *Pestic. Biochem. Physiol.* **32**, 135-145.

Knipple, D.C., Payne, L.L., and Soderlund, D.M. (1991). PCR-generated conspecific sodium channel gene probe for the house fly homologue of the *para* locus of *Drosophila melanogaster*. *Arch. Insect Biochem. Physiol.* **16**, 45-53.

Knipple, D.C., Doyle, K.E., Marsella-Herrick, P.A., and Soderlund, D.M. (1994). Tight genetic linkage between the *kdr* insecticide resistance trait and a voltage-sensitive sodium channel gene in the house fly. *Proc. Natl. Acad. Sci. USA* **91**, 2483–2487.

LaLonde, D.I.V., and Brown, A.W.A. (1954). The effects of insecticides on the action potentials of insect nerve. *Can. J. Zool.* **32**, 74-81.

Lawrence, L.J., and Casida, J.E. (1982). Pyrethroid toxicology: mouse intracerebral structure-toxicity relationships. *Pestic. Biochem. Physiol.* **18**, 9-14.

Leake, L.D. (1977). The action of (*S*)-3-allyl-2-methyl-4-oxocyclopent-2-enyl (1*R*)-*trans*-chrysanthemate, (*S*)-bioallethrin, on single neurones in the central nervous system of the leech, *Hirudo medicinalis*. *Pestic. Sci.* **8**, 713-721.

Lombet, A., Bidard, J.-N., and Lazdunski, M. (1987). Ciguatoxin and brevetoxins share a common receptor site on the neuronal voltage-dependent Na^+ channel. *FEBS Lett.* **219**, 355-359.

Lombet, A., Mourre, C., and Lazdunski, M. (1988). Interactions of insecticides of the pyrethroid family with specific binding sites on the voltage-dependent sodium channel from mammalian brain. *Brain Res.* **459**, 44-53.

Loughney, K., Kreber, R., and Ganetzky, B. (1989). Molecular analysis of the *para* locus, a sodium channel gene in *Drosophila*. *Cell* **58**, 1143-1154.

Lowenstein, O. (1942). A method of physiological assay of pyrethrum extracts. *Nature* **150**, 760-762.

Lund, A.E., and Narahashi, T. (1982). Dose-dependent interaction of the pyrethroid isomers with sodium channels of squid axonal membranes. *Neurotoxicology* **3**, 11-24.

Lund, A.E., and Narahashi, T. (1983). Kinetics of sodium channel modification as the basis for the variation in the nerve membrane effects of pyrethroids and DDT analogs. *Pestic. Biochem. Physiol.* **20**, 203-216.

Miller, T.A., Kennedy, J.M., and Collins, C. (1979). CNS insensitivity to pyrethroids in the resistant *kdr* strain of house flies. *Pestic. Biochem. Physiol.* **12**, 224-230.

Murayama, K., Abbott, N.J., Narahashi, T., and Shapiro, B.I. (1972). Effects of allethrin and condylactis toxin on the kinetics of sodium conductance of crayfish giant axon membranes. *Comp. Gen. Pharmacol.* **3**, 391-400.

Narahashi, T. (1962a). Effects of the insecticide allethrin on membrane potentials of cockroach giant axons. *J. Cell. Comp. Physiol.* **59**, 61-65.

Narahashi, T. (1962b). Nature of the negative afterpotential increased by the insecticide allethrin in cockroach giant axons. *J. Cell. Comp. Physiol.* **59**, 67-76.

Narahashi, T. (1969). Mode of action of DDT and allethrin on nerve: cellular and molecular mechanisms. *Residue Rev.* **25**, 275-288.

Narahashi, T. (1984). Nerve membrane sodium channels as the target of pyrethroids. *In* "Cellular and Molecular Neurotoxicology" (T. Narahashi, ed.), pp. 85-108. Raven Press, New York.

Narahashi, T. (1992). Nerve membrane Na^+ channels as targets of insecticides. *Trends Pharmacol. Sci.* **13**, 236-241.

Narahashi, T., and Anderson, N.C. (1967). Mechanism of excitation block by the insecticide allethrin applied externally and internally to squid giant axons. *Toxicol. Appl. Pharmacol.* **10**, 529-547.

Oppenoorth, F.J. (1985). Biochemistry and genetics of insecticide resistance. *In* "Comprehensive Insect Physiology Biochemistry and Pharmacology" (G.A. Kerkut and L.I. Gilbert, eds.), pp. 731-773. Pergamon Press, Oxford.

Pauron, D., Barhanin, J., Amichot, M., Pralavorio, M., Berge, J.-B., and Lazdunski, M. (1989). Pyrethroid receptor in the insect Na^+ channel: alteration of its properties in pyrethroid-resistant flies. *Biochemistry* **28**, 1673-1677.
Payne, G.T., and Soderlund, D.M. (1989). Allosteric enhancement by DDT of the binding of [^{3}H]batrachotoxinin A-20-a-benzoate to sodium channels. *Pestic. Biochem. Physiol.* **33**, 276-282.
Roche, M., Frelin, C., Bruneau, P., and Meinard, C. (1985). Interaction of tralomethrin, tralocythrin, and related pyrethroids in Na^+ channels of insect and mammalian cultured cells. *Pestic. Biochem. Physiol.* **24**, 306-316.
Rossignol, D.P. (1988). Reduction in number of nerve membrane sodium channels in pyrethroid resistant house flies. *Pestic. Biochem. Physiol.* **32**, 146-152.
Rubin, J.G., Payne, G.T., and Soderlund, D.M. (1993). Sructure-activity relationships for pyrethroids and DDT analogs as modifiers of [^{3}H]batrachotoxinin A 20-a-benzoate binding to mouse brain sodium channels. *Pestic. Biochem. Physiol.* **45**, 130–140.
Salgado, V.L., Irving, S.N., and Miller, T.A. (1983). Depolarization of motor nerve terminals by pyrethroids in susceptible and *kdr*-resistant house flies. *Pestic. Biochem. Physiol.* **20**, 100-114.
Sattelle, D.B., and Yamamoto, D. (1988). Molecular targets of pyrethroid insecticides. *Adv. Insect Physiol.* **20**, 147-213.
Sawicki, R.M. (1978). Unusual response of DDT-resistant houseflies to carbinol analogues of DDT. *Nature* **275**, 443-444.
Schallek, W., and Wiersma, C.A.G. (1948). The influence of various drugs on a crustacean synapse. *J. Cell. Comp. Physiol.* **31**, 35-47.
Scott, J.G., and Matsumura, F. (1983). Evidence for two types of toxic actions of pyrethroids on susceptible and DDT-resistant German cockroaches. *Pestic. Biochem. Physiol.* **19**, 141-150.
Soderlund, D.M., and Bloomquist, J.R. (1989). Neurotoxic actions of pyrethroid insecticides. *Annu. Rev. Entomol.* **34**, 77-96.
Soderlund, D.M., and Bloomquist, J.R. (1990). Molecular mechanisms of insecticide resistance. *In* "Pesticide Resistance in Arthropods" (R.T. Roush and B.E. Tabashnik, eds.), pp. 58-96. Chapman and Hall, New York, NY.
Soderlund, D.M., Ghiasuddin, S.M., and Helmuth, D.W. (1983). Receptor-like stereospecific binding of a pyrethroid insecticide to mouse brain membranes. *Life Sci.* **33**, 261-267.
Soderlund, D.M., Bloomquist, J.R., Ghiasuddin, S.M., and Stuart, A.M. (1987). Enhancement of veratridine-dependent sodium channel activation by pyrethroids and DDT analogs. *In* "Sites of Action for Neurotoxic Pesticides" (R.M. Hollingworth and M.B. Green, eds.), pp. 251-261. American Chemical Society, Washington, DC.
Starkus, J.G., and Narahashi, T. (1978). Temperature dependence of allethrin-induced repetitive discharges in nerves. *Pestic. Biochem. Physiol.* **9**, 225-230.
Stern, M., Kreber, R., and Ganetzky, B. (1990). Dosage effects of a *Drosophila* sodium channel gene on behavior and axonal excitability. *Genetics* **124**, 133-143.
Suzuki, D.T., Grigliatti, T., and Williamson, R. (1971). Temperature-sensitive mutations in *Drosophila melanogaster*. VII. A mutation ($para^{ts}$) causing reversible adult paralysis. *Proc. Natl. Acad. Sci. USA* **68**, 890-893.
Tamkun, M.M., and Catterall, W.A. (1981). Ion flux studies of voltage-sensitive sodium channels in synaptic nerve ending particles. *Mol. Pharmacol.* **19**, 78-86.
Verschoyle, R.D., and Aldridge, W.N. (1980). Structure-activity relationships of some pyrethroids in rats. *Arch. Toxicol.* **45**, 325-329.
Verschoyle, R.D., and Barnes, J.M. (1972). Toxicity of natural and synthetic pyrethrins to rats. *Pestic. Biochem. Physiol.* **2**, 308-311.
Vijverberg, H.P.M., van der Zalm, J.M., and van den Bercken, J. (1982). Similar mode of action of pyrethroids on sodium channel gating in myelinated nerves. *Nature* **295**, 601-603.
Vijverberg, H.P.M., van der Zalm, J.M., van Kleef, R.G.D.M., and van den Bercken, J. (1983). Temperature- and structure-dependent interaction of pyrethroids with the sodium channels in frog node of Ranvier. *Biochim. Biophys. Acta* **728**, 73-82.
Vincent, J.P., Balerna, M., Barhanin, J., Fosset, M., and Lazdunski, M. (1980). Binding of sea anemone toxin to receptor sites associated with the gating system of sodium channel in synpatic nerve endings *in vitro*. *Proc. Natl. Acad. Sci. USA* **77**, 1646-1650.
Wang, C.M., Narahashi, T., and Scuka, M. (1972). Mechanism of negative temperature coefficient of nerve blocking action of allethrin. *J. Pharmacol. Exp. Ther.* **182**, 442-453.
Welsh, J.H., and Gordon, H.T. (1947). The mode of action of certain insecticides in the arthropod nerve axon. *J. Cell. Comp. Physiol.* **30**, 147-172.
Williamson, M.S., Denholm, I., Bell, C.A., and Devonshire, A.L. (1993). Knockdown resistance (*kdr*) to DDT and pyrethroid insecticides maps to a sodium channel gene locus in the housefly (*Musca domestica*). *Mol. Gen. Genet.* **240**, 17–22.

Wouters, W., van den Bercken, J., and van Ginneken, A. (1977). Presynaptic action of the pyrethroid insecticide allethrin in the frog motor end plate. *Eur. J. Pharmacol.* **43**, 163-171.
Wu, C.-F., Ganetzky, B., Jan, L.Y., Jan, Y.-N., and Benzer, S. (1978). A *Drosophila* mutant with a temperature-sensitive block in nerve conduction. *Proc. Natl. Acad. Sci. USA* **75**, 4047-4051.
Yamamoto, D., Quandt, F.N., and Narahashi, T. (1983). Modification of single sodium channels by tetramethrin. *Brain Res.* **274**, 344-349.

11

Insect Resistance to Pyrethrins and Pyrethroids

DONALD G. COCHRAN

I. INTRODUCTION

Insecticide resistance was scarcely mentioned in a book on pyrethrum of about 20 years ago (Casida, 1973). Indeed, one contributor to the 1973 treatise espoused the position that there were numerous reasons to suggest that resistance to pyrethrins would not develop into a major problem (Mrak, 1973). There were only limited cases of resistance to natural pyrethrins reported at that time, even though resistance to other classes of insecticides was already well advanced (Georghiou, 1990). Although that prediction was not entirely without merit, resistance to pyrethrins has become of sufficient importance that it warrants thorough consideration at this time.

Insect resistance to insecticides is an example of evolution on a time scale that is sufficiently brief to allow its easy detection and quantification. It is a population phenomenon that results from repeated exposure over a number of generations to an insecticide that kills the most susceptible portion of the population. Survivors possess genetically-derived capabilities that allow them to withstand the effects of the insecticide. Thus, resistance can be defined as a change in an insect population over time that results in their ability to withstand dosages of a given insecticide that were previously effective in killing them. It was first evident early in this century (Melander, 1914) and has become extensive since the introduction of the synthetic organic insecticides at mid-century. Currently, there are over 500 species of insects and closely-related arthropods that are resistant to one or more insecticides and/or acaricides (Georghiou, 1990). The problem appears to become more extensive with each passing year.

Resistance is a complex phenomenon that can result from one of several mechanisms. The most frequently encountered of them are: (1) metabolic resistance in which the resistant insects have the enzymatic capability to degrade the insecticide before it can kill them (Mouches *et al.*, 1987; Georghiou, 1987; Soderlund and Bloomquist, 1990), (2) target-site insensitivity, e.g., knockdown resistance (*kdr*), in which the nervous system has been modified in such a way that the insecticide no longer has a major adverse effect on it (Chiang and Devonshire, 1982; Salgado *et al.*, 1983; Soderlund and Bloomquist, 1990), and (3) a decreased rate of penetration of the insecticide into the insect's body (Plapp and Hoyer, 1968; Sawicki and Farnham, 1968; Georghiou, 1987). The first two mechanisms often convey such a high-level of resistance that insects possessing either of them cannot be killed by the insecticide involved. The third mechanism, reduced penetration, typically produces only a small effect, but may greatly enhance the overall effect of the other two mechanisms (Georghiou, 1987). In addition, other mechanisms, such as insecticide storage in fatty tissue or rapid excretion, are known, but are uncommon. Some populations have more than one of these independent mechanisms and can be referred to as multi-resistant (Scott, 1990).

The situation can also be complicated by cross resistance in which tolerance to one insecticide confers resistance to other insecticides even though the insects have not been exposed to the other compounds. These latter cases can usually be explained on the basis of a similar chemistry among the insecticides or a resistance mechanism that can deal with diverse chemicals. Cross resistance between two organophosphates is an example of the first (Georghiou, 1972) and that between DDT and pyrethroids examplifies the second type (Scott and Matsumura, 1981; 1983). However, it is clear that the mechanisms may be more complex than was originally thought (Fournier *et al.*, 1987; Raymond *et al.*, 1989; Scott, 1990).

II. USE PATTERNS AND THE DEVELOPMENT OF RESISTANCE

The development of resistance is generally related to insecticide use patterns. In other words, one does not expect to find resistance to compounds that have not been used to control a particular pest species. This situation can be complicated by cross resistance, as mentioned above, but usually an analysis of the chemicals applied or the mechanisms involved will reveal the cause and type of cross resistance. Other situations may be more difficult to explain. For example, resistance to sulfluramide, a fluorinated sulfonamide, occurred in the German cockroach prior to any significant use of this material for cockroach control (Schal, 1992). That resistance did not correlate with resistance to any other class of insecticides. However, it may correlate with exposure to compounds in certain household products that have a similar chemistry. Other cases of unexpected resistance may result from a species being present where insecticides are used to control other insects at the same location.

Current and new uses for pyrethrins are discussed elsewhere in this book, however, it is important to mention past use patterns here because the development of resistance to pyrethrins has been affected by those use patterns. For example, pyrethrins have not been used extensively in the control of agricultural insect pests since the mid-1940s (Lange and Akesson, 1973). This is true because of their photoinstability, cost, and, perhaps most importantly, because of the availability of more competitive insecticides. As a result, reports of resistance to pyrethrins among these pests are very limited (Attia and

Hamilton, 1978). This, of course, contrasts markedly with the situation for the photostable synthetic pyrethroids against which resistance is becoming an increasingly important problem (Georghiou, 1990; Plapp *et al.*, 1990; Pree, 1990). Pyrethrins have been used over long periods of time for the control of household pests (Burden and Smittle, 1961), stored products pests (Gillenwater and Burden, 1973), in dairy barns for housefly control (Fisher, 1955; Keiding, 1977), for medical and veterinary pest control (Smith, 1973), and in similar environments that are protected from direct exposure to sunlight. Obviously, it is among pests found in these situations where one would expect to find cases of resistance to pyrethrins.

III. DOCUMENTED CASES OF RESISTANCE TO PYRETHRINS

The modern era of insect control with chemical insecticides began in the 1940s with the advent of DDT (Brown, 1951). While pyrethrins were being used prior to that date (Lange and Akesson, 1973), their use also intensified about the same time as a result of a heightened awareness that higher levels of control were possible through the use of chemicals. Because resistance is fostered by intensive insecticide use, it is not suprising that documented cases of resistance to pyrethrins bear dates subsequent to 1950 (Sawicki, 1985).

A listing of the known cases of resistance to pyrethrins is presented in Table 11-1. Several points emerge from these data. First, it is apparent that the insects

Table 11-1. Documented Cases of Resistance to Pyrethrins in Chronological Order of First Reports

Species	RR[a]	Reference
Pediculus humanus	Mod.	Nicoli and Sautet, 1955
	9×	Cole and Clarke, 1961
Blattella germanica	20×	Keller *et al.*, 1956
	140×	Cochran, 1989
Sitophilus granarius	7×	Holborn, 1957
	148×	Lloyd, 1969
Tribolium castaneum	3.5×	Holborn, 1957
Cimex lectularius	4×	Busvine, 1958
Cimex hemipterus	10×	Busvine, 1958
Boophilus decoloratus	18×	Whitehead, 1959
Musca domestica	10×	Plapp and Hoyer, 1968
	13×	Keiding, 1977
	>450×	Farnham, 1971
Culex tarsalis	10×	Plapp and Hoyer, 1968
Sitophilus oryzae	6×	Cichy, 1971
Plodia interpunctella	2.5×	Zettler *et al.*, 1973
Ephestia cautella	3.5×	Zettler *et al.*, 1973
Myzus persicae	13×	Attia and Hamilton, 1978
Oryzaephilus surinamensis	Low	Attia and Frecker, 1984
Fannia canularis	100×	Myers *et al.*, 1989

[a] RR = resistance ratio.

involved are those associated with the use patterns described above. Only one species, the green peach aphid *Myzus persicae*, is an agricultural pest. Second, it is clear that resistance to pyrethrins is not widespread in the insect world. Only 15 species are involved, which is about 3% of the total number of species known to be resistant to insecticides (Georghiou, 1990). Third, the levels of resistance, as reflected by resistance ratios (RR), are not high. Over half of the RR values reported were 10 or less. While it is well-known that the method of testing for resistance may influence the RR (Scott *et al.*, 1986; Milio *et al.*, 1987), it seems unlikely that resistance has reached a high level in most of the species listed. Fourth, RRs of >100 were reported for the German cockroach, the granary weevil, the house fly, and the little house fly. Hence, insects from diverse taxa are capable of developing high-level resistance to pyrethrins. In those species or populations of a given species where RRs are low, it appears that selection pressure has not been sufficient to foster the development of high-level resistance. From a practical point of view, this probably means that pyrethrins were used less intensively against such pests. Alternatively, pyrethrins may have been superceded by other insecticides and what remains can be viewed as residual resistance (Farnham *et al.*, 1984).

IV. PYRETHRINS RESISTANCE IN THE GERMAN COCKROACH

As stated above, the German cockroach is one of four species known to have developed high-level resistance to pyrethrins (Keller *et al.*, 1956; Cochran, 1973, 1989). This phenomenon has been studied in detail for the house fly and the German cockroach but not for the other two species. Because the information on the house fly was reviewed recently (Sawicki 1985), emphasis here will be placed on pyrethrins resistance in the German cockroach. This case history illustrates what can happen when pyrethrins are used extensively in a control program.

Pyrethrins have been used as a flushing agent in cockroach-control programs since at least the 1940s (Mallis, 1969). This use depends on the rapid excitatory response caused by pyrethrins. The response produces movement that brings cockroaches out of their hiding places so they will come in contact with the residual insecticide that is normally applied along with pyrethrins. Undoubtedly because of this use, there is a great deal of resistance to pyrethrins (Table 11-2).

Table 11-2. Status of Pyrethrins Resistance in the German Cockroach

No. strains tested	RR[a]
5	0.0–1.0
18	1.0–2.0
8	2.0–3.0
4	3.0–5.0
51	>140

[a]RR = resistance ratio (LT_{50} test strain/LT_{50} susceptible strain).

Table 11-3. Estimated Gene Frequency of the Pyrethrins-Resistance Gene in Susceptible Strains of German Cockroaches

Strain	GF[a]	LT_{50} RR[b]
VPI[c]	0.02	—
Blacksburg	<0.01	1.3
Orlando	<0.01	1.2
Hazard	0.07	0.7
Lincoln Terrace	0.09	1.6
HRDC	0.16	1.1
Navy # 3	0.40	2.6

[a]GF = gene frequency of the pyrethrins-resistance gene calculated according to Falconer (1960).
[b]RR = resistance ratio based on mortality data.
[c]VPI, Orlando, and Hazard are laboratory susceptible strains.

Of the 86 strains tested, nearly two thirds were not killed by a 24 hour exposure to a dose (0.3 nl of 20% pyrethrins/cm^2) that killed 100% of a susceptible strain. Many of these field populations are resistant to pyrethrins, but not to any other pyrethroid. Of the 51 strains with high-level resistance, shown in Table 11-2, 29 were resistant only to pyrethrins.

What appears to be involved here is the long history of pyrethrins use against this insect and the fact that the gene frequency for the pyrethrins-resistance gene in some "susceptible" strains is already finite. The data in Table 11-3 provide estimates of gene frequency for a series of populations that would typically be deemed susceptible on the basis of their times for 50% lethality (LT_{50}) RRs. Yet in five of them gene frequency was high enough to be measurable from samples ranging in size from 100–1,500 insects. In the Navy # 3 strain about 15% of the insects were already homozygous resistant even though the RR was only 2.6. The relatively high frequency of this gene in such populations means that selection for resistance should proceed at a rapid pace if they are regularly exposed to pyrethrins. It appears that this is what has happened in many instances. Information of this kind led Cochran (1994a) to speculate that the oldest type of resistance in the German cockroach was probably to pyrethrins, even though it was not detected until the mid-1950s (Keller *et al.*, 1956).

It is generally recognized that pyrethrins have two types of action, i.e., knockdown and kill. Knockdown usually occurs within minutes, while it may take several hours to kill an insect. It is the knockdown that is of importance in cockroach control because the flushing action, referred to above, immediately precedes knockdown. It takes only about twice as long for pyrethrins-resistant cockroaches to be knocked down as susceptible ones (Table 11-4). The main difference is that insects from resistant populations recover from knockdown in 2-4 hours, and appear to be completely normal at 24 hours. Thus, pyrethrins have not lost their effectiveness as a flushing agent, even in cockroaches that are resistant to kill by pyrethrins.

Table 11-4. Knockdown Times for Pyrethrins-Resistant Strains of German Cockroaches

Strain	KD_{50} (min)	RR[a]	
		KD_{50}	LT_{50}
VPI (susceptible)	13	—	—
Muncie	21	1.6	>140
Alachua # 146	23	1.8	>140
Gary	34	2.6	>140
Alachua # 108	35	2.6	>140

[a]RR = resistance ratio based on knockdown time or mortality.

Of fundamental importance to an understanding of pyrethrins resistance is gaining an insight into the mechanisms by which resistance is achieved. This may be accomplished by electrophysiological means for examining the nerve insensitivity mechanism (Umeda *et al.*, 1988) or by studying the enzymes involved in metabolic resistance (Wirth *et al.*, 1990). One simple way of approaching the latter is to study the effect of synergists on resistance. If synergists with a known mode of action block resistance, it is assumed that resistance is metabolic in nature and is attributable to an enzyme or enzyme system that is inhibited by that synergist. The synergists piperonyl butoxide (PB) and *N*-octyl bicycloheptenedicarboximide (MGK 264) act as inhibitors of microsomal mixed-function oxidases (Casida, 1970). The synergist DEF (*S*,*S*,*S*-tributyl phosphorotrithioate) (normally used as a defoliant) is usually described as an esterase inhibitor (Bell and Busvine, 1967; Apperson and Georghiou, 1975; Horowitz *et al.*, 1988) although it can also be an oxidase inhibitor (Casida, 1970). These two types of enzyme action are known to be important pathways for degrading pyrethroid insecticides (Matsumura, 1985). Thus, if synergists fail to block resistance, it is more likely that the mechanism is nonmetabolic.

Table 11-5. Effects of Synergists on Pyrethrins-Resistant Strains

Strain	RR alone	RR with synergist		
		PB	MGK 264	DEF
Clarke	>140[a]	1.8	2.4	1.7
Muncie	>140	1.1	1.2	1.5
Gary	>140	0.5	0.9	0.9
Cooper	>140	1.0	1.1	0.9
Alachua # 108	>140	1.4	1.2	2.4
Hawthorne	>140	1.5	1.3	2.1
H-360	>140	2.3	3.6	1.5
Montg. Gardens	>140	1.5	1.8	1.7

[a]Values are LT_{50} resistance ratios against pyrethrins alone and in the presence of the indicated synergist. The pyrethrins concentration was 0.3 nl of 20% pyrethrins/cm^2. The synergist-insecticide ratio was 5:1.

The effects of PB, MGK 264, and DEF on pyrethrins resistance in the German cockroach are shown in Table 11-5. They were applied in combination with pyrethrins at a dosage that produced no mortality when the synergists were applied alone. The synergist:insecticide ratio was 5:1. Each of these synergists reduced the RR of the strains tested to a very low level. Furthermore, those insects did not recover after a 24 hour exposure, as resistant cockroaches normally do. They were clearly rendered susceptible again by the presence of a synergist. While this type of evidence is circumstantial, it seems clear that resistance to pyrethrins in these strains is metabolic in nature.

An interesting aspect of the results shown here is that synergists generally considered to have different types of action produced essentially the same result. A possible explanation derives from the statement by Casida (1973) that pyrethrin I is metabolized more readily by oxidases, while pyrethrin II is degraded by esterases. If this is correct, then in the presence of PB or MGK 264, the metabolism of pyrethrin I would be inhibited, and, similarly, the presence of DEF would spare pyrethrins II. Perhaps under either condition there would be enough of one or the other intact pyrethrin to kill the otherwise resistant cockroaches. Alternatively, the synergism by DEF may be due to its action as an oxidase inhibitor. The latter suggestion appears more plausible, in view of the monofactorial inheritance mechanism for pyrethrins resistance (Cochran, 1973, 1994b).

V. PYRETHRINS RESISTANCE IN POPULATIONS RESISTANT TO OTHER PYRETHROIDS

A. Allethrin

German cockroaches are resistant not only to pyrethrins but also to synthetic pyrethroids (Cochran, 1989). The one that has the greatest impact on pyrethrins resistance is *S*-bioallethrin. Twenty-two of the 51 strains with high-level resistance to pyrethrins, shown in Table 11-2, were also highly resistant to allethrin. Among these 22 allethrin-resistant strains, knockdown was greatly

Table 11-6. Effects of Synergists on Allethrin-Resistant Strains

Strain	RR alone	RR with synergist: PB	RR with synergist: DEF
San Diego	>140[a]	>140	>140
MDSU	>140	>140	>140
Navy # 1	>140	>140	>140
Reddick	>140	>140	>140
Ft. Knox	>140	>140	>140
Forest Green	>140	>140	>140
Las Palms	>140	>140	>140

[a] See footnote for Table V. The (*S*)-bioallethrin concentration was 0.3 nl/cm^2 (93.1% active ingredient).

Table 11-7. Cross Resistance in Pyrethrins-Resistant Strains

Insecticide	Hawthorne	Coretta	Navy # 4	New Opelika
Pyrethrins	>140[a]	>140	>140	>140
Allethrin	>140	>140	>140	>140
Permethrin	0.5	>100	1.8	>120
Phenothrin	0.6	>140	2.4	>140
Fenvalerate	0.9	2.0	>50	>60
Cyfluthrin	1.8	2.9	3.7	>70
Cypermethrin	1.6	2.1	2.6	>50
Esfenvalerate	0.8	1.7	2.6	7.0
Cyhalothrin	0.7	1.8	1.7	5.6

[a]Values are LT_{50} resistance ratios. Insecticide concentrations were as reported previously (Cochran, 1989).

reduced; in some of them it did not occur and in others a portion of the test insects were briefly knocked down. Furthermore, in these strains none of the three synergists tested negated resistance to allethrin (Table 11-6) nor to pyrethrins. This indicates that allethrin resistance is due to a second mechanism, probably a *kdr*-type (Scott *et al.*, 1990), that also protects these insects from pyrethrins. Scott and Matsumura (1981, 1983) have previously reported a *kdr*-type mechanism in the German cockroach, but it originated from a DDT-selected strain and its effect was mainly on permethrin and cypermethrin. Preliminary electrophysiological evidence supports the contention that a *kdr*-type mechanism is involved in allethrin resistance as well (Bloomquist, 1993). In many of these strains there was no resistance or cross resistance to other pyrethriods, as illustrated by the Hawthorne strain in Table 11-7. It remains to be clarified whether this represents a second *kdr*-type mechanism in this species (see below).

B. Other Pyrethroids

Many populations of German cockroaches are resistant to pyrethroids other than pyrethrins and allethrin. This fact is illustrated by the data from the Coretta, Navy # 4, and New Opelika strains shown in Table 11-7. It is obvious that each of these strains has a somewhat different resistance profile. Indeed, examination of the profiles of over 80 populations has revealed that resistance to various combinations of pyrethroids is possible, including across-the-board resistance to as many as 9 or 10 pyrethroids. It can be argued that this is evidence in support of the notion that several pyrethroid-resistance mechanisms exist in this species, in addition to those for pyrethrins and allethrin.

In all cases reported here where resistance to other pyrethroids has been detected, the populations were also resistant to pyrethrins and allethrin. This probably means that the gene frequencies for the genes conferring resistance to the latter two insecticides were initially significantly higher than those for the newer pyrethroids. This concept is supported by data from laboratory selections for pyrethroid resistance (Table 11-8). In this case, fenvalerate was the selecting agent and the strain initially had no resistance to any pyrethroid. By the F_4

Table 11-8. Response of the Fenvalerate-Selected Bowl Strain to Several Pyrethroids

	Generations of selection			
Insecticide	F_6	F_{10}	F_{14}	F_{17}
Pyrethrins	>140[a]	>140	>140	>140
Allethrin	>140	>140	>140	>140
Permethrin	1.3	3.8	1.6	>140
Phenothrin	2.6	>60	>140	>140
Fenvalerate	1.1	2.0	2.6	>75
Cyfluthrin	0.7	3.2	2.0	30.6
Cypermethrin	—	—	1.6	20.2
Esfenvalerate	—	—	—	3.0
Cyhalothrin	—	—	—	4.8

[a]Values are LT_{50} resistance ratios.

generation (not shown), the strain was resistant to pyrethrins. Resistance to allethrin appeared by the F_6 generation, but little if any resistance to the other pyrethroids tested was present. However, based on the gene frequency evidence presented in Table 11-3 and events in subsequent generations, it is likely that the RR of 2.6 for phenothrin indicated a rapidly-increasing gene frequency for this resistance by the F_6 generation of selection. While a similar argument could be made for permethrin and cyfluthrin at F_{10}, it is evident that high-level resistance to them and other pyrethroids, including the selecting agent fenvalerate, was not apparent until the F_{17} generation. The early appearance of resistance to phenothrin here and the presence of the singular resistance to fenvalerate in the Navy # 4 strain (Table 11-7) lend additional support to the contention that several independent pyrethroid-resistance mechanisms exist in this species. It is also clear from the F_{17} generation (Table 11-8) and the data from the New Opelika strain (Table 11-7) that the German cockroach is capable of developing resistance to all of the pyrethroids tested, as well as to others not included in the tables.

It was of interest to determine the effects of synergists on resistance to permethrin and phenothrin in the German cockroach. As a general statement, synergists do negate resistance to these materials as they do with pyrethrins. Data showing the effects of PB and MGK 264 on resistance to permethrin and phenothrin are presented in Table 11-9. In most cases the RRs for permethrin are quite low, but those for the Seasons (Sel.) strain indicate that resistance was only partially negated. A similar, but more pronounced, effect is seen when these synergists were used against phenothrin-resistant strains. In some cases (e.g., Jacksonville and Coretta), resistance was largely negated by both PB and MGK 264. In other instances (especially K 851 and Long Island), one or the other or both synergists failed to reduce the RRs to a level that would restore the effectiveness of phenothrin. In these latter instances, it appears that more than one resistance mechanism may be functioning. Thus, with the newer pyrethroids the effectiveness of synergists in negating resistance varies with the insecticide and population being tested.

Table 11-9. Effects of Synergists on Permethrin and Phenothrin Resistance in the German Cockroach

	Permethrin			Phenothrin		
		RR with synergist			RR with synergist	
Strain	RR alone	PB	MGK 264	RR alone	PB	MGK 264
Jacksonville	>120[a]	1.7	2.7	>140	1.7	2.5
Forest Green	>90	1.9	2.6	>140	2.7	4.6
Coretta	>100	1.9	2.9	>140	2.9	2.4
Jones	>120	2.5	2.0	>140	4.8	2.7
New Opelika	>120	1.9	1.9	>140	3.0	3.5
Seasons (Sel.)	>140	5.3	4.1	>140	7.5	3.8
K 851	>120	3.9	3.1	>140	>140	8.0
Long Island	>120	2.6	2.6	>140	7.0	>70

[a]Values are resistance ratios against permethrin and phenothrin alone and in the presence of the indicated synergist. The permethrin concentration was 1.5 nl/cm^2 (92.0% active ingredient) and the phenothrin concentration was 3.0 nl/cm^2 (93.5% active ingredient). The synergist:insecticide ratio was 5:1.

VI. GENETICS OF PYRETHRINS RESISTANCE

Another important facet of understanding the pyrethrins-resistance phenomenon is providing information on the mechanism of inheritance of this trait, and how it compares with other closely-related traits. Pyrethrins resistance in the house fly is controlled by a series of factors located on at least four different chromosomes (2, 3, 4, and 5) (Tsukamoto, 1969). Each factor apparently conveys only low-level resistance by itself. For example, a gene located on chromosome 3 was reported to convey resistance in the order of 10X (Plapp and Hoyer, 1968). It is only when all or at least most of these factors are present together that higher levels of resistance occur. For example, Farnham (1971) reported a strain that was "immune" to pyrethrins. Subsequent studies confirmed that several factors were involved, the most important of which appeared to control a *kdr*-type mechanism located on chromosome 3 (Farnham 1973). It is probably the same factor reported earlier by Plapp and Hoyer (1968). In spite of the remaining uncertainties, this is the most completely studied example of how pyrethrins resistance is inherited in any insect.

In the German cockroach high-level resistance to pyrethrins was reported to be controlled by a simple, autosomal, incompletely-dominant trait that is linked with the Group VI trait pallid eye at a distance of about 33 linkage units (Cochran, 1973). A recent reevaluation of this trait, using a different method of testing, has confirmed that it is a simple, autosomal, dominant trait (Cochran, 1994b). The synergists PB, MGK 264, and DEF are all effective in negating resistance in the heterozygotes, as would be expected and as they are in homozygous resistant individuals. When allethrin resistance is also present, no knockdown occurs in highly resistant strains. As indicated above, the nervous system of these insects does not respond to the presence of either pyrethrins or allethrin. However, the heterozygotes are knocked down as are insects that have resistance only to pyrethins, and, as is the case with the latter, they subsequently recover. That recovery is blocked by synergists such as PB

(Cochran, unpublished data). These facts provide evidence that there are, indeed, two mechanisms involved, and that they are both functioning in insects that are resistant to pyrethrins and allethrin. One is the simple, autosomal, dominant metabolic resistance that is blocked by the synergists. The other is the *kdr*-type mechanism that confers resistance to allethrin. It is a simple, autosomal, recessive trait that renders the heterozygotes susceptible to knockdown. However, because the dominant metabolic resistance trait is also present, these insects recover from knockdown, and that recovery can be blocked by synergists. A similar conclusion that two independent factors are involved in resistance to pyrethrins (metabolic) and allethrin (*kdr*) was also reached by Dong and Scott (1991) and Cochran (1994b).

In conducting genetic analyses, like those discussed here, it is important to realize that in the German cockroach resistant strains are seldom completely homozygous for the resistance gene. In other words, the gene frequency of the resistance gene is somewhat less than 1.0. Also, as shown in Table 11-3, susceptible strains often have the resistance gene present at a level that is high enough to produce significant numbers of heterozygotes and even a few homozygous-resistant individuals. Thus, the gene frequencies of the resistance gene have to be determined for the parental lines. The values obtained must be taken into account when conducting crosses because the expected ratios in the F_2 and backcrosses will not be 3:1 or 1:1, respectively. Rather, calculations must be made to determine the expected ratios based on known gene frequencies. These measurements and calculations have proven to be relatively easy for the strains reported here, and expected and observed ratios have been in close agreement (Cochran, 1994b).

The results discussed in this section require a return to the question of whether there is more than one *kdr*-type resistance mechanism in the German cockroach. The example described by Scott and Matsumura (1981, 1983) occurred as a consequence of selection with DDT that also conveyed a significant measure of cross resistance to a series of pyrethroids, including pyrethrins, allethrin, permethrin, and fenvalerate. The parental strain had RRs of >2 for all of these insecticides, as well as for DDT. Three generations of laboratory selection with DDT produced a *kdr*-type resistance mechanism that resulted in RRs of approximately 50 for DDT, 25 for pyrethrins, 20 for allethrin and permethrin, and 10 for fenvalerate. That resistance declined rapidly in the absence of further selection with DDT. Clearly, this is a very different outcome from laboratory selections with fenvalerate (Table 11-8) or as occurred in field populations in which the selecting agent was a pyrethroid, most probably cypermethrin (Table 11-7, Hawthorne strain). DDT was not a factor here because it was never used in the field to control German cockroaches in the USA, and the laboratory-selected strain was never exposed to DDT. Furthermore, additional selections of the DDT-resistant strain with permethrin resulted in a stabilization of resistance to DDT and several pyrethroids (Dong and Scott, 1991). Thus, it appears that there are at least two types of *kdr* mechanisms that can occur in this insect. The determining factor apparently is whether DDT has been involved as a selecting agent. The nature of the differences remains to be defined.

VII. PYRETHRINS RESISTANCE IN THE FUTURE

The future of pyrethrins use will depend on the impact of resistance, and the use patterns will continue to play a major role in which species and to what extent resistance will occur. The question of importance then becomes what new use patterns will develop? What additional insect species can be controlled with pyrethrins? What new situations will allow the successful employment of pyrethrins? If such species and situations exist, resistance may well develop in some of them as has been the case in the past. Whether resistance will develop becomes a question of the biology of the species involved and how intensively pyrethrins are used against those species.

It is particularly important to consider the possible impact the development of resistance to other pyrethroids will have on pyrethrins resistance. It is evident that resistance to pyrethroids is a rapidly emerging problem in several areas of insecticide use. Most prominant are the areas involving control of agricultural insect pests (Myers *et al.*, 1987; Metcalf, 1989; Plapp *et al.*, 1990; Pree, 1990) and disease vectors (Shidrawi, 1990). In most of these cases, resistance to pyrethrins was not even considered because, for whatever reason, they were not effective against those species. Yet from Table 11-8 it is clear that selection of a pyrethroid-susceptible population of German cockroaches with fenvalerate resulted in the development of pyrethrins resistance before resistance to any other pyrethroid appeared. A similar result occurred when permethrin was the selecting agent (Cochran, unpublished data). Of course, the mechanism of resistance and the frequency of the gene or genes conferring resistance to pyrethrins may be very different in other insect species. Nevertheless, it appears likely that, in at least some of them, resistance to pyrethrins may already be well established. The importance of such an occurrence will depend on whether pyrethrins has any potential future role as a control insecticide for those species. In any case, it is of scientific interest to know whether the German cockroach response is a general one, that has already resulted in many cases of undetected resistance to pyrethrins, or whether it is specific for that species.

VIII. CONCLUSION

On the basis of existing evidence it appears that resistance to natural pyrethrins is not currently a serious or widespread problem in the insect world. However, this situation may be a reflection of past use patterns more than the lack of an ability by insects to develop resistance. From studies with the house fly and the results presented here with the German cockroach, there is no reason to expect that resistance to pyrethrins will not develop if they are used extensively against a specific pest. Future use patterns will undoubtedly play a major role in determining the extent to which pyrethrins resistance develops.

A major concern for the future of pyrethrins use is that control efforts with other pyrethroids may have inadvertently fostered resistance to pyrethrins in many insect species. Evidence from the German cockroach indicates that this is a strong possibility. Because of the extensive recent use of pyrethroids in insect control, this point should have a high priority in future research whenever pyrethrins are being considered for control of an insect species that has been exposed to other pyrethroids.

REFERENCES

Apperson, C.S., and Georghiou, G.P. (1975). Mechanisms of resistance to organophosphorus insecticides in *Culex tarsalis*. *J. Econ. Entomol.* **68**, 153-157.

Attia, F.I., and Hamilton, J.T. (1978). Insecticide resistance in *Myzus persicae* in Australia. *J. Econ. Entomol.* **71**, 851-852.

Attia, F.I., and Frecker, T. (1984). Cross-resistance spectrum and synergism studies in organophosphorus-resistant strains of *Oryzaephilus surinamensis* (L.) (Coleoptera: Cucugidae) in Australia. *J. Econ. Entomol.* **77**, 1367-1370.

Bell, J.D., and Busvine, J.R. (1967). Synergism of organophosphates in *Musca domestica* and *Chrysomya putoria*. *Entom. Exp. Appl.* **10**, 263-269.

Bloomquist, J.R. (1993). Personal communication.

Brown, A.W.A. (1951). "Insect Control by Chemicals," 817 pp. Wiley, New York.

Burden, G.S., and Smittle, B.J. (1961). New and current insecticides for German cockroach control. *Pest Contr.* **29**(6), 30.

Busvine, J.R. (1958). Insecticide resistance in bed bugs. *Bull. W.H.O.* **19**, 1042-1052.

Casida, J.E. (1970). Mixed-function oxidase involvement in the biochemistry of insecticide synergists. *J. Agric. Food Chem.* **18**, 753-772.

Casida, J.E. (1973). Biochemistry of the pyrethrins. *In* "Pyrethrum, The Natural Insecticide" (J.E. Casida, ed.), pp. 101–120. Academic Press, New York.

Chiang, C., and Devonshire, A.L. (1982). Changes in membrane phospholipids, identified by Arrhenius plots of acetylcholinesterase and associated with pyrethroid resistance (*kdr*) in houseflies (*Musca domestica*). *Pestic. Sci.* **13**, 156-160.

Cichy, D. (1971). The role of some ecological factors in the development of pesticide resistance in *Sitophilus oryzae* L. and *Tribolium castaneum* Herbst. *Ekol. Pol.* **19**, 563-616.

Cochran, D.G. (1973). Inheritance and linkage of pyrethrins resistance in the German cockroach. *J. Econ. Entomol.* **66**, 27-30.

Cochran, D.G. (1989). Monitoring for insecticide resistance in field-collected strains of the German cockroach (Dictyoptera: Blattellidae). *J. Econ. Entomol.* **82**, 336-341.

Cochran, D.G. (1994a). Insecticide Resistance. *In* "Understanding and Controlling the German Cockroach" (M.K. Rust, J.M. Owens, and D.A. Reierson, eds.). Oxford University Press, in the press.

Cochran, D.G. (1994b). Resistance to pyrethrins in the German cockroach: inheritance and gene frequency estimates in field-collected populations (Dictyoptera: Blattellidae). *J. Econ. Entomol.* **87**, 280-284.

Cole, M.M., and Clarke, P.H. (1961). Development of resistance to synergized pyrethrins in body lice and cross-resistance to DDT. *J. Econ. Entomol.* **54**, 649-651.

Dong, K., and Scott, J.G. (1991). Neuropharmacology and genetics of *kdr*-type resistance in the German cockroach, *Blattella germanica* (L.). *Pestic. Biochem. Physiol.* **41**, 159-169.

Falconer, D.S. (1960). "An Introduction to Quantitative Genetics," 416 pp. Oliver and Boyd Ltd., Edinburgh.

Farnham, A.W. (1971). Changes in the cross-resistance patterns of houseflies selected with natural pyrethrins or resmethrin. *Pestic. Sci.* **2**, 138-143.

Farnham, A.W. (1973). Genetics of resistance of pyrethroid selected houseflies *Musca domestica* L. *Pestic Sci.* **4**, 513-520.

Farnham, A.W., O'Dell, K.E., Denholm, I., and Sawicki, R.M. (1984). Factors affecting resistance to insecticides in houseflies, *Musca domestica* L. (Diptera: Muscidae). III. Relationship between the level of resistance to pyrethroids, control failure, and the frequency of gene *kdr*. *Bull. Entomol. Res.* **74**, 581-589.

Fisher, E.H. (1955). A dairy-barn fogging method for fly control. *J. Econ. Entomol.* **48**, 330.

Fournier, D., Bride, J.-M., Mouches, C., Raymond, M., Magnin, M., Berge, J.-B., Pasteur, N., and Georghiou, G.P. (1987). Biochemical characterization of the esterases A1 and B1 associated with organophosphate resistance in the *Culex pipiens* L. complex. *Pestic. Biochem. Physiol.* **27**, 211-217.

Georghiou, G.P. (1972). The evolution of resistance to pesticides. *Ann. Rev. Ecol. System.* **3**, 133-168.

Georghiou, G.P. (1987). Insecticides and pest resistance: the consequences of abuse. 36th Ann. Fac. Res. Lect., pp. 1-27. University of California, Riverside.

Georghiou, G.P. (1990). Overview of insecticide resistance. *In* "Managing Resistance to Agrochemicals" (M.B. Green, H.M. LeBaron, and W.K. Moberg, eds.), pp. 18–41. *ACS Symp. Ser.* **421**. American Chemical Society, Washington, DC.

Gillenwater, H.B., and Burden, G.S. (1973). Pyrethrum for control of household and stored-product insects. *In* "Pyrethrum, The Natural Insecticide" (J.E. Casida, ed.), pp. 243–259. Academic Press, New York.

Holborn, J.M. (1957). The susceptibility to insecticides of laboratory cultures of an insect species. *J. Sci. Food Agric.* **8**, 182-188.

Horowitz, A.R., Toscano, N.C., Youngman, R.R., and Georghiou, G.P. (1988). Synergism of insecticides with DEF in sweetpotato whitefly (Homoptera: Aleyrodidae). *J. Econ. Entomol.* **81**, 110-114.

Keiding, J. (1977). Resistance in the housefly in Denmark and elsewhere. *In* "Pesticide Management and Insecticide Resistance" (D.L. Watson and A.W.A. Brown, eds.), pp. 261–302. Academic Press, New York.

Keller, J.C., Clark, P.H., and Lofgren, C.S. (1956). Susceptibility of insecticide resistant cockroaches to pyrethrins. *Pest Contr.* **24**, 14-15, 30.

Lange, W.H., and Akesson, N.B. (1973). Pyrethrum for control of agricultural insects. *In* "Pyrethrum, The Natural Insecticide" (J.E. Casida, ed.), pp. 261–279. Academic Press, New York.

Lloyd, C.J. (1969). Studies on the cross-tolerance to DDT-related compounds of a pyrethrin-resistant strain of *Sitophilus granarius* (L.) (Coleoptera, Curculionidae). *J. Stored Prod. Res.* **5**, 337-356.

Mallis, A. (1969). "Handbook of Pest Control," 1158 pp. MacNair-Dorland Co., New York.

Matsumura, F. (1985). "Toxicology of Insecticides," 2nd Ed., 598 pp. Plenum Press, New York.

Melander, A.L. (1914). Can insects become resistant to sprays? *J. Econ. Entomol.* **7**, 167-173.

Metcalf, R.L. (1989). Insect resistance to insecticides. *Pestic. Sci.* **26**, 333-358.

Milio, J.F., Koehler, P.G., and Patterson, R.S. (1987). Evaluation of three methods for detecting chlorpyrifos resistance in the German cockroach (Orthoptera: Blattellidae). *J. Econ. Entomol.* **80**, 44-46.

Mouches, C., Magnin, M., Berge, J.-B., de Silvestri, M., Beyssat, V., Pasteur, N., and Georghiou, G.P. (1987). Overproduction of detoxifying esterases in organophosphate-resistant Culex mosquitoes and their presence in other insects. *Proc. Natl. Acad. Sci. USA.* **84**, 2113-2116.

Mrak, E.M. (1973). Advantages and disadvantages of pyrethrum. *In* "Pyrethrum, The Natural Insecticide" (J.E. Casida, ed.), pp. 307–311. Academic Press, New York.

Myers, J.A., Georghiou, G.P., and Hawley, M.K. (1987). House fly (Diptera: Muscidae) resistance to permethrin on Southern California dairies. *J. Econ. Entomol.* **80**, 636-640.

Myers, J.A., Georghiou, G.P., Bradley, F.A., and Tran, H. (1989). Filth fly resistance to pyrethrins associated with automated spray equipment in poultry houses. *Poult. Sci.* **69**, 736-740.

Nicoli, R.M., and Sautet, J. (1955). Papport sur la fréquence et la sensibilité aux insecticides de *Pediculus humanus humanus* L. dans le sud-est de la France. *Monogr. Inst. Nat. Hyg.* **8**, 1-78.

Plapp, Jr., F.W., and Hoyer, R.F. (1968). Possible pleiotropism of a gene conferring resistance to DDT, DDT analogs and pyrethrins in the housefly and *Culex tarsalis*. *J. Econ. Entomol.* **61**, 761-765.

Plapp, Jr., F.W., Campanhola, C., Bagwell, R.D., and McCutchen, B.F. (1990). Management of pyrethroid-resistant tobacco budworms on cotton in the United States. *In* "Pesticide Resistance in Arthropods" (R.T. Roush and B. Tabashnik, eds.), pp. 237-260. Chapman and Hall, New York and London.

Pree, D.J. (1990). Resistance management in multiple-pest apple orchard ecosystems in eastern North America. *In* "Pesticide Resistance in Arthropods" (R.T. Roush and B. Tabashnik, eds.), pp. 261–276. Chapman and Hall, New York and London.

Raymond, M., Beyssat-Arnaouty, V., Sivasubramanian, N., Mouches, C., Georghoiu, G.P., and Pasteur, N. (1989). Amplification of various esterase B's responsible for organophosphate resistance in Culex mosquitoes. *Biochem. Genet.* **27**, 417-423.

Salgado, V.L., Irving, S.M., and Miller, T.A. (1983). Depolarization of nerve terminals by pyrethroids in susceptible and *kdr*-resistant house flies. *Pestic. Biochem. Physiol.* **20**, 100-114.

Sawicki, R.M. (1985). Resistance to pyrethroid insecticides in arthropods. *In* "Insecticides" (D.H. Hutson and T.R. Roberts, eds.), pp. 143–192. John Wiley and Sons, London.

Sawicki, R.M., and Farnham, A.W. (1968). Genetics of resistance to insecticides in the SKA strain of *Musca domestica*. III. Location and isolation of the factors of resistance to dieldrin. *Entomol. Exp. Appl.* **11**, 133-142.

Schal, C. (1992). Sulfluramid resistance and vapor toxicity in field-collected German cockroaches (Dictyoptera: Blattellidae). *J. Med. Entomol.* **29**, 207-215.

Scott, J.G. (1990). Investigating mechanisms of insecticide resistance: methods, strategies, and pitfalls. *In* "Pesticide Resistance in Arthropods" (R.T. Roush and B. Tabashnik, eds.), pp. 39–57. Chapman and Hall, New York and London.

Scott, J.G., and Matsumura, F. (1981). Characteristics of a DDT-induced case of cross-resistance to permethrin in *Blattella germanica*. *Pestic. Biochem. Physiol.* **16**, 21-27.

Scott, J.G., and Matsumura, F. (1983). Evidence for two types of toxic actions of pyrethroids on susceptible and DDT-resistant German cockroaches. *Pestic. Biochem. Physiol.* **19**, 141-150.

Scott, J.G., Ramaswamy, S.B., Matsumura, F., and Tanaka, K. (1986). Effect of method of application on resistance to pyrethroid insecticides in *Blattella germanica* (Orthoptera: Blattellidae). *J. Econ. Entomol.* **79**, 571-575.

Scott, J.G., Cochran, D.G., and Siegfried, B.D. (1990). Insecticide toxicity, synergism, and resistance in the German cockroach (Dictyoptera: Blattellidae). *J. Econ. Entomol.* **83**, 1698-1703.

Shidrawi, G.R. (1990). A WHO global programme for monitoring vector resistance to pesticides. *Bull. W.H.O.* **68**, 403-408.

Smith, C.N. (1973). Pyrethrum for control of insects affecting man and amimals. *In* "Pyrethrum, The Natural Insecticide" (J.E. Casida, ed.), pp. 225–241. Academic Press, New York.

Soderlund, D.M., and Bloomquist, J.R. (1990). Molecular mechanisms of insecticide resistance. *In* "Pesticide Resistance in Arthropods" (R.T. Roush and B. Tabashnik, eds.), pp. 58–96. Chapman and Hall, New York and London.

Tsukamoto, M. (1969). Biochemical genetics of insecticide resistance in the housefly. *Residue Rev.* **25**, 289-314.

Umeda, K., Yano, T., and Hirano, M. (1988). Pyrethroid-resistance mechanism in the German cockroach, *Blattella germanica* (Orthoptera: Blattellidae). *Appl. Ent. Zool.* **23**, 373-380.

Whitehead, G.B. (1959). Pyrethrum resistance conferred by resistance to DDT in the blue tick. *Nature* **184**, 378-379.

Wirth, M.C., Marquine, M., Georghiou, G.P., and Pasteur, N. (1990). Esterases A2 and B2 in *Culex quinquefasciatus* (Diptera: Culicidae): role in organophosphate resistance and linkage. *J. Med. Entomol.* **27**, 202-206.

Zettler, J.L., McDonald, L.L., Redlinger, L.M., and Jones, R.D. (1973). *Plodia interpunctella* and *Cadra cautella* resistance in strains to malathion and synergised pyrethrins. *J. Econ. Entomol.* **66**, 1049-1050.

12

Mammalian Toxicology of Pyrethrum Extract

GERALD P. SCHOENIG

I. INTRODUCTION

In 1985, a consortium was formed to respond to the data requirements of the United States Environmental Protection Agency (EPA) associated with the reregistration of pesticides. This consortium, the Pyrethrin Joint Venture, consists of all the major manufacturers and formulators of pyrethrum extract. Included in the data requirements was the development of a comprehensive, state-of-the-art mammalian toxicology database on pyrethrum extract. This review presents the results of the studies conducted as part of this data development program. No attempt is made to discuss the prior studies which were summarized earlier (Williams, 1973).

All studies in this program were conducted on a single composite sample of pyrethrum extract. This composite was made up of equal portions of typical manufacturing use product from three manufacturers: the Pyrethrum Board of Kenya, McLaughlin Gormley King Co. (Papua New Guinea) and OPYRWA, Rwanda. Active ingredients (total pyrethrins) were 57.6% of this composite sample with the remaining material mainly other plant extractives, e.g., fatty acids, smaller amounts of isoparaffinic solvents added during the refining process and the antioxidant butylated hydroxytoluene. The ratio of pyrethrins I to pyrethrins II in this sample was 1.85. All studies met or exceeded the Pesticide Assessment Guidelines (EPA, 1984) and were conducted in accordance with Good Laboratory Practice Standards (EPA, 1989). They were carefully designed and overseen by the Technical Committee of the Pyrethrin Joint Venture and an independent toxicology consultant.

II. ACUTE TOXICITY, SENSITIZATION, AND IRRITATION STUDIES

A. Acute Oral Toxicity

Acute oral toxicity was evaluated in Charles River CD® rats (five males and five females per group). The pyrethrum extract was administered without dilution and the rats were observed for 14 days following dose administration. The LD_{50}s for male and female rats were 2.37 and 1.03 g/kg, respectively (as total pyrethrins). Clinical signs consisted of hyperactivity and tremors with no-effect levels for these symptoms of 0.71 and 0.32 g/kg for males and females, respectively. Surviving rats recovered 1–2 days after dose administration.

B. Acute Dermal Toxicity

Acute dermal toxicity was evaluated in New Zealand albino rabbits. Five male and five female rabbits were administered undiluted pyrethrum extract at the limit dose of 2.0 g/kg after which time they were observed for 14 days. No clinical signs or mortality were observed and skin irritation was limited to slight erythema.

C. Acute Inhalation Toxicity

Acute inhalation toxicity was evaluated in Charles River CD® rats which were exposed to aerosols of pyrethrum extract for 4 hours (five males and five females per group). Acetone was used as a vehicle in order to produce a liquid aerosol of suitable particle size and an acetone control group was included in the study. Chamber concentrations were determined gravimetrically and analytically. The mean analytical exposures were 0, 0.69, 2.1, and 4.6 mg/L total pyrethrins. The average mass median aerodynamic diameter of the aerosol particles was 2.6 microns with 96% of the particles $\leqslant$10 microns and 11% $<$1 micron. Under these conditions, there were no differences in response between males and females and the combined-sex LC_{50} in terms of total pyrethrins was calculated to be 3.4 mg/L. Clinical signs consisted of various respiratory and secretory responses which were also observed in the acetone control, but tremors and mortality were observed at the two higher concentrations. Surviving animals appeared normal within 6 days after exposure. The no-effect level for pyrethrins-specific effects was 0.69 mg/L.

D. Skin Sensitization

Delayed contact hypersensitivity was evaluated in guinea pigs by the method of Buehler (1965). A group of 10 Hartley-strain guinea pigs received three applications of undiluted pyrethrum extract each week over 3 weeks (nine induction applications). The challenge application was made 2 weeks after the last induction and treated animals were compared to naive and positive-control groups. The known sensitizer 1-chloro-2,4-dinitrobenzene (DNCB) was applied to the test animals in the positive-control group. No evidence of sensitization was observed in the animals administered pyrethrum extract, but a positive response was obtained with DNCB.

E. Skin Irritation

Skin irritation was evaluated in six albino rabbits which were administered 0.5 ml of undiluted pyrethrum extract for a 4-hour period. The application sites were occluded during the exposure period. Mild skin irritation in the form of erythema was observed 24 hours after application, but the skin of all animals appeared normal within 72 hours after application.

F. Eye Irritation

Eye irritation was evaluated in six albino rabbits which received 0.1 ml ocular instillations of undiluted pyrethrum extract. Mild conjunctival irritation was observed during the first 48 hours after application; however, the eyes of all animals appeared normal within 72 hours.

III. SUBCHRONIC TOXICITY STUDIES

In most of the studies discussed in Sections III to VI the parameters evaluated were: observations for clinical signs, body weight and food consumption measurements, hematology, clinical chemistry, organ weight measurements, and gross and microscopic pathology. This set of observations is referred to as the "standard parameters."

A. Rat 90 Day Oral Toxicity

Pyrethrum extract was incorporated into the diet and administered to Charles River CD® rats at 0, 300, 1,000, 3,000, 10,000, and 20,000 ppm (as total pyrethrins). Fifteen male and 15 female rats were evaluated at each concentration for the "standard parameters" plus urinalyses. Clinical signs, including increased respiration, tremors, hyperactivity, convulsions, and death, were observed at 10,000 and/or 20,000 ppm. Decreased body weight gains and food consumption, evidence of anemia, increased liver and kidney weights, and tubular degeneration in the kidneys were observed at $\geqslant$3,000 ppm. No treatment-related effects were observed at 300 or 1,000 ppm. On the basis of the results of this study, dietary concentrations of 100, 1,000 and 3,000 ppm were selected for a 2 year rat chronic toxicity/oncogenicity study and a rat two-generation reproduction study.

B. Mouse 90 Day Oral Toxicity

Pyrethrum extract was incorporated into the diet and administered to Charles River CD-1® mice at 0, 300, 3,000, 10,000, and 30,000 ppm (as total pyrethrins). Fifteen male and 15 female mice were evaluated at each concentration for the "standard parameters" except for hematology and clinical chemistry.

Clinical signs, including labored breathing, tremors, dilated pupils, and death, were observed at 10,000 and 30,000 ppm while liver congestion and increased liver weights were observed at $\geqslant$1,000 ppm. Hepatocellular hypertrophy was observed microscopically at $\geqslant$3,000 ppm. No treatment-related effects were noticed in mice fed diets containing 300 ppm total pyrethrins. On the basis of the results of this study, dietary levels of 100, 2,500 and 5,000 ppm were selected for an 18-month mouse oncogenicity study.

C. Dog 8 Week Oral Toxicity

Pyrethrum extract was incorporated into the diet and administered to beagle dogs at 0, 600, 1,000, 3,000 and 6,000 ppm (as total pyrethrins). Two male and two female dogs were evaluated at each dose for the "standard parameters."

Clinical signs including inappetence, ataxia, and tremors were observed at 3,000 and 6,000 ppm while three dogs at the highest concentration died. Decreased body weight and food consumption, anemia, alterations in electrolytes, and increases in two serum enzymes (glutamic pyruvic and oxaloacetic transaminases) (SGPT and SGOT) were observed at 6,000 ppm. Liver weight increases and testes weight decreases occurred at $\geqslant$1,000 ppm. No treatment-related effects were observed in dogs fed diets containing 600 ppm total pyrethrins. On the basis of the results of this study, dietary concentrations of 100, 500 and 2,500 ppm were selected for a 1-year chronic dog toxicity study.

D. Rabbit 21 Day Dermal Toxicity

Pyrethrum extract was applied to the shaven backs of New Zealand white albino rabbits 5 days per week for 3 weeks at dose levels of total pyrethrins corresponding to 0, 100, 300 and 1,000 mg/kg/day. Five male and five female rabbits were evaluated in each group. In a preliminary dose range-finding dermal irritation study, undiluted pyrethrum extract produced more than slight dermal irritation with repeated application over 5 days. Therefore, the pyrethrum extract was diluted and administered as a 25% (w/v) mixture in corn oil in the 21 day study. A vehicle control group also was included in the study and used vegetable oil because of its low potential for skin irritation and because it most closely resembled the inert plant material in pyrethrum extract. The application sites were occluded during the 6–8 hour exposure. Evaluations included observations for clinical signs of skin irritation and the "standard parameters" as above. With the exception of minor skin irritation which was noted with equal or greater severity in control animals treated with vegetable oil, no treatment-related effects were observed in this study.

E. Rat 90 Day Inhalation Toxicity

Pyrethrum extract was evaluated in Charles River CD® rats as a liquid aerosol at total pyrethrins concentrations of 0, 11, 30, 100 and 356 mg/m^3 which were selected on the basis of findings from an acute, 4 hour aerosol inhalation exposure in which the LC_{50} was 3.4 mg/L (3,400 mg/m^3). Fifteen male and 15 female rats were exposed to each concentration for 6 hours per day, 5 days per week for 13 weeks. Chamber concentrations were determined both gravimetrically and analytically. The average mass median aerodynamic diameter of the aerosol particles was 2.7 microns. The mean diameter of 99% of the particles was $\leqslant$10 microns and 3% were $\leqslant$1.0 micron. Evaluation included observations for the "standard parameters" plus ophthalmology.

Clinical signs indicative of systemic toxicity (labored breathing, hyperactivity, and tremors) were observed at the 356 mg/m^3 concentration. Clinical signs of irritation of the respiratory tract were observed at $\geq$30 mg/m^3. Decreased body weight gains and evidence of anemia were observed at $\geq$100 mg/m^3 and some evidence of anemia also was observed in the 30 mg/m^3 group. Increased liver weights occurred at 356 mg/m^3. Extremely sensitive methods were used to process and microscopically examine the tissues of the respiratory tract. As a result, microscopic changes indicative of irritation were observed in the respiratory tract from animals at all test concentrations and the air controls. The irritation observed in the animals from the pyrethrins groups was more pronounced than that from the air-control group, especially at 356 mg/m^3. The no-effect level for systemic toxicity was 11 mg/m^3.

IV. TERATOLOGY STUDIES

A. Rat Teratology

Dose range-finding and definitive teratology studies were conducted with Charles River CD® rats. In both studies, pyrethrum extract was suspended in 0.5% methyl cellulose to facilitate dosing and was administered by oral gavage at a constant volume of 3 ml/kg on gestation days 6–15. Control animals received vehicle (0.5% methyl cellulose) only. In the dose range-finding study, five mated female rats per group were evaluated at dose levels of total pyrethrins corresponding to 0, 37.5, 75, 150, 300, and 600 mg/kg/day. In the definitive study, 25 mated females per group were evaluated at levels of total pyrethrins corresponding to 0, 5, 25, and 75 mg/kg/day. Parameters evaluated in both studies included observations for clinical signs, body weight, food consumption, maternal ovarian and uterine exams, and fetal external examination. In addition, in the definitive study the fetuses were given detailed internal soft-tissue and skeletal examinations.

In the dose range-finding study, maternal toxicity in the form of mortality, convulsions and/or tremors was observed at $\geq$150 mg/kg/day; tremors also were observed at 75 mg/kg/day. No maternal toxicity was observed at 37.5 mg/kg/day and no evidence of developmental toxicity was observed at any dose level. On the basis of these results, dose levels of 0, 5, 25, and 75 mg/kg/day were selected for the definitive study where no evidence of maternal or developmental toxicity was observed at any dose. While it was unexpected that the tremors, observed in the pregnant females during the dose range-finding study at 75 mg/kg/day, did not occur in the definitive study, the results of the range-finding study indicate that the high dose level evaluated in the definitive study was very close to being maternally toxic.

B. Rabbit Teratology

Dose range-finding and definitive teratology studies were conducted in female New Zealand White SPF rabbits. In both studies, pyrethrum extract suspended in 0.5% methyl cellulose was administered by oral gavage at 3 ml/kg on gestation days 7–19. Control animals received vehicle (0.5% methyl cellulose) only. In the dose range-finding study, five mated female rabbits per group were evaluated at levels of total pyrethrins corresponding to 0, 37.5, 75, 150, 300, and 600 mg/kg/day. In the definitive study, 16 mated females per group were evaluated at doses of total pyrethrins of 0, 25, 100, and 250 mg/kg/day. Parameters evaluated in both studies included observations for clinical signs, body weights, and maternal uterine and ovarian examinations. In addition, in the definitive study, the fetuses were weighed and externally examined, and given detailed internal soft-tissue and skeletal examinations.

In the dose range-finding study, maternal toxicity in the form of mortality, tremor/convulsions and weight loss, and fetotoxicity in the form of high postimplantation loss were observed at 600 mg/kg/day. Maternal toxicity included weight loss during the treatment period and tremors were also observed at 300 mg/kg/day. No treatment-related effects were observed at 37.5, 75, or 150 mg/kg/day. On the basis of these results, dose levels of 25, 100, and 250 mg/kg/day were selected for the definitive study where maternal toxicity in the form of weight loss or reduced weight gain during the treatment period, excessive salivation and arched head were observed in both the 100 and 250 mg/kg/day groups. No maternal toxicity was observed at 25 mg/kg/day and developmental toxicity was not observed in the study.

V. REPRODUCTIVE TOXICOLOGY STUDY

A two-generation, two-litter-per-generation reproduction study was conducted in Charles River CD® rats. Pyrethrum extract was incorporated into the diet and administered to the rats at 0, 100, 1,000, and 3,000 ppm (as total pyrethrins) while the control group received basal diet only. The F_0 parental generation consisted of 28 males and 28 females per group which were fed treated or control diet for at least 77 days prior to the first mating. Following a minimum of 10 days after weaning their first litters, the animals were mated a second time. Twenty-eight male and 28 female offspring per group from the second mating were selected randomly to become parents of the F_1 generation. These animals were treated for at least 95 days prior to being mated twice as above. For both parental groups, treatment was continued through gestation and lactation. Parameters evaluated in the parental rats included clinical signs, growth, food consumption, gross pathology, organ weight, and microscopic pathology. Reproductive and litter parameters evaluated included male and female fertility indices, events at parturition, gestation length, litter size, numbers of viable and stillborn pups, and pup survival and growth during lactation.

No parental toxicity in the F_0 generation was observed at any dietary concentration. In the F_1 generation, parental toxicity in the form of decreased body weights and food consumption was noted for males and females in the 1,000 and 3,000 ppm groups. Neonatal toxicity as evidenced by reduced pup body weights in both generations was observed in the 1,000 and 3,000 ppm groups. No treatment related effects were observed at 100 ppm and reproductive toxicity was not observed in the study.

VI. CHRONIC TOXICITY AND ONCOGENICITY STUDIES

A. Dog Chronic Toxicity

Pyrethrum extract was incorporated into the diet and administered for 1 year to purebred beagle dogs at 0, 100, 2,500, and 5,000 ppm (as total pyrethrins). Four male and four female dogs were evaluated at each dietary concentration and control animals were fed basal diet. Evaluation included the "standard parameters" plus ophthalmology.

Diet aversion and decreased food consumption during the first 2 weeks of the study were observed in the 500 and 2,500 ppm groups. In the 2,500 ppm group, evidence of anemia, increased SGPT levels and increased liver weights also were observed. No toxicologic effects occurred in the 100 or 500 ppm groups.

B. Mouse Oncogenicity Study

Pyrethrum extract was incorporated into the diet and administered for 18 months to Charles River CD-1® mice at 100, 2,500, and 5,000 ppm (as total pyrethrins). Sixty male and 60 female mice were evaluated at each dietary concentration. In addition, two independent untreated control groups, each consisting of 60 male and 60 female mice, were included in the study. Mice in the untreated control groups were fed basal diet. Evaluation included observations for palpable masses and the "standard parameters" except clincial chemistry.

Hyperactivity was observed and two deaths occurred during the first week of the study in the 5,000 ppm group. In both the 2,500 and 5,000 ppm groups, discolored dark livers, increased liver weights, and microscopically observed vacuolar fatty changes in liver tissue were present. No treatment- related effects were observed in the 100 ppm group and oncogenicity was not observed in the study.

C. Rat Chronic Toxicity/Oncogenicity

Pyrethrum extract was incorporated into the diet and administered for 104 weeks to Charles River CD® rats at 100, 1,000, and 3,000 ppm (as total pyrethrins). Sixty male and 60 female rats were evaluated at each dietary concentration. In addition, two independent untreated control groups were included, each consisting of 60 male and 60 female rats and fed basal diet. Evaluation included the "standard parameters," plus observations for palpable mass and ophathalmology.

Decreased body weight, increased SGOT and SGPT levels, and a small (but statistically significant) increase in the incidence of keratoacanthomas of the skin in male rats were observed in the 3,000 ppm group. In addition, a small increase in hyperplasia and follicular cell adenoma in the thyroid gland was observed in male rats treated at 1,000 and 3,000 ppm and in females at 3,000 ppm. No treatment-related effects were observed in the 100 ppm group. The increased incidence of keratoacanthomas in the skin of male rats was not considered to be of any toxicological significance because of the site and the

Table 12-1. Follicular Cell Tumors in the Thyroid Gland

		Dietary concentration (ppm)				
Sex	Lesion	0	0	100	1,000	3,000
Male[a]	Adenoma	2	0	3	5	5
	Carcinoma	0	1	1	2	2
	Total	2	1	4	7	7
Female[a]	Adenoma	0	0	2	3	5
	Carcinoma	1	2	0	0	1
	Total	1	2	2	3	6

[a] $n = 60$ except for 1,000 ppm male where $n = 59$.

self-limiting nature of this lesion. The follicular cell tumor incidence in the thyroid gland is given in Table 12-1.

Because the increased incidences over controls were small and because the presence of follicular cell tumors in the treated rats approximated or were within the range of historical control data for the Charles River CD® rat, the toxicological significance of this finding as being indicative of a treatment-related effect is not very convincing. In addition, it is now becoming increasingly clear that chemicals like the natural pyrethrins, which have a marked stimulatory effect on liver metabolism, can interfere with the normal biofeedback mechanism that controls the circulating levels of thyroid hormone (TH) and thyroid stimulating hormone (TSH) (EPA, 1986). This stimulatory effect on metabolism results in lower than normal TH levels and higher than normal TSH levels. This in turn is known to induce hyperplasia and follicular cell tumors in the thyroid gland. Because the stimulatory effect on metabolism is observed only at pyrethrins concentrations much higher than those ever expected to be encountered from ingestion of foodstuffs containing residues of pyrethrum extract, the consensus of the scientific and regulatory communities is that tumors that arise via this mechanism have no relevance to human health (EPA, 1986).

VII. MUTAGENICITY STUDIES

A. *Salmonella*/Mammalian-Microsome Plate Incorporation Assay (Ames Test)

Five strains of *Salmonella typhimurium*, TA98, TA100, TA1535, TA1537, and TA1538 were used. Each strain was tested in the presence and absence of metabolic activation by a rat liver S-9 system induced with Aroclor 1254. The concentrations of pyrethrum extract evaluated in these studies (with and without metabolic activation) were 0, 292, 585, 877, 2,924, 5,848, and 8,772 μg/plate (as total extract). Mutagenic frequency did not increase in any of the tester strains. Results from the initial assay were confirmed independently.

B. Chromosomal Aberrations in Chinese Hamster Ovary (CHO) Cells

This assay was conducted in Chinese Hamster Ovary (CHO) cells in the presence and absence of metabolic activation with the rat liver S-9 fraction after induction by Aroclor 1254. In the absence of metabolic activation, the concentrations of pyrethrum extract evaluated were 0.005, 0.01, 0.02, 0.04, and 0.08 μl/ml (as total extract) while with activation, 0.02, 0.04, 0.08, 0.16, and 0.32 μl/ml were used. No increase in chromosomal aberrations was observed with or without metabolic activation.

C. Unscheduled DNA Synthesis in Rat Primary Hepatocytes

This assay was conducted in rat primary hepatocytes. Concentrations of pyrethrum extract of 0.03, 0.1, 0.3, 0.6, and 1.0 μl/ml were evaluated both in an initial and in an independent confirmatory assay. No increase in DNA synthesis was observed in either assay.

VIII. CONCLUSION

In these recent state-of-the-art studies, pyrethrum extract was shown to have a low order of acute toxicity and very little potential to cause eye or skin irritation and skin sensitization. The longer-term studies have demonstrated a low systemic toxicity and that pyrethrum extract is not a teratogen or a reproductive toxin. The results of the mutagenicity studies as well as an 18 month mouse oncogenicity investigation and a 2 year rat toxicity/oncogenicity study indicate that pyrethrum extract has minimal potential to induce tumors in mammalian systems. An apparent small increase in benign thyroid tumors was observed in the 2 year rat toxicity/oncogenicity study; however, this observation can be explained by a mechanism that both scientists and regulators agree has little or no relevance to human health (EPA, 1986). The results of the studies conducted in this large industry-sponsored data development program have confirmed the findings from earlier studies that the use of pyrethrum extract as an insecticide poses little risk to humans.

REFERENCES

Buehler, E.V. (1965). Delayed contact hypersensitivity in the guinea pig. *Arch. Dermat.* **92**, 171–175.

Environmental Protection Agency (1984). "Pesticide Assessment Guidelines, Subdivision F, Hazard Evaluation, Human and Domestic Animals (Revised)," Washington, D.C., November.

Environmental Protection Agency (1986). "Standard Evaluation Procedure, Neoplasia Induced by Inhibition of Thyroid Gland Function (Guidance for Analysis and Evaluation)," Washington, D.C., draft document dated June 9.

Environmental Protection Agency (1989). "Federal Insecticide, Fungicide and Rodenticide Act (FIFRA)"; Good Laboratory Practice Standards; Final Rule. 40 CFR Part 160. *Federal Register* **54**, No. 158, 34067 – 34074.

Williams, C.H. (1973). Tests for possible teratogenic, carcinogenic, mutagenic, and allergenic effects of pyrethrum. In "Pyrethrum, The Natural Insecticide" (J.E. Casida, ed.), pp. 167–176. Academic Press, New York.

13

Metabolism and Synergism of Pyrethrins

JOHN E. CASIDA and GARY B. QUISTAD

I. INTRODUCTION

The active ingredients of pyrethrum extract, the pyrethrins, are short-residual and biodegradable insecticides. These characteristics provide an advantage in safety but also restrict the conditions under which it is most effective. The synergist piperonyl butoxide (PB) is added to enhance the potency by retarding pyrethrins detoxification in insect pests. Accordingly, metabolism and synergism are intimately interrelated. This review emphasizes results since 1973 on the biochemistry of the pyrethrins (Casida, 1973) and the action of synergists (Casida, 1970; Yamamoto, 1973).

Rapid metabolism of the pyrethrins is evident from several features of their action in insects and mammals. With insects they are more effective for knockdown than for kill, implying efficient detoxification and recovery. Low-oxidase species such as the mustard beetle are more sensitive than high-oxidase species, such as the house fly (Elliott and Janes, 1973). More importantly, PB is added to inhibit oxidative detoxification enzymes in pest species. With mammals the low oral toxicity is in marked contrast to the high intravenous toxicity, i.e., LD_{50}s of >600 vs 1 mg/kg, respectively, for pyrethrin II in rats, which is probably due to detoxification of the slowly absorbed oral dose. Finally, the low tissue residues, or lack thereof in mammals, indicate rapid biodegradation.

The six pyrethrins each contain chemical substituents sensitive to oxidation and the pyrethrate methoxycarbonyl group is easily hydrolyzed (Fig. 13-1). Some metabolites are formed by two or more sites of metabolic attack. Accordingly, there are many possible metabolites of relatively complex structure requiring thorough studies on their identification and significance.

II. ANALYTICAL TECHNIQUES FOR METABOLISM STUDIES

Metabolism studies are normally made with ^{14}C preparations, or less frequently with ^{3}H compounds, for sensitivity and particularly for total accountability. Radiolabeled pyrethrins used for such investigations are shown in Fig. 13-2

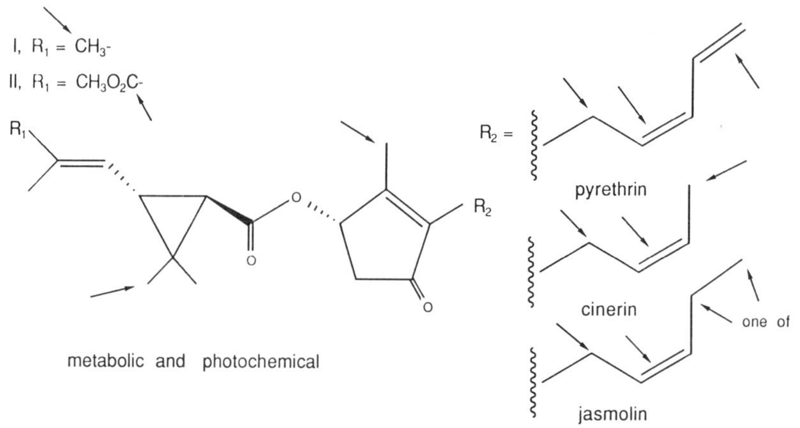

I, $R_1 = CH_3$-
II, $R_1 = CH_3O_2C$-
R_2 =
pyrethrin
cinerin
jasmolin
one of
metabolic and photochemical

Figure 13-1. Structures of the six pyrethrins showing chemical substituents sensitive to oxidation and hydrolysis.

(Elliott and Casida, 1972; Elliott *et al.*, 1969; Pellegrini *et al.*, 1952; Yamamoto and Casida, 1968). Care must be taken in radiosynthesis to reproduce the stereochemistry of the natural material and therefore determine the fate of the botanical insecticide free of irrelevant isomers.

Sensitivity in metabolism studies with nonradioactive materials can be achieved for the pyrethrins by HPLC with an ultraviolet (UV) detector (see Crosby, 1994, for relevant spectra) and by gas chromatography (GC) with an electron capture detector (the cyclopentenolone moiety provides excellent detector response). Metabolite characterization is best achieved by ^{1}H NMR as illustrated in Fig. 13-3 for a pyrethrin I metabolite in rat urine. This NMR spectrum was reported in 1972 for a sample of *ca.* 3 mg at 60 MHz with 16 scans. Of course, today's higher-field instruments require vastly less metabolite. The chemical ionization/mass spectrometry (CI/MS) fragmentation pattern allows an interpretation of the structure of a pyrethrins metabolite in a very

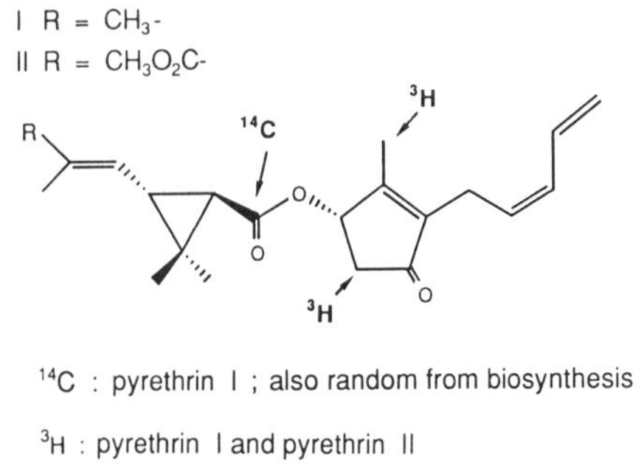

I R = CH_3-
II R = CH_3O_2C-

^{14}C : pyrethrin I ; also random from biosynthesis
^{3}H : pyrethrin I and pyrethrin II

Figure 13-2. Radiolabeled preparations of [^{14}C]pyrethrin I, [^{3}H]pyrethrin I, and [^{3}H]pyrethrin II.

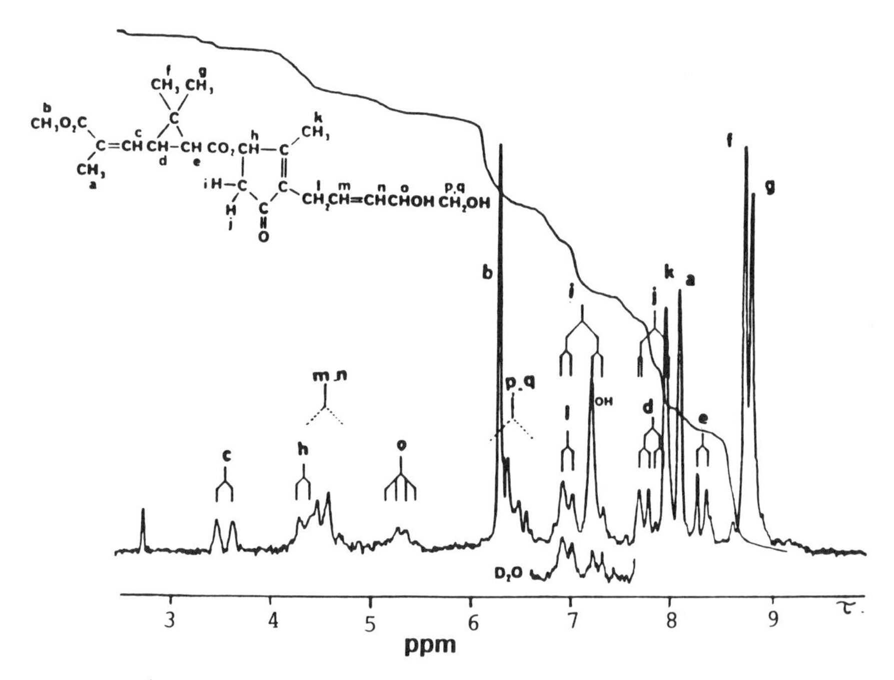

Figure 13-3. ^{1}H NMR spectrum with assignments for a pyrethrin II dihydrodiol derived from methylation of a rat urinary metabolite of pyrethrin I (Elliott *et al.*, 1972).

small amount (Fig. 13-4). Both techniques have been important to the current understanding of the metabolism of the six pyrethrins.

Additional steps are required to identify polar metabolites and conjugates. The polar metabolites can be analyzed directly by MS using soft ionization techniques (e.g., fast-atom bombardment) or derivatized (e.g., methyl esters for carboxylic acids and trimethylsilyl derivatives for alcohols, Fig. 13-4). The conjugates are cleaved enzymatically, e.g., for those in urine by glucuronidase and sulfatase. The deconjugated metabolites are then analyzed directly or following derivatization.

III. INSECT METABOLISM

The first study on insect metabolism of the pyrethrins used randomly-labeled esters from biosynthesis which were administered to the American cockroach. The only identified metabolite was $^{14}CO_2$ (up to 12%) while up to 75% of the label was retained in the body and "hydrolysis products" were extracted and analyzed by paper chromatography (Zeid *et al.*, 1953). Early studies (Winteringham *et al.*, 1955; Chang and Kearns, 1964) were compromised by low-specific-activity [^{14}C]pyrethrins and relatively poor separation methods. The techniques were sufficiently advanced by 1966 for more meaningful studies (Yamamoto and Casida, 1966). An NADPH-dependent enzyme from house flies converted [^{14}C-*acid*]pyrethrin I to a metabolite identified as *O*-demethyl-

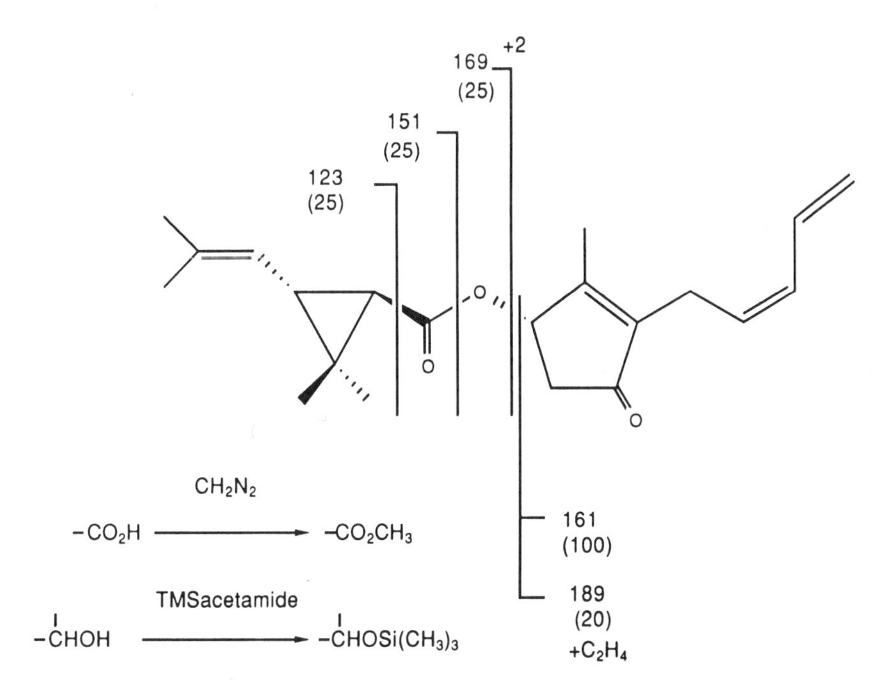

Figure 13-4. Chemical ionization-mass spectrometry fragmentation of pyrethrin I and derivatization reactions for metabolite acids and alcohols. Numbers in parentheses are relative intensities.

pyrethrin II by (1) hydrolysis to chrysanthemum dicarboxylic acid and (2) methylation to pyrethrin II (Fig. 13-5). The intermediate alcohol and aldehyde were also observed and identified by TLC cochromatography of the labeled metabolites with unlabeled standards. A large number of house fly metabolites were unidentified and no other insect species has been examined for metabolite characterization (Yamamoto *et al.*, 1969).

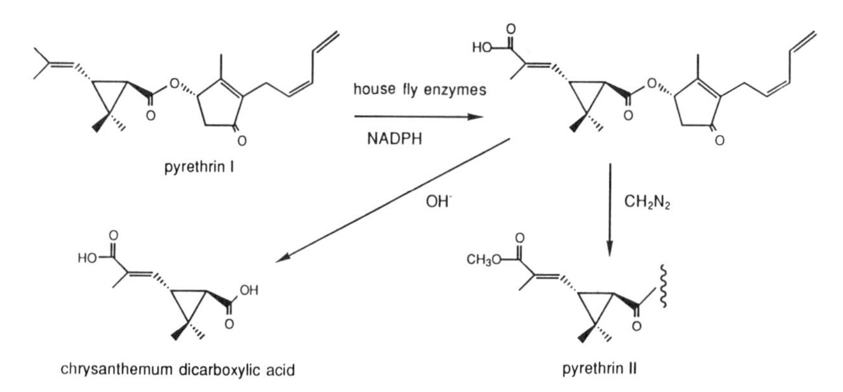

Figure 13-5. First identified metabolite of the pyrethrins (Yamamoto and Casida, 1966).

IV. RAT METABOLISM

A. Metabolism in Liver Microsomal Oxidase Systems

The metabolic reactions carried out by the house fly microsome-NADPH system (Fig. 13-5) are also apparent in the analogous rat liver system involving oxidation of the *trans*-methyl substituent of the isobutenyl moiety based on ^{14}C and ^{3}H studies and TLC cochromatography (Elliott *et al.*, 1969, 1972). ^{13}C NMR

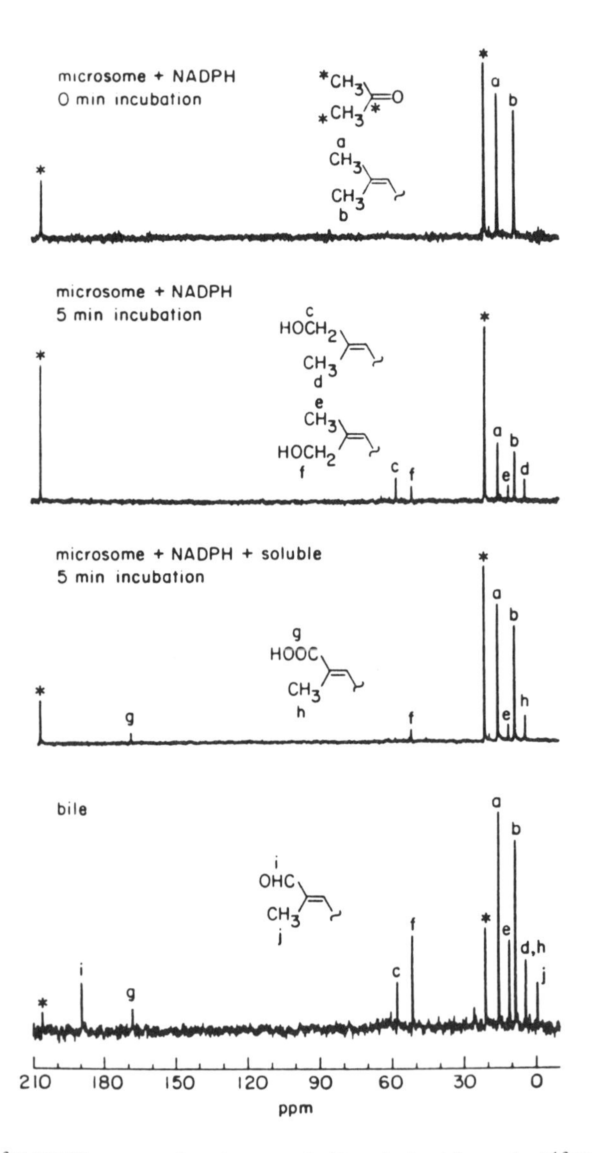

Figure 13-6. ^{13}C NMR spectra showing metabolites derived from the [$^{13}CH_3$]isobutenyl substituent of the chrysanthemate moiety in the rat liver microsome-NADPH system and in bile. Acetone is used for reference signals (Brown *et al.*, 1985).

spectroscopic studies on [$^{13}CH_3$]chrysanthemate metabolism show that either methyl group of the isobutenyl substituent can be oxidized (Fig. 13-6) (Brown *et al.*, 1985). Studies of these types are limited by the availabililty of ^3H-, ^{14}C- and ^{13}C-labeled compounds.

The need for information on metabolism of each component of the pyrethrins is partially fulfilled by a study designed to (1) isolate all six pure pyrethrins, (2) develop GC/MS techniques for each metabolite involving capillary GC with H_2 as the carrier gas and derivatization with Me_3Si or CH_3 moieties, and (3) identify microsomal metabolites by GC/MS without the requirement for ^{14}C or ^3H labeling or synthetic standards. Metabolites were recognized and identified by CI fragmentation patterns. Modifications in the chrysanthemate isobutenyl group involved epoxidation of the double bond or oxidation of the *trans*-methyl substituent to the alcohol, aldehyde, and carboxylic acid. The pyrethric acid portion undergoes hydrolysis and oxidation at one of the *gem*-dimethyl substituents (Fig. 13-7). The rethronyl moieties undergo epoxidation and epoxide cleavage to dihydrodiols for the *cis*-pentadienyl, *cis*-pentenyl, and *cis*-butenyl substituents of the pyrethrins, jasmolins, and cinerins, respectively (Fig. 13-8). In addition the terminal methyl substituents of the jasmolins and cinerins are oxidized to the corresponding alcohols. Combinations of these reactions result in complicated overall pathways when all 63 of the metabolites are considered for the pyrethrins I (Fig. 13-9) and the pyrethrins II (Fig. 13-10). The relative rates of microsomal oxidation are similar for the four major pyrethrins, i.e., pyrethrin I, pyrethrin II, cinerin I, and cinerin II (Soderland and Casida, 1977). It is clear that the pyrethrins are readily oxidized by cytochrome P-450-dependent oxidases and that multiple sites are involved for each of the pyrethrum constituents.

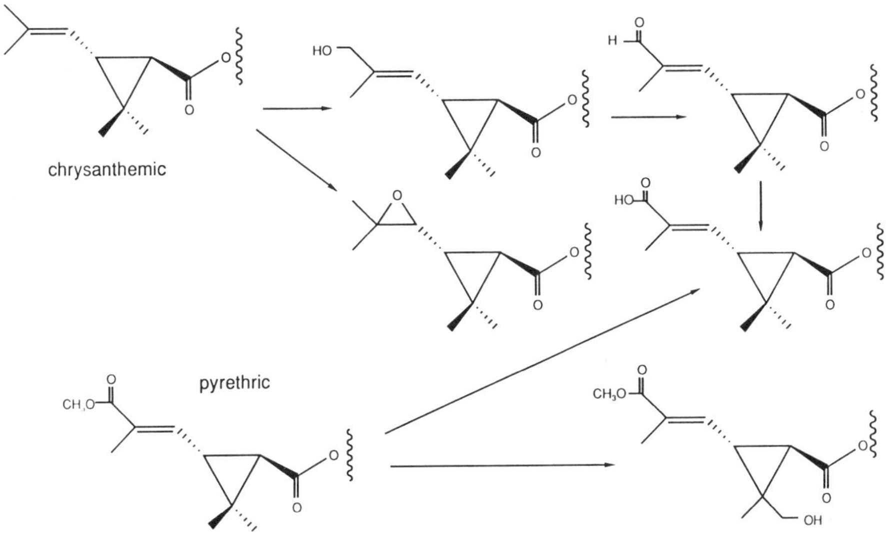

chrysanthemic

HO

H

O

pyrethric

CH_3O

OH

Figure 13-7. Metabolic modifications in the chrysanthemic and pyrethric acid substituents.

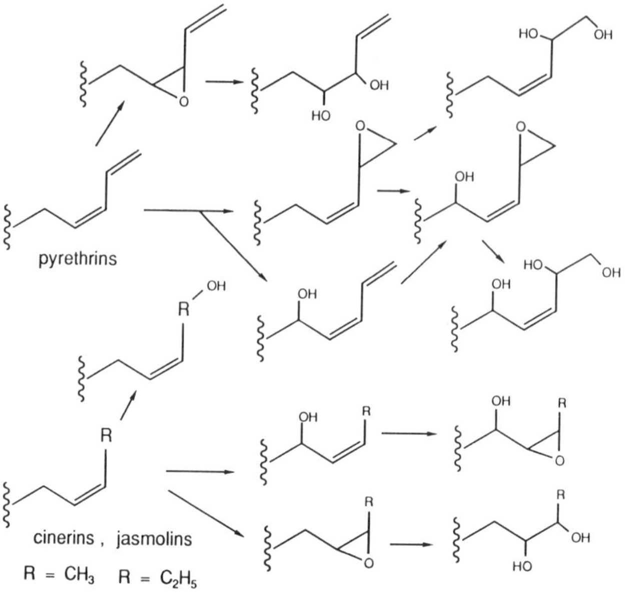

Figure 13-8. Metabolic modifications in the rethronyl side-chains.

B. Rat Urinary Metabolites

The applicability of the enzyme studies to *in vivo* metabolism is established by identification of a series of metabolites of pyrethrin I and pyrethrin II in the urine of orally-treated rats (Casida *et al.*, 1971; Elliott *et al.*, 1972) (Fig. 13-11). The major metabolites consist of the alcohol, aldehyde, and carboxylic acid modifications from oxidation of the isobutenyl methyl group of pyrethrin I, the carboxylic acid from hydrolysis of the carbomethoxy group of pyrethrin II, and the dihydrodiols from pentadienyl epoxidation and hydration. The structure of one of the dihydrodiols from previous work (Elliott *et al.*, 1972) was reassigned based on studies specifically focused on this type of modification (Ando *et al.*, 1991). By analogy with a GC/MS study on allethrin, several of the urinary metabolites of pyrethrin I and pyrethrin II are expected to be glucuronides of the corresponding carboxylic acid and dihydrodiols (Class *et al.*, 1990) (Fig. 13-11).

C. Toxicity of Metabolites

The acute and chronic toxicity of the pyrethrins are attributable to a combination of the parent esters and the metabolites they generate. Although only a few of the metabolites have been separately tested (e.g., chrysanthemic acid), the favorable toxicology of the pyrethrins (Schoenig, 1994) indicates that the metabolites as well as the pyrethrins themselves must be of relatively low toxicity.

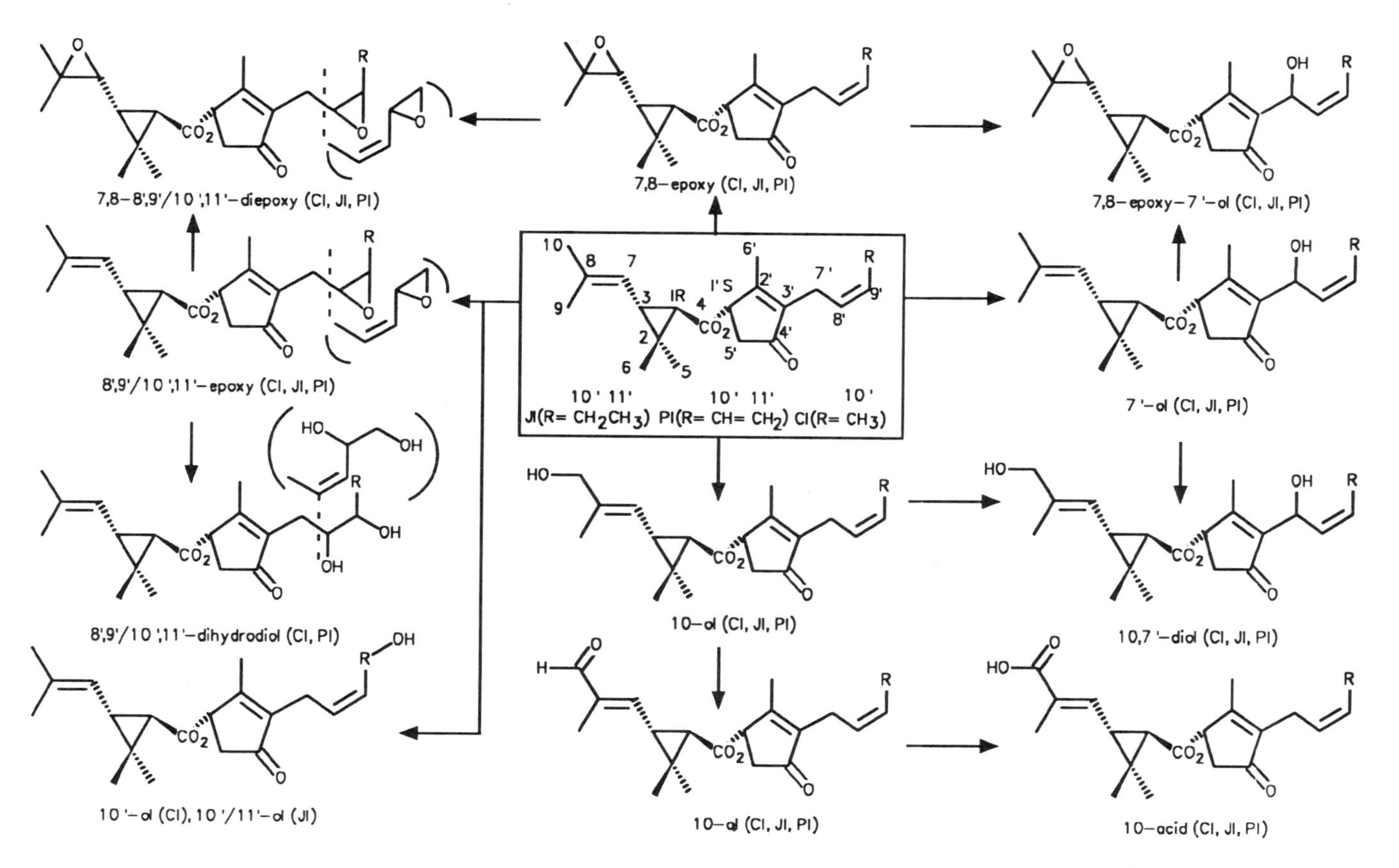

Figure 13-9. Complete pathways for formation of microsomal oxidase metabolites of pyrethrin I, cinerin I and jasmolin I (Class *et al.*, 1990).

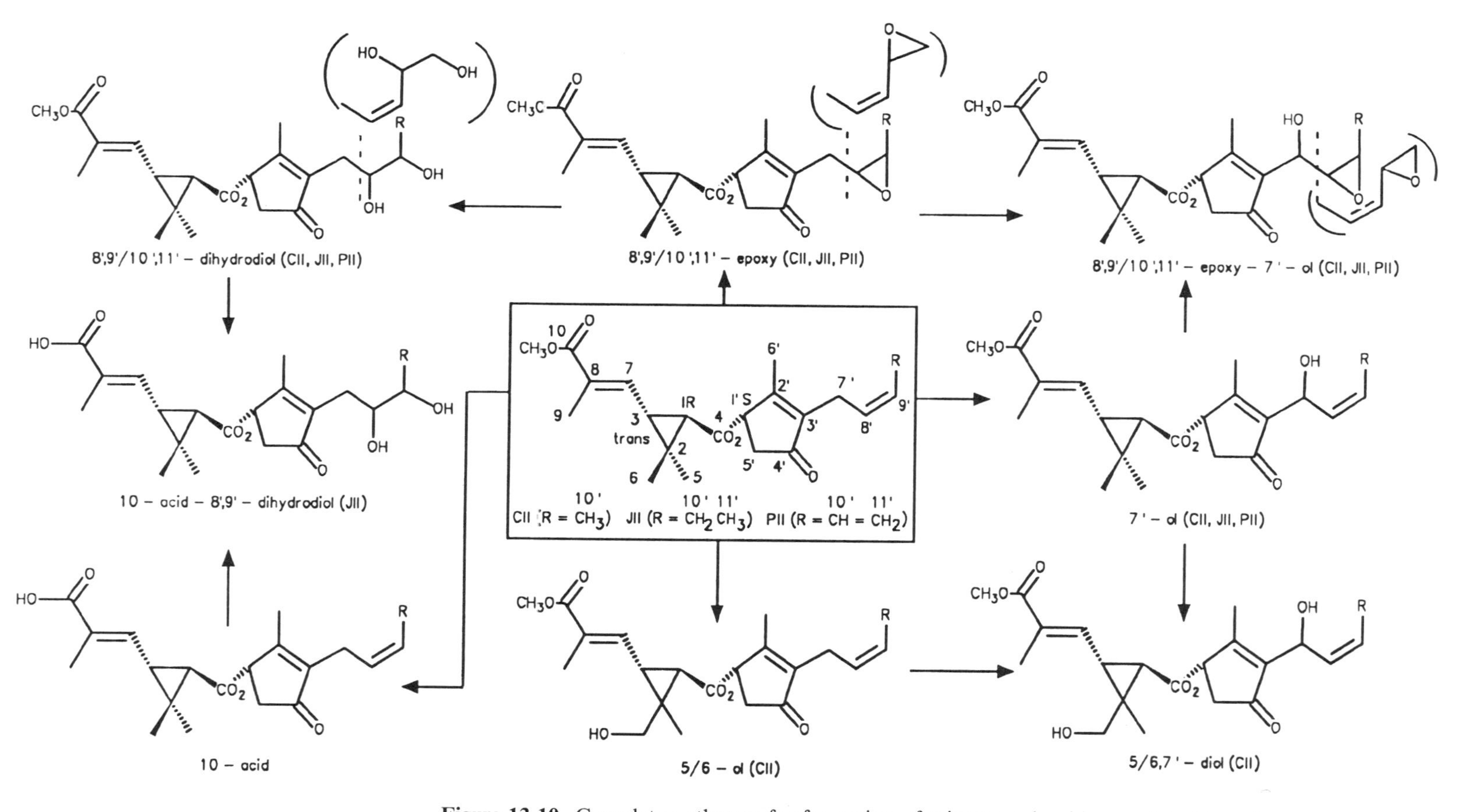

Figure 13-10. Complete pathways for formation of microsomal oxidase metabolites of pyrethrin II, cinerin II, and jasmolin II (Class *et al.*, 1990).

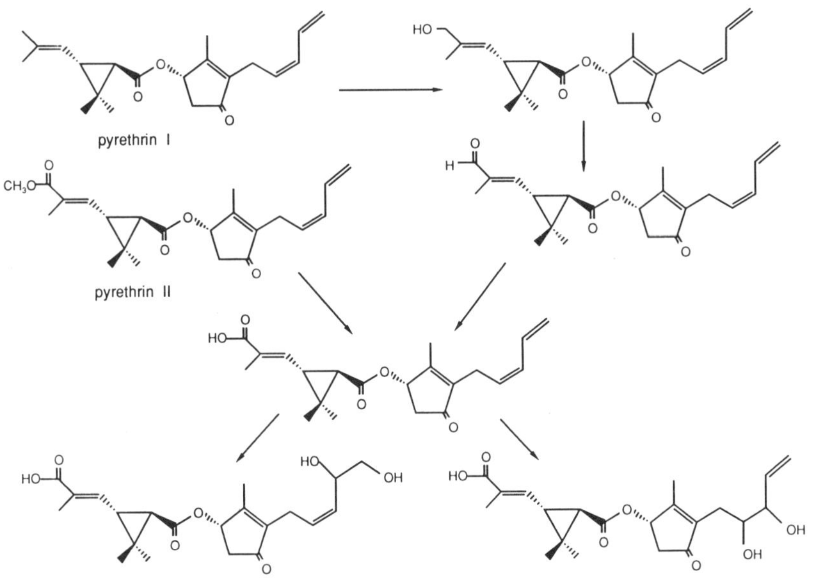

Figure 13-11. Pathways for formation of metabolites in the urine of rats treated orally with pyrethrin I and pyrethrin II.

D. Further Studies

Current knowledge of the metabolism of the pyrethrins in rats is very extensive but the studies were designed to understand the metabolic fate not to fit the study protocols of the Environmental Protection Agency. These protocols are being followed in studies currently underway in rats using [^{14}C-*acid*]pyrethrin I.

V. SYNERGISM

A. History and Types of Synergists

The history of insecticide synergists originates with attempts to enhance the potency of the pyrethrins (B-Bernard and Philogène, 1993). A 1938 observation that *N*-isobutylundecyleneamide enhanced the insecticidal activity initiated the use of insecticide synergists and the search for better compounds. The discovery of methylenedioxyphenyl synergists started with defining that the synergistic activity of sesame oil was due to the sesamin and sesamolin components. Synthesis and testing of related compounds led to sulfoxide, propyl isome, tropital, and PB (Fig. 13-12) (Yamamoto, 1973). Propynyl phosphonates are also effective as are certain amides such as MGK 264. Octachlorodipropyl ether was extensively used in Japan. The sifting of compounds for effectiveness, economics, and toxicology has led to only two major synergists for practical utility, PB and MGK 264. The latter compound may act somewhat like a cosynergist and is not specifically considered here.

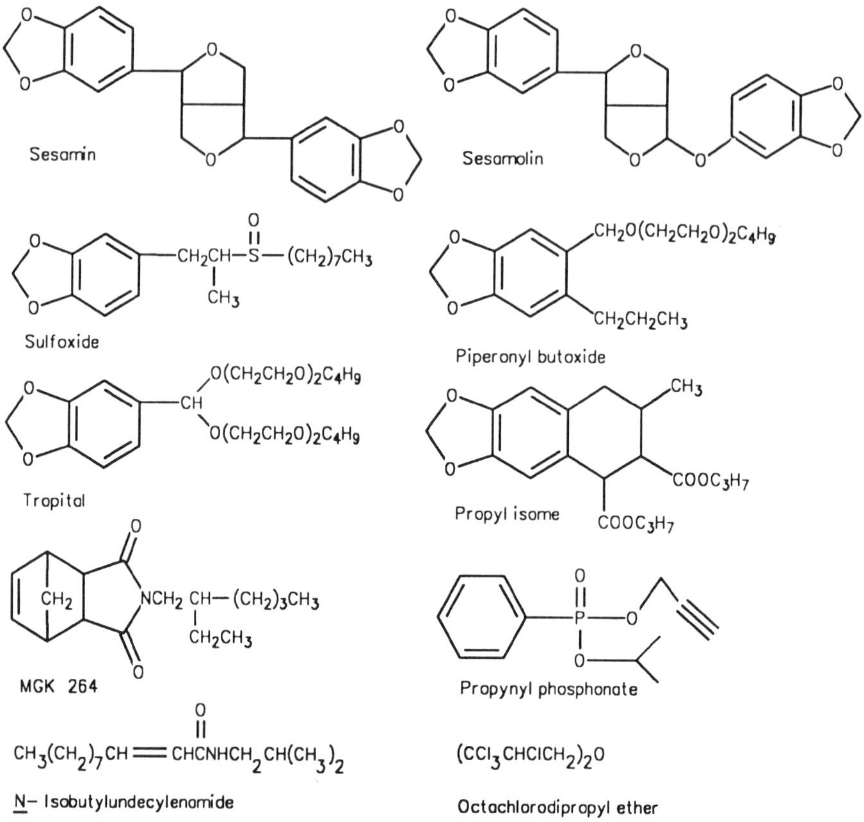

Figure 13-12. Structures of piperonyl butoxide, MGK 264, and earlier commercial or candidate synergists.

The pyrethrum industry benefited greatly from the possibility of using inexpensive additives for lowering the required concentration of the expensive pyrethrins, allowing expanded uses at favorable cost. It is only with the pyrethrins and pyrethroids that extensive use of synergists is involved, due to both economics and the ease of synergism.

B. Piperonyl Butoxide — Source, Synthesis and Bioactivity

PB is prepared by hydrogenation of safrole, chloromethylation, and addition of the butylcarbityl side chain (Wachs, 1947) (Fig. 13-13). The current limitation is the supply of safrole from the ocotea tree in Brazil. Synthetic safrole and dihydrosafrole cannot compete economically with the natural material for now. Radiosynthesis of PB has involved the [^{14}C]methylenedioxy label via $^{14}CH_2I_2$ (Kuwatsuka and Casida, 1965) (Fig. 13-13) and the α-carbon-^{14}C label has also been prepared (New England Nuclear Corp.).

PB is a synergist for any insecticide undergoing detoxification by the cytochrome P-450-dependent microsomal oxidases (Casida, 1970; Yamamoto,

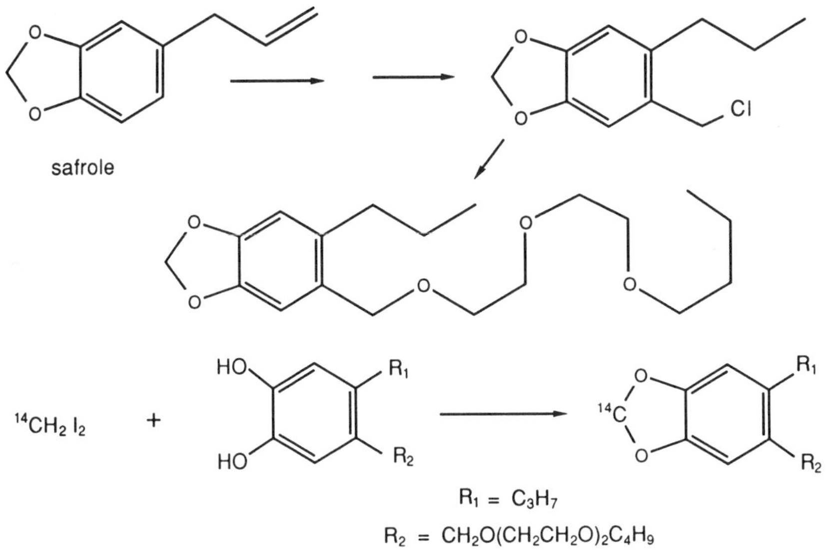

safrole

$^{14}CH_2I_2$ +

$R_1 = C_3H_7$

$R_2 = CH_2O(CH_2CH_2O)_2C_4H_9$

Figure 13-13. Synthesis of piperonyl butoxide and [^{14}C]piperonyl butoxide.

1973). It inhibits metabolism of allethrin maximally during the first 8 hours after topical application to house flies (Fig. 13-14) (Yamamoto *et al.*, 1969). The enhancement of house fly toxicity for the pyrethrins is up to 140- to 325-fold compared with 43-fold for *S*-bioallethrin (Ando *et al.*, 1983) (Fig. 13-15). PB also inhibits xenobiotic metabolism 3–12 hours after ip administration to mice followed by induction of cytochrome P-450 and microsomal oxidase activity (Skrinjaric-Spoljar *et al.*, 1971; Matthews *et al.*, 1970).

C. Piperonyl Butoxide — P-450 Interaction

Early observations suggested that metabolism of PB was significant for synergistic action (Casida *et al.*, 1966; Wilkinson and Hicks, 1969) and that PB binds to P-450 at a single site (Philpot and Hodgson, 1972). PB and other methylenedioxybenzenes have been studied extensively as both inhibitors and inducers of cytochrome P-450. Treatment of rats with these compounds results in a biphasic change in P-450 activity. The initial rapid inhibition of P-450-mediated metabolic activity is attributed to formation of a metabolite: P-450 complex. Although other structures have been proposed, currently the best evidence suggests that the metabolite is a carbene (Ortiz de Montellano and Reich, 1986; Levi and Hodgson, 1989). Since both O_2 and NADPH are required for formation of the complex *in vitro*, oxidation of the methylenedioxy group probably precedes carbene formation (Fig. 13-16). Spectral support for the carbene comes from synthesis of a model complex between methylenedioxy-benzene and iron:*meso*-tetraphenylporphyrin (Fe:TPP) (Fig. 13-17) (Mansuy *et al.*, 1979). The reduced metabolite:P-450 complex displays a double Soret difference spectrum (Type III) with absorption at 455 and 427 nm. Studies on

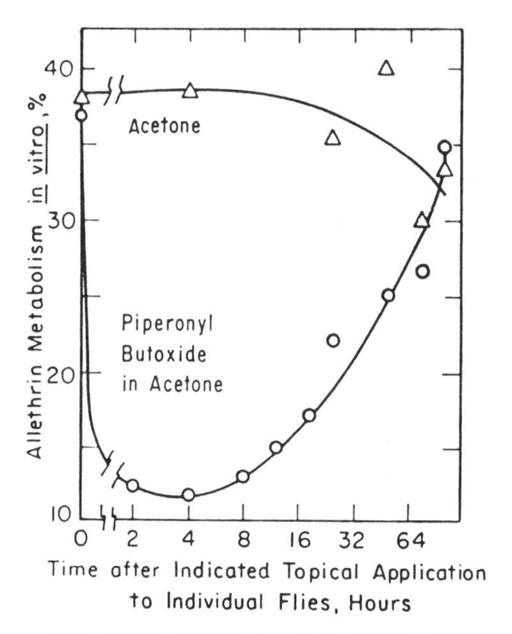

Figure 13-14. Inhibition of cytochrome P-450 in house flies treated with piperonyl butoxide. Allethrin metabolism *in vitro* is compared for enzyme from control insects (acetone) and synergist-treated insects (5 μg piperonyl butoxide in acetone) (Yamamoto *et al.*, 1969).

rats treated with safrole demonstrated the relative stability of the corresponding metabolite:P-450 complex since it survived microsome isolation, including dialysis and detergent treatment (Elcombe *et al.*, 1975). Displacement investigations with nucleophiles have shown that the metabolite:P-450 complexes from several methylenedioxybenzenes vary considerably in stability

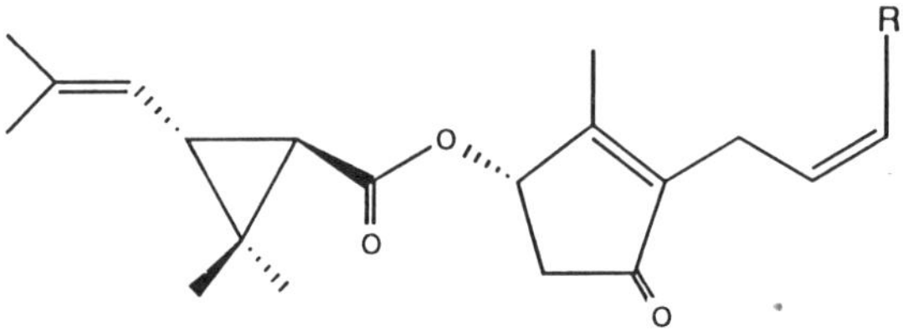

Cmpd.	R	topical LD_{50}, μg/g alone	PB	alone/PB
pyrethrin I	$CH=CH_2$	26	0.08	325
jasmolin I	CH_2CH_3	95	0.68	140
cinerin I	CH_3	85	0.27	315
S - bioallethrin	H	29	0.68	43

Figure 13-15. Piperonyl butoxide synergism of the toxicity of pyrethrins I and allethrin to house flies.

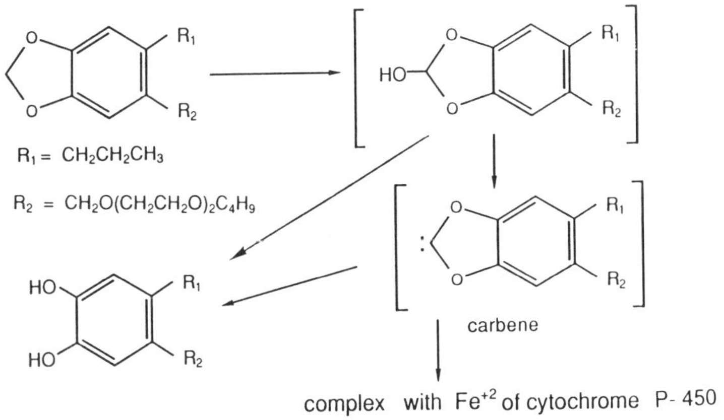

$R_1 = CH_2CH_2CH_3$

$R_2 = CH_2O(CH_2CH_2O)_2C_4H_9$

carbene

complex with Fe^{+2} of cytochrome P-450

Figure 13-16. Metabolism of piperonyl butoxide by cytochrome P-450 oxidases.

(Dickins *et al.*, 1979; Murray *et al.*, 1983). In rat liver the initial inhibition of P-450 is followed by increased catalytic activity which is attributed to enhanced synthesis of selective P-450 holoenzymes (Marcus *et al.*, 1990). The methylene carbon of methylenedioxybenzenes is essential to induction of P-450 and substitution of one or two methyl groups blocks induction activity (Cook and Hodgson, 1983).

Similar carbene adducts of P-450 have been proposed for metabolites of CCl_4, trichloromethylthio fungicides, and DDT (Mansuy *et al.*, 1978; Mansuy, 1980). Model complexes have been prepared using Fe:TPP with structural confirmation by X-ray crystal analysis (Mansuy, 1980).

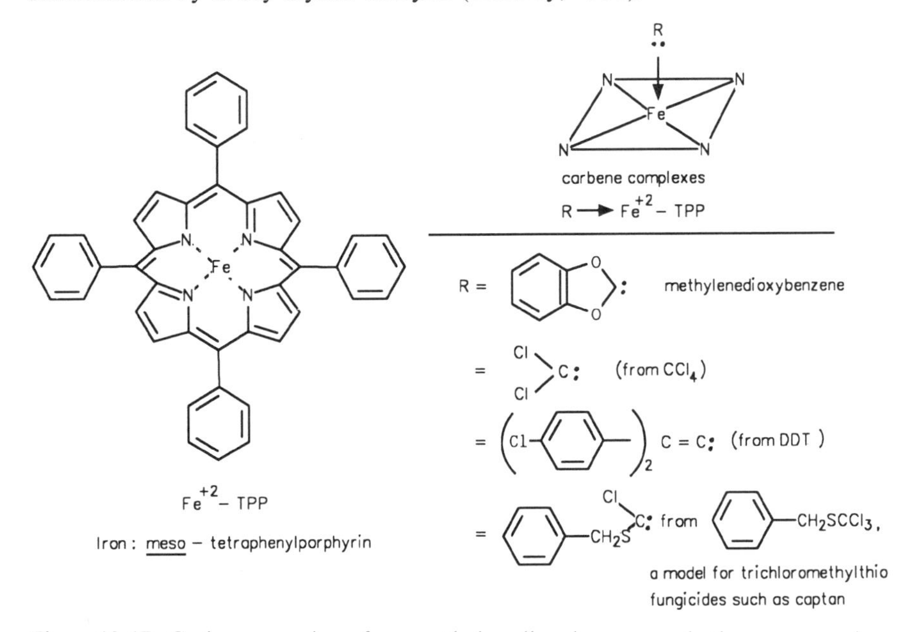

Figure 13-17. Carbene complexes from methylenedioxybenzene and other compounds as models for the corresponding complexes from cytochrome P-450.

D. Piperonyl Butoxide — Metabolism

Radiosynthesis of [*methylenedioxy*-^{14}C]PB and several related pyrethrins synergists (Kuwatsuka and Casida, 1965) allowed preliminary elucidation of metabolic pathways in mammals and insects. Studies using mouse liver microsomes and house fly abdomen homogenates demonstrated that the methylene-^{14}C group is oxidized to ^{14}C-formate, the major water-soluble product (Casida *et al.*, 1966). *In vivo*, [*methylenedioxy*-^{14}C]PB is converted predominantly to $^{14}CO_2$ in mice (76% yield) but to a lesser extent in house flies (11%).

More extensive metabolic studies in mammals utilized PB radiolabeled with ^{14}C in both the methylenedioxy group as well as the α-carbon of the polyether side chain (R_2 in Fig. 13-18) (Kamienski and Casida, 1970). After oral dosage with [*methylenedioxy*-^{14}C]PB, 66–76% of the recovered radiocarbon was expired as $^{14}CO_2$ for both mice and rats while only 6% was excreted in urine whereas with [*α-carbon*-^{14}C]PB, 65–73% of the recovered ^{14}C was in urine. The only metabolites tentatively identified in urine were 6-propylpiperonylic acid (R_1=propyl, R_3=carboxyl, Fig. 13-18) and 6-propylpiperonylglycine. Since collectively these metabolites represented less than 0.5% of the applied dose and there were 10–16 unknown metabolites in urine, it appears that PB is metabolized extensively by mammals, but the identity of individual metabolites is poorly understood. The catechol derived on demethylenation of PB does not occur in urine or mixed-function oxidase enzyme systems, suggesting that it undergoes further metabolism.

Metabolism of [*methylenedioxy*-^{14}C]- and [*α-methylene*-^{14}C]PB in rats after intravenous injection yields numerous metabolites but no intact PB determined by thin-layer chromatography in both bile and urine (Fishbein *et al.*, 1969). PB was the major residue (>90% of tissue radiocarbon) in lungs and perineal fat which contained 15–25 and 9–18% of the applied dose, respectively.

PB was one of ten ^{14}C-labeled methylenedioxyphenyl compounds studied in house flies (Esaac and Casida, 1969). With [*methylenedioxy*-^{14}C]- and

R_1 = $CH_2CH_2CH_3$
R_2 = $CH_2O(CH_2CH_2O)_2C_4H_9$

R_3 = $CH_2OCH_2CH_2OCH_2CH_2OH$
= $CH_2OCH_2CH_2OH$
= CO_2H
= $CONHCH_2CO_2H$ (mammal)
= CO_2 - glucosyl (house fly)

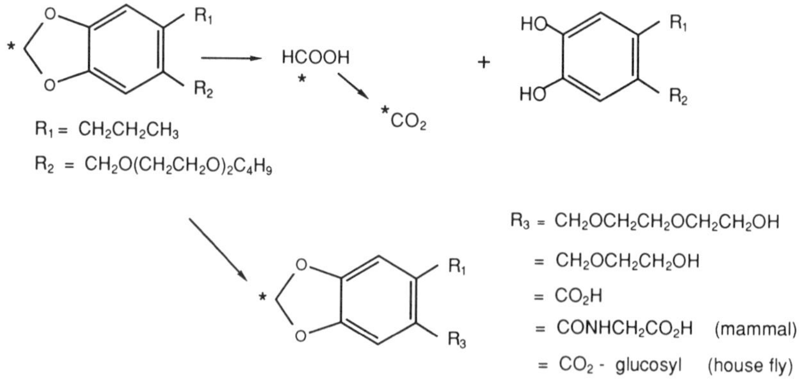

Figure 13-18. Metabolism of piperonyl butoxide in mice and house flies.

[*α-carbon*-^{14}C]PB, 69 and 81% of the recovered radiocarbon were in excrement, respectively. Since only 11% of the dose was recovered as $^{14}CO_2$ for the methylenedioxy-^{14}C sample, most of the excreted products are likely to retain the methylenedioxyphenyl moiety. The catechol resulting from excision of the methylene group is excreted without conjugation. Another major pathway involves oxidation of the polyether side-chain to give, ultimately, 6-propylpiperonylic acid, which is conjugated prior to excretion as the glucoside and other conjugates. In contrast to [^{14}C]piperonylic acid administered to house flies alone (Esaac and Casida, 1968), amino acid conjugates of 6-propylpiperonylic acid are not recovered from [^{14}C]PB-treated house flies (Essac and Casida, 1969).

The metabolism of several PB analogs is relevant to P-450 inhibition and possible toxicity. Under anaerobic conditions rat liver microsomes and NADPH convert methylenedioxybenzenes to carbon monoxide which has been trapped by hemoglobin and confirmed by GC (Yu *et al.*, 1980). Since normally the liver is aerobic, this pathway probably has little toxicological relevance. Isotope studies using ^{18}O and ^{13}C demonstrated that the methylenic group is the source of the carbon in carbon monoxide suggesting that hydroxylation of the methylene is the initial metabolic step, but overall, carbon monoxide is a minor product (Anders *et al.*, 1984). As noted previously, oxidative demethylenation is a major metabolic pathway for PB in mammals so the catechol(s) must be transitory. Catechols can be converted to quinones which react with cellular nucleophiles (such as GSH) as has been shown for the hallucinogen, methylenedioxyphenylmethamphetamine (Kumagai *et al.*, 1991).

E. Piperonyl Butoxide — Toxicology

The normal uses of PB are considered to be relatively risk-free. Even in the more sensitive mammals (rabbits and mice) PB has low acute toxicity (oral $LD_{50} > 2{,}500$ mg/kg) (Brown, 1971). The most significant effects of PB in mammals involve its interaction at high doses with drugs (and pesticides) to modify therapeutic effects or initiate adverse biological responses (Conney *et al.*, 1972). PB on injection will prolong the activity of barbituates and a variety of insecticides in mice (Skrinjaric-Spoljar *et al.*, 1971). This synergist is not teratogenic in rats when administered by gavage at 63 to 1,000 mg/kg on days 6 to 15 of gestation (Kennedy *et al.*, 1977; Khera *et al.*, 1979). Some behavioral and developmental effects were seen in mice from PB administered continuously at 0.1 to 0.8% of the diet in two- and three-generation studies (Tanaka, 1992; Tanaka *et al.*, 1992). The weights of pups were significantly reduced in higher-dose groups and the survival index at postnatal day 21 was reduced in each generation for the group dosed at 0.8%. Several behavioral parameters such as olfactory orientation, surface righting, and cliff avoidance were adversely affected in treated groups. The results suggested that PB modified reproductive, developmental, and behavioral responses, with increasing effects in subsequent generations of mice. However, this study in Japan used exaggerated dose levels and the authors predicted no effects in humans at levels of actual dietary intake.

VI. CONCLUSION

Pyrethrin I is rapidly metabolized by cytochrome P-450-dependent microsomal oxidases in house flies and probably in many other insects. The insecticidal activity of the six individual pyrethrins and their mixture in pyrethrum extract is enhanced by PB and other synergists which retard the rate of oxidative detoxification. The pyrethrins are generally of low toxicity to mammals due in part to their rapid metabolism by oxidative and hydrolytic pathways. The 63 identified metabolites of the six pyrethrins in mammals and insects and their microsomal enzyme systems illustrate the ease of conversion by attack at multiple molecular sites. As with the pyrethrins, PB is also rapidly metabolized in mammals. The pyrethrins-PB combination is preferred for both effectiveness and safety.

ACKNOWLEDGMENT

This publication was made possible by grant number P01 ES00049 from the National Institute of Environmental Health Sciences, NIH.

REFERENCES

Anders, M.W., Sunram, J.M., and Wilkinson, C.F. (1984). Mechanism of the metabolism of 1,3-benzodioxoles to carbon monoxide. *Biochem. Pharmacol.* **33**, 577–580.

Ando, T., Ruzo, L.O., Engel, J.L., and Casida, J.E. (1983). 3-(3,3-Dihalo-2-propenyl) analogues of allethrin and related pyrethroids: synthesis, biological activity, and photostability. *J. Agric. Food Chem.* **31**, 250–253.

Ando, T., Toia, R.F., and Casida, J.E. (1991). Epoxy and hydroxy derivatives of (*S*)-bioallethrin and pyrethrins I and II: synthesis and metabolism. *J. Agric. Food Chem.* **39**, 606–611.

B-Bernard, C., and Philogène, B.J.R. (1993). Insecticide synergists: role, importance and perspectives. *J. Toxicol. Environ. Health* **38**, 199–223.

Brown, M.A., Holden, I., Glickman, A.H., and Casida, J.E. (1985). Biooxidation of chrysanthemate isobutenyl methyl groups directly examined by carbon-13 nuclear magnetic resonance spectroscopy. *J. Agric. Food Chem.* **33**, 8–13.

Brown, N.C. (1971). A review of the toxicology of piperonyl butoxide. *Pyrethrum Post* **11**, 66–68.

Casida, J.E. (1970). Mixed-function oxidase involvement in the biochemistry of insecticide synergists. *J. Agric. Food Chem.* **18**, 753–772.

Casida, J.E. (1973). Biochemistry of the pyrethrins. *In* "Pyrethrum, The Natural Insecticide" (J. E. Casida, ed.), pp. 101–120. Academic Press, New York.

Casida, J.E., Engel, J.L., Esaac, E.G., Kamienski, F.X., and Kuwatsuka, S. (1966). Methylene- C^{14}-dioxyphenyl compounds: metabolism in relation to their synergistic action. *Science* **153**, 1130–1133.

Casida, J.E., Kimmel, E.C., Elliott, M., and Janes, N.F. (1971). Oxidative metabolism of pyrethrins in mammals. *Nature* **230**, 326–327.

Chang, S.C., and Kearns, C.W. (1964). Metabolism *in vivo* of C^{14}-labelled pyrethrin I and cinerin I by house flies with special reference to the synergistic mechanism. *J. Econ. Entomol.* **57**, 397–404.

Class, T.J., Ando, T., and Casida, J.E. (1990). Pyrethroid metabolism: microsomal oxidase metabolites of (*S*)-bioallethrin and the six natural pyrethrins. *J. Agric. Food Chem.* **38**, 529–537.

Conney, A.H., Chang, R., Levin, W.M., Garbut, A., Munro-Faure, A.D., Peck, A.W., and Bye, A. (1972). Effects of piperonyl butoxide on drug metabolism in rodents and man. *Arch. Environ. Health* **24**, 97–106.

Cook, J.C., and Hodgson, E. (1983). Induction of cytochrome P-450 by methylenedioxyphenyl compounds: importance of the methylene carbon. *Toxicol. Appl. Pharmacol.* **68**, 131–139.

Crosby, D. (1994). This volume.

Dickins, M., Elcombe, C.R., Moloney, S.J., Netter, K.J., and Bridges, J.W. (1979). Further studies on the dissociation of the isosafrole metabolite-cytochrome P-450 complex. *Biochem. Pharmacol.* **28**, 231–238.

Elcombe, C.R., Bridges, J.W., Gray, T.J.B., Nimmo-Smith, R.H., and Netter, K.J. (1975). Studies on the interaction of safrole with rat hepatic microsomes. *Biochem. Pharmacol.* **24**, 1427–1433.

Elliott, M., and Casida, J.E. (1972). Optically pure pyrethroids labeled with deuterium and tritium in the methylcyclopentenonyl ring. *J. Agric. Food Chem.* **20**, 295–299.

Elliott, M., and Janes, N.F. (1973). Chemistry of the Natural Pyrethrins. *In* "Pyrethrum, The Natural Insecticide" (J.E. Casida, ed.), pp. 55–100. Academic Press, New York.

Elliott, M., Kimmel, E.C., and Casida, J.E. (1969). ^{3}H-Pyrethrin I and -pyrethrin II: preparation and use in metabolism studies. *Pyrethrum Post* **10**(2), 3–8.

Elliott, M., Janes, N.F., Kimmel, E.C., and Casida, J.E. (1972). Metabolic fate of pyrethrin I, pyrethrin II, and allethrin administered orally to rats. *J. Agric. Food Chem.* **20**, 300–313.

Esaac, E.G., and Casida, J.E. (1968). Piperonylic acid conjugates with alanine, glutamate, glutamine, glycine, and serine in living houseflies. *J. Insect Physiol.* **14**, 913–925.

Esaac, E.G., and Casida, J.E. (1969). Metabolism in relation to mode of action of methylenedioxyphenyl synergists in house flies. *J. Agric. Food Chem.* **17**, 539–550.

Fishbein, L., Falk, H.L., Fawkes, J., Jordan, S., and Corbett, B. (1969). The metabolism of piperonyl butoxide in the rat with ^{14}C in the methylenedioxy or α-methylene group. *J. Chromatr.* **41**, 61–79.

Kamienski, F.X., and Casida, J.E. (1970). Importance of demethylenation in the metabolism *in vivo* and *in vitro* of methylenedioxyphenyl synergists and related compounds in mammals. *Biochem. Pharmacol.* **19**, 91–112.

Kennedy, Jr., G.L., Smith, S.H., Kinoshita, F.K., Keplinger, M.L., and Calandra, J.C. (1977). Teratogenic evaluation of piperonyl butoxide in the rat. *Food Cosmet. Toxicol.* **15**, 337–339.

Khera, K.S., Whalen, C., Angers, G., and Trivett, G. (1979). Assessment of the teratogenic potential of piperonyl butoxide, biphenyl, and phosalone in the rat. *Toxicol. Appl. Pharmacol.* **47**, 353–358.

Kumagai, Y., Wickham, K.A., Schmitz, D.A., and Cho, A.K. (1991). Metabolism of methylenedioxyphenyl compounds by rabbit liver preparations. Participation of different cytochrome P-450 isozymes in the demethylenation reaction. *Biochem. Pharmacol.* **42**, 1061–1067.

Kuwatsuka, S., and Casida, J.E. (1965). Synthesis of methylene-C^{14}-dioxyphenyl compounds: radioactive safrole, dihydrosafrole, myristicin, piperonyl butoxide, and diastereoisomers of sulfoxide. *J. Agric. Food Chem.* **13**, 528–533.

Levi, P.E., and Hodgson, E. (1989). Metabolites resulting from oxidative and reductive processes. *In* "Intermediary Xenobiotic Metabolism in Animals: Methodology, Mechanisms, and Significance" (D.H. Hutson, J. Caldwell, and G.D. Paulson, eds.), pp. 119–138. Taylor and Francis, London.

Mansuy, D. (1980). New iron-porphyrin complexes with metal-carbon bond-biological implications. *Pure Appl. Chem.* **52**, 681–690.

Mansuy, D., Lange, M., and Chottard, J.C. (1978). 2,2-Bis(*p*-chlorophenyl)-1,1,1-trichloroethane (DDT) with iron (II) porphyrins. Isolation of the vinylidene carbene complex, tetraphenylporphyriniron(II) ($C{=}C(p\text{-}Cl\text{-}C_6H_4)_2$). *J. Amer. Chem. Soc.* **100**, 3213–3214.

Mansuy, D., Battioni, J.-P., Chottard, J.-C., and Ullrich, V. (1979). Preparation of a porphyrin-iron-carbene model for the cytochrome P-450 complexes obtained upon metabolic oxidation of the insecticide synergists of the 1,3-benzodioxide series. *J. Amer. Chem. Soc.* **101**, 3971–3973.

Marcus, C.B., Wilson, N.M., Jefcoate, C.R., Wilkinson, C.F., and Omiecinski, C.J. (1990). Selective induction of cytochrome P-450 isozymes in rat liver by 4-*n*-alkylmethylenedioxybenzenes. *Arch. Biochem. Biophys.* **277**, 8–16.

Matthews, H.B., Skrinjaric-Spoljar, M., and Casida, J.E. (1970). Insecticide synergist interactions with cytochrome P-450 in mouse liver microsomes. *Life Sci.* **9**, 1019–1048.

Murray, M., Wilkinson, C.F., Marcus, C., and Dubé, C.E., (1983). Structure-activity relationships in the interactions of alkoxymethylenedioxybenzene derivatives with rat hepatic microsomal mixed-function oxidases *in vivo*. *Mol. Pharmacol.* **24**, 129–136.

Ortiz de Montellano, P.R., and Reich, N.O. (1986). Inhibition of cytochrome P-450 enzymes. *In* "Cytochrome P-450: Structure, Mechanism and Biochemistry" (P.R. Ortiz de Montellano, ed.), pp. 273–314. Plenum Press, New York.

Pellegrini, Jr., J.P., Miller, A.C., and Sharpless, R.V. (1952). Biosynthesis of radioactive pyrethrins using $C^{14}O_2$. *J. Econ. Entomol.* **45**, 532–536.

Philpot, R.M., and Hodgson, E. (1972). The effect of piperonyl butoxide concentration on the formation of cytochrome P-450 difference spectra in hepatic microsomes from mice. *Mol. Pharmacol.* **8**, 204–214.

Schoenig, G. (1994). This volume.
Skrinjaric-Spoljar, M., Matthews, H.B., Engel, J.L., and Casida, J.E. (1971). Response of hepatic microsomal mixed-function oxidases to various types of insecticide chemical synergists administered to mice. *Biochem. Pharmacol.* **20**, 1607–1618.
Soderlund, D.M., and Casida, J.E. (1977). Effects of pyrethroid structure on rates of hydrolysis and oxidation by mouse liver microsomal enzymes. *Pestic. Biochem. Physiol.* **7**, 391–401.
Tanaka, T. (1992). Effects of piperonyl butoxide on F_1 generation mice. *Toxicol. Lett.* **60**, 83–90.
Tanaka, T., Takahashi, O., and Oishi, S. (1992). Reproductive and neurobehavioral effects in three-generation toxicity study of piperonyl butoxide administered to mice. *Food Chem. Toxicol.* **30**, 1015–1019.
Wachs, H. (1947). Synergistic insecticides. *Science* **105**, 530–531.
Wilkinson, C.F., and Hicks, L.J. (1969). Microsomal metabolism of the 1,3-benzodioxole ring and its possible significance in synergistic action. *J. Agric. Food Chem.* **17**, 829–836.
Winteringham, F.P.W., Harrison, A., and Bridges, P.M. (1955). Absorption and metabolism of [^{14}C]pyrethroids by the adult housefly, *Musca domestica* L., *in vivo*. *Biochem. J.* **61**, 359–367.
Yamamoto, I. (1973). Mode of action of synergists in enhancing the insecticidal activity of pyrethrum and pyrethroids. *In* "Pyrethrum, The Natural Insecticide" (J.E. Casida, ed.), pp. 195–210. Academic Press, New York.
Yamamoto, I., and Casida, J.E. (1966). *O*-Demethyl pyrethrin II analogs from oxidation of pyrethrin I, allethrin, dimethrin and phthalthrin by a house fly enzyme system. *J. Econ. Entomol.* **59**, 1542–1543.
Yamamoto, I., and Casida, J.E. (1968). Syntheses of ^{14}C-labeled pyrethrin I, allethrin, phthalthrin, and dimethrin on a submillimole scale. *Agric. Biol. Chem.* **32**, 1382–1391.
Yamamoto, I., Kimmel, E.C., and Casida, J.E. (1969). Oxidative metabolism of pyrethroids in houseflies. *J. Agric. Food Chem.* **17**, 1227–1236.
Yu, L.-S., Wilkinson, C.F., and Anders, M.W. (1980). Generation of carbon monoxide during the microsomal metabolism of methylenedioxyphenyl compounds. *Biochem. Pharmacol.* **29**, 1113–1122.
Zeid, M.M.I., Dahm, P.A., Hein, R.E., and McFarland, R.H. (1953). Tissue distribution, excretion of $C^{14}O_2$ and degradation of radioactive pyrethrins administered to the American cockroach. *J. Econ. Entomol.* **46**, 324–336.

14

Environmental Toxicology of Pyrethrum Extract

KARL L. GABRIEL and RAYMOND MARK

I. INTRODUCTION

Although pyrethrum has been used for over a century without evidence of deleterious effects to humans, wildlife, plants, or the environment, the current call-in for updated toxicologic and safety data by the United States Environmental Protection Agency (EPA) has, in effect, imposed requirements for additional proof of its safety. Under the Federal Food, Drug, and Cosmetic Act of 1954, accepted tolerances and exemptions from tolerances were provided for pyrethrum and a list of synergists, and the Food Additives Amendment of 1958 established tolerances for almost every suitable use (Moore, 1973). Development and submission of the data to the EPA is required to support the registration of a pesticide product such as pyrethrum. In addition, states such as California and Arizona have their own regulatory agencies and requirements for data submission.

The EPA's objective in calling for new pyrethrum test data is to develop a data base comparable to that of other newer pesticides. Existing reports on pyrethrum toxicity consist mainly of studies performed from 1950 to 1970 in various laboratories. During that period, EPA regulations, such as those outlining the required tests and procedures and Good Laboratory Practice standards, were not in place. The call for reregistration of pyrethrum is now based on a defined data set applicable to all pesticide products. The test methods and procedures are standardized and provide for observations covering a wide spectrum of potential effects. In addition, there have been technical advances for pesticide detection and a greater understanding of environmental concerns over the past two decades. There is also greater knowledge of mechanisms of action and metabolic pathways which are useful for evaluation of toxicity. Finally, the past 30 or more years of extensive use of various pesticidal agents have provided a broad data base of ecotoxicological effects within which pyrethrum can be positioned.

The call for data by EPA includes studies on acute and chronic toxicity, effects on wildlife and aquatic organisms, terrestrial and aquatic field exposures, and environmental fate. These studies form the basis for development of toxicity information to provide a starting point for pesticide hazard assessments, including: (1) determination of the levels of the active ingredient which will produce toxic effects on selected test organisms; (2) measurements of the pesticide residues for estimating their potential impact to fish and wildlife; (3) determining the need for precautionary label statements with respect to minimizing any potential adverse environmental effects; and (4) determining if any need exists for further laboratory and/or field studies. This review considers toxicity data from recently completed studies and annotates those which are still in progress.

II. GENERAL CONSIDERATIONS

The direct and indirect environmental impacts of a pesticide are related to its toxicological effects and ultimate fate. A host of possibilities exists for wide-ranging effects, arising from combinations or permutations of the pesticide's mode of action, metabolic fate, and degradability. Although the primary toxicity of the pesticide is conferred by its chemical structure, its mechanism of action on living processes (Casida, 1973) may be modified by metabolism into more or less active by-products and may result in different effects on various classes of organisms. Hence, EPA directives call for direct toxicity data on insects, fish, birds, and mammals. Persistence of a pesticide or its by-products would generally have greater adverse effects on wildlife and the environment, whereas rapid degradation and dissipation would minimize its environmental impact. Pyrethrum has been especially notable for its lack of persistence. From its long history of use, its impact on the environment has been considered to be minimal in contrast to that of many other chemical pesticides. There has been no evidence of pyrethrins entering into the food chain through environmental residues, runoff, or leaching.

The EPA environmental impact study categories are comprised of toxicological studies and environmental fate investigations. Toxicological studies include (1) acute toxicity to selected species of mammals, birds, aquatic organisms, and invertebrate organisms; (2) phytotoxicity; (3) reproductive and life-cycle studies; and (4) field investigations. Environmental fate studies cover (1) terrestrial field dissipation; (2) aquatic field dissipation; (3) bioaccumulation and bioconcentration; and (4) residue testing.

III. HISTORICAL BACKGROUND

Available information on the toxicity of pyrethrum extract to mammals was reviewed by Barthel (1973) and included studies reported through 1972. Substantial variations existed in the oral LD_{50} of pyrethrins for rats, ranging from 100 to $>2,600$ mg/kg. However, the test materials used in these various studies differed in type, purity, and concentration and, in some studies, certain assumptions were necessary to calculate the LD_{50}. Malone and Brown (1968) showed differences in susceptibility between the rat, mouse, and chick, and that the degree of purity of pyrethrum extract was an important factor. The preparation containing the highest concentration of pyrethrins (77.8%) was least toxic to all three species. Chronic toxicity of pyrethrum extract was studied by Lehman (1952) in a 104 week feeding study. The no-effect level was 1,000 ppm pyrethrins and the chronic oral LD_{50} was 250 mg/kg/day for rats. Inhalation toxicity data are also limited with the most extensive study reported by Carpenter *et al.* (1950). Rats exposed to 0.5 g of pyrethrins in peanut oil per 1,000 cubic feet of air for 27 to 85 half-hour exposures showed no excess mortality nor any histopathological changes above that of the vehicle-control groups. Dermal toxicity studies in rabbits (Carpenter *et al.*, 1950) and rats (Malone and Brown, 1968) showed LD_{50} values of 2,060 mg/kg and $>1,350$ mg/kg, respectively.

The environmental toxicology of pyrethrum extract was reviewed by Pillmore (1973), who coordinated the 1970 cooperative field appraisals. An experimental area in Colorado was selected in which twelve 160 acre treatment blocks were delineated. Within each treatment block, the central 40 acres were developed for intensive study. A stabilized formulation in mineral oil with antioxidant and ultraviolet-screening materials was applied at 0.5 gal/acre. Field appraisal included estimates of bird numbers and survival, fish survival, sampling of terrestrial and aquatic insects, and sampling for pyrethrins deposit. Based on trend counts, the spraying at 0.1 and 0.2 lb pyrethrins/acre did not appear to result in any discernable changes in species composition or relative abundance of birds. No effect was observed on fish life.

IV. REGULATORY REQUIREMENTS

The test methods used are based upon the following EPA Pesticide Assessment Guidelines for hazard evaluation and environmental fate studies (1982): (1) Subdivision F—Hazard Evaluation: Human and Domestic Animals; (2) Subdivision E—Hazard Evaluation: Wildlife and Aquatic Organisms; (3) Subdivision L—Hazard Evaluation: Nontarget Insects; and (4) Subdivision N—Chemistry: Environmental Fate. Registration of pesticides under the Federal Insecticide, Fungicide, and Rodenticide Act (FIFRA) is based on tests performed in accordance with sections of the above guidelines (1989). The mammalian toxicological studies under Subdivision F are reported by Schoenig (1994). Studies completed or scheduled to be performed under Subdivisions E, N, and L are listed in Tables 14-1 and 14-2 based on Schoenig (1992). Since many of the ecotoxicological studies require the use of radiolabeled pyrethrins, performance of these tests was delayed pending the preparation of appropriate radiolabeled material. [^{14}C]Pyrethrins prepared by Amersham International (Buckinghamshire, England) which met purity and stability standards (acid-labeled Pyrethrin I, 98% radiochemical purity) was finally available in late 1992 and the environmental fate studies are now in progress.

Table 14-1. Wildlife Studies

FIFRA guideline	Test
	Using pyrethrum extract—completed
71-1a	Acute oral toxicity in bobwhite quail
71-2a	Avian dietary LC_{50} in bobwhite quail
71-2b	Avian dietary LC_{50} in mallard ducks
141-1	Honeybee acute toxicity LD_{50}
	Using radiolabeled pyrethrin I—in progress or scheduled
72	Fish toxicology: 1a bluegill; 1b bluegill, typical end product (TEP); 1c rainbow trout; 1d rainbow trout, TEP
72	Invertebrate toxicology: 2a invertebrate toxicology; 2b invertebrate toxicology, TEP
72	Estuarine/marine: 3a fish toxicology—sheepshead minnow; 3b mollusk toxicology—American oyster; 3c mollusk toxicology—shrimp; 3d fish toxicology—sheepshead minnow, TEP; 3e mollusk toxicology—American oyster, TEP; 3f mollusk toxicology—shrimp, TEP
72-4a	Fish early life stage
72-4b	Invertebrate life cycle
164-2	Aquatic field dissipation
165-4	Bluegill bioconcentration

Table 14-2. Environmental Fate Studies

FIFRA guideline	Test
	Using pyrethrum extract—completed
164-1	Terrestrial field dissipation
164-2	Aquatic field dissipation
	Using radiolabeled pyrethrin I—in progress or scheduled
161-1	Hydrolysis
161-2	Photodegradation—water
161-3	Photodegradation—soil
162	Metabolism: 1 soil; 3 anaerobic aquatic; 4 aerobic aquatic
163-1	Leach, adsorption, desorption
163-2	Volatility—lab
165-4	Bioaccumulation in fish
171-4	Nature of residue: a plants; b livestock

V. ECOTOXICOLOGY STUDIES

For the following studies, the test material used was pyrethrum extract, Task Force Blend FEK-99 with a reported purity of 57.6% active ingredient (combined pyrethrins I and II).

A. Acute Oral Toxicity in Bobwhite Quail

Ten northern bobwhite quail (*Colinus virginanus*), five males and five females, were randomly assigned to dosage groups of 125, 250, 500, 1,000, and 2,000 mg/kg of active ingredient (ai) or to a corn oil vehicle control group. All birds were fasted for at least 15 hours prior to dosing and all received a constant volume of 4 ml/kg. There were no mortalities at any of the dosages tested. No signs of toxicity were observed at 125 or 250 mg ai/kg. At 500 mg ai/kg, toxic signs occurred within 3 hours after dosing. Signs of toxicity noted at this dosage were prostrate posture, wing droop, lower limb weakness, shallow and rapid respiration, twitching, hyperexcitability, loss of coordination, and lethargy. Toxic signs persisted through day 4, but all birds were normal in appearance and behavior by day 5. Similar findings were observed for the groups dosed with 1,000 or 2,000 mg ai/kg, but onset of signs occurred slightly sooner. At 2,000 mg ai/kg, additional signs of toxicity included reduced reaction to sound and movement and a ruffled appearance. A reduction in body weight gain was observed among all birds at 1,000 mg ai/kg and a loss in body weight among all birds at 2,000 mg ai/kg for days 0–3. The acute oral LD_{50} value and the no-mortality level for northern bobwhite quail exposed to pyrethrum extract as a single oral dose was $>2,000$ mg ai/kg. The no-observed-effect level was 250 mg ai/kg (Campbell and Lynn, 1991).

B. Dietary Toxicity in Bobwhite Quail

Ten northern bobwhite quail (*C. virginanus*) chicks, 10 days of age, were randomly assigned to dietary concentrations of 562, 1,000, 1,780, 3,160, and 5,620 ppm of ai. The pyrethrum extract was mixed into the ration with corn oil. Each group was fed the appropriate test or control diet for 5 days, following which all groups were given untreated feed for 3 days. There were no mortalities at any of the concentrations tested and all birds were normal in appearance and behavior throughout the study. A slight reduction in body weight gain was observed at 5,620 ppm ai during exposure days 0 to 5. No effect on feed consumption was observed at any of the test concentrations. The dietary LC_{50} value and no-mortality level for northern bobwhite quail exposed to pyrethrum extract was $>5,620$ ppm ai. The no-observed-effect level was 3,160 ppm ai based on the reduction in body weight gain at 5,620 ppm ai (Grimes *et al.*, 1991a).

C. Dietary Toxicity in the Mallard Duck

Ten mallard ducklings (*Anas platyrhynchos*), 10 days of age, were randomly assigned to dietary concentrations of 562, 1,000, 1,780, 3,160, and 5,620 ppm of ai. The pyrethrum extract was mixed into the ration with corn oil. Each group was fed the appropriate test or control diet for 5 days, following which all groups were given untreated feed for 3 days. There were no mortalities at any of the concentrations tested. At 562, 1,000, and 1,780 ppm ai, all birds were normal in appearance and behavior throughout the study. At 3,160 and 5,620 ppm ai, all birds were slightly lethargic on the third day and this lethargy persisted in the 5,620 ppm ai group for an additional 12 hours. All birds were then normal in appearance and behavior throughout the remainder of the study. There was a slight reduction in body weight gain at 5,620 ppm ai during exposure days 0 to 5. No effect on feed consumption was observed at any of the test concentrations. The dietary LC_{50} value and the no-mortality level for the mallard exposed to pyrethrum extract was >5,620 ppm ai. The no-observed-effect level was 1,780 ppm ai based on the lethargy noted at 3,160 ppm ai (Grimes *et al.*, 1991b).

D. Acute Contact Toxicity in the Honey Bee

Honey bees (*Apis mellifera*) were tested at five treatment levels representing 586, 58.6, 5.86, 0.586 and 0.0586 ng ai/bee. The pyrethrum extract was prepared in acetone and dosed topically on the thorax and/or abdomen of at least 25 bees per group. At test termination on day 2, mortality in the untreated group was 4%, and mortalities in the acetone solvent control group and the 0.0586 and 0.586 ng ai/bee groups were 8%. Mortalities at 5.86, 58.6, and 586 ng ai/bee were 14%, 78%, and 98%, respectively. All the bees at the two highest dose levels were immobile within the first hour of dosing. According to the toxicity categories of Atkins *et al.* (1976), pyrethrins was classified as highly toxic, i.e., $LD_{50} < 2$ μg/bee. From the 5.86, 58.6, and 586 ng ai/bee data, the honey bee 48 hour contact LD_{50} was approximately 22 ng (0.022 μg) ai/bee. The no-observed-effect dose was 0.586 ng ai/bee (Lynn and Hoxter, 1991).

VI. ENVIRONMENTAL FATE STUDIES

Terrestrial field dissipation studies were performed at three locations in California, Michigan, and Georgia. These studies evaluated the extent and rate of dissipation, and the vertical mobility of pyrethrum in the soil profile under typical use conditions following an application of Pyrenone® Crop Spray (Roussel-Uclaf, Montvale, New Jersey, 20% pyrethrins) at the maximum annual rate. The maximum annual application rate of 0.46 lb ai/acre for pyrethrum was evenly applied on bareground at these three field locations at a time that was typical for the control of insect pests in cropping situations. Soil samples were taken to a depth of 36″. Maximum duration of sampling varied between 97 to 179 days after treatment for the three locations. Extracts of soil samples were analyzed by gas chromatography. Pyrethrins had a half-life of approximately 1 hour when applied to bareground in California and Georgia and approximately 2 hours in Michigan. In all three locations, no residues of pyrethrins above the 0.1 ppm limit of quantitation were detectable in any 0–6″ soil core sample by 1 day after treatment and beyond, or in any samples below the 0–6″ soil horizon (Hatterman, 1992). Aquatic field-dissipation studies are in progress. As shown in Table 14-2, the other environmental fate studies are also underway following the preparation of radiolabeled pyrethrins, which became available for use in late 1992.

VII. CONCLUSION

The findings of the ecotoxicological and environmental fate studies completed to date are totally consistent with previously-reported data. The potential hazard of pyrethrins for warm-blooded species is low. The oral LD_{50} of >2,000 mg pyrethrins/kg and the dietary LC_{50} of >5,620 ppm pyrethrins for both the bobwhite quail and the mallard duck indicates that pyrethrum spraying is very unlikely to produce any toxic effects on avian wildlife. The impact of terrestrial spraying of pyrethrum extract on nontarget organisms other than sensitive insects is also minimized by extremely rapid dissipation. Although data acquisition continues for toxicological and fate investigations which require [^{14}C]pyrethrins, the available database from current and past studies indicates that pyrethrum extract, applied at insecticidal dosages, has only a minimal direct impact on wildlife and no long-term effects on various ecosystems.

REFERENCES

Atkins, Jr., E.L., Anderson, L.D., Kellum, D., and Neuman, K.W. (1976). Protecting honey bees from pesticides. Univ. of California, Div. of Agricultural Sciences, Leaflet 2883.

Barthel, W.F. (1973). Toxicity of pyrethum and its constituents to mammals. *In* "Pyrethrum, The Natural Insecticide" (J.E. Casida, ed.), pp. 123-142. Academic Press, New York.

Campbell, S., and Lynn, S.P. (1991). Unpublished report from Wildlife International Ltd. to Pyrethrum Task Force. An acute oral toxicity study with pyrethrum extract in the northern bobwhite quail, August 30.

Carpenter, C.P., Weil, C.S., Pozzani, U.C., and Smythe, Jr., H.C. (1950). Comparative acute and subacute toxicities of allethrin and pyrethrin. *Arch. Ind. Hyg. Occup. Med.* **2**, 420-432.

Casida, J.E. (1973). Biochemistry of the pyrethrins. *In* "Pyrethrum, The Natural Insecticide" (J.E. Casida, ed.), pp. 123-142. Academic Press, New York.

Grimes, J., Lynn, S.P., and Smith, G.J. (1991a). Unpublished report from Wildlife International Ltd. to Pyrethrum Task Force. A dietary LC_{50} study with pyrethrum extract in the northern bobwhite quail, July 31.

Grimes, J., Lynn, S.P., and Smith, G.J. (1991b). Unpublished report from Wildlife International Ltd. tò Pyrethrum Task Force. A dietary LC_{50} study with pyrethrum extract in the mallard, July 31.

Hatterman, D.R. (1992). Unpublished report from Landis International Inc. to Pyrethrum Task Force. Terrestrial field dissipation of LX1180-02 (pyrethrum + piperonyl butoxide) applied to bareground.

Lehman, A.J. (1952). Chemicals in food. Part II. Pesticides. Section III. Subacute and chronic toxicity. Association of Food and Drug Officials of the United States, Quarterly Bulletin **16**, 47-53.

Lynn, S.P., and Hoxter, K.A. (1991). Unpublished report from Wildlife International Ltd. to Pyrethrum Task Force. An acute contact toxicity study with pyrethrum extract with the honey bee, July 30.

Malone, J.C., and Brown, N.C. (1968). Toxicity of various grades of pyrethrum to laboratory animals. *Pyrethrum Post* **9**, 3-8.

Moore, J.B. (1973). Residue and tolerance considerations with pyrethrum, piperonyl butoxide, and MGK 264. *In* "Pyrethrum, The Natural Insecticide" (J.E. Casida, ed.), pp. 293-311. Academic Press, New York.

Pillmore, R.E. (1973). Toxicity of pyrethrum to fish and wildlife. *In* 'Pyrethrum, The Natural Insecticide" (J.E. Casida, ed.), pp. 143-165. Academic Press, New York.

Schoenig, G.P. (1992). Unpublished material prepared for the Pyrethrum Joint Venture; Overview and summary of pyrethrum extract data development program.

Schoenig, G.P. (1994). This volume.

US Environmental Protection Agency, Pesticide Assessment Guidelines (1982). Subdivision E, Addendum 1, Hazard Evaluation: Wildlife and Aquatic Organisms; Subdivision F, Hazard Evaluation: Human and Domestic Animals; Subdivision L, Hazard Evaluation: Nontarget Insects; Subdivision N, Chemistry: Environmental Fate, Washington, DC.

US Environmental Protection Agency (1989). FIFRA Accelerated Reregistration Phase 3 Technical Guidance, Washington, DC.

V

Pyrethrum for Pest Control

15

Pyrethrum for Control of Pests of Agricultural and Stored Products

CHARLES A. SILCOX and EDWIN S. ROTH

I. INTRODUCTION

The use of pyrethrum in agricultural and stored products insect control was extensively reviewed as individual chapters in "Pyrethrum, The Natural Insecticide" (Casida, 1973). At that time, the development of pyrethrum into agricultural markets was diametrically opposed to its commercialization in stored products markets. Pyrethrum had been displaced from agricultural insect control by the introduction of synthetic organic insecticides, which were more cost-effective, but was used extensively in a variety of applications for stored products insect control.

This review describes the significant advances since 1973 in reviving the agricultural uses of pyrethrum and in redefining one of its applications for stored products insect control. It details the unique features of pyrethrum that have been exploited during the past decade to reintroduce it to American agriculture and that have traditionally made it an ideal product for stored products insect control in a wide variety of use areas. Finally, it describes a new formulation developed to overcome some of the inherent problems associated with pyrethrum-based, Ultra-Low Volume (ULV) space spray applications. This unique formulation addresses both insecticide application technology and contemporary regulatory issues to expand the use of pyrethrum for stored products insect control in food processing plants.

II. AGRICULTURAL INSECT CONTROL

Lange and Akesson (1973) documented the extensive use of pyrethrum for agricultural and horticultural insect control prior to World War II. However, the development of synthetic organic insecticides during the mid-1940s greatly reduced the amount of pyrethrum used in these areas because a succession of insecticide classes were developed that provided improved and more economical insect control. They suggested that research on formulations and application methods could, once again, increase the role of pyrethrum in agriculture. This research did not materialize, perhaps due to the introduction of photostable pyrethroid insecticides for agricultural insect control.

The increased agricultural use of pyrethrum during the 1980s resulted from a commercial development program that related the regulatory status and physical and toxicological properties of pyrethrum to specific niches of agricultural insect control. Pyrethrum is unique in being exempt from the establishment of tolerances when applied to growing crops, having a zero-day preharvest interval that allows it to be used up to and including the day of harvest, and having a broad spectrum of activity as well as a renowned capacity for rapid knockdown of insects. There is no doubt that pyrethrum is not cost-effective on major (large acreage) crops where alternative insecticides may be selected to control pests. However, the characteristics listed above make pyrethrum an ideal material for insect control on minor (small acreage) crops. These are generally high value crops, which can be profitably produced despite substantial insect control costs, for which there are few or no registered insecticides. The lack of alternative insecticides decreases the level of insect control that is demanded from a pyrethrum-based product.

A. Label Expansion

The major factor responsible for the increased agricultural use of pyrethrum during the 1980s was the development of an expanded label that permitted it to be applied to virtually any growing crop. The label was based on a crop grouping scheme that was devised by the United States Environmental Protection Agency (EPA) to expedite minor-use pesticide registrations (Anonymous, 1983). This scheme (Table 15-1) provided a vehicle for a group tolerance to be established that would allow an active ingredient to be registered for use on all crops within a specific group after tolerances were established on several crops that were selected to be representative of the group. The fact that pyrethrum (and the synergist piperonyl butoxide) is exempt from tolerances permitted the crop grouping format to be incorporated into a product label that allowed pyrethrum to be applied to an extremely wide variety of crops.

B. Primary Uses in Agriculture

Although there are a number of agricultural situations that are ideally suited for pyrethrum-based insecticide applications, the primary positioning of

Table 15-1. United States Environmental Protection Agency Crop Grouping Scheme[a]

Vegetables: root and tuber; leaves of root and tuber; bulb (*Allium* spp.); leafy: brassica (cole) leafy; legume; leaves of legume; fruiting; cucurbit

Fruits: citrus (*Citrus* spp., *Fortunella* spp.); pome; stone; small fruits; berries; tree

Grains and Hay: cereal grains; forage, fodder, and straw of cereal grains; grasses for seed, forage, fodder, and hay; nongrass animal feeds

Herbs and Spices

[a]40 CFR 180.34 gives the list of crops included but not limited to in each category.

pyrethrum in the agricultural market is as a broad-spectrum insecticide for use on minor crops. In this role, pyrethrum fills the void created by the relative lack of insecticide registrations for most of these crops. This role will undoubtedly expand in the future because a large number of minor use insecticide registrations will be terminated as a result of the current reregistration of active ingredients by the EPA (Anonymous, 1989). Pyrethrum products are also used on major crops, although this is generally at lower, more economical application rates that are directed at pests that are particularly susceptible to pyrethrins. One of the traditional agricultural uses of pyrethrum is as a tank-mixture, at low application rates (0.004 to 0.008 lb pyrethrins/acre), with conventional insecticides. This application exploits the inherently rapid action of pyrethrins on the insect nervous system (Camougis, 1973) by increasing insect motility immediately after treatment (Shepard, 1951), which improves control by exposing the insect to greater quantities of both pyrethrum and the companion insecticide.

The combination of low mammalian toxicity (Barthel, 1973) and rapid degradation by ultraviolet radiation (Tattersfield, 1932; Elliott and Janes, 1973; Bullivant and Pattenden, 1976) allows pyrethrum to have a zero-day preharvest interval. Thus, pyrethrum is one of the few materials that may be applied to crops up to, and including, the day of harvest. As a crop nears harvest, pyrethrum can replace conventional insecticides, which cannot be applied at this time without exceeding residue tolerances, to control populations of insects that would directly damage the crop. It may also be applied as a quick knockdown spray on the day of harvest either to control insects that would be annoying to the persons harvesting the crop or to greatly reduce the possibility of rejection by processors for those crops that will be inspected for insects prior to processing. With the increasing encroachment of urban development into agricultural areas, the potential for increased pyrethrum use is substantial because it is an ideal material for insect control in areas where agricultural land borders homes, schools, shopping centers, and other locations where there is potential for inadvertent contact by people who are not familiar with the precautions necessary to avoid exposure to agricultural chemicals.

The short persistence of pyrethrins and the moderate to low toxicity to honey bees and beneficial arthropods under normal use conditions provide two additional advantages for agricultural pest control. Pollination by honey bees is an important factor in the production of many agricultural crops. Although pyrethrins are intrinsically toxic to honey bees (Atkins, 1975), the safety to bees can be enhanced by applying pyrethrum early in the morning or late in the evening when they are not actively foraging. This avoids direct contact at the time of application and also allows time for the photodegradation of pyrethrins from plant surfaces before the bees resume their activity. Thus, there is an agricultural niche for pyrethrum as a bloom period spray to control pests, yet not to adversely affect honey bees or decrease pollination. Pyrethrum may also see expanded use in integrated control programs that are based on selective insecticides that do not disrupt beneficial arthropod populations. Although pyrethrum is a broad spectrum contact insecticide, it can be applied as a selective

Table 15-2. Control of Apple Aphids and a Cecidomyid Predator by Pyrethrum, Endosulfan, and Fluvalinate (Data from Forsythe and Gardner, 1986)

		Average control (%)[a]		
Treatment	lb ai/acre	Apple aphids	Cecidomyid predators	Aphids/ terminal[b]
Untreated		0	0	39
Pyrethrum	0.0125	80	58	5
Endosulfan	0.5	95	100	1.3
Fluvalinate	0.06	93	100	0.1

[a]Average of samples taken 1, 4, and 7 days after treatment.
[b]14 days after treatment.

toxicant, particularly at low application rates, to provide a degree of insect control and not disrupt beneficial arthropod populations (Wilkinson *et al.*, 1975). Table 15-2 illustrates the effect of pyrethrum and conventional insecticides on both apple aphids and a cecidomyid predator in a Maine apple orchard (Forsythe and Gardner, 1986). These results show that, although pyrethrum was not quite as effective as the other insecticides towards apple aphids, it was not as disruptive to the predator population as the more persistent, synthetic insecticides. This allowed the predators to enact an additional measure of control and the apple aphid population was below the treatment threshold level in all test plots at 14 days after treatment.

C. Cooperative Development Program

The tolerance exemption and crop grouping scheme that eventually led to expansive agricultural label claims for pyrethrum products also created a problem for the manufacturers of the product. Essentially, the breadth of the label was such that insect specific control recommendations could not be made for the vast majority of crops listed on the label. Of course, the manufacturer could independently generate this information, but this was to revisit the minor use crisis where a company could not justify expending resources to generate data to support specific claims for crops with little potential return on investment. A unique solution to this problem involved an informal cooperative development program that was initiated to define the spectrum of activity of pyrethrum towards agricultural insects. This program recognized that there are more than 400 entomologists with extension responsibilities in the USA (Anonymous, 1990) and that many of these scientists are responsible for insecticide use recommendations on minor crops. The program was based on the need for these entomologists to be familiar with the activity of pyrethrum towards a variety of agricultural insect pests because the updated labeling made pyrethrum a potentially important insecticide for use on minor crops. It involved the manufacturer providing research samples to interested entomologists in exchange for reports that would establish the data base from which specific use recommendations could be incorporated into product labeling.

D. Spectrum of Activity

The continuing informal cooperative development program described above has been successful in defining the spectrum of activity of pyrethrum towards agricultural pests (Table 15-3). Results to date show that pyrethrum is most effective against certain lepidopterous larvae, whiteflies, aphids, leafhoppers, flea beetles, and thrips. It should be kept in mind, however, that even a fair level of control (a rating of 2 or 3) may be acceptable on minor crops where few or no other insecticides are registered.

III. STORED PRODUCTS INSECT CONTROL

A. Review of Product Types

Pyrethrum is one of the most widely-used insecticides for stored products insect control because its low mammalian toxicity and short residual life allow application in numerous situations, its broad spectrum of activity is suitable for control of a wide variety of pests, and established tolerances for raw agricultural commodities and processed foods are acceptable for both direct grain treatments and in food handling establishments. There are literally hundreds of pyrethrum products with label claims for stored products insect control. These range from consumer liquid and aerosol products for control of these insects in private homes to a wide variety of professional products for use by commercial pesticide applicators either to protect commodities through direct applications to grain, as repellent treatments to packaging material, or for space, contact, surface, crack-and-crevice, and spot treatments within food handling establishments. These establishments are defined by the EPA as "an area or place other than a private residence in which food is held, processed, prepared, and/or served." They include, but are not limited to, bakeries, cafeterias, canneries, commercial airplanes, hospitals, mobile caterers, restaurants, schools, supermarkets, and taverns. This section will address the use of pyrethrum for stored products insect control in one type of food handling establishment—the food processing plant. It will not address the use of pyrethrum in consumer products or repellent packaging because there have been no substantial changes in these use areas since the previous review (Gillenwater and Burden, 1973). Likewise, it will not address the use of pyrethrum as a grain treatment because, although there are many products registered for this use, very little pyrethrum is actually used in this manner since more cost-effective insecticides have displaced it from this market.

B. Use in Food Processing Plants

The use of insecticides in food processing plants is regulated by a number of federal agencies including the EPA, the Food and Drug Administration (FDA) and the United States Department of Agriculture (USDA). All insecticide products in the United States must be registered with the EPA. In addition, those used in food areas of FDA-inspected facilities (for receiving, serving,

Table 15-3. Spectrum of Activity of Pyrethrum (in Combination with Piperonyl Butoxide) Towards Agricultural Insect Pests

Species	Crop	Dose relative to 0.025 lb/A	Relative activity[a] (*n*)[b]
	LEPIDOPTERA		
Tomato pinworm, *Keiferia lycopersicella*	tomato	1 ×	2 (1)
Beet armyworm, *Spodoptera exigua*	tomato	1 ×	3 (1)
	celery	1 ×	4 (1)
Black cutworm, *Agrotis ipsilon*	celery	1 ×	4 (1)
Cabbage looper, *Trichoplusia ni*	cabbage	1 or 1.3 ×	3 (1 and 3)
Fall armyworm, *Spodoptera frugiperda*	sweet corn	0.3 ×	1 (1)
		1 ×	2 (3)
	sorghum	2 ×	5 (2)
Tomato fruitworm, *Helicoverpa zea*	tomato	1 ×	3 (1)
Corn earworm, *Helicoverpa zea*	sweet corn	1, 2 or 4 ×	1 (1 each)
	sorghum	2 ×	5 (2)
Imported cabbageworm, *Pieris rapae*	cabbage	1 ×	3 (1)
		1.3 ×	5 (2)
European corn borer, *Ostrinia nubilalis*	sweet corn	1 ×	1 (1)
Navel orangework, *Amyelois transitella*	walnut	2 ×	1 (1)
Fruittree leafroller, *Archips argyrospila*	oak	1 ×	5 (1)
Blueberry spanworm, *Itame argillacearia*	blueberry	1 ×	5 (1)
Spiny elm caterpillar, *Nymphalis antiopa*	willow	0.3 ×	5 (2)
	HOMOPTERA		
Greenhouse whitefly, *Trialeurodes vaporariorum*	tomato	1.2 ×	3 (1)
		2.4 ×	5 (1)
	cucumber	1.2 ×	4 (1)
		2.4 ×	5 (1)
	lettuce	1.2 or 2.4 ×	5 (1 each)
Apple aphid, *Aphis pomi*	apple	0.5 ×	3 (2)
Cabbage aphid, *Brevicoryne brassicae*	cabbage	1.3 ×	5 (2)
Green peach aphid, *Myzus persicae*	apple	1 ×	5 (1)
Pea aphid, *Acyrthosiphon pisum*	alfalfa	0.7 ×	3 (1)
		1 or 2x	5 (1 each)
Potato aphid, *Macrosiphum euphorbiae*	potato	1 ×	4 (7)
Sunfloweer aphid, *Aphis helianthi*	celery	2 ×	4 (1)
Grape leafhopper, *Erythroneura elegantula*	grape	0.3 or 1 ×	5 (1 each)
Potato leafhopper, *Empoasca fabae*	potato	1 ×	5 (17)
	alfalfa	1 or 2 ×	4 (3 and 4)
Variegated leafhopper, *Erythroneura variabilis*	grape	0.5, 1 or 2 ×	4 (1 each)

Table 15-3. *continued*

Species	Crop	Dose relative to 0.025 lb/A	Relative activity[a] (*n*)[b]
	COLEOPTERA		
Blueberry flea beetle, *Altica sylvia*	blueberry	1 ×	5 (2)
Colorado potato beetle, *Leptinotarsa decemlineata*	potato	1 ×	5 (1)
Potato flea beetle, *Epitrix cucumeris*	potato	1 ×	5 (11)
Alfalfa weevil, *Hypera postica*	alfalfa	0.67 ×	2 (1)
		1 ×	3.(1)
Cowpea cucurlio, *Chalcodermus aeneus*	southern peas	1 or 2 ×	2 (1 each)
	THYSANOPTERA		
Blueberry thrips, *Frankliniella vaccinii*	blueberry	1 or 2 ×	5 (1 each)
Greenhouse thrips, *Heliothrips haemorrhoidalis*	avocado	1 ×	5 (1)
Onion thrips, *Thrips tabaci*	onion	2 ×	4 (1)
Privet thrips, *Dendrothrips ornatus*	privet	2.7 × [c]	5 (1)
	HEMIPTERA		
Minute pirate bug, *Orius tristicolor*	sorghum	2 ×	2 (2)
Tarnished plant bug, *Lygus lineolaris*	potato	1 ×	3 (17)
	DIPTERA		
Vegetable leafminer, *Liriomyza sativae*	tomato	1 ×	1 (1)
	celery	1 ×	1 (1)
Birch leafminer, *Fenusa pusilla*	birch	0.5 ×	1 (1)
		1 ×	4 (1)
		2 ×	5 (1)
Blueberry maggot, *Rhagoletis mendax*	blueberry	1 or 1.4 ×	2 (2 and 1)
	HYMENOPTERA		
Blueberry sawfly, *Neopareophora litura*	blueberry	1 ×	5 (2)
	ACARINA		
Citrus flat mite, *Brevipalpus lewisi*	lemon	0.3 or 1 ×	4 (1 each)
Spider mites, *Tetranychus* sp.	canteloupe	0.33 or 0.67 ×	2 (1 each)

[a]Relative activity based on a 1–5 scale where 1 = 0–20% control; 2 = 21–40% control; 3 = 41–60% control; 4 = 61–80% control; and 5 = 81–100% control. % control calculated by the following formula: [number of insects_{UNT} − number of insects_{TRT}/number of insects_{UNT}] × 100.
[b]Number of trials.
[c]lb pyrethrins/100 gallons; applied to runoff.

storage, packaging, preparing, edible-waste, and enclosed food processing equipment) must have established food additive tolerances and be applied in a manner that ensures the tolerance will not be exceeded. The use of insecticides in meat and poultry plants is more complicated than in other types of food processing plants because some insecticides that are approved by the EPA for use in general food handling establishments may not be used in facilities operating under the USDA meat and poultry inspection program. This is a case where the insecticide label does not reflect the only applicable law and the current "List of Chemical Compounds Authorized for Use under USDA Inspection and Grading Programs" must be consulted to determine whether a particular product may be applied in these facilities. Insecticides are categorized by the EPA as either residual or nonresidual products based on their relative persistence after application. The EPA also recognizes five types of indoor applications: general, spot, and crack-and-crevice applications of residual insecticides as well as space and contact applications of nonresidual insecticides. Although pyrethrum is classified as a nonresidual insecticide, it may also be applied in general, spot, and crack-and-crevice applications. Thus, pyrethrum is an ideal insecticide for use in food processing plants because it is approved for a variety of applications to control a broad spectrum of pests within all areas of both FDA and USDA inspected facilities.

The primary use of pyrethrum in food processing plants is as a space spray. These applications are an integral part of insect control programs in these facilities (Corrigan and Klotz, 1992) and they may only be made when the plant is not in operation. This is usually at night and corresponds with the time when insects are generally most active and exposed to contact by the insecticide. Space spraying is an important method of insecticide application within food plants because the time when the plant is not in operation must be kept to a minimum and the internal structure of these facilities is diverse and contains many microhabitats that are suitable for incipient infestations of both flying and crawling insects. Space sprays allow a large volume of space (and surface area) to be treated in a relatively short time and also deliver insecticide to many locations that would not be treatable using other application methods. Most of the pyrethrum space sprays that are applied in food processing plants are oil-based products applied in ULV applications to deliver 0.03 to 0.3 g of pyrethrins (depending on the targeted pest) per 1,000 cubic feet of space (Bennett *et al.* 1988; Gillenwater and Burden, 1973; Silcox, unpublished). Treatments usually occur at monthly intervals, but more frequent applications are made during the summer months when insects are more active (Corrigan and Klotz, 1992).

C. Advanced ULV Space-Spray Technology

At the time of the Gillenwater and Burden review (1973) of pyrethrum for stored products insect control, ULV insecticide application was an emerging technology (Anonymous, 1972) that subsequently became the standard space-spray application technique within food processing plants. Stationary or mobile ULV applicators generally replaced thermal foggers and mist applicators and were used to apply oil-based products, either undiluted or diluted with oil, at the rate of 0.5 to 2 fluid ounces per 1,000 cubic feet to deliver 0.03 to 0.3 g of pyrethrins within this volume. Although this technique was superior to previous methods of application and became widely accepted within the industry, it had inherent problems that included oil deposition, inaccurate dosing, and nonuniform distribution of spray droplets.

There is a new space-spray technology that utilizes liquid carbon dioxide both as a solvent and propellent to greatly reduce the volume of oil applied while maintaining or improving the traditional performance of pyrethrum products. The key technical factors that were considered as the product was developed included droplet size, spray distribution, and biological efficacy while the assessment of commercial viability addressed environmental, liability, and regulatory concerns. The concept of utilizing liquid carbon dioxide both as a solvent and propellent was originally investigated in Australia and several formulations have been commercialized in that country since 1976 (Slatter *et al.*, 1981). In the United States, a similar product (Turbocide® Pest Control System, Fairfield American Corp., subsidiary of Roussel-Uclaf) was registered with the EPA and has been marketed since 1985. The current Turbocide formulation contains 0.5% pyrethrins, 4.0% piperonyl butoxide, 7.9% aliphatic hydrocarbon solvent, and 87.6% liquid carbon dioxide and is marketed in 70 lb net-weight cylinders. A relatively "dry" fog is generated at the time of application because the liquid carbon dioxide becomes gaseous as soon as it leaves the nozzle and the active ingredients are delivered in approximately one-eighth of the volume of oil that is applied in conventional ULV applications. The system employs a grid of nozzles that are connected to the insecticide cylinder by various sized aluminum tubing. Precise dosages are applied because the flow of pressurized formulation through the tubing is balanced to ensure uniform discharge from each nozzle and an exact volume of material is applied to the treatment area. Depending on the volume of the facility being treated, the desired application rate is achieved by either totally discharging one or multiple 70 lb net-weight cylinders or, if relatively small spaces are being treated or the targeted pest is particularly susceptible to pyrethrins (i.e., house flies, fruit flies, Indianmeal moths, etc.), incorporating a reduced-size dosing cylinder into the system. This cylinder is generally located near the 70 lb cylinder and is connected to the system using a three-way valve that allows it to be charged from a 70 lb cylinder and, by turning the valve, discharged to the nozzles.

Technical development of the Turbocide system addressed the physical parameters of space-spray applications and documented the biological performance of the system towards a variety of insects (Groome *et al.*, 1987). The system was evaluated in a 240,000 cubic foot warehouse (120 × 100 × 20 feet) that was equipped with a single four-nozzle block (Fig. 15-1). Five sampling sites were established relative to the nozzle block to assess the uniformity of application within one quadrant of the building (Fig. 15-2). One site was located directly under the nozzle block while the other four sites were located at two distances along two angles from one of the nozzles. The distances were 30 and 60 feet from the nozzle block and the angles were both in-line and 45° out-of-line with one of the nozzles. Physical measurements were made at each site at various times after application. Droplet size was determined using a 10-stage, quartz-crystal microbalance cascade impactor (Berkeley Controls Model C1000A). Spray cloud distribution and duration were determined by measuring the aerial concentration of the nonliquid carbon dioxide components of the formulation at 1 m above floor level using a respirable mass monitor (Thermal Systems Inc.).

Figure 15-1. Turbocide four-nozzle block.

Droplet size is one of the most important factors influencing the performance of aerosol insecticide application. Droplets must be small enough to ensure maximum dispersion and thorough coverage of the treated area, but must remain large enough to readily impinge upon the target insect. A number of studies have addressed the issue of optimum droplet size for ULV insecticide application and it is generally agreed that maximum efficiency is achieved within the range of 2–16 microns for the mass mean diameter (Lofgren *et al.*, 1973; Corrigan and Klotz, 1992). Droplets generated by the Turbocide system remain within this size range for more than 90 minutes and the 1 micron droplets remain airborne for 3–6 hours after application (Table 15-4). Droplets of this

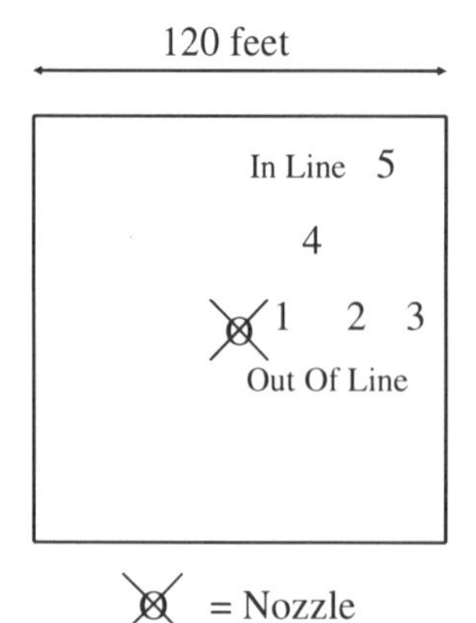

Figure 15-2. Schematic of Turbocide test site showing sampling sites relative to nozzle-block location and orientation.

Table 15-4. Droplet Size Through Time After Application of 12.1 g of Turbocide Formulation per 1,000 Cubic Feet (4.3 mg Pyrethrins per Cubic Meter)

Time after application (min)	Mass mean diameter[a] (microns)
18	8.1
23	6.1
63	3.7
96	2.8
185	1.2
237	1.1
294	1.0
347	0.9

[a]Mean droplet size from five sampling sites and three applications.

size would be expected to readily disperse and provide uniform distribution throughout the treated area and this was documented by aerial concentration measurements (Table 15-5). Although the aerial concentration was consistently greatest at the test site that was located 30 feet from and directly in-line witha nozzle, the relative concentration between sites was remarkably uniform, especially for samples taken 60 or more minutes after application. The maximum aerial concentration (assuming absolutely uniform dispersion) for the application rate employed in these treatments was 86 mg/m^3 and it is interesting to note that 19% of this material remained airborne at 60 minutes after application.

Biological assays were conducted using a series of application rates and by exposing three replicates of 10 to 20 insects (per species) at each of the five sites. Groups of insects were exposed for 2, 4, and 6 hours and were then transferred to clean holding cages, provided with food and water (where

Table 15-5. Aerial Concentration Through Time After Application of 12.1 g of Turbocide Formulation per 1,000 Cubic Feet (4.3 mg Pyrethrins per Cubic Meter)

	Aerial concentration (mg/m^3)[a]					
	Sampling position[b]					
Time after application (min)	1	2	3	4	5	Ave.
30	33	36	35	42	30	35.1
60	16	14	15	18	17	16.2
140	4	3	4	5	4.8	4.5
240	2	1.5	1.5	2.2	2	1.2
360	0.8	1	0.5	0.5	0.5	0.5

[a]Mean aerial concentration from three applications.
[b]See Fig. 2.

appropriate) and transported to the laboratory where they were maintained at $80 \pm 2°F$. Mortality was recorded at 48 or 72 hours after treatment. The biological performance of the system was determined with nine insect species as summarized in Table 15-6. The means presented are pooled for each of the test sites because an analysis of the factorial arrangement of treatment variables showed that sample location did not significantly ($P = 0.05$) influence the results (which, again, demonstrates the uniform distribution of droplets from the Turbocide system). With the exception of the *Tribolium* species, high levels of mortality were achieved, for at least one application rate, for each insect species. It should be noted, however, that both *Tribolium* species were very quickly "knocked down" when exposed to even the lowest application rate for the shortest exposure period. Recovery was not complete until 72 hours after treatment and this raises the invariable question of the relevance of laboratory or, as in the case of these tests, semilaboratory tests to performance under actual use conditions. Despite the low efficacy towards *Tribolium* in these tests, the Turbocide system is an important component of management programs for these species in practical use situations within operating food plants and warehouses (T. Osterberg and D. Pirrung, personal communications).

The commercial viability of all insecticide products is impacted by both their inherent technical capabilities and the external regulatory constraints that ultimately establish the limits of their use. An essential element of new product development is the continued and unbiased assessment of the potential to market an improved technology in light of current or anticipated regulations. The commercial development of the Turbocide system was unique in that the dramatically new technology for pyrethrum-based space sprays not only enhanced the physical and biological performance of these applications, but also made the technology more commercially acceptable from regulatory, liability, and environmental perspectives. In comparison to traditional oil-based ULV space sprays of pyrethrum, the time required for application is greatly reduced with the Turbocide system because it is activated by merely opening a valve. Dermal and inhalation exposure by applicators is totally eliminated because there is no diluting of product or filling of application equipment and the applicator is not in the area being treated at the time of application. In addition, the potential fire hazard from the application of an oil-based product is substantially reduced in direct proportion to the eight-fold reduction in the volume of oil applied through the Turbocide system as well as the fire retardant properties of carbon dioxide.

A number of federal, state, and local agencies are charged with the responsibility of regulating the transportation, hazard assessment, and application of insecticides as well as the disposal of empty insecticide containers. The Turbocide formulation is classified by the US Department of Transportation as a nonflammable gas and this favorably impacts both the cost of interstate transportation and the insurance premiums of customers who store insecticide products within their facility. There has been a distinct trend within the pesticide industry towards reduced application of volatile organic compounds (VOCs) and increased use of returnable containers. Inherent to the Turbocide system

Table 15-6. Efficacy of Turbocide Formulation with Pyrenone Towards Various Insects in Tests Conducted in a 240,000 Cubic Foot Warehouse

Pyrethrins applied (mg/m^3)	Exposure period (hrs)	Species[a] and % Mortality[b]								
		HF	FF	IMM	GCR	STGB	CGB	CFB	RFB	WHB
0.9	2	60	38	100	19	10	22	3	0	46
	4	65	30	100	25	8	18	1	0	49
	6	69	45	100	64	15	28	6	0	60
1.5	2	99	83	100	54	28	26	13	5	75
	4	99	92	100	66	34	35	11	5	96
	6	99	97	100	73	32	39	10	4	96
3.1	2	100	95	100	79	59	65	49	13	95
	4	100	94	100	86	65	74	45	14	96
	6	100	100	100	94	75	88	45	20	94
4.3	2	100	99	100	90	91	70	48	17	92
	4	100	100	100	89	90	98	47	31	96
	6	100	100	100	97	94	84	50	34	96

[a]HF = house fly (*Musca domestica*); FF = fruit fly (*Drosophila melanogaster*); IMM = Indianmeal moth (*Plodia interpunctella*); GCR = German cockroach (*Blatella germanica*); STGB = sawtoothed grain beetle (*Oryzaephilus surinamensis*); CGB = cigarette beetle (*Lasioderma serricorne*); CFB = confused flour beetle (*Tribolium confusum*); RFB = red flour beetle (*Tribolium castaneum*); WHB = warehouse beetle (*Trogoderma variabile*).
[b]Mortality at 72 hours after treatment (except HF, FF and IMM = 24 hours after treatment).

is a nearly eight-fold reduction in VOCs as well as a mandatory return policy for empty cylinders. The system also eliminates localized environmental contamination and potential pesticide enforcement actions that may result from the misapplication of an insecticide because it is designed to uniformly deliver an exact insecticide dose within an area of known dimension.

IV. CONCLUSION

In response to heightened environmental awareness, the commercial viability of insecticide products in the future will depend as much on their environmental compatibility as on their ability to control pest populations. Pyrethrum was an important agricultural insecticide until the introduction of the chlorinated hydrocarbons nearly 50 years ago. A succession of insecticide classes was subsequently developed and has provided the core products for agricultural insect control since that time. Although these materials will remain as important tools for agricultural insect control, the industry is undeniably moving away from persistent and highly toxic insecticides and towards materials that offer a balance of selective performance and reduced environmental hazard. During the past decade, pyrethrum has been reintroduced to American agriculture as a material for use on minor crops, as a preharvest spray, for bloom-period applications and in areas that are sensitive to potential exposure to insecticides. Its agricultural importance will, undoubtedly, increase during the coming years as tolerances are established to maintain its use on growing crops and as more and more of the currently existing uses of other insecticides are lost during reregistration of these products with the EPA. Pyrethrum will also see increased use in integrated control programs where it is shown to be nondisruptive to beneficial arthropod populations. A potential use that is currently being investigated is the combination of pyrethrum with *Bacillus thuringiensis* or other similar active ingredients that have complementary labeling. The combined application of biorational insecticides may be the wave of the future due to the relatively limited spectrum of activity of these products. Finally, the federal LISA (Low Input Sustainable Agriculture) program will inevitably identify a number of uses for pyrethrum in agricultural insect control.

In contrast with its agricultural use, pyrethrum has never been displaced from its position as one of the primary insecticides for stored products insect control. Its physical and toxicological properties allow it to be applied by a wide variety of methods to a great number of sites for the control of a broad spectrum of stored products insects. The prominent role that pyrethrum has played in space-spray applications will expand further with the greatly reduced, and possibly eliminated, use of dichlorvos space sprays and as further improvements in application technology are introduced to the food processing industry. A new Turbocide system is currently under development and will further enhance the physical and biological performance while increasing its commercial viability in light of anticipated environmental legislation. The recent introduction of Actisol® application equipment (Roussel-Uclaf) brings a new concept of pyrethrum application to stored products insect control. This system provides a directed, ULV, crack-and-crevice application of pyrethrum into harborages that were previously inaccessible to insecticide treatment. As these systems foretell, future stored products insect control programs will rely as much on new application technologies as on the introduction of novel chemistries and pyrethrum will undoubtedly remain as a material of choice for the control of these insects.

REFERENCES

Anonymous (1972). ULV insecticide application indoors. Technical Release No. 10-72, National Pest Control Association, pp. 1–6.

Anonymous (1983). EPA crop grouping scheme under 40 CFR 180.34(F). *Federal Register* **20**(20), 290.

Anonymous (1989). Federal Insecticide, Fungicide, and Rodenticide Act amendments of 1988; schedule of implementation. *Federal Register* **54**(79), 18076–18086.

Anonymous (1990). State extension entomology specialists. US Government Printing Office. pp. 1–48.

Atkins, E.L. (1975). Injury to honey bees by poisoning. *In* "The Hive and the Honey Bee" (Dadant and Sons, eds.), pp. 663–696. Dadant and Sons, Hamilton, Illinois.

Barthel, W.F. (1973). Toxicity of pyrethrum and its constituents to mammals. *In* "Pyrethrum, The Natural Insecticide" (J.E. Casida, ed.), pp. 123–142. Academic Press, New York.

Bennett, G.W., Owens, J.M., and Corrigan, R.M. (1988). "Truman's Scientific Guide to Pest Control Operations." Edgell Communications, Duluth, Minnesota.

Bullivant, M.J., and Pattenden, G. (1976). Photodecomposition of natural pyrethrins and related compounds. *Pestic. Sci.* **7**, 231–235.

Camougis, G. (1973). Mode of action of pyrethrum in arthropod nerves. *In* "Pyrethrum, The Natural Insecticide" (J.E. Casida, ed.), pp. 211–222. Academic Press, New York.

Casida, J.E. (ed.) (1973). "Pyrethrum, The Natural Insecticide." Academic Press, New York.

Corrigan, R.M., and Klotz, J.H. (1992). "Food Plant Pest Management." Purdue University, West Lafayette, Indiana.

Elliott, M., and Janes, N.F. (1973). Chemistry of natural pyrethrins. *In* "Pyrethrum, The Natural Insecticide" (J.E. Casida, ed.), pp. 56–100. Academic Press, New York.

Forsythe, H.Y., and Gardner, W.E. (1986). Insect and mite control experiments on apples in Maine. Unpublished report.

Gillenwater, H.B., and Burden, G.S. (1973). Pyrethrum for control of household and stored products insects. *In* "Pyrethrum, The Natural Insecticide" (J.E. Casida, ed.), pp. 243–259. Academic Press, New York.

Groome, J.M., Slatter, R., Bassler, M.A., Silcox, C.A., and Swanson, D.E. (1987). A biological and physical evaluation of Fairfield American Corporation's Turbocide fixed installation spray system. Wellcome Research Laboratories. Unpublished report.

Lange, H., and Akesson, N.B. (1973). Pyrethrum for control of agricultural insects. *In* "Pyrethrum, The Natural Insecticide" (J.E. Casida, ed.), pp. 261–279. Academic Press, New York.

Lofgren, C.S., Anthony, D.W., and Mount, G.A. (1973). Size of aerosol droplets impinging on mosquitoes as determined with a scanning electron microscope. *J. Econ. Entomol.* **66**, 1085–1088.

Shepard, H.H. (1951). "The Chemistry and Action of Insecticides." McGraw-Hill, New York.

Slatter, R., Stewart, D.C., Martin, R., and White, A.W.A. (1981). An evaluation of Pestigas BB — a new system for applying synthetic pyrethroids as space sprays using pressurized carbon dioxide. *International Pest Control* **23**, 162–164.

Tattersfield, F. (1932). The loss of toxicity of pyrethrum dusts on exposure to light and air. *J. Agric. Sci.* **22**, 396–417.

Wilkinson, J.D., Biever, K.D., and Ignoffo, C.M. (1975). Contact toxicity of some chemical and biological pesticides to several insect parasitoids and predators. *Entomophaga* **20**, 113–120.

16

Pyrethrum for Control of Pests of Medical and Veterinary Importance

EUGENE J. GERBERG

I. HISTORY

Pyrethrum was first used at the beginning of the 19th Century, when finely ground flowers known as "Persian Insect Powder" were marketed for control of vermin. The first commercial production was in Dalmatia (Yugoslavia). Pyrethrum powder was used for the control of bedbugs as early as 1875, to control mosquitoes in the 1880s, to protect horses from "buffalo gnats" in the 1890s, and was burned to repel mosquitoes in the early 1900s. Pyrethrum was used with kerosene in household fly sprays about 1918 and as a mist of 0.1% pyrethrins in kerosene for spray-killing adult mosquitoes in houses and buildings in villages in malarious areas during World War II. Pyrethrum sprays were also used to control bedbugs in military barracks. A pyrethrum "vanishing" cream was used as a mosquito repellent, but was not very long lasting, and was therefore replaced by other repellents.

The literature on pyrethrum for the control of insects affecting man and animals was last reviewed by Smith (1973) and so the present emphasis is on work conducted after 1970.

II. PYRETHRUM FOR CONTROL OF PESTS OF MEDICAL IMPORTANCE

A. House Flies and Cluster Flies

Water-based pressurized sprays containing 0.25% pyrethrins, 0.80% piperonyl butoxide, 0.40% *N*-octyl bicycloheptenedicarboximide, and 8.05% petroleum distillate are used against flies and other flying insects. Synergized pyrethrum is also effective for the control of cluster flies (Chadwick, 1972).

B. Mosquitoes

The use of pyrethrum as a larvicide originated with application of powdered pyrethrum flowers to mosquito breeding sites in 1918. Pyrethrum as a 0.006%

emulsion was used as a larvicide at 55 gal/acre. The New Jersey mosquito larvicide, developed in 1941, consisted of 0.07% pyrethrins, 0.5% sodium lauryl sulphate and 33.5% water; 3–4 gal of the pyrethrum oil emulsion replaced the 35–50 gal/acre treatment. Darwazeh and Mulla (1981) reported excellent control of mosquito larvae in dairy waste lagoons using pyrethrum "Tossits," at the rate of one Tossit/3 m^2.

Pyrethrum was used for adult mosquito control in 1911 with a spray consisting of a tincture of pyrethrum flowers, potash soap, and glycerine diluted with water 20 times by volume. A mixture of pyrethrum products in kerosene was used to kill malaria and yellow fever mosquitoes. In Africa, houses and huts were sprayed with pyrethrins to control *Anopheles* mosquitoes. Kerosene-based fly and mosquito sprays now contain 0.025–0.075% pyrethrins and 0.2–0.6% piperonyl butoxide. There are a number of synergized pyrethrins fogging concentrates available. A ready-to-use ultralow volume (ULV) mosquito adulticiding formulation may contain pyrethrins, piperonyl butoxide, petroleum distillate, and mineral oil. Hobbs (1976) used ULV spraying of a pyrethrins formulation for the control of *An. albimanus*. Malaria surveillance indicated that the ULV spraying had a marked impact on the transmission of the disease during the main season. ULV formulations containing 0.0013–0.008% pyrethrins and 1.5% piperonyl butoxide have been used for the control of *Aedes nigromaculis*, *Ae. taeniorhynchus*, *Ae. vexans*, *Culex pipiens*, *Mansonia perturbans*, *Psorophora confinnis*, and *An. quadrimaculatus*. Schuyler and Massing (1974) using a Micro-Gen, applying 2.6–2.8 gal/ha of a pyrethrins formulation, obtained 100% kill in 30 minutes at 30–60 m from the vehicle line of travel. Aerial application of 1% synergized pyrethrins in diesoline at 10 fl. oz/acre gave satisfactory control of *Ae. taeniorhynchus* on Grand Cayman Island (Lee, 1975). Water-based pressurized sprays containing pyrethrins, piperonyl butoxide, *N*-octyl bicycloheptenedicarboximide, and petroleum distillate have also been used against mosquitoes.

Aircraft disinsectization required an aqueous-based spray of pyrethrum. In 1938 a carbon tetrachloride pyrethrum extract in a water base, dispersed by means of carbon dioxide, was used in aircraft. After Goodhue and Sullivan (1943) developed the aerosol "bomb," aerosols containing pyrethrins and later pyrethroids were widely used for disinsectization of airplanes. Pressurized aerosols now usually contain 0.2–0.4% pyrethrins and 1.6–3.0% piperonyl butoxide.

Mosquito coils, made from 1.3% pyrethrum powder, pyrethrum "marc" (residue left over after pyrethrum flowers have been extracted with a solvent), filler, binder, dye, and fungistatic agent, are widely used in tropical areas. The slow-burning coils are ignited and the smoke acts both as a killing agent and repellent. Pyrethrins irritate and then activate mosquitoes to fly from the source of the stimuli. The irritation dose is estimated at 1.2×10^{-7} mg while the activation dose lies between 1.2 and 6.0×10^{-7} mg (Anonymous, 1989). At the higher dosage, the insect may be knocked down and death may occur. Hudson (1974) conducted field tests in Canada of mosquito coils containing 0.3% pyrethrins, resulting in a 93% reduction of *Aedes* spp. To avoid the smoke emission, which may be uncomfortable to some people, mats and chips impregnated with pyrethrins have been developed. Electric vaporizers deliver the active ingredient, e.g., at 40 to 60 mg pyrethrins per mat knockdown was 70–76%. A mosquito mat using a catalytic evaporator and lighter fluid has been used in the field.

C. Other Biting Flies

Tsetse flies. Lee *et al.* (1968) claimed a 95% reduction of *Glossina pallipides* in Tanzania by aerial application of 0.04% pyrethrins and 2.0% piperonyl butoxide at a rate of 0.016 gal/acre at 3 week intervals. Tarimo *et al.* (1971a,b) obtained 85% control of *Glossina swynnertoni* and 95% control of *G. pallipides* in tests conducted 2 years later using 0.5% pyrethrins, 2.0% piperonyl butoxide, and 5% DDT. Kuria and Bwogo (1986) obtained 60–72% reduction of *G. pallidipes* using an aerial ULV spray of 0.65% pyrethrins. In another test using ground equipment and two applications of 0.25% pyrethrins and 1.0% piperonyl butoxide, they obtained 98% control of *G. swynnertoni.*

Ceratopogonid flies. Woodward *et al.* (1985) used a pyrethrins larvicide to control *Culicoides variipennis*, a vector of bluetongue viral disease in sheep, in an alkaline lake in California. The peripheral area of the breeding site was treated at a rate of 701 g pyrethrins/ha, resulting in 99.3% reduction in the density of *C. variipennis* larvae along the shoreline.

D. Fleas

An aerosol formulation containing 0.056% pyrethrins and 0.05% permethrin was effective for flea and tick control on cats. Pyrethrins dips and total release aerosol sprays have also been used for flea control.

E. Bedbugs

Water-based pressurized sprays containing pyrethrins, piperonyl butoxide, *N*-octyl bicycloheptenedicarboximide, and petroleum distillate were effective for 9 days against bedbugs (Twinn, 1945). In Kenya, bed frames were sprayed with 0.1% oil solutions of pyrethrins, which provided fairly good protection for 1 month.

F. Lice

A body louse powder called MYL containing pyrethrins, phenol-S as an antioxidant, *N*-isobutylundecylenamide as a synergist, and 2,4-dinitroanisole as an ovicide was used during World War II, but it was later replaced by a DDT powder. For head lice, Twinn and MacNay (1943) suggested oils containing pyrethrins. Pyrethrins in shampoo containing 0.03% pyrethrins and 3.0% piperonyl butoxide are still widely used (Robinson and Shepherd, 1980). Other formulations containing 0.15% pyrethrins and 1.5% piperonyl butoxide gave 100% control of head lice after 2 weeks (Svanasbakken *et al.*, 1985).

G. Mites

Ointments containing pyrethrins are used for control of *Sarcoptes scabiei*.

II. PYRETHRUM FOR CONTROL OF PESTS OF VETERINARY IMPORTANCE

A. Flies

Flies that attack cattle include the stable fly *Stomoxys calcitrans*, the horn fly *Haematobia irritans*, the screw-worm fly *Cochliomyia macellaria*, the black blow fly *Phormia regina*, the green bottle fly *Lucilia sericata*, the heel fly or warble fly *Hypoderma lineatum*, horse flies *H. bovis*, and deer flies (Tabanidae). Cattle sheds, barns, stables, and dairies are often sprayed or misted with pyrethrins formulations synergized with piperonyl butoxide. Animals may be treated with pyrethrins plus synergist sprays. The treatment may be used to destroy or repel flies. Wettable powder formulations have been used for tabanid control on cattle. Control of *S. calcitrans* on a dairy farm in Germany was obtained using a pyrethrins aerosol at a rate of 0.1 ml/m^3. Butterfat production by dairy cattle was increased by applying an emulsion containing 0.1% pyrethrins and 1.0% piperonyl butoxide at a rate of 10–15 oz/animal/week. Oil sprays containing 25 mg pyrethrins, 200 mg piperonyl butoxide, and 200 ml odorless oil have been used with hand operated equipment. Blume *et al.* (1973) reported 7 days of protection from stable flies and 10 days from horn flies, using 58.8 ml of a 25% EC formulation of pyrethrins and piperonyl butoxide (0.1:1.0) applied as a fine mist. Axtell and Dukes (1974) used ULV nonthermal aerosols containing 5% pyrethrins and 15% piperonyl butoxide at the rate of 3.5–7 fl. oz/minute against tabanids. At 100 feet from the line of travel the tabanid mortality was 27–78%. Dukes and Axtell (1975) evaluated a synergized pyrethrins against *Stomoxys* and *Tabanus* applied as ULV aerosols.

A refillable ear tag (Morgan tag) charged with 8% pyrethrins was tested on cattle exposed to horn flies (Hogsette *et al.*, 1991). The pyrethrins significantly reduced horn flies 4 days post treatment and there was still significant reduction after 7 days but not sufficient for commercial acceptance. A permethrin formulation significantly reduced horn fly populations for a total of 12 weeks.

Flies in poultry houses were controlled by ULV fogging of a concentrate containing 5% pyrethrins, 25% piperonyl butoxide, and 70% petroleum distillate. Other formulations may contain an additional synergist, MGK 264. An automatic piped aerosol system that dispersed synergized pyrethrins throughout a poultry breeding facility provided excellent control of house flies.

B. Fleas

Pressurized flea and tick sprays containing 0.056% pyrethrins and 0.050% permethrin have been very effective against fleas and ticks on dogs and cats. Area treatment, using a 0.22% microencapsulated pyrethrins formulation in a 1 gal B & G sprayer, resulted in a 52% reduction of adult fleas in 24 hours and 95.6% reduction after 30 days. MacDonald and Miller (1986) found that pyrethrins (0.15%), piperonyl butoxide (1.5%), another synergist, and a repellent in isopropyl alcohol wetted the coats of the dogs and penetrated better. They concluded that there was some residual activity, but that frequent applications were necessary to control fleas and ticks.

C. Ticks

An early control of lice and ticks on cattle and sheep consisted of a mixture of pyrethum and flour. Pyrethrins and piperonyl butoxide (1:10) was the most effective of 26 widely-used insecticides against the brown dog tick (*Rhipicephalus sanguineus*) in laboratory tests (Gladney *et al.*, 1972). In Kenya, 0.008% pyrethrins in a water-based emulsion gave good control of the blue tick *Boophilus decoloratus*. In another test an aqueous (EC) formulation of 0.05% pyrethrins and 0.15% piperonyl butoxide sprayed on animals at 5 l/animal gave residual control of *B. decoloratus* and the red tick, *Rhipicephalus appendiculatus*, the vector of East Coast fever. A "tick grease" consisting of 0.5% pyrethrins with pyrethrum vegetable waxes and petroleum jelly has also been used. Hoffman (1986) sprayed and fogged a mixture of pyrethrins and 2% permethrin as a new method to eliminate the brown dog tick and avoid air contamination. Effective control was obtained for the fowl tick *Argus persicus* using a pyrethrum solution, applied by means of a knapsack sprayer, on walls of the mud and wattle houses in Kenya.

Repellency has been reported against Ixodid and Argasid ticks. Bar-Zeev and Gothilf (1973) using *Ornithodorus tholozani* evaluated 539 chemicals for repellent effect. Pyrethrins was outstanding and far superior to all the other chemicals tested in protection from tick bites. There was also a delayed toxic effect on the ticks.

D. Mites

The chicken mite, *Dermanyssus gallinae*, can be controlled in poultry houses by two treatments 7 days apart with 0.03% pyrethrins. Synergized pyrethrins controlled an infection of *Mycoptes musculinus* and *Myobia musculi* in mice (Constantin, 1972). Lowenstine *et al.* (1979) treated a cat for trombiculosis by topical applications of carbaryl and pyrethrins. In 1972, the Food and Drug Administration registered a pharmaceutical preparation containing 0.05% pyrethrins, 0.5% piperonyl butoxide, and 25% squalene for the treatment of ear mites in dogs and cats.

IV. CONCLUSION

There are many established uses of pyrethrins for the control of medical and veterinary pests, but there is a paucity of new information. Hopefully there will be an upsurge in research on this useful insecticide.

REFERENCES

* Note: References designated by asterisks are not cited in the text but are included here for easy access to relevant literature.

Anonymous (1989). News and features. *Pyrethrum Post* **17**, 78–79.

Arther, R.G., and Young, N. (1985). Efficacy of a rotenone/pyrethrin dip for control of fleas and ticks. *Vet. Med.* **80**, 53,56–57.*

Axtell, R.C., and Dukes, J.C. (1974). ULV chemical control of mosquitoes, culicoides and tabanids in coastal North Carolina. *Proc. Pap. 42nd Annu. Conf. Calif. Mosq. Control Assoc.*, 99–101.

Baker, G.J. (1973). Mosquito adulticiding with pyrethrum. *Pyrethrum Post* **12**, 12–13.*

Bar-Zeev, M., and Gothilf, S. (1973). Laboratory evaluation of tick repellents. *J. Med. Entomol.* **10**, 71–74.

Bennett, G.W., and Lund, R.D. (1977). Evaluation of encapsulated pyrethrins (Sectrol) for German cockroach and cat flea control. *Pest Contr.* **45**, 44,46,48–50.*

Betke, P., and Schultka, H. (1980). Studies into behavior of house fly (*Musca domestica* L.) under conditions of swine fattening unit and attempts for control by means of pyrethrin aerosol (in German). *Monatsh. Veterinarmed.* (*Jena*) **35**, 850–852.*

Betke, P., Schultka, H., and Ribbeck, K. (1986). *Stomoxys calcitrans* as a pest on a dairy farm (in German). *Angew. Parasitol.* **27**, 39–44.*

Bills, G.T. (1980). Effective housefly control in British piggeries. *International Pest Control* **22**, 84–85, 88.*

Billstein, S., and Laone, P. (1979). Demographic study of head lice infestations in Sacramento County school children. *Int. J. Dermatol.* **18**, 301–304.*

Blume, R.R., Matter, J.J., and Eschle, J.L. (1973). Biting flies (Diptera: Muscidae) on horses: laboratory evaluation of five insecticides for control. *J. Med. Entomol.* **10**, 596–598.

Chadwick, P.R. (1972). The problem of cluster flies. *Environ. Health* (*London*) **80**, 9–11.

Constantin, M.L. (1972). Effects of insecticides on acariasis in mice. *Lab. Anim.* **6**, 279–286.

Curtis, C.F. and Hill, N. (1988). Comparison of methods of repelling mosquitoes. *Entomol. Exp. Appl.* **49**, 175–179.*

Darwazeh, H.A., and Mulla, M.S. (1981). Pyrethrin tossits against mosquito larvae and their effects on mosquito fish and selected nontarget organisms. *Mosq. News* **41**, 650–655.

Dukes, J.C., and Axtell, R.C. (1975). Chemical control of coastal biting flies and gnats. *Proc. N. J. Mosq. Control Assoc.*, 323–333.

Fales, J.H., Bodenstein, O.F., and Bowers, W.S. (1970). Seven juvenile hormone analogues as synergists for pyrethrins against house flies. *J. Econ. Entomol.* **63**, 1379–1380.*

Fales, J.H., Bodenstein, O.F., Waters, R.M., and Fields, E.S. (1970). Development of synergized pyrethrin aerosols for use against public health insects on aircraft and ships. *Aerosol Age*, May, 41–60.*

Fox, J.G. (1982). Outbreak of tropical rat mite dermatitis in laboratory personnel. *Arch. Dermat.* **118**, 676–678.*

Fridinger, T.L. (1984). Designing the ultimate weapon against fleas. *Vet. Med. Small Anim. Clin.* **79**, 1151–1155.*

Gaaboub, I.A., and Hammad, S.M. (1970). Susceptibility levels of the Egyptian body louse, *Pediculus humanis corporis* De Geer, in Alexandria city (U.A.R.) to DDT, gamma BHC and pyrethrins (Anoplura: Pediculidae). *Entomol. Soc. Egypt Bull. Econ. Ser.* **4**, 189–195.*

Gallegos, G.P., Lagunes-Tejeda, A., and Bravo-Mojica, H. (1984). Analisis de accion conjunta de insecticidas en el mosquito *Culex quinquefasciatus* Say (Diptera: Culicidae). *Agrociencia* **57**, 37–48.*

Galun, R. (1975). Protection of livestock from tsetse (*Glossina morsitans*) bites by means of repellents. *Pyrethrum Post* **13**, 2–4.*

Gladney, W.J., Dawkins, C.C., and Drummond, R.O. (1972). Insecticides tested for control of nymphal brown dog ticks by the "tea-bag" technique. *J. Econ. Entomol.* **65**, 174–176.

Golder, T.K., Otieno, L.H., Patel, N.Y., and Omyango, P. (1984). Increased sensitivity to a natural pyrethrum extract of *Trypanosoma*-infected *Glossina morsitans. Acta. Trop.* (*Basel*) **41**, 77–79.*

Goodhue, L.D., and Sullivan, W.H. (1943). Parasiticidal aerosols suitable for combating flies, roaches, mosquitoes, ... U.S. Patent 2,321,023.

Hobbs, J.H. (1976). A trial of ultralow volume pyrethrin spraying as a malaria control measure in El Salvador. *Mosq. News* **36**, 132–137.

Hoffman, G. (1986). New methods to eliminate the brown dog tick (*Rhipicephalus sanguineus* L.) avoiding indoor air contamination. *Dtsch. Tieraer Ztlwolchenschr.* **93**, 418–424.

Hogsette, J.A., Prichard, D.L., Ruff, J.P., and Jones, C.J. (1991). Development of a refillable ear tag for control of horn flies (Diptera: Muscidae) on beef cattle. *J. Controlled Release* **15**, 167–176.
Hudson, J.E. (1974). Field tests of mosquito coils near Edmonton, Canada. *Pyrethrum Post* **12**, 123–127.
Hudson, J.E., and Esozed, S. (1971). The effects of smoke from mosquito coils on *Anopheles gambiae* Giles and *Mansonia uniformis* (Theo.) in verandah-trap huts at Magugu, Tanzania. *Bull. Entomol. Res.* **61**, 247–265.*
Joseph, S.R., Sagle III, W.B., and Mallack, J. (1973) Evaluation of ground ULV applications for adult mosquito control in Maryland. *Proc. N. J. Mosq. Exterm. Assoc.*, 40–44.*
Kissam, J.N., and Query, G.W. (1976). Accudose aerosol—an effective housefly (Diptera: Muscidae: *Musca domestica* L.) adulticide control system for cage type poultry houses. *Poult. Sci.* **55**, 1906–1913.*
Kuria, J., and Bwogo, R.K. (1986). Aerial application of pyrethrum ULV formulation for control of *Glossina pallidipes* Aust. (Diptera: Glossinidae) in Lambwe Valley, Kenya. *Pyrethrum Post* **16**, 52–60.
Lee, C.W. (1975). Pyrethrum air sprays control mosquitoes in the Caribbean. *Pyrethrum Post* **13**, 5–6.
Lee, C.W., Irving, N.S., and Parker, J.D. (1968) Aerial applications of pyrethrum aerosol to control Tsetse fly. *Pyrethrum Post* **9**, 37–40.
Lee, J.E. (1972). Mosquito adulticiding with synergized pyrethrum. *Proc. Pap. 40th Annu. Conf. Calif. Mosq. Control Assoc.*, 52–54.*
Loren, V.R. (1974). Procedure for mixing pyrethrum adulticiding materials used by the metropolitan mosquito control district. *Mosq. News* **34**, 235–236.*
Lowenstine, L.J., Carpenter, J.L., and O'Connor, B.M. (1979). Trombiculosis in a cat. *J. Am. Vet. Med. Assoc.* **175**, 289–292.
Lynfield, Y.L., and O'Donoghue, M.N. (1982). Pediculosis therapy. *J. Am. Acad. Dermatol.* **6**, 949–950.*
Maas, W., and Zindel, E.E. (1974). The use of insecticides in ultra low volume applications. *Pyrethrum Post* **12**, 110–112, 115.*
MacDonald, J., and Miller, T.A. (1986). Dynamics of natural flea infestation and evaluation of a control program. *Pyrethrum Post* **16**, 84–88.
Maclver, D.R. (1975). Stability of pyrethrins in mosquito coils under long storage—a note. *Pyrethrum Post* **13**, 15–16.*
Mader, D.R., Houston, R.S. and Frye, F.L. (1986). *Hirstiella trombidiiformis* infestation in a colony of chuckwallas. *J. Am. Vet. Med. Assoc.* **189**, 1138–1139.*
Mankowska, H., Styczynska, B., and Krzeminska, A. (1976). Evaluation of the effectiveness of the preparation CX99 in the control of insect pests in sanitary hygiene (in Polish). *Rocz. Panstw. Zaki. Hig.* **25**, 569–574.*
Mansour, N.A., Gaaboub, I.A., and Kamel, F.M. (1972). Evaluation of some insecticides with three pyrethrin synergists against the mosquito *Culex pipiens fatigans* Wied. in Egypt. II. Adult stage. *Z. Angew. Entomol.* **70**, 375–378.*
Meyer, J.A., and Bradley, F. (1986). Field evaluation of Synerid as a house fly larvicide in a northern California caged-layer facility. *Prog. Poult. Through Res. Calif. Coop. Ext.* (*Univ. Calif.*) **34**, 1–3.*
Miller, T.A. (1986). Maximizing the potency of nature's own flea and tick insecticide-pyrethrin. *Pyrethrum Post* **16**, 89–90.*
Mosha, F.W., Njau, R.J.A., and Myamba, J. (1989). Biological efficacy of new formulations of mosquito coils and a critical review of test methods. *Pyrethrum Post* **17**, 47–51.*
Muirhead-Thomson, R.C. (1970). The potentiating effect of pyrethrins and pyrethroids on the action of organophosphorus larvicides in *Simulium* control. *Roy. Soc. Trop. Med. Hyg. Trans.* **64**, 895–906.*
Muirhead-Thompson, R.C., and Merryweather, J. (1970). Ovicides in *Simulium* control. *Bull. W.H.O.* **42**, 174–177.*
Mulla, M.S., Arias, J.R., Sjogren, R.D., and Akesson, N.B. (1973). Aerial application of mosquito adulticides in irrigated pastures. *Proc. Pap. 41st Annu. Conf. Calif. Mosq. Control Assoc.*, 51–56.*
Page, K.W. (1974). Automatic control of guinea-pig lice with a synergised pyrethrins aerosol. *Vet-Rec* **94**, 254–255.*
Parmar, B.S. (1974). Some studies using mixed synergists with pyrethrins. *Pyrethrum Post* **12**, 121–122.*
Pittman, M.R., Silveira, S.M., and Gillies, P.A. (1974). Flit MLO with pyrethrins for mosquito adulticiding. *Proc. Pap. 42nd Conf. Calif. Mosq. Control Assoc.*, 185–187.*
Preiss, F.J. (1973). Mosquito adulticiding with synergized pyrethrins. *Soap Chem. Specialities* **49**, 72–74, 106.*

Rathburn, Jr., C.B., and Boike, Jr., A.H. (1975). Ultra low volume tests of several insecticides applied by ground equipment for the control of adult mosquitoes. *Mosq. News* **35**, 28-29.*
Roberts, R.H., and Pund, W.A. (1974). Control of biting flies on beef steers: effect on performance in pasture and feedlot. *J. Econ. Entomol.* **67**, 232–234.*
Robinson, D.H., and Shepherd, D.A. (1980) Control of head lice in school-children. *Current Therap. Res.* **27**, 1–6.
Rust, M.K., and Reierson, D.A. (1988). Performance of insecticides for control of cat fleas (Siphonaptera: Pulicidae) indoors. *J. Econ. Entomol.* **81**, 236–240.*
Schultz, H. (1975). Human infestation by *Ophionyssus natricis* snake mite. *Br. J. Dermatol.*, (*London*) **93**, 695–697.*
Schuyler, Jr., K.C., and Massing, R.A. (1974). Observations on ground ULV applications of synergised pyrethrins on non-target insects and mosquitoes in Centre County, Pennsylvania. *Pyrethrum Post* **12**, 142–144.
Sham, F.T.K. (1971). On mosquito coils. *Pyrethrum Post* **11**, 50–54.*
Smith, A., Obudho, W.O., Esozed, S., and Myamba, J. (1972). Verandah-trap hut assessments of mosquito coils with a high pyrethrin I/pyrethrin II ratio against *Anopheles gambiae* Giles. *Pyrethrum Post* **11**, 138–140.*
Smith, A., Hudson, J.E., and Esozed, S. (1972). Trials with mosquito coils against *Anopheles gambiae* Giles, *Mansonia uniformis* Theo., and *Culex fatigans* Wied. entering verandeh trap huts. *Pyrethrum Post* **11**, 111–115.*
Smith, C.N. (1973). Pyrethrum for control of insects affecting man and animals. *In* "Pyrethrum, The Natural Insecticide" (J.E. Casida, ed.), pp. 225–241. Academic Press, New York.
Sullivan, W.N., Gebhart, W.A., Chaniotis, B.N., and Whitlaw, J.T. (1976). The effectiveness of pyrethrins and pyrethroid aerosols against mosquitoes endemic to Panama. *Mosq. News* **36**, 316–320.*
Svanasbakken, K.T., von Hanno, T., Kollbaer, R., Savonsnick, R., Skagseth, P.A., and Skjerven, O. (1985). Treatment of headlice *Pediculus humanus* with Rinsoderm shampoo — a multicentric investigation. *Pyrethrum Post* **16**, 21–23.
Tarimo, C.S. (1974). The control of *Glossina swynnertoni* Aust. with synergised pyrethrum from the ground. *Pyrethrum Post* **12**, 116–118.*
Tarimo, C.S., Lee, C.W., Parker, J.D., and Matechi, H.T. (1970). Aircraft applications of insecticides in East Africa. XIX. A comparison of two sampling techniques for assessing the effectiveness of pyrethrum application on *Glossina pallipedes* Austen. *Bull. Entomol. Res.* **608**, 221–223.*
Tarimo, C.S., Parker, J.D., and Kahumbura, J.M. (1971a). Aircraft applications of insecticides in East Africa. XX. The control of *Glossina pallidipes* Aust. in savannah woodland with a pyrethrum/DDT mixture. *Pyrethrum Post* **11**, 18–20.
Tarimo, C.S., Parker, J.D., and Kahambura, J.M. (1971b). Aircraft applications of insecticides in East Africa. XXI. The control of *Glossina swynnertoni* Aust. and *Glossina pallidipes* Aust. in savannah woodland with pyrethrum. *Pyrethrum Post* **11**, 21–23.
Tarimo, C.S., Parker, J.D., and Kahumbura, J.M. (1972). An attempt to reduce the cost of pyrethrum aerial sprays against *Glossina swynnertoni* Aust. *East Afr. Agr. Forest J.* **38**, 47–55.*
Twinn, C.R. (1945). The bedbug and its control. *Dept. Agr. Can. Div. Ent. Proc. Publ.*, 33.
Twinn, C.R., and MacNay, G.C. (1943). Insecticides for head and crab lice. *Can. Entomol.* **75**, 4–13.
Voiculescu, A., and Ichim, A. (1976). Possibilities for using emulsions of natural pyrethrins in sanitary disinsection. *Igenia* (*Bucharest*) **25**, 351–353.*
Washino, R.K., Fukushima, C.K., Mount, G.A., and Wolmendorf, D.J. (1977). Control of adult *Aedes nigromaculis* (Ludlow) with aerosols of pyrethrins and synthetic pyrethroids. III. Comparison of several techniques of evaluating effects upon mosquitoes and nontarget organisms. *Proc. Pap. 45th Annu. Conf. Calif. Mosq. Control Assoc.*, 157–159.*
Whitehorse, S.J.O. (1975). The use of pyrethrin as an insect repellent in dingo baits. *Pyrethrum Post* **13**, 43–44.*
Williams, R.E., Knapp, F.W., and Clarke, Jr., J.L. (1979). Aerial insecticide applications for control of adult mosquitoes in the Ohio River Basin. *Mosq. News* **39**, 622–626.*
Winney, R. (1971). The biological activity of mosquito coils with a high pyrethrin I content. *Pyrethrum Post* **11**, 55–57.*
Winney, R. (1975). Pyrethrins and pyrethroids in coils — a review. *Pyrethrum Post* **13**, 17–22.*
Winney, R., and Kuria, R.M. (1975). Some observations in the biological performance of pyrethrum based mosquito coils. *Pyrethrum Post* **13**, 7–14.*
Womeldorf, D.J., and Mount, G.A. (1977). Control of adult *Aedes nigromaculis* (Ludlow) with aerosols of pyrethrins and synthetic pyrethroids. II. Field evaluations. *Proc. Pap. 45th Annu. Conf. Calif. Mosq. Control Assoc.*, 154–156.*
Woodward, D.L., Colwell, A.E., and Anderson, N.L. (1985). Use of a pyrethrin larvicide to control

Culicoides variipenis (Diptera: Ceratopogonidae) in an alkaline lake. *J. Am. Mosq. Control Assoc.* **1**, 363–368.

Wright, R.E. (1977). Evaluation of Air Guard actuators and insecticide application aerosols for control of flies in dairy barns. *Pyrethrum Post* **14**, 2–9.*

Yap, H.H., Thiruvengadam, V., and Yap, K.H. (1975). A preliminary field evaluation of ULV synergized pyrethrins for the control of *Aedes aegypti* (Linnaeus) and *Aedes albopictus* (Skuse) in urban area of West Malaysia. *Pyrethrum Post* **14**, 98–102.*

Zburay, E.P., and Mount, G.A. (1977). Control of adult *Aedes nigromaculis* (Ludlow) with aerosols of pyrethrins and synthetic pyrethroids. I. Laboratory assays. *Proc. Pap. 45th Annu. Conf. Calif. Mosq. Control. Assoc.*, 152–153.*

Ziv, M., Hadani, A., and Abada, M. (1976). Spraying synergized pyrethrins for the control of house flies (*Musca domestica*, var. *vicina*) on dairy cattle. *Refu. Vet. (Tel Aviv)* **33**, 139–141.*

17

Pyrethrum for Control of Insects in the Home

M. KEITH KENNEDY and ROBERT L. HAMILTON

I. INTRODUCTION

Natural pyrethrins has been used continuously since its insecticidal properties were discovered in Persia in the early 1800s (Gnadinger, 1936; McLaughlin, 1973). First introduced into the United States about 1848 (Gullickson, 1994), pyrethrins is still in wide use today. It is unique in the history of chemical pest control in the United States—it has survived almost 145 years of use without being: (1) rendered totally ineffective by insecticide resistance; (2) restricted or banned by federal regulations; or (3) fatally attacked in the news media by antichemical groups because of health and/or environmental concerns.

II. BENEFITS

Four characteristics of the pyrethrins account for its extraordinary history of use, i.e., knockdown (KD), natural, safety, and nonpersistence.

A. Knockdown

The single most important feature that dictates the selection of pyrethrins as a key formulation component is the rapid insect KD it produces on a wide variety of insect pests (Elliott, 1990, 1994). Until the recent advent of synthetics which can equal its performance on selected insects, pyrethrins' KD activity had been unrivaled for the majority of its commercial history.

B. Natural

A second important characteristic is that pyrethrins are naturally-occurring compounds extracted from the flowers of chrysanthemum plants (Gnadinger, 1936). Although counterintuitive to experts knowledgeable about secondary plant compounds, the descriptor "natural" connotes and/or provides a certain level of implied safety and environmental comfort to today's general public.

The "natural" positioning of insect control products containing pyrethrins has been emphasized by the advertising of many consumer products companies in an effort to broaden the insecticidal product's appeal. Although the use of pyrethrins appears to be especially compelling to environmentally-conscious consumers, it is not altogether clear whether this strategy has provided an overall increase in consumer purchase interest as measured by product sales.

C. Safety

An equally-important factor contributing to pyrethrins prevalent use is its relative safety. Coincident with its natural derivation, the low mammalian toxicity of this material (rat oral LD_{50} 1,500 mg/kg; dermal LD_{50} 1,800 mg/kg) (Ware, 1989; Anonymous, 1993) does, in fact, provide a moderate level of user safety and thus exemplifies the "natural-equates-to-safety" concept mentioned above. It is one of the few chemicals that currently has Environmental Protection Agency (EPA) clearance for use in and around commercial food handling establishments. However, some of the pyrethroids currently used in consumer products also have a high level of user safety in terms of dermal exposure (Fig. 17-1).

D. Non-Persistence

The nonpersistence of pyrethrins (Allan and Miller, 1990; Elliott, 1990) makes them widely acceptable as safe and environmentally-friendly alternatives to other "hard pesticides" (Otieno, 1983). Pyrethrins' lack of residual activity (generally measured in hours) has eschewed the well-publicized issue of household contamination by persistent pesticides such as chlordane. Thus, this natural, fast-acting, KD insecticide is often the active ingredient of choice for

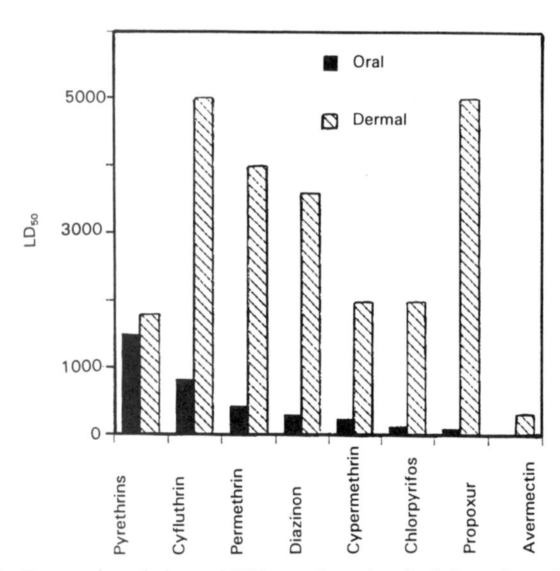

Figure 17-1. Rat oral and dermal LD_{50} values (mg/kg) for selected insecticides.

use in sensitive areas such as food handling establishments, hospitals, food storage areas, vector control, and on pets.

III. LIMITATIONS

A. Cost/Availability

Cost and availability are the most critical factors affecting the global use of pyrethrins. Because chrysanthemum plants are grown as a labor-intensive agricultural crop, the supply, and to some extent quality, are dependent on a variety of biotic and abiotic factors which have a direct impact on yield. Not only is weather important, but the political climate of some of the major pyrethrins-producing countries can dramatically influence the size of the chrysanthemum acreage planted each year and ultimately the cost (Otieno, 1983; Wainaina, 1994). Many of these factors were influential in effecting the large shortfall of pyrethrins in the late 1970s (Gullickson, 1994). They also represent important limitations to the predictability of global pyrethrins supplies which is deemed requisite by manufacturers of insect control products for long-term business development planning. Moreover, cost and availability also constitute critical factors in short-term decisions on the level of pyrethrins utilized in various insecticide products given the type of performance that is desired. Due to the vagaries of the world's pyrethrins supply, manufacturers of pyrethrins-containing, insect-control products have been forced to carefully manage their use of pyrethrins and to develop contingency plans (especially registration of alternate formulas) in case of another world shortage. The 1979 shortage specifically forced many manufacturers to turn to synthetics which had acceptable performance and more enticing cost/benefit ratios. Current alternatives to pyrethrins include propoxur and many of the synthetic pyrethroids. The trend of pyrethroid substitution in lieu of pyrethrins continued through the 1980s and early 1990s in the US consumer insecticide market. This tendency of decreased reliance on pyrethrins is unlikely to be reversed without a major improvement in the long-term stability and supply of the global pyrethrum crop.

B. Photostability

The photoinstability of pyrethrins, while considered a benefit as previously discussed, limits applications where residual activity, in addition to KD, is equally important and/or required (Elliott, 1983). Most frequently, pyrethrins is used in combination with slower-acting, but longer-residual compounds such as propoxur or various pyrethroids to provide quick knockdown with longer lasting control.

C. Need for Synergist

The pyrethrins are always used with a synergist such as piperonyl butoxide (PB) to ensure the performance (mortality) of the formulation against pyrethrins-tolerant or -resistant insects (Cochran, 1987a). However, just as the mercurial nature of the pyrethrins supply is cause for concern, the future global supply of PB is also ambiguous. Deforestation in Brazilian rainforests has led to an environmental and political controversy. As a result, harvest of ocotea trees (from which the PB precursor safrole is extracted) has been restricted in Brazil even though they are very vigorous trees and are not considered to be endangered (Hartshorn, 1993). While Brazil currently supplies a significant proportion of the global safrole, the continued availability will largely depend on the country's developing environmental policy, an unknown at this time (Allred, 1993). The lack of a commercially-suitable alternative synergist that has the same broad activity profile (*N*-octyl bicycloheptanedicarboximide or MGK 264 is much less active on flies) (Section V.C) creates a specter of uncertainty about the viability of currently-registered pyrethrins-PB formulations.

IV. REGISTRATION

A. Reregistration

Because of gaps and potential problems with supporting data, all insecticides (>300 chemicals) registered with the Federal EPA prior to November 1, 1984, are undergoing reregistration. This requires the generation of new data (i.e., toxicology, product chemistry, environmental fate, and fish, wildlife, and crop residue data) supporting all proposed uses and their acceptance by the EPA for the registration to remain active. Because pyrethrins was registered prior to 1984, it too, is subject to reregistration. Since synergized pyrethrins has so many EPA-approved uses (Gillenwater and Burden,1973), the data requirements are massive. To support development of the necessary data package on pyrethrins, a consortium of interested parties was formed to share the costs of reregistration which is administered through the Chemical Specialties Manufacturers Association (CSMA). The data generation is estimated to cost approximately $4 million and is scheduled for completion in December, 1994. Allowing time for EPA review, it is anticipated that pyrethrins will be reregistered for the supported uses in 1996. Until the reregistration process is complete, current registrations remain active and products may be sold to the public.

B. Minor Uses

Data to support pyrethrins registrations for the major uses such as food, livestock, and agriculture are being generated by the consortium. However, relatively "minor" usages such as application on domestic animals or use in pet sprays is not being supported. This means that after 1996, companies will be required to generate a substantial data package for EPA review prior to registration and sale of a pet spray containing pyrethrins. Because the cost of generating these data may outweigh the potential profits involved, it is possible that pyrethrins will become unavailable in the near future for one of its exemplary uses: direct application on pets and domestic animals. However, major formulators and/or suppliers may elect to conduct the necessary studies to obtain the pet registrations. They could then offer these registrations to their customers as part of a supply agreement. Regardless, it will take several years before the issue of pyrethrins registrations on pets is sorted out.

V. PYRETHRINS FORMULATIONS

Formulators are faced with many choices when selecting insecticides for use in consumer products. Attributes of each chemical such as KD performance, persistence, odor, safety, availability, cost, and public perception must be evaluated and understood prior to inclusion in any formulation. Pyrethrins formulations are considered here on the basis of house flies as models for flying insects and German cockroaches as models for crawling insects. Our data and discussion of insecticide preparations are focused on solvent-based formulations, the traditional vehicle for household consumer products. However, increasing pressure from regulatory authorities to reduce VOCs (Volatile Organic Compounds) and consumer demand for "safer," more aesthetic products is driving the industry to develop water-based alternatives. Advantages of water-based aerosols include: lower flammability, decreased VOCs, better smell, and a reduction in oily residue. Many consumer products are currently water-based [Flying Insect Killers (FIKs) and total release foggers]. Reformulation of these products has been driven primarily by the safety concerns with airborne solvents and the aesthetics of the residue left after treatment. Because current solvent-based Crawling Insect Killers (CIKs) generally meet consumer demands and expectations, commercialization of aqueous formulations has been slow. With CIKs, the primary concern of consumers is instant action; they literally want insects to stop in their tracks after being treated. Under the increasing consumer and regulatory demands for more environmentally-friendly products, the consumer product industry will be forced to deliver water-based CIKs. Pyrethrins is left as one of the few active ingredients that can meet consumer expectations of rapid KD in aqueous aerosols.

Most insecticidal pyrethrins formulations in the consumer market are not composed of single chemicals; rather, they are mixtures of different compounds, each adding unique properties to the formulation. The basic insecticidal components of aerosol formulations are: a KD agent, a killing agent, and a synergist. The KD agent is used to immobilize the insect quickly and the killing agent adds lethal efficiency (crawling and flying insects) or persistence (crawling insects). The synergist aids in kill by pyrethrins and selected pyrethroids and can help overcome resistance to insecticides.

A. Knockdown

Table 17-1 gives the KD efficiency of several commonly-used insecticides against house flies and German cockroaches when assayed in systems that mimic consumer usage patterns [space spray with houseflies (CSMA susceptible strain) and direct spray with German cockroaches (SCJ susceptible strain)], comparing each chemical to the performance of 0.25% pyrethrins: 1.0% PB in kerosene,

Table 17-1. Relative KD_{50} for Selected Insecticides in Consumer Products Compared to 0.25% Pyrethrins: 1.0% PB. All Insecticides in Kerosene and Assayed at 0.25% Unless Otherwise Noted[a]

	KD_{50} relative to pyrethrins/PB	
Insecticide	Space spray[b] *Musca domestica* female	Direct spray[c] *Blatella germanica* male
Pyrethrins/PB[d]	1.0	1.0
Propoxur (1%)	—	0.91
Bioallethrin/PB	0.75	4.00[e]
Cyfluthrin	0.86	4.76
Cyfluthrin/PB	0.81	5.26
Cypermethrin	0.96	4.55
Cypermethrin/PB	0.77	3.13
Permethrin	1.64	10.00
Permethrin/PB	1.45	10.00
Resmethrin/PB	1.09	8.33
Tetramethrin/PB	0.92	1.79
Tralomethrin	1.02	7.14
Tralomethrin/PB	1.03	7.14

[a]D. Broadbent, S.C. Johnson Wax unpublished data.
[b]Space spray = standard PEET-GRADY methodology with 0.66 g of insecticide dissolved in kerosene.
[c]Direct spray = direct spray of groups of cockroaches with insecticide dissolved in kerosene.
[d]Insecticide:synergist ratio 1:4 for all compounds.
[e]$KD_{50} > 1$ is slower than pyrethrins.

chosen as the standard because of excellent KD properties for both flying and crawling insects. For house flies treated with solvent-based formulations, there are synthetic pyrethroids which, on an equal percentage basis, exceed the KD performance of pyrethrins, but (on the same basis) for German cockroaches only propoxur and tetramethrin approach the activity of pyrethrins. KD is generally slower in water-based formulatons and only tetramethrin is left as the alternative to pyrethrins in providing acceptable KD.

For FIKs, many cost-effective potential replacements for pyrethrins exist (Fig. 17-2). Tetramethrin (Neo-Pynamin®) costs three to four times less than pyrethrins and is used widely in FIKs as a KD agent. For CIKs, propoxur and tetramethrin can equal the KD of pyrethrins at a lower cost. Propoxur is used to a great extent around the world in CIKs because it is inexpensive and has excellent knockdown at 0.5% or greater. However, at this time, propoxur is under review by the EPA and its future availability is unclear. This leaves only pyrethrins and tetramethrin as fast KD agents for roach sprays. Because of its cost and performance, tetramethrin is a good alternative for CIKs.

One potential problem with using pyrethrins or some of the pyrethroids (such as tetramethrin) as KD agents in consumer products is the negative temperature coefficient associated with their activity. At higher temperatures, their effectiveness as KD and killing agents decreases (Blum and Kearns, 1956). Insect populations and sales of consumer aerosol products are generally greatest in the warmer southern climates. Formulators need to be aware of the potential

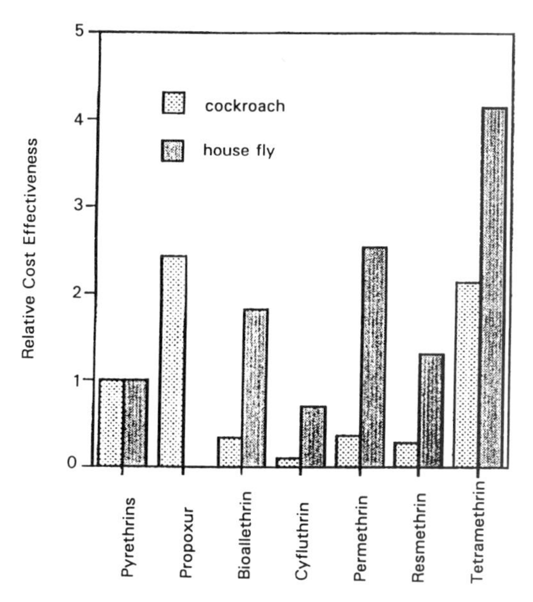

Figure 17-2. Knockdown cost effectiveness of pyrethroids relative to pyrethrins. Values >1 reflect equal performance at lower cost.

interaction of warm southern climates and KD performance of CIKs and FIKs containing either pyrethrins or selected pyrethroids.

B. Killing agent

Although pyrethrins is used primarily as a KD agent, the addition of a synergist does contribute to the killing efficiency of a formulation. Other compounds are more efficient insecticidal agents than pyrethrins and are generally included to assure the product performs as expected. For FIKs these include tetramethrin, resmethrin, bioallethrin, and phenothrin. Persistence is generally not required in FIKs; however, it is an important component of CIKs where consumers want a product that will not only KD and kill an insect quickly, but also provide long-lasting control. Pyrethrins can accomplish the first task, but it has very little residual activity. Cost-efficient killing and residual agents available include chlorpyrifos, propoxur, permethrin, and cypermethrin.

C. Synergists

Synergists are used with natural pyrethrins to increase their killing efficiency even with susceptible insects (Table 17-2) but have little effect upon KD (Table 17-1). The two widely-accepted synergists for use in consumer products are PB and MGK 264. The ratio of synergist to pyrethrins or pyrethroid is often 5:1 or 10:1 in consumer products. With susceptible house flies the toxicity of pyrethrins can be increased 12.5 times by addition of PB. PB is an excellent general purpose synergist, but can cause nasal irritation in high concentrations.

Table 17-2. Relative LD_{50} for Selected Insecticides in Consumer Products Compared to 0.25% Pyrethrins:1.0% PB[a]

	LD_{50} relative to pyrethrins/PB	
Insecticide	Space spray *Musca domestica* female	Direct spray *Blattella germanica* male
Pyrethrins/PB[b]	1.0	1.0
Pyrethrins	12.5[c]	1.54
Bioallethrin	4.54	2.63
Bioallethrin/PB	1.33	2.00
Permethrin	0.26	0.59
Permethrin/PB	0.14	0.55
Resmethrin	0.35	0.81
Resmethrin/PB	0.18	0.79
Tetramethrin	7.69	12.5
Tetramethrin/PB	1.82	4.76
Chlorpyrifos	0.48	0.61
Propoxur	6.66	1.25

[a] D. Broadbent, S.C. Johnson Wax unpublished data. See Table I for conditions.
[b] Insecticide:synergist ratio 1:4 for all compounds using topical application.
[c] $LD_{50} > 1$ requires more chemical to kill than pyrethrins.

Although MGK 264 is generally regarded as a less-effective, broad-spectrum synergist, it has less potential for irritation. Therefore, PB is often combined with MGK 264 in the same formulation to provide broad-spectrum synergistic activity with reduced irritation. However, the main use of synergists is not to aid KD, but rather to help overcome metabolic resistance in many species including flies and roaches.

D. Resistance

Although resistance is thoroughly reviewed by Cochran (1994), it is briefly addressed here from the perspective of consumer products. German cockroaches and house flies have shown the potential to develop high levels of resistance to most insecticides used for their control. In a survey of resistance in field-collected strains of German cockroaches, Cochran (1989) found high pyrethrins resistance in half of the strains tested. Apparently, resistance to pyrethrins develops rather quickly when populations are selected, i.e., in as little as six generations or about 2 years under field conditions (Cochran, 1987b). Resistance is generally thought not to be a problem for consumer products because of consumer usage patterns. Although consumers consistently rate persistence and long-term killing efficiency in CIKs as an important attribute of a quality product, their patterns of aerosol use (either CIKs or FIKs) suggest that persisting residues are less important than KD. Consumers sometimes use aerosols as "liquid fly swatters," i.e., not as residual treatments, but rather, they spray a few insects at a time and apply such a heavy dosage of a solvent-based product that the insect may die from the solvent exposure alone.

VI. PYRETHRINS IN US CONSUMER PRODUCTS

The general categories used in Table 17-3 describe all but a handful of the consumer insecticide products sold in the US and are used by many consumer products companies.

As part of a periodic review, the Research Services Department at S.C. Johnson Wax conducted a survey (unpublished) of the insect-control products found on store shelves in major cities across the US in 1992. The 348 products that were collected are by no means meant to be inclusive but provide a reasonable approximation of the general availability of product types and active ingredients which are being marketed to the public through grocery stores, mass merchandisers, drugstores, and hardware stores. The data presented in Tables 17-4 to 17-7 have been extracted from this survey.

The pet-care category has by far more pyrethrins-containing products (40%) than any other segment (Table 17-4). This number is offset by the fact that none of the 32 flea collars sampled contained pyrethrins. Flea killers and wasp/hornet sprays had pyrethrins in 30 and 27% of the products sampled, respectively, while house-and-garden, a multipurpose spray, was fourth at 24%. Few of the FIKs, indoor foggers, and ant-and-roach products contained pyrethrins (16, 15, and 13%, respectively). Overall, only 25% of the insecticide products sampled in this survey utilized some level of pyrethrins.

A. Ant and Roach Sprays

These products are formulated as aerosols (both solvent and aqueous), liquid sprays (triggers, pumps, etc.), and crack-and-crevice sprays. Of the 348 products collected in the survey, 70 (or 20%) were designated as ant-and-roach products (Table 17-4). While this is a seemingly small representation, the ant-and-roach category still represents the largest volume of sales for the household insecticide market according to industry sales and consumer surveys. Regardless of its volume, the ant-and-roach category has the smallest percentage (13%) of products that utilize pyrethrins (Table 17-4). The exact reasons for this phenomenon are probably multiple, but cost and availability to manufacturers are likely key-contributing factors. Moreover, given the market share dominance of a small number of brands (Raid, Black Flag, Combat, and Hot Shot), fewer than seven products account for the majority of pyrethrins use in this category. The majority of the aerosol ant-and-roach sprays rely on bioallethrin and/or propoxur for KD of roaches and other crawling insects (Table 17-5). In ant-and-roach nonaerosol liquids (trigger sprayers), pyrethrins are used in even a smaller number of products (2 of 8) compared to aerosols (5 of 44). Additionally, only 2 of the 9 crack/crevice products contained pyrethrins. However, since the latter products are designed as residual sprays for long-term crawling insect control, the paucity of pyrethrins in this type of products is not surprising. Home Insect Controls (HICs) are large (0.5–1 gal) tankards usually containing only a residual insecticide such as chlorpyrifos or diazinon. Since they are made for longlasting control for both indoor and outdoor use, they typically would not be formulated with pyrethrins. Of the 19 HICs sampled, none contained pyrethrins (Table 17-5).

Table 17-3. Commonly Available Consumer Insecticide Products Containing Pyrethrins (Py) in the United States, 1992

Category	% in formula		% of total actives
	Py	Synergist	Py + Synergist
Ant and roach killer			
Raid	0.20	0.5[a]	77.7
Raid Max	0.05	1.0[a]	48.8
Enforcer A & R Killer 3	0.05	0.25[b]	57.7
Hot Shot Pro Strength	0.075	0.25[b]	22.8
Flying insect killer			
Raid	0		
Bengal FIK	0.30	1.6[c]	100
Purina Fly-A-Rest	0.5	5.0[a]	100
TAT Fly and Mosquito Killer	0.75	0.375[a]	100
Flea killer aerosol			
Raid	0.14	1.98[c]	90.7
Raid Flea Killer Plus	0.14	2.00[c]	95.8
Bengal Flea Killer	0.025	0.05[a]	11.3
Zodiac 120 Day Carpet	0.20	2.00[c]	99.3
Wasp and hornet killer			
Raid	0		
Gro-well Hornet and Wasp	0.1	0.53[c]	55.8
Purina Frost'em W & H	0.05	0	11.1
Indoor foggers			
Raid	0.5	2.67[c]	77.1
Raid Max	0.075	2.0[c]	95.4
Enforcer Four Hour Fogger X	0.05		
Outdoor foggers	Not used in this category		
House and garden			
Raid H & G	0		
Black Flag H & G	0		
Real Kill H & G II	0.25	0	38.5
Pet care aerosols			
Double Duty Cat Flea and Tick Spray	0.30	2.4[a]	100
Happy Jack Flea-Tick Spray	0.05	0.1[a]	18.4
Hartz Blockade	0		
Sargent's Flea and Tick Repellent for cats	0.15	1.5[a]	84.6
Mosquito coils			
Catch-France	0.25	0	
SCJ-Kenya	0.2	0	
Lake-Kenya	0.3	0	
Swan Thailand	0		
Personal repellent			
Prevent-UK	1.0	?	

[a]Uses PB as synergist.
[b]Uses MGK 264 as synergist.
[c]Uses combination of PB and MGK 264.

Table 17-4. Incidence of Pyrethrins in Commonly Available Consumer Insecticide Products by Category

Category	Total no. products (% of Total)	No. with Py (% with Py)
Ant/roach	70 (20)	9 (13)
Flying insect killers	25 (7)	4 (16)
Indoor foggers	52 (15)	8 (15)
House/garden	21 (6)	5 (24)
Wasp/hornet	30 (9)	8 (27)
Flea killers	44 (13)	13 (30)
Pet care	99 (28)	40 (40)
Other	7 (2)	0 (0)
Total	348	87 (25)

Data courtesy of Research Services, S.C. Johnson Wax.

Table 17-5. Incidence of Pyrethrins in Commonly Available Ant and Roach Consumer Insecticide Products

	Number of products (%)			
	Aerosols		Liquids	
Actives	Ant and roach	Crack/crevice	Trigger sprays	HIC[a]
Bioallethrin	24 (55)	1 (11)	1 (13)	0 (0)
Chlorpyrifos	20 (45)	1 (11)	1 (13)	5 (26)
Propoxur	10 (23)	2 (22)	2 (25)	0 (0)
Pyrethrins	**5 (11)**	**2 (22)**	**2 (25)**	**0 (0)**
Permethrin	5 (11)	0 (0)	0 (0)	0 (0)
Diazinon	0 (0)	3 (33)	3 (38)	3 (16)
Phenothrin	0 (0)	9 (100)	0 (0)	0 (0)
Total No. sampled	44[b]	9	8	19

[a]HIC = Home Insect Control.
[b]Note: many of the products contain more than one active; thus, the total number of products sampled is not a sum of the actives column.
Data courtesy of Research Services, S.C. Johnson Wax.

B. Flying Insect Killers

FIKs constitute only 7% of the total products found on the shelf in the survey (Table 17-4). Of the 25 FIKs found, only four (or 16%) contained pyrethrins while tetramethrin, phenothrin, bioallethrin, and resmethrin were the predominant active ingredients utilized in this category (Table 17-6). As discussed above, the cost effectiveness of the pyrethroids and their availability have made these synthetics the actives of choice for this segment.

C. Indoor Foggers (Total Release Aerosols)

This category (52 products) comprises 15% of the total insecticides found in the survey (Table 17-4) and represents the largest percentage of broad-spectrum (primarily roaches and fleas) aerosol products consisting of one delivery form.

Table 17-6. Incidence of Pyrethrins in Commonly Available Flying Insect Killer, Indoor Fogger (Total Release Aerosol), House and Garden (Multipurpose) and Wasp and Hornet Consumer Insecticide Products

	Number of products (%)			
Actives	Flying insect killer	Indoor fogger	House and garden	Wasp hornet
Tetramethrin	12 (48)	19 (37)	6 (29)	
Phenothrin	12 (48)	14 (27)	6 (29)	
Bioallethrin	9 (36)	14 (27)	10 (48)	6 (20)
Resmethrin	5 (20)	17 (33)	8 (38)	6 (20)
Permethrin		14 (27)		
Fenvalerate		13 (25)		
Chlorpyrifos		4 (8)		5 (17)
DDVP		5 (10)		
Cyfluthrin		1 (2)		
Propoxur				9 (30)
Pyrethrins	**4 (16)**	**8 (15)**	**5 (24)**	**8 (27)**
Total sampled[a]	25	52	21	30

[a]Note: many of the products contain more than one active; thus, the total number of products sampled is not a sum of the actives column.
Data courtesy of Research Services: S.C. Johnson Wax.

Despite the large number of indoor fogger brands sold, only 15% utilize some level of pyrethrins (Table 17-6). Given its lack of persistence, but fast, broad-spectrum activity, pyrethrins is a logical choice as the primary active for an environmentally-friendly, total-release aerosol. However, of the eight indoor foggers containing pyrethrins, only one (Raid Indoor Fogger) had a level greater than 0.05%.

D. House and Garden (Multipurpose)

This is a relatively small category with only 21 products (6% of the total) found on the shelf (Table 17-4). These aerosols are formulated to be used on plants as well as on hard surfaces to control a myriad of arthropod problems. Only five (24%) of these multipurpose sprays contained pyrethrins (Table 17-6), despite the potential benefits provided by inclusion.

E. Wasp and Hornet

Wasp-and-hornet (W&H) aerosols are specialty products aimed at providing rapid KD and kill of wasps either flying or on their exposed nests. Thirty W&H products were found on store shelves representing 9% of the total products found in the survey (Table 17-4). Of these, eight (or 27%) contained some level of pyrethrins. However, other active ingredients such as propoxur, chlorpyrifos, or bioallethrin are usually included to provide the necessary insecticidal action (Table 17-6).

F. Flea Killers

Flea killers are available as aerosols and liquids which are formulated for indoor use as a premise spray and/or for spot treatment. In addition, some formulations may be used directly on pets while powders are applied to carpets and pet bedding. Overall there were 44 flea killer products representing *ca.* 13% of the total insecticide products collected (Table 17-4). The number of aerosols (23) was slightly higher than the liquid sprays (18) that were sampled (Table 17-7). Regardless, pyrethrins was represented equally between the two main delivery forms (26 vs. 28%) while it was present in two of the three powders found in the survey. Other actives such as phenothrin, tetramethrin, and chlorpyrifos were also represented while the largest percentage of products contained bioallethrin. Overall, the use of pyrethrins in flea control products for the home (30%) is second only to products specifically formulated for pet use.

G. Pet Care

This is by far the largest category in terms of number of products (99) with 28% of the total found on the shelves (Table 17-4). Liquids account for the largest segment (35 products) followed closely by pet collars (32) with powders and aerosols rounding out the category (18 and 14, respectively, Table 17-7). Pyrethrins use in these products is two to three times more prevalent than in any of the other product categories surveyed. Some level of pyrethrins was present in approximately 74, 50, and 39% of the liquid pet products, aerosols, and powders, respectively. None of the 32 pet collars sampled contained pyrethrins.

Numerous other actives are used in combination with pyrethrins to provide the residual kill of the adult fleas on the pet, but none are used as extensively as pyrethrins (Table 17-7). Clearly, this is one area where pyrethrins use has been maximized. This may be due, in the most part, to a large number of EPA registrations which allow pyrethrins on pets. Additionally, many of the brand products are undoubtedly sub-registrations on behalf of small regional companies.

VII. PYRETHRINS AND ENVIRONMENTAL ATTITUDES

Although people use insecticide products to rid their home of "bugs," there is an increased public concern over indoor air quality and chemicals in the home environment. As mentioned earlier, natural pyrethrins is an environmentally-friendly, nonpersistent pesticide with an excellent safety profile. However, given its cost, will consumers preferentially select for an environmentally-friendly natural insecticide product or even pay a premium?

A. Roper Study

The Roper Organization (1990) found in a suvey of consumer attitudes and behavior towards environmental issues that of the people surveyed: (1) more

Table 17-7. Incidence of Pyrethrins in Commonly Available Pet Care Consumer Insecticide Products

	Number of products (%)						
	Aerosols		Liquids		Powders		Collars
Actives	Flea killer	Pet care	Flea killer	Pet care	Flea killer	Pet care	Pet care
Propoxur		2 (14)		2 (6)			6 (19)
Resmethrin		3 (21)		1 (3)			
Permethrin		3 (21)		6 (17)			
Chlorpyrifos	6 (26)		7 (39)	4 (11)		1 (6)	10 (31)
Carbaryl		2 (14)		2 (6)		10 (56)	3 (9)
DDVP							2 (6)
Fenvalerate		2 (14)					
Bioallethrin	8 (35)				1 (33)		
Phenothrin	7 (30)				1 (33)		
Tetramethrin	5 (22)						
Pyrethrins	**6 (26)**	**7 (50)**	**5 (28)**	**26 (74)**	**2 (67)**	**7 (39)**	
Total No. sampled[a]	23	14	18	35	3	18	32

[a]Note: many of the products contain more than one active; thus, the total number of products sampled is not a sum of the actives column.
Data courtesy of Research Services, S.C. Johnson Wax.

than 90% thought that companies should solve any problems that develop from insecticide use; (2) 72% favored stronger government regulations of insecticide products; and (3) 70% were more interested in convenience. However, only 29% of those Americans surveyed said they bought a product because an advertisement or the label said it was environmentally safe. In general, the study concluded that Americans do not seem to spend much for environmentally-safe "green" products.

B. Organic Gardening Paradigm

Organic gardening products now represent *ca.* 10% of the $1 billion consumer fertilizer market, but only 2.6% of the $1.6 billion consumer pesticide market (Cuneo, 1993). Despite apparent popularity, Cuneo (1993) notes that sales of organic gardening products have leveled off, leaving companies "wondering how deep the 'green' movement really is." The following quote attributed to an official of the National Gardening Association epitomizes the disconnect between the dogma so zealously espoused by organic gardeners and their purchasing behavior dictated by economy: "consumers vote with their wallet instead of their heart and head" (Cuneo, 1993). This latter information coupled with similar findings by the Roper Organization suggests that cost will likely influence a consumer's decision to purchase an organic insecticide product such as pyrethrins and that relying solely on an environmental and/or natural position will not be overly convincing to many customers.

VIII. FUTURE OF PYRETHRINS USE

A. Impetus

The compelling reason for continued use of natural pyrethrins in consumer insecticides is their fast KD performance coupled with a strong safety profile, natural positioning, broad-target range, and extensive list of registrations. Additionally, this active has an exceedingly long history of use that has contributed to a strong equity of both consumer and formulator confidence.

B. Constraints

The key limiting factors for expanded use of pyrethrins are cost (vs. synthetic pyrethroids), the stability of global supply, and possible regulatory restrictions. The supply shortfall in 1978-9 dealt a major blow to the pyrethum industry from which it has not fully recovered after 15 years. The substitution of cost-competitive synthetics in many of the heretofore standard pyrethins formulas has continued to increase and has allowed many companies to market formulas highly competitive to pyrethrins.

Regulatory restrictions and/or elimination of uses (e.g., pets) and label claims (i.e., natural or safe) will significantly impact the future use of natural pyrethrins. Although natural claims have been leveraged in the past by various marketing advertisements for both consumer and professional products, the US EPA has recently disallowed the use of these types of claims on the product label. Thus, marketing strategies which rely on the natural positioning of pyrethrins-containing products may be dropped either for regulatory reasons or because of an inability to connect with the consumer.

In general, the future of pyrethins usage may largely depend on the marketing strategies of a few major companies who control access to the majority of the US supply. Should one of these companies decide to drop their pyrethrins-based strategy because of price, supply, regulatory changes, or consumer attitudes, the inability of other companies to pick up the excess quickly would create an enormous overabundance of pyrethrins on the world market. The impact of this excess could have serious consequences on the industry in the short and long term.

IX. CONCLUSION

Natural pyrethrins is a versatile insecticide with a long history of use. It is construed as a safe, nonpersistent, environmentally-friendly insecticide with a broad target range and fast KD activity. It has a strong presence in growth categories of consumer insecticide products such as flea control, especially for on-pet use and has advantages in water-based formulations. However, its relatively higher cost vis-a-vis synthetic pyrethroids and its lack of an adequate, stable long-term supply have resulted in pyrethrins being poorly represented (in terms of numbers of different products/brands) in the largest category of consumer insecticides, the ant-and-roach control products. The greatest threats to the future use of pyrethrins are: (1) cheaper, effective synthetics; (2) regulatory issues; and (3) vagaries in marketing strategies of several key consumer-products companies.

REFERENCES

Allan, G.G., and Miller, T.A. (1990). Long-acting pyrethrin formulations. *In* "Pesticides and Alternatives: Innovative Chemical and Biological Approaches to Pest Control" (J.E. Casida, ed.), pp. 357–364. Elsevier Science Publishers, New York.

Allred, J. (1993). CSMA PB Task Force Personal Communication.

Anonymous (1993). "Farm Chemicals Handbook '93," Meister Publishing Co., Willoughby, OH.

Blum, M.S., and Kearns, C.W. (1956). Temperature and the action of pyrethrum in the American cockroach. *J. Econ. Entomol.* **49**, 862–865.

Cochran, D.G. (1987a). Effects of synergists on bendiocarb and pyrethrins resistance in the German cockroach (Dictyoptera: Blattellidae). *J. Econ. Entomol.* **80**, 728–732.

Cochran, D.G. (1987b). Selection for pyrethroid resistance in the German cockroach (Dictyoptera: Blattellidae). *J. Econ. Entomol.* **80**, 1117–1121.

Cochran, D.G. (1989). Monitoring for insecticide resistance in field-collected strains of the German cockroach (Dictyoptera: Blattellidae). *J. Econ. Entomol.* **82**, 336–341.

Cochran, D.G. (1994). This volume.

Cuneo, A.Z. (1993). Marketing organic just comes naturally for some. *Advertising Age*, p. 6, June 28.

Elliott, M. (1983). Developments in the chemistry of action of pyrethroids. *In* "Natural Products for Innovative Pest Management" (D.L. Whitehead and W.S. Bowers, eds.), pp. 127–150. Pergamon Press, New York.

Elliott, M. (1990). Pyrethroid insecticides and human welfare. *In* "Pesticides and Alternatives: Innovative Chemical and Biological Approaches to Pest Control" (J.E. Casida, ed.), pp. 345–364. Elsevier Science Publishers, New York.

Elliott, M. (1994). This volume.
Gillenwater, H.B., and Burden, G.S. (1973). Pyrethrum for control of household and stored-product insects. *In* "Pyrethrum, The Natural Insecticide" (J.E. Casida, ed.), pp. 243–259. Academic Press, New York.
Gnadinger, C.B. (1936). "Pyrethrum Flowers." 2nd Ed. McLaughlin Gormley King Co., Minneapolis, Minnesota.
Gullickson, Sr., W.D. (1994). This volume.
Hartshorn, G. (1993). World Wildlife Fund. Personal communication.
McLaughlin, G.A. (1973). History of pyrethrum. *In* "Pyrethrum, The Natural Insecticide" (J.E. Casida, ed.), pp. 3 15. Academic Press, New York.
Otieno, D.A. (1983). Natural pyrethrin as an insecticide: problems of chemical activity, industrialization and use. *In* "Natural Products for Innovative Pest Management" (D.L. Whitehead and W.S. Bowers, eds.), pp. 93–107. Pergamon Press, New York.
Roper Organization Inc. (1990). "The Environment: Public Attitudes and Individual Behavior," 86 pp. New York, NY.
Wainaina, J. (1994). This volume.
Ware, G.W. (1989). "The Pesticide Book," 3rd Ed., 340 pp. Thompson Publications, Fresno, CA.

18

Pyrethrins — Residues and Tolerances

FREDERICK J. PREISS

I. INTRODUCTION

Pyrethrum, used alone or synergized, is registered by the United States Environmental Protection Agency (EPA) for the control of a large number of pests under diverse conditions. These uses range from human pediculocide preparations to fly control to protection of stored fabric from insect attack. Some uses of pyrethrum, namely those associated with protection of food, feed and food animals from insect attack, require the establishment of residue tolerances.

II. RESIDUE TOLERANCES AND EXEMPTIONS FROM TOLERANCES

Currently applicable residue tolerances and exemptions for pyrethrins are given in Table 18-1 in the areas of: growing crops (pyrethrins are exempt from tolerance requirements); raw agricultural commodities; and food processing and storage.

Table 18-1. Pyrethrins Residue Tolerances and Exemptions from Tolerances

Pyrethrins: exemptions from tolerances

Section 180.1001 Exemptions from the requirement of a tolerance.

(a) An exemption from a tolerance shall be granted when it appears that the total quantity of the pesticide chemical in or on all raw agricultural commodities for which it is useful under conditions of use currently prevailing or proposed will involve no hazard to the public health.

(b) When applied to growing crops, in accordance with good agricultural practice, the following pesticide chemicals are exempt from the requirement of a tolerance:
(7) Pyrethrum and pyrethrins.
These pesticides are not exempted from the requirement of a tolerance when applied to a crop at the time of or after harvest.

Pyrethrins: tolerances for residues

Section 180.128 Tolerances for residues of the insecticide pyrethrins (insecticidally active principles of *Chrysanthemum cinerariaefolium*) are established in or on raw agricultural commodities, as follows:

From postharvest application: 3 ppm in or on barley, birdseed mixtures, buckwheat, corn (including popcorn), rice, rye, and wheat.

Table 18-1. *continued*

From postharvest application: 1 ppm in or on almonds, apples, beans, blackberries, blueberries (huckleberries), boysenberries, cherries, cocoa beans, copra, cottonseed, crabapples, currants, dewberries, figs, flaxseed, gooseberries, grain sorghum, grapes, guavas, loganberries, mangoes, muskmelons, oats, oranges, peaches, peanuts (with shell removed), pears, peas, pineapples, plums (fresh prunes), raspberries, tomatoes, and walnuts.

0.5 ppm in milk fat reflecting negligible residues in milk.

0.2 ppm in meat, fat, and meat byproducts of poultry.

0.1 ppm (negligible residue) in eggs and meat, fat, and meat byproducts of cattle, goats, hogs, horses, and sheep.

0.05 ppm in or on potatoes and sweet potatoes from postharvest application.

Pyrethrins: the Food Additives

Section 185.5200 The food additive pyrethrins may be safely used in accordance with the following prescribed conditions:

(a) It is used or intended for use in combination with piperonyl butoxide for control of insects:
  (1) In certain grain mills and in storage areas for milled cereal grain products, whereby the amount of pyrethrins is from 10% to 100% of the amount of piperonyl butoxide in the formulation.
  (2) On the outer ply of multiwall paper bags of 50 lb or more capacity in amounts not exceeding 6 mg per square foot, whereby the amount of pyrethrins is equal to 10% of the amount of piperonyl butoxide in the formulation. Such treated bags are to be used only for dried foods.
  (3) On cotton bags of 50 lb or more capacity in amounts not exceeding 5.5 mg per square foot of cloth, whereby the amount of pyrethrins is equal to 10% of the amount of piperonyl butoxide in the formulation. Such treated bags are constructed with waxed paper liners and are to be used only for dried foods that contain 4% fat or less.
  (4) In two-ply bags consisting of cellophane/polyolefin sheets bound together by an adhesive layer when it is incorporated in the adhesive. The treated sheets shall contain not more than 10 mg of pyrethrins per square foot (107.6 mg per square meter). Such treated bags are to be used only for packaging dried prunes, raisins, and other dried fruits and are to have a maximum ratio of 0.31 mg of pyrethrins per ounce of fruit (0.01 mg of pyrethrins per g of product).
  (5) In food processing areas and food storage areas (provided that the food is removed or covered prior to such use).

(b) It is used or intended for use in combination with piperonyl butoxide and *N*-octyl bicycloheptenedicarboximide for insect control in accordance with Section 185.4500.

(c) A tolerance of 1 ppm is established for residues of pyrethrins in or on:
  (1) Milled fractions derived from cereal grains when present as a result of its use in cereal grain mills and in storage areas for milled cereal grain products.
  (2) Dried foods when present as the result of migration from its use on the outer ply of multiwall paper bags of 50 lb or more capacity.
  (3) Foods treated in accordance with Section 185.4500.
  (4) Dried foods that contain 4% fat, or less, when present as a result of migration from its use on the cloth of cotton bags of 50 lb or more capacity constructed with waxed paper liners.
  (5) Foods treated in accordance with paragraphs (a)(4) and (a)(5) of this section.

(d) To assure safe use of the additive, its label and labeling shall conform to that registered with the US Environmental Protection Agency, and it shall be used in accordance with such label and labeling.

(e) Where tolerances are established under sections 408 and 409 of the Act on both raw agricultural commodities and processed foods made therefrom, the total residues of pyrethrins in or on the processed food shall not be greater than that permitted by the larger of the two tolerances.

(40 FR 14156, March 29, 1975. Redesignated at 41 FR 26568, June 28, 1976, and further redesignated and amended at 53 FR 24666, June 29, 1988)

Reprinted, in part, from the Code of Federal Regulations 40, Parts 150 to 189, Revised July 1, 1990. Published by the Office of the Federal Register National Archives and Records Administration.

The tolerances and exemptions from tolerances were established before 1972 using less sensitive analytical methods than are available today. Therefore, the EPA (1986), as part of its data call-in program, requested that residue studies be conducted to support all areas of uses requiring tolerances. This issue was addressed by the Pyrethrin Steering Committee comprised of member companies McLaughlin Gormley King Company (MGK), Roussel-Uclaf, S.C. Johnson & Son, Commonwealth Industrial Gases (CIG), Pyrethrum Board of Kenya (PBK), Tanganyika Pyrethrum Board (TPB), and Office Du Pyrethre Au Rwanda (OPYRWA). The Pyrethrin Steering Committee's Technical Sub-committee reviewed existing data, found it unsatisfactory, and with approval from the Steering Committee, initiated studies on residues resulting from use of pyrethrum in food processing, growing crops (which covers tolerances on raw agricultural commodities), milk, eggs, and meat animals.

This review considers pyrethrins residue studies conducted to date, those ongoing, and those to be conducted.

III. GENERAL COMMENTS ON RESIDUE ANALYSIS

Pyrethrum contains both pyrethrins I (consisting of pyrethrin I, cinerin I, and jasmolin I) and pyrethrins II (consisting of pyrethrin II, cinerin II, and jasmolin II). Pyrethrins I and II are typically present in similar amounts and represent the major insecticidal components of the mixture. The EPA requested that analysis be conducted for both pyrethrins I and II, but the Technical Sub-committee petitioned the EPA to accept analysis of pyrethrins I and to calculate pyrethrins II to determine total pyrethrins residues (Novak, 1992). The quantitation of small amounts of pyrethrins residues has historically been based on pyrethrins I due to thermal instability and decomposition of pyrethrins II under gas chromatographic analysis conditions. Attempts were made by the Technical Sub-committee to use other methods of analysis, but they were unsuccessful. Subsequently, the EPA agreed with the position. Therefore, all residues are based on analysis of pyrethrins I and calculation of pyrethrins II with the assumption that pyrethrins II are essentially equal in amounts to pyrethrins I.

IV. MEAT, MILK, EGG, AND POULTRY RESIDUE

Pyrethrum sprays, wipe-on, and dip products are used for control of flies and ectoparasites on meat and milk animals and poultry. A residue program has been initiated in these areas after a long delay in obtaining ^{14}C-labeled pyrethrin I which was needed to first determine the nature of the residues when applied to animals (parent material and metabolites). Based on this information, tissues, milk, and eggs of various animals will be analyzed after treatment with a typical product containing pyrethrum. Protocols are currently being developed for review by EPA. Because of less toxicological concern over the metabolites, it is anticipated that only the parent compound will be analyzed by gas chromatography.

V. INCIDENTAL FOOD ADDITIVE

Because of its flushing action, quick knockdown, and kill of arthropods, and its nonresidual nature, synergized pyrethrum is undoubtedly the product of choice in commercial food-processing and storage areas. Product label directions for use of pyrethrum in these areas mandate that an establishment be closed during treatment and that food be removed or covered. Food-handling surfaces must also be covered and washed before reuse. Even with these precautions, the possibility exists that residues might result from pyrethrum use—or misuse.

Five separate residue studies were conducted to cover all use patterns in commercial food processing and preparation areas. They were: (1) single contact spray—food processing; (2) multiple contact spray—food storage warehouse; (3) single space spray—food processing; (4) multiple space spray—food storage warehouse; and (5) intermittent aerosol—simulated restaurant. MGK was contracted to conduct the studies.

A. Single Contact Spray – Food Processing

Method. Food commodities were representative of the different types commonly processed. These consisted of cream pies, butter, hard candy (lemon drops), meat (hamburger), bread, and white potatoes. The simulated food processing area was a room measuring 7 by 16 feet (112 square feet). Four samples of each commodity were placed on a table in the center of the room on paper plates (Fig. 18-1). Two of each were completely uncovered while the remaining two were covered with a double layer of brown paper. The test material was based on the maximum amount of active ingredients that would be found in an EPA registered, synergized pyrethrum formulation. It contained 0.75% pyrethrins,

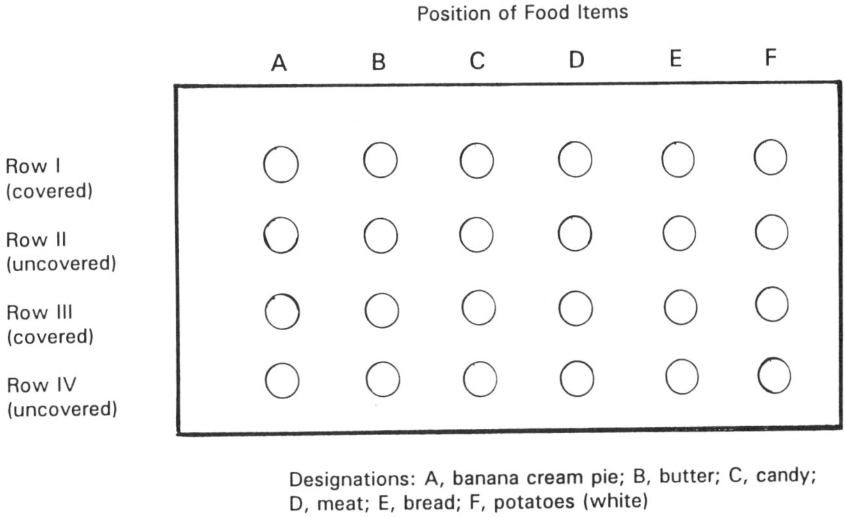

Figure 18-1. Simulated food processing table (4 by 8 feet) showing various food commodities.

the synergists (piperonyl butoxide 7.50% and MGK® 264 3.00%), and 88.75% petroleum distillate. The synergists are also the subject of a residue data call-in and were analyzed separately (data not included here). The test formulation was prepared under Good Laboratory Practices protocols and analyzed by gas-liquid chromatography — flame ionization detector (GLC-FID) before and after completion of all applications. The test material was applied using a compressed air sprayer with a fan type nozzle at the standard label rate of 1 gal per 1,000 square feet to the walls and floor of the simulated processing room. One hour posttreatment all samples were placed in plastic storage containers and frozen at −28°C for later analysis. To determine the storage stability of pyrethrins, several representative commodities were treated with 0.2 ppm of pyrethrins and stored at −28°C. Analysis was done after 1 year of storage. Additionally, standard recoveries of pyrethrins were conducted from the same foods used in the study.

Meinen and Bergman (1991a–e), MGK Senior Research Associate and Analytical Chemist, respectively, developed and used the procedure in Fig. 18-2 for residue analysis of pyrethrins in food with the following gas chromatographic conditions: instrument — Varian, electron capture; column — 120 cm × 4 mm id, 3% Xe-60 on Chromosorb W (HP), 80–100 mesh; carrier gas–argon/methane, 50 ml/min; temperatures–column 180°C, injector 300°C, detector 300°C; and injection 2–3 μl. The limits of detection are 0.05 ppm for total pyrethrins I.

Results. The range of standard pyrethrins recoveries from the foods used in the study is shown in Table 18-2. The percent pyrethrins recovered ranged from 60–136, depending on the commodity. These recovery data were used for all five residue studies conducted. The recovery of pyrethrins on various commodities after 1 year of freezer storage at −28°C is shown in Table 18-3. The recovery of 58-116% indicates that pyrethrins are relatively stable under standard storage conditions. Because of the nature of the application (a coarse, wet contact spray applied to the floor, bottom portion of the walls, and around the base of the food processing table), no residues of pyrethrins were expected and no residues above 0.1 ppm were detected. Trace amounts (residue calculated at 0.05 to 0.10 ppm) were found in uncovered bread samples.

Food or Crop Sample → Acetonitrile Extraction →

Acetonitrile/Petroleum Ether Partition → Florisil Cleanup →

Electron Capture/Gas Liquid Chromatography

Figure 18-2. Schematic method for analysis of pyrethrins residues in processed food commodities and in crops.

Table 18-2. Range of Standard Recoveries of Total Pyrethrins (I and II) Residues from Various Food Commodities

Commodity	Recovery (%)
Bread	70–93
Candy	70–110
Meat	80–118
Potato	93–110
Cream pie	70–136
Butter	74–85
Lettuce	68–108
Flour	60–98
Sugar	58–65
Beans	80–107
Peanuts	65–85
Prunes	67–106
Chocolate cake	90–130

Table 18-3. Storage Stability of Total Pyrethrins (I and II) Residues at −28°C for Approximately 1 Year

	Recovery (%)	
Commodity[a]	Storage sample[b]	Typical range
Candy	66	58–107
Meat	54	80–116
Bread	50	70–93
Sugar	63	58–65
Peanuts	69	65–85
Beans	81	80–107

[a]Commodities were spiked with approximately 0.2 ppm of pyrethrins and stored in a freezer at −28°C.
[b]Spiked samples stored for 1 year at −28°C.

B. Multiple Contact Spray – Food Storage Warehouse

Method. In this study, the food commodities used were representative of those which would be stored in warehouses for extended times and would potentially be exposed to multiple applications of insecticidal sprays. They were navy beans, Spanish peanuts, dried prunes, granulated sugar, and chocolate cake. Each commodity was left in the package in which it was purchased. The exception was navy beans, which were placed in cotton bags. Ten samples of each commodity were placed on a pallet in each of the rooms utilized. The test formulation was the same as used in the single contact spray. Ten applications per week of the standard rate (1X) (1 gal/1,000 square feet) and exaggerated rate (4X) (4 gal/1,000 square feet) were made over a 5 week period. After each

Table 18-4. Total Pyrethrins (I and II) Residues in a Food Storage Warehouse Following Multiple Applications (Contact Spray) at Standard (1 ×) and Exaggerated Rate (4 ×)

	Pyrethrins (ppm)[a]				
Spray treatment	Navy beans	Spanish peanuts	Dried prunes	Granulated sugar	Chocolate cake
1	0[b]/0	0/0	TR[c]/0	0/0	0/0
2 and 3	0/0	0/0	0/0	0/0	0/0
4	0/0	0/0	TR/0	0/0	0/0
5 through 10	0/0	0/0	0/0	0/0	0/0

[a]Data given for the standard rate (1 ×) above the line and the exaggerated rate (4 ×) below the line.
[b]Residues <0.05 ppm are reported as 0 ppm.
[c]Residues between 0.05 and 0.10 ppm are reported as a trace (TR).

treatment, samples were removed and stored at −28°C for later analysis. Sample handling, storage stability, and analysis were the same as for the single application previously discussed.

Results. Since this application in a simulated warehouse was made as a coarse, wet spray, little or no pyrethrins residues were expected in food stored in the rooms. The treatments at the 1X rate resulted in no detectable residues in any commodities except prunes, which had a trace amount (0.05 ppm to 0.10 ppm). No residues were detected in any commodities exposed at the 4X rate (Table 18-4).

C. Single Space Spray – Food Processing

Space spraying of unoccupied food processing and serving facilities is a major method of control for flying and crawling insects. Synergized pyrethrins sprays typically contain from 0.075% to 5.0% pyrethrins depending on the type of product, i.e., mist, thermal fog, or ultra low volume (ULV).

Method. The floor diagram for the study was discussed in detail earlier (Fig. 18-1). The representative spray was chosen to cover all levels of synergized pyrethrum space sprays currently used for this type of application. The ULV spray contained 5.0% pyrethrins, 25.0% piperonyl butoxide, 10.0% MGK® 264, and 60.0% petroleum distillate. The test formulation was applied using a cold, ULV aerosol generator at a rate of one (1X) and four ounces (4X) per thousand cubic feet of room space. A single application was made over the food, the room closed for one-half hour and then opened and ventilated thoroughly. The remainder of the study was conducted in the same way as previously described in the section on contact spray–food processing.

Results. Since space sprays are applied into the air and settle over time, significant residues were expected on the uncovered food, while covered food would be protected. Table 18-5 shows the results of residue analyses of covered and uncovered food at 1X and 4X applications. As expected, pyrethrins residues after 1 half hour exposure were significant in the uncovered food. Ranges found were 2.3 ppm (meat) to 22 ppm (bread) at 1X and 11 ppm (potato) to 90 ppm (bread) at 4X. Covered commodities in both treatments had no pyrethrins residues, thus supporting the statement on pesticide labels "to cover or remove food before treatment."

Table 18-5. Total Pyrethrins (I and II) Residues in a Food Processing Facility Following Single Application (Space Spray) at Standard (1 ×) and Exaggerated Rate (4 ×)

	Pyrethrins (ppm)[a]	
Commodity	Covered	Uncovered
Bread	0/0[b]	22/90
Meat	0/0	2.3/16
Candy	0/0	11/22
Potato	0/0	5.9/11
Banana cream pie	0/0	7.1/25
Butter	0/0	5.4/13

[a]Data are given for the standard rate (1 ×) above the line and exaggerated rate (4 ×) below the line.
[b]Residues calculated at <0.05 ppm are reported as 0 ppm.

D. Multiple Space Spray – Food Storage Warehouse

Packaged food commodities typically remain in warehouses for some time and can potentially be exposed to numerous insecticidal sprays. This study was conducted to determine pyrethrins residues that could be present in food from multiple applications of a space spray in a warehouse.

Method. Food commodities and room set up used for this study were the same as those used for the food storage warehouse study outlined earlier. The test formulation was the same as used in the single space spray–food processing study. Ten applications were made at one ounce (1X) and four ounces (4X) per one thousand cubic feet of space. Spraying was conducted twice weekly for 5 weeks. The formulation was space sprayed over the food using a ULV cold fog aerosol generator. After each treatment, one sample of each commodity was removed from the room, bagged, and stored at −28°C until analyzed.

Results. Pyrethrins residues from 10 applications of 1X and 4X space sprays are given in Table 18-6. With the 1X treatment, pyrethrins residues greater than a trace (0.05 to 0.10 ppm) were found after seven treatments in Spanish peanuts but not in other treated foods. This may be attributed to either the food commodity packaging being impervious to pyrethrins or the pyrethrins degrading rapidly between treatments. Pyrethrins residues above trace levels were greater in the 4X treatment in all food items except sugar and chocolate cake. These two commodities most likely had better barriers to pyrethrins migration than the peanuts (cellophane) and navy beans (cotton bag). Residues found in the prunes were unexpected since they were in foil-lined bags. Residues found were probably due to contamination when opening the prune bags for processing and analysis. In no case did the pyrethrins residues exceed 0.75 ppm.

E. Intermittent Aerosol Spray – Simulated Restaurant

Intermittent, or timed-release aerosols, are used primarily to control house flies in restaurants and other food-processing or serving areas during operation. The

Table 18-6. Total Pyrethrins (I and II) Residues in a Food Storage Warehouse Following Multiple Applications (Space Spray) at Standard (1 ×) and Exaggerated (4 ×) Rates

	Pyrethrins (ppm)[a]				
Spray treatment	Navy beans	Spanish peanuts	Dried prunes	Granulated sugar	Chocolate cake
1	0[b]/0	0/0	TR[c]/TR	0/0	0/0
2	0/0	0/0	0/TR	0/0	0/0
3	0/0	0/TR	0/0	0/0	0/0
4	0/TR	TR/0.10	0/TR	0/0	0/0
5	0/TR	TR/0.18	0/0.18	0/0	0/0
6	0/0.10	TR/0.53	0/0	0/0	0/0
7	0/TR	0.16/0.24	0/0	0/0	0/0
8	0/TR	0.12/0.33	0/TR	0/0	0/0
9	0/0.10	0.23/0.28	0/TR	0/TR	0/0
10	0/0.17	0.21/0.72	0/0.14	0/TR	0/0

[a]Data given for the standard rate (1 ×) above the line and exaggerated (4 ×) rate below the line.
[b]Residues <0.05 ppm are reported as 0 ppm.
[c]Residues between 0.05 and 0.10 ppm are reported as a trace (TR).

aerosol operates on a battery and timer system that activates the sprayer every 7.5 or 15 minutes. Depending on the aerosol valve, from 50 to 150 mg of total product are released into the air at each activation of the unit.

Method. The residue evaluation was conducted in a 5,400 cubic foot room using four tables set at varying distances from the wall-mounted aerosol, which was 6 feet off the floor (Fig. 18-3). Each table held one covered and one uncovered commodity and triplicate food handling utensils/surfaces. The food commodities used were bread, candy, meat, potatoes, banana cream pie, butter, lettuce, and flour. The food handling utensils/surfaces were knives, aluminum foil, and plates. The intermittent aerosol contained (w/w): 2.0% pyrethrins, 20.0% piperonyl butoxide (technical), 5.0% MGK® 264, 13.0% petroleum distillate, and 60.0% propellent A-70 (isobutane/propane). The test formulation was applied at a rate of 50 mg every 15 minutes for 24 hours. The samples were collected, bagged, and held at −28°C for analysis.

Results. Pyrethrins residues ranging from a trace (0.05 to 0.10 ppm) to 2.1 ppm were found in uncovered food (Table 18-7). Basically, no pyrethrins residues (one sample had a trace) were found in covered food after 24 hours exposure to the aerosol. The amount of residue was greatest in the uncovered food closest to the intermittent aerosol. Food handling surfaces/utensils all contained pyrethrins residues. The greatest amounts were detected closest to the aerosol and the least farthest from the dispensers.

VI. CROP RESIDUE

Pyrethrum and pyrethrins are currently exempt from the requirements of finite tolerances when used on growing crops but are not exempt from tolerance requirements when applied to a crop at the time of harvest or at postharvest

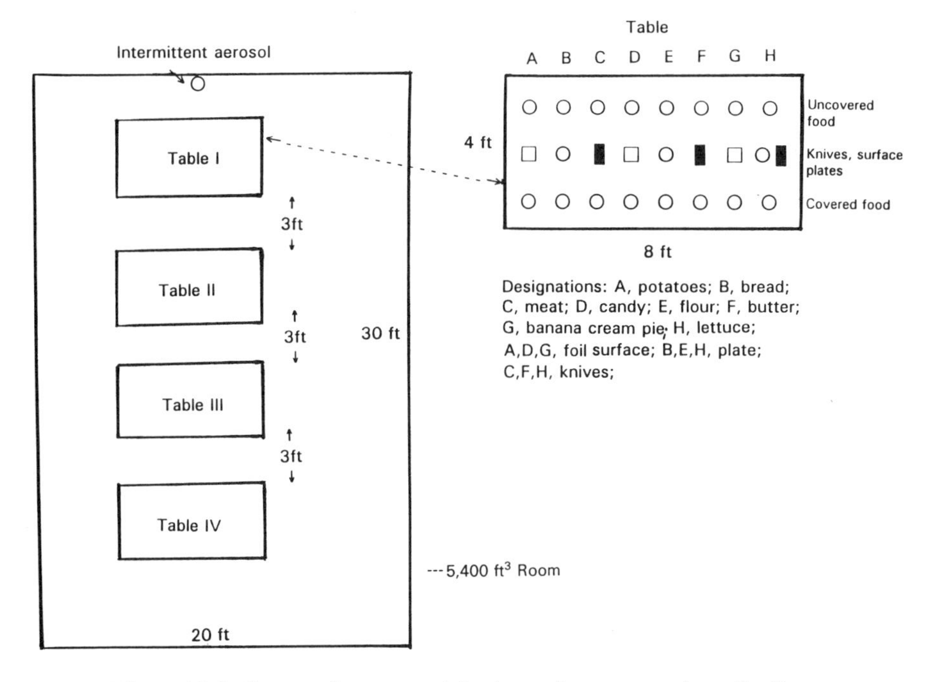

Figure 18-3. Set-up of room used for intermittent aerosol application.

Table 18-7. Total Pyrethrins (I and II) Residues in Intermittent Aerosol Spray Simulated Restaurant

	Table Number[a]							
	I		II		III		IV	
Item	C[b]	UC[b]	C	UC	C	UC	C	UC
Commodity				*Pyrethrins (ppm)*				
Bread	0[c]	1.7	0	0.78	0	0.13	0	0.72
Candy	0	2.1	0	0.16	0	0	0	TR[d]
Meat	TR	1.2	0	0.17	0	0	0	0
Potatoes	0	0.14	0	TR	0	0	0	0
Banana cream pie	0	0.28	0	TR	0	TR	0	TR
Butter	0	0.24	0	TR	0	TR	0	TR
Lettuce	0	0	0	0	0	0	0	0
Flour	0	0.26	0	0	0	0	0	0
Utensil				*Surface average (μg pyrethrins/in²)*				
Knife	9.0		1.4		0.45		0.95	
Foil	12		0.82		0.37		0.26	
Plate	12		0.92		0.37		0.27	

[a]Table number I closest to aerosol; IV farthest from aerosol (3 feet intervals).
[b]C = covered; UC = uncovered.
[c]Residues < 0.05 ppm are reported as 0 ppm.
[d]Residues between 0.05 and 0.10 ppm are reported as a trace (TR).

(Table 18-1). Pyrethrum products are registered by the EPA for use on a large variety of crops and can be applied up to the time of harvest. The EPA data call-in on pyrethrum of 1986 mandated that crop residue data be generated and postharvest tolerances be supported through residue data on each crop on which pyrethrum is used. This program would lead to establishing tolerances for agricultural crops and processed food commodities, in addition to determining the postapplication interval at which residues would be acceptable. The Pyrethrin Steering Committee Technical Sub-committee contracted with Landis International, Valdosta, Georgia to set up, coordinate, and conduct the required residue studies in accordance with EPA guidelines.

A. Program

A major issue confronting the Sub-committee was the broad agricultural crop use pattern of pyrethrum. It is labeled for use on crops bridging a spectrum from rice to sweet potatoes. Dr. W.R. Landis, Landis International, was able to negotiate the requirements for the crop field trials with EPA. This program, initiated in 1992, will be complete in 1995. Due to the high cost of pyrethrum, lack of residual insect control, and requirements for eco-fate studies beyond those which the Pyrethrin Steering Committee had endorsed (Gabriel and Mark, 1994), a decision was made to limit the residue program to selected crops. This action will eventually reduce the crops on which pyrethrum can be used. The crop residue program consists of 71 studies comprised of 64 raw agricultural commodities and seven processed commodities. Eight crop groups made up of a number of representative crops of each group will be evaluated for residues (Table 18-8). The field trials will be conducted in geographical regions where the representative crops are grown. The residue studies will cover all crops that fall into the crop group category according to EPA guidelines.

The representative formulation selected for the crop field trials is PYRENONE® Crop Spray (Roussel-Uclaf) containing 6.0% pyrethrins, 60.0% piperonyl butoxide, and 34.0% emulsifiers and solvents. Ten applications (0.05 lb actual pyrethrins per application) will be made by commercial applicator to each crop before harvest using agricultural equipment. Each crop will then be harvested and frozen until residue analysis can be conducted.

B. Analysis

Since the metabolites of pyrethrins I and II are likely to be less toxic than pyrethrins themselves (Elliott *et al.*, 1972), the Pyrethrin Steering Committee Technical Sub-committee decided to analyze plant tissue for residues of pyrethrins only. The residue determination in treated crops is being conducted by an independent contract laboratory. A schematic diagram for the method of analysis by GLC-EC for pyrethrins I is given in Fig. 18-2 while Fig. 18-4 gives a standard curve and Fig. 18-5 a chromatogram for the three pyrethrins standards which are easily separated. Work has been initiated to validate the analytical method on each crop. As an example, Fig. 18-6 is a chromatogram from lettuce

Table 18-8. Pyrethrum Crop Residue Trials Program

Crop group	Representative crops	Type of study	Trial location (state)
Brassica	Broccoli	RAC[a]	CA, TX, OR
	Mustard	RAC	GA, TX
	Cabbage	RAC	FL, CA, NY
Leafy vegetable	Leaf lettuce	RAC	FL, AZ
	Head lettuce	RAC	CA, FL
	Spinach	RAC	TX, CO
	Celery	RAC	CA, MI
Legume vegetable	Dry pea	RAC	TX, WA
	Succulent pea	RAC	MN, CA
	Succulent bean	RAC, PC[b]	NY, FL, WI
	Dry bean	RAC	ND, CO
Fruiting vegetable	Tomato	RAC, PC	FL, MI, NJ, CA
	Pepper	RAC	CA, TX, NC
Root/tuber	Carrot	RAC	TX
	Sugar beet	RAC, PC	CA, ND, MN
	Potato	RAC, PC	CO, ID, ME, WA
	Radish	RAC	FL
Cucurbita	Summer squash	RAC	FL, TX, GA, NJ
	Cucumber	RAC	NC, MI
	Cantaloupe	RAC	AZ, CA
Citrus	Orange	RAC, PC	FL, CA, TX
	Lemon	RAC	CA, AZ
	Grapefruit	RAC	FL, TX
Small fruit	Strawberry	RAC	OR, FL
	Blueberry	RAC	MI, NC
	Cranberry	RAC	MA
	Grape	RAC, PC	NY, CA
	Blackberry	RAC	TX

[a] Raw agricultural commodity.
[b] Processed commodity.

that was fortified with 1.0 ppm pyrethrins I. The recovery of pyrethrins using the method was 97.5%, well above the EPA acceptable level of 70%. Additionally, studies of freezer storage stability are underway. Samples of treated crops will be held in the freezer in the event that analysis for metabolites is necessary.

VII. CONCLUSION

A residue program has been initiated for pyrethrins in meat, milk, poultry, and eggs. Studies in support of a pyrethrins food additive have been conducted. In all cases, with the exception of the intermittent aerosol application, residues in uncovered foods were less than the established 1.0 ppm. Several of the uncovered commodities on the table closest to the dispenser (3 feet) had pyrethrins residues ranging from 1.2 to 2.1 ppm. A crop residue program is underway on eight crop groups to establish finite tolerances on growing crops preharvest and to

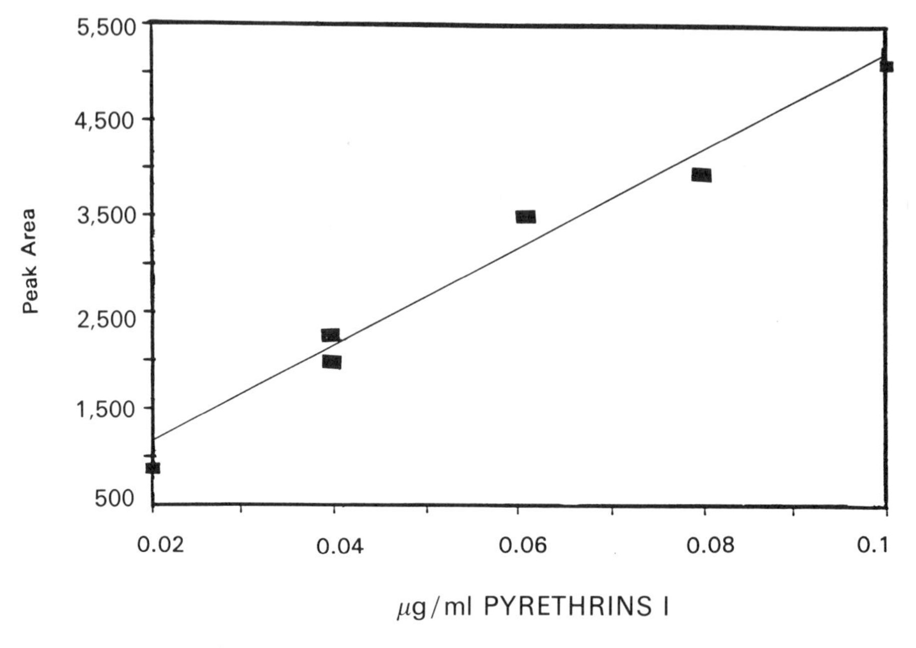

Figure 18-4. Pyrethrins I crop residue analysis standard curve.

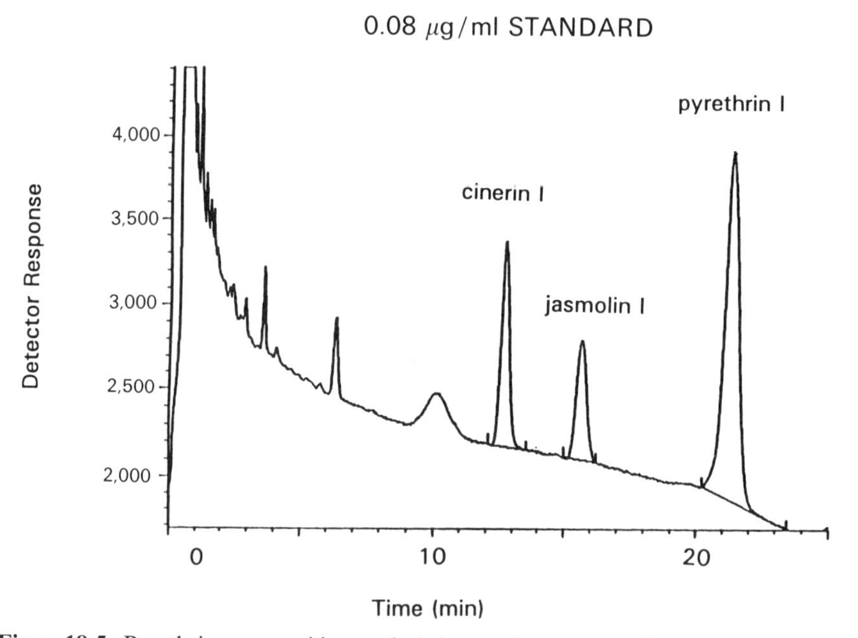

Figure 18-5. Pyrethrins crop residue analysis by gas chromatography. Instrument—H/P 5880, electron capture; column—J & W Megabore DB-225, 30 m × 0.53 mm, 1.0 μm film; carrier gas—helium at 30 ml/min; make-up gas—argon/methane at 60 ml/min; temperatures—column 190°C, injector 300°C, detector 300°C; injection vol—1 μl (0.08 μg/ml standard).

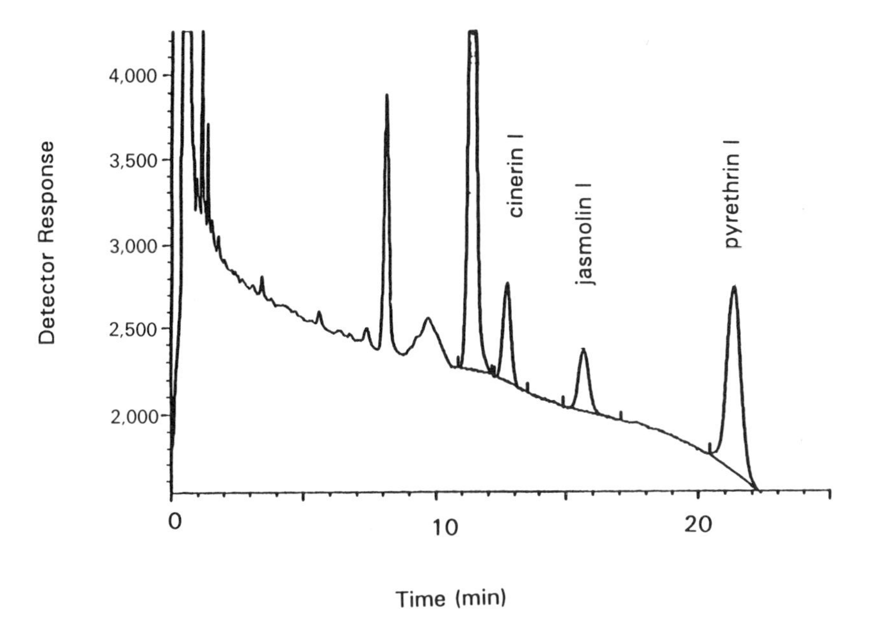

Figure 18-6. Analysis of pyrethrins in lettuce fortified with 1.0 ppm pyrethrins I (97.5% recovery).

confirm the established postharvest tolerances. These studies should be completed in 1995. Due to limited use and high cost to develop residue data for pyrethrins, the Pyrethrin Steering Committee elected not to generate data on stored grains and multiwall bags used to prevent insect infestation.

ACKNOWLEDGMENT

I acknowledge the Pyrethrin Steering Committee for allowing me to use the residue data presented here. The author thanks Dr. R.W. Landis (Landis International), Dr. Roger Novak (NPC, Incorporated), and Vernon J. Meinen (Senior Research Associate, MGK) for reviewing and making recommendations for this manuscript and Mary Brennan, MGK, for typing and editing.

REFERENCES

Elliott, M., Janes, N.F., Kimmel, E.C., and Casida, J.E. (1972). Metabolic fate of pyrethrin I, pyrethrin II, and allethrin administered orally to rats. *J. Agric. Food Chem.* **20**, 300-313.

Environmental Protection Agency (1986). Data call-in notice for product chemistry and residue chemistry data for pyrethrins. Unpublished document.

Gabriel, K.L., and Mark, R. (1994). This volume.

Meinen, V.J. and Bergman, J.T. (1991a). Residue study of pyrethrins, piperonyl butoxide, and MGK® 264 in certain food commodities resulting from use as a contact spray in a simulated feed and food processing situation, 146 pp. Unpublished report.

Meinen, V.J. and Bergman, J.T. (1991b). Residue study of pyrethrins, piperonyl butoxide, and MGK® 264 in certain food commodities resulting from use as a contact spray in a simulated warehouse situation, 114 pp. Unpublished report.

Meinen, V.J. and Bergman, J.T. (1991c). Residue study of pyrethrins, piperonyl butoxide, and MGK® 264 in certain food commodities resulting from use as a space spray in a simulated feed and food processing situation, 145 pp. Unpublished report.

Meinen, V.J. and Bergman, J.T. (1991d). Residue study of pyrethrins, piperonyl butoxide, and MGK® 264 in certain food commodities resulting from use as a space spray in a warehouse situation, 116 pp. Unpublished report.

Meinen, V.J. and Bergman, J.T. (1991e). Residue study of pyrethrins, piperonyl butoxide, and MGK® 264 in certain food commodities and surfaces resulting from use of an intermittent aerosol in a simulated restaurant situation, 199 pp. Unpublished report.

Novak, R.A. (1992). Justification for quantitation of pyrethrins residues in crops based on analysis of pyrethrin I esters, 4 pp. Unpublished report.

VI

Summary

19

Pyrethrum — A Benefit to Human Welfare

JOHN E. CASIDA and GARY B. QUISTAD

I. INTRODUCTION

This multiauthor treatise considers all aspects of pyrethrum flowers and their insecticidal constituents, the pyrethrins. It updates the available knowledge of the production, chemistry, toxicology and uses for the past 20 years since the publication "Pyrethrum, The Natural Insecticide" (Casida, 1973).

Insecticides must combat pests at a low dose and be safe and cost-effective; this requires an increasingly unusual set of biological, chemical, and economic features. The compound in the form and manner applied must control the pest without harming beneficial organisms including both closely-related species (predators, parasites, and pollinators) and higher organisms (wildlife and humans). It must degrade sufficiently fast that residues are not a problem and there is no accumulation through food chains. The insecticide must be adequate in supply, competitive in price and easily formulated for use. Most important, it must be of low toxicity to humans exposed directly to the compound or to its residues on a short- or long-term basis.

This book defines the extent to which pyrethrum meets these goals of an ideal insect control agent. It provides the best up-to-date information on the role of pyrethrum in the spectrum of insecticides.

II. BOTANICAL INSECTICIDES

Botanical and inorganic insecticides predated the synthetics by many centuries. Their known toxic properties were applied to pest control with variable effectiveness. Inorganics based on arsenic and fluorine were the dominant stomach poisons for chewing insects prior to the advent of DDT and other chlorinated hydrocarbons. Pyrethrum, rotenone, and nicotine were the three primary botanicals. Nicotine has been totally replaced in most parts of the world and rotenone is now used primarily for control of garden pests, ectoparasites, and undesirable fish in lakes and reservoirs. Other botanicals such as ryania and sabadilla never went beyond the minor-use stage. A more recent botanical insecticide, neem with azadirachtin as its active ingredient, is being given increasing consideration on both an academic and practical basis.

Pyrethrum is the only botanical insecticide from before World War II that continues to be of major importance today. It is therefore appropriate to consider the unusual properties and circumstances that make pyrethrum so unique.

Botanicals until recently had a favored status. If they had been used for decades with no evidence of problems, they were not subjected to critical reexamination as is the case for synthetic insecticides. In some cases botanicals were categorized as "exempt from the requirement for a tolerance" without further scrutiny. This changed with further amendments to the Federal Insecticide, Fungicide, and Rodenticide Act, particularly the 1988 amendment which required reregistration of older pesticides and contained a data call-in provision to fill all gaps in the evaluations. This has a special meaning for botanicals because it requires: defining and consistently producing a standard, reproducible plant extract and formulation; identifying all major and significant components; defining the fate of the active ingredients; and carrying out full toxicological and environmental fate studies. These requirements are made more difficult by the often complex structures of the active ingredients, other diverse bioactive materials that are present, and often a need for specific additives. Economics also play a major factor since botanicals have classically been low-volume, multiproducer commodity items. The cost of the toxicology and environmental studies might often exceed all potential sales income for years ahead, totally ruling out investment in data development.

It was only with pyrethrum that these major investments, both scientific and economic, justified an effort to update the information to meet current standards and requirements. It was apparent to individual pyrethrum producers that the economic burden to satisfy the reregistration process was unjustified if addressed alone. Therefore, the major participants in the pyrethrum industry formed a coalition, the Pyrethrin Steering Committee, to ensure continued use of their products.

An initial problem was to accurately define pyrethrum which was necessary for all required studies. This was no easy task since pyrethrum has six active ingredients, and the composition is dependent on many factors. An important additional complication is that the Environmental Protection Agency (EPA) is seldom confronted with registration of pesticidal mixtures and uncertainties in procedural requirements usually translate into multiple negotiations with the Agency, often resulting in considerable delay. A third important hurdle was to define which residues required analytical methods—the more chemicals analyzed, the higher the cost. And finally, with most botanical insecticides, as with many dietary constituents (Gold *et al.*, 1992; Ames *et al.*, 1990), the probability of passing through the battery of toxicological tests unscathed is probably no better than that of the synthetics. However, with pyrethrum there was already a massive literature on toxicological tests that provided confidence it would pass the rigorous criteria for reregistration and continued use.

III. PROTOTYPES FOR SYNTHETIC PYRETHROIDS

Botanical insecticides are important for two reasons: they are applied directly for pest control as indicated above; and they are used as prototypes for synthetics that are improved in particular characteristics of effectiveness, safety, economics, or proprietary status over pyrethrins. The testing of the myriad of plants, microbes, or other biologicals as insecticides amounts in effect to screening millions of natural products for biological activity. Once recognized, the active component(s) are isolated and identified. The active ingredient may then be synthesized for direct use or for structural modification, i.e., as a prototype for synthetics. The most effective and safest natural products serve as the best prototypes. If their structure is complicated, e.g., the avermectins, they may be used for reconstructive synthesis and commercialization of an analog using the microbial or botanical product for modification. If less complicated in structure, the compound could conceivably (but rarely) be generated by synthesis for use directly. More often the lessons from the chemistry and structure-activity relationships of the natural material and its derivatives are applied to optimization of the synthetic substitutes. The pyrethrins were particularly suitable prototypes because of their only moderately complex structure and their very high insecticidal activity combined with low oral and dermal mammalian toxicity.

Pyrethrin I can be synthesized by routes that make it a practical, although perhaps not economically-feasible, replacement for the natural pyrethrins, but it would probably lack some of the knockdown (KD) effectiveness of pyrethrin II combined with pyrethrin I. The synthetic material would presumably have to go through the entire toxicology-environmental impact package for EPA registration. *S*-Bioallethrin, a synthetic homolog of pyrethrin I, replaced some of its uses as did tetramethrin and phenothrin. The success of the synthetic pyrethroids is a direct result of the lessons of pyrethrins chemistry developed so well by Ryo Yamamoto, Hermann Staudinger, Leopold Ruzicka, Leslie Crombie, Michael Elliott, and Milton Schechter from 1919 until 1973 (Elliott and Janes, 1973) and a host of others following since then. Despite the intrusions of the synthetics, the use of pyrethrum extract continues such that this prototype survived the synthetics, at least for now.

IV. ROLE IN PEST CONTROL

The use of pyrethrum over the past 150–200 years evolved at first slowly and then rapidly in its role in pest control. It was always the first and foremost contact insecticide among the botanicals. It was often threatened, but never totally surpassed and replaced, as a major component in pest management strategies.

Pyrethrum through the 1930s was a major component of agricultural insect control. Its continued use was strongly threatened by first DDT and the chlorinated hydrocarbons and later the synthetic pyrethroids. The continuance of pyrethrum for agriculture depends primarily on two properties: selective performance and reduced environmental hazard. Four uses have been reintroduced in the past decade: pest control on minor crops; a preharvest spray; applications during the period of blooms; and for areas sensitive to potential insecticide exposure. The future in agriculture will depend on the expanded registrations on growing crops balanced against the loss of other products through failure to reregister with EPA. Special combinations are also becoming important with biological control agents, biorational insecticides, and in low-input sustainable agriculture.

Stored-products insect control continues to be a major market for pyrethrum. This is dependent on new application technology including: space sprays in the food processing industry; the new Turbocide® system of enhanced environmental acceptability; and the Actizol® application equipment for crack-and-crevice treatments. Pyrethrum is the material of choice in these uses. It is also important in protection of stored fabric.

Medical and veterinary pests were the first target of the pyrethrins. The flea, body louse, and bed bug powder of the 19th century became the mosquito and fly control agent of the 20th century. Pyrethrins are usually considered the safest choice in light of their exceptionally low mammalian toxicity and lack of environmental persistence.

The use in protection of food, feed, and farm animals from insect attack requires the establishment of residue tolerances. A residue program on the pyrethrins now under way will help fill this gap. This includes food handling facilities and selected crops in the field. The rapid decomposition of the pyrethrins in light and air insures that there will not be a persisting residue which is a favorable factor in the risk-benefit equation.

Insects in the home are major targets for the pyrethrins. Major uses are on-pet treatments and water-based formulations for fly control. The pyrethrins are of lesser importance in ant and roach control products. In choosing an insecticide, cheap, effective synthetics are balanced against the long history of safe use and rapid KD of the pyrethrins.

The pyrethrins are almost unique in their KD efficacy. They bind to a distinct but ill-defined site in the neuronal voltage-sensitive sodium channel. Some selected strains of *Drosophila*, *Musca*, and other pests have a low-affinity target, a type of resistance that can be a serious problem with no short-term solution once it appears.

Resistance is a constant and major threat to the use of any insecticide. It is fortunately for now not a serious challenge to the continued used of the pyrethrins. This favorable situation reflects the current use patterns, e.g., for house fly and German cockroach control. Synthetic pyrethroids are a threat to the pyrethrins because of their persistence and ability to select for strains with more resistance to the pyrethrins than to the synthetics. This would be less of a problem or not relevant if only the pyrethrins but not the synthetic pyrethroids were registered for use on a particular pest complex.

V. SAFE USE

Pyrethrum was at one time labeled as "natural" and "nontoxic to humans and pets," designations that can no longer legally be used as either guides to the purchaser or slogans by the producer. It is important but no longer adequate that several generations of users have been convinced of the safety from their own experience. It is now necessary to apply the same standards to both a natural material with the long history of pyrethrum and a new synthetic organic insecticide of unknown mechanism of action and fate. Pyrethrum is undergoing reregistration with EPA which has required standardization and definition of pyrethrum extract and a large data-development program on its mammalian and environmental toxicology and its fate in areas of intentional and unintentional exposure.

The environmental fate and toxicology of the pyrethrins have not been areas of concern because of their known ease of chemical, photochemical, and metabolic transformations. Studies on their movement and breakdown under actual environmental conditions—those prescribed in the EPA guidelines—are currently underway. [^{14}C]Pyrethrin I is being used for precise definition of the chemical identity, amount, and distribution of the products. These data combined with analyses on residues are essential in defining the direct impact on wildlife and ecosystems. All findings to date indicate minimal disruptions and no long-term effects when used according to label instructions.

Pyrethrum extract has a very favorable toxicological profile—one that can be matched by only a few other insecticides. It has a low order of acute oral toxicity and little potential to cause skin irritation or sensitization. It is not a reproductive toxin, teratogen, or mutagen, and has minimal potential to induce tumors in mammalian systems by state-of-the-art studies. Thus, these investigations confirm earlier studies that this insecticide poses little risk to humans.

There are three mechanistic reasons for the unusual safety of the pyrethrins. First, due to their lipophilicity, they penetrate rapidly into insects but mammalian barriers are more substantial. Second, they are exceptionally active in insect voltage-dependent sodium channels relative to the mammalian nervous system, allowing very low application rates. Third, they are metabolized and detoxified very rapidly so they do not reach the nervous system of mammals in active form. They are also metabolized quickly in some insects, requiring synergists such as piperonyl butoxide to slow down the detoxification and maximize the effectiveness. Fortunately, piperonyl butoxide also has a favorable toxicological profile by current tests and the use of synergized pyrethrins continues as one of the safest forms of pest control.

VI. CONCLUSION

Pyrethrum, the most economically important botanical insecticide, is used in the same form and manner as the synthetic materials—both closely-related (pyrethroids) and of other types (organophosphorus compounds, methyl-carbamates, and chlorinated hydrocarbons). It has been used for a longer time than any other type and has survived frequent challenges to its continued significance. The most important production areas have changed with periods of wartime and economic uncertainties. Efforts continue to improve the pyrethrins content in flowers through breeding programs and thereby expand the pyrethrum business. It has survived years of scarcity and overproduction. It can now be grown economically under labor-intensive conditions or using a fully mechanized production system. The major and minor ingredients are fully characterized and their metabolic and environmental fates are known or studies are underway to fill data gaps. The toxicological tests to maintain and to potentially expand current registrations were passed without significant problems. Even the challenge of synthetic analogs of high effectiveness, consistent quality and supply, and reasonable cost has not reduced the demand other than temporarily.

Pyrethrum users and researchers have long been fascinated with and convinced of its importance, effectiveness, and safety. This volume provides the available information on pyrethrum and its active ingredients, the pyrethrins, pointing out their advantages and disadvantages, the breadth and depth of knowledge, and the deficiencies or data gaps when they occur. Pyrethrum is an amazing material not only to those who have "kept the faith" but also to all people who have benefitted from its use.

ACKNOWLEDGMENT

This publication was made possible by grant number PO1 ES00049 from the National Institute of Environmental Health Sciences, NIH.

REFERENCES

Ames, B.N., Profet, M., and Gold, L.S. (1990). Dietary pesticides (99.99% all natural). *Proc. Natl. Acad. Sci. USA* **87**, 7777–7781.

Casida, J.E. (ed.). (1973). "Pyrethrum, The Natural Insecticide." Academic Press, New York.

Elliott, M., and Janes, N.F. (1973). Chemistry of the natural pyrethrins. *In* "Pyrethrum, The Natural Insecticide" (J.E. Casida, ed.), pp. 55–100. Academic Press, New York.

Gold, L.S., Slone, T.H., Stern, B.R., Manley, N.B., and Ames, B.N. (1992). Rodent carcinogens: setting priorities. *Science* **258**, 261–265.

Index